Lecture Notes in Computer Science 14836

The series Lecture Notes in Computer Science (LNCS), including its subseries Lecture Notes in Artificial Intelligence (LNAI) and Lecture Notes in Bioinformatics (LNBI), has established itself as a medium for the publication of new developments in computer science and information technology research, teaching, and education.

LNCS enjoys close cooperation with the computer science R & D community, the series counts many renowned academics among its volume editors and paper authors, and collaborates with prestigious societies. Its mission is to serve this international community by providing an invaluable service, mainly focused on the publication of conference and workshop proceedings and postproceedings. LNCS commenced publication in 1973.

Leonardo Franco · Clélia de Mulatier ·
Maciej Paszynski · Valeria V. Krzhizhanovskaya ·
Jack J. Dongarra · Peter M. A. Sloot
Editors

Computational Science – ICCS 2024

24th International Conference
Malaga, Spain, July 2–4, 2024
Proceedings, Part V

Editors
Leonardo Franco
University of Malaga
Malaga, Spain

Clélia de Mulatier
University of Amsterdam
Amsterdam, The Netherlands

Maciej Paszynski
AGH University of Science and Technology
Krakow, Poland

Valeria V. Krzhizhanovskaya
University of Amsterdam
Amsterdam, The Netherlands

Jack J. Dongarra
University of Tennessee
Knoxville, TN, USA

Peter M. A. Sloot
University of Amsterdam
Amsterdam, The Netherlands

ISSN 0302-9743 ISSN 1611-3349 (electronic)
Lecture Notes in Computer Science
ISBN 978-3-031-63774-2 ISBN 978-3-031-63775-9 (eBook)
https://doi.org/10.1007/978-3-031-63775-9

This Springer imprint is published by the registered company Springer Nature Switzerland AG
The registered company address is: Gewerbestrasse 11, 6330 Cham, Switzerland

Preface

Welcome to the proceedings of the 24th International Conference on Computational Science (https://www.iccs-meeting.org/iccs2024/), held on July 2–4, 2024 at the University of Málaga, Spain.

In keeping with the new normal of our times, ICCS featured both in-person and online sessions. Although the challenges of such a hybrid format are manifold, we have always tried our best to keep the ICCS community as dynamic, creative, and productive as possible. We are proud to present the proceedings you are reading as a result.

ICCS 2024 was jointly organized by the University of Málaga, the University of Amsterdam, and the University of Tennessee.

Facing the Mediterranean in Spain's Costa del Sol, Málaga is the country's sixth-largest city, and a major hub for finance, tourism, and technology in the region.

The University of Málaga (Universidad de Málaga, UMA) is a modern, public university, offering 63 degrees and 120 postgraduate degrees. Close to 40,000 students study at UMA, taught by 2500 lecturers, distributed over 81 departments and 19 centers. The UMA has 278 research groups, which are involved in 80 national projects and 30 European and international projects. ICCS took place at the Teatinos Campus, home to the School of Computer Science and Engineering (ETSI Informática), which is a pioneer in its field and offers the widest range of IT-related subjects in the region of Andalusia.

The International Conference on Computational Science is an annual conference that brings together researchers and scientists from mathematics and computer science as basic computing disciplines, as well as researchers from various application areas who are pioneering computational methods in sciences such as physics, chemistry, life sciences, engineering, arts, and the humanities, to discuss problems and solutions in the area, identify new issues, and shape future directions for research.

The ICCS proceedings series have become a primary intellectual resource for computational science researchers, defining and advancing the state of the art in this field.

We are proud to note that this 24th edition, with 17 tracks (16 thematic tracks and one main track) and close to 300 participants, has kept to the tradition and high standards of previous editions.

The theme for 2024, "Computational Science: Guiding the Way Towards a Sustainable Society", highlights the role of Computational Science in assisting multidisciplinary research on sustainable solutions. This conference was a unique event focusing on recent developments in scalable scientific algorithms; advanced software tools; computational grids; advanced numerical methods; and novel application areas. These innovative novel models, algorithms, and tools drive new science through efficient application in physical systems, computational and systems biology, environmental systems, finance, and others.

ICCS is well known for its excellent lineup of keynote speakers. The keynotes for 2024 were:

- David Abramson, University of Queensland, Australia
- Manuel Castro Díaz, University of Málaga, Spain
- Jiří Mikyška, Czech Technical University in Prague, Czechia
- Takemasa Miyoshi, RIKEN, Japan
- Coral Calero Muñoz, University of Castilla-La Mancha, Spain
- Petra Ritter, Berlin Institute of Health & Charité University Hospital Berlin, Germany

This year we had 430 submissions (152 to the main track and 278 to the thematic tracks). In the main track, 51 full papers were accepted (33.5%); in the thematic tracks, 104 full papers (37.4%). The higher acceptance rate in the thematic tracks is explained by their particular nature, whereby track organizers personally invite many experts in the field to participate. Each submission received at least 2 single-blind reviews (2.6 reviews per paper on average).

ICCS relies strongly on our thematic track organizers' vital contributions to attract high-quality papers in many subject areas. We would like to thank all committee members from the main and thematic tracks for their contribution to ensuring a high standard for the accepted papers. We would also like to thank Springer, Elsevier, and Intellegibilis for their support. Finally, we appreciate all the local organizing committee members for their hard work in preparing this conference.

We hope the attendees enjoyed the conference, whether virtually or in person.

July 2024

Leonardo Franco
Clélia de Mulatier
Maciej Paszynski
Valeria V. Krzhizhanovskaya
Jack J. Dongarra
Peter M. A. Sloot

Organization

Conference Chairs

General Chair

Valeria Krzhizhanovskaya	University of Amsterdam, The Netherlands

Main Track Chair

Clélia de Mulatier	University of Amsterdam, The Netherlands

Thematic Tracks Chair

Maciej Paszynski	AGH University of Krakow, Poland

Thematic Tracks Vice Chair

Michael Harold Lees	University of Amsterdam, The Netherlands

Scientific Chairs

Peter M. A. Sloot	University of Amsterdam, The Netherlands
Jack Dongarra	University of Tennessee, USA

Local Organizing Committee

Leonardo Franco (Chair)	University of Malaga, Spain
Francisco Ortega-Zamorano	University of Malaga, Spain
Francisco J. Moreno-Barea	University of Malaga, Spain
José L. Subirats-Contreras	University of Malaga, Spain

Thematic Tracks and Organizers

Advances in High-Performance Computational Earth Sciences: Numerical Methods, Frameworks & Applications (IHPCES)

Takashi Shimokawabe	University of Tokyo, Japan
Kohei Fujita	University of Tokyo, Japan
Dominik Bartuschat	FAU Erlangen-Nürnberg, Germany

Artificial Intelligence and High-Performance Computing for Advanced Simulations (AIHPC4AS)

Maciej Paszynski	AGH University of Krakow, Poland

Biomedical and Bioinformatics Challenges for Computer Science (BBC)

Mario Cannataro	University Magna Graecia of Catanzaro, Italy
Giuseppe Agapito	University Magna Graecia of Catanzaro, Italy
Mauro Castelli	Universidade Nova de Lisboa, Portugal
Riccardo Dondi	University of Bergamo, Italy
Rodrigo Weber dos Santos	Federal University of Juiz de Fora, Brazil
Italo Zoppis	University of Milano-Bicocca, Italy

Computational Diplomacy and Policy (CoDiP)

Roland Bouffanais	University of Geneva, Switzerland
Michael Lees	University of Amsterdam, The Netherlands
Brian Castellani	Durham University, UK

Computational Health (CompHealth)

Sergey Kovalchuk	Huawei, Russia
Georgiy Bobashev	RTI International, USA
Anastasia Angelopoulou	University of Westminster, UK
Jude Hemanth	Karunya University, India

Computational Optimization, Modelling, and Simulation (COMS)

Xin-She Yang	Middlesex University London, UK
Slawomir Koziel	Reykjavik University, Iceland
Leifur Leifsson	Purdue University, USA

Generative AI and Large Language Models (LLMs) in Advancing Computational Medicine (CMGAI)

Ahmed Abdeen Hamed	State University of New York at Binghamton, USA
Qiao Jin	National Institutes of Health, USA
Xindong Wu	Hefei University of Technology, China
Byung Lee	University of Vermont, USA
Zhiyong Lu	National Institutes of Health, USA
Karin Verspoor	RMIT University, Australia
Christopher Savoie	Zapata AI, USA

Machine Learning and Data Assimilation for Dynamical Systems (MLDADS)

Rossella Arcucci	Imperial College London, UK
Cesar Quilodran-Casas	Imperial College London, UK

Multiscale Modelling and Simulation (MMS)

Derek Groen	Brunel University London, UK
Diana Suleimenova	Brunel University London, UK

Network Models and Analysis: From Foundations to Artificial Intelligence (NMAI)

Marianna Milano	Università Magna Graecia of Catanzaro, Italy
Giuseppe Agapito	University Magna Graecia of Catanzaro, Italy
Pietro Cinaglia	University Magna Graecia of Catanzaro, Italy
Chiara Zucco	University Magna Graecia of Catanzaro, Italy

Numerical Algorithms and Computer Arithmetic for Computational Science (NACA)

Pawel Gepner	Warsaw Technical University, Poland
Ewa Deelman	University of Southern California, Marina del Rey, USA
Hatem Ltaief	KAUST, Saudi Arabia

Quantum Computing (QCW)

Katarzyna Rycerz	AGH University of Krakow, Poland
Marian Bubak	Sano and AGH University of Krakow, Poland

Simulations of Flow and Transport: Modeling, Algorithms, and Computation (SOFTMAC)

Shuyu Sun	King Abdullah University of Science and Technology, Saudi Arabia
Jingfa Li	Beijing Institute of Petrochemical Technology, China
James Liu	Colorado State University, USA

Smart Systems: Bringing Together Computer Vision, Sensor Networks and Artificial Intelligence (SmartSys)

Pedro Cardoso	University of Algarve, Portugal
João Rodrigues	University of Algarve, Portugal
Jânio Monteiro	University of Algarve, Portugal
Roberto Lam	University of Algarve, Portugal

Solving Problems with Uncertainties (SPU)

Vassil Alexandrov	Hartree Centre – STFC, UK
Aneta Karaivanova	IICT – Bulgarian Academy of Science, Bulgaria

Teaching Computational Science (WTCS)

Evguenia Alexandrova	Hartree Centre – STFC, UK
Tseden Taddese	UK Research and Innovation, UK

Reviewers

Ahmed Abdelgawad	Central Michigan University, USA
Samaneh Abolpour Mofrad	Imperial College London, UK
Tesfamariam Mulugeta Abuhay	Queen's University, Canada
Giuseppe Agapito	University of Catanzaro, Italy
Elisabete Alberdi	University of the Basque Country, Spain
Luis Alexandre	UBI and NOVA LINCS, Portugal
Vassil Alexandrov	Hartree Centre – STFC, UK
Evguenia Alexandrova	Hartree Centre – STFC, UK
Julen Alvarez-Aramberri	Basque Center for Applied Mathematics, Spain
Domingos Alves	Ribeirão Preto Medical School, University of São Paulo, Brazil
Sergey Alyaev	NORCE, Norway
Anastasia Anagnostou	Brunel University London, UK
Anastasia Angelopoulou	University of Westminster, UK
Rossella Arcucci	Imperial College London, UK
Emanouil Atanasov	IICT – Bulgarian Academy of Sciences, Bulgaria
Krzysztof Banaś	AGH University of Krakow, Poland
Luca Barillaro	Magna Graecia University of Catanzaro, Italy
Dominik Bartuschat	FAU Erlangen-Nürnberg, Germany
Pouria Behnodfaur	Curtin University, Australia
Jörn Behrens	University of Hamburg, Germany
Adrian Bekasiewicz	Gdansk University of Technology, Poland
Gebrail Bekdas	Istanbul University, Turkey
Mehmet Ali Belen	Iskenderun Technical University, Turkey
Stefano Beretta	San Raffaele Telethon Institute for Gene Therapy, Italy
Anabela Moreira Bernardino	Polytechnic Institute of Leiria, Portugal
Eugénia Bernardino	Polytechnic Institute of Leiria, Portugal
Daniel Berrar	Tokyo Institute of Technology, Japan
Piotr Biskupski	IBM, Poland
Georgiy Bobashev	RTI International, USA
Carlos Bordons	University of Seville, Spain
Bartosz Bosak	PSNC, Poland
Lorella Bottino	University Magna Graecia of Catanzaro, Italy

Roland Bouffanais	University of Geneva, Switzerland
Marian Bubak	Sano and AGH University of Krakow, Poland
Aleksander Byrski	AGH University of Krakow, Poland
Cristiano Cabrita	Universidade do Algarve, Portugal
Xing Cai	Simula Research Laboratory, Norway
Carlos Calafate	Universitat Politècnica de València, Spain
Victor Calo	Curtin University, Australia
Mario Cannataro	University Magna Graecia of Catanzaro, Italy
Karol Capała	AGH University of Krakow, Poland
Pedro J. S. Cardoso	Universidade do Algarve, Portugal
Eddy Caron	ENS-Lyon/Inria/LIP, France
Stefano Casarin	Houston Methodist Hospital, USA
Brian Castellani	Durham University, UK
Mauro Castelli	Universidade Nova de Lisboa, Portugal
Nicholas Chancellor	Durham University, UK
Thierry Chaussalet	University of Westminster, UK
Sibo Cheng	Imperial College London, UK
Lock-Yue Chew	Nanyang Technological University, Singapore
Pastrello Chiara	Krembil Research Institute, Canada
Su-Fong Chien	MIMOS Berhad, Malaysia
Marta Chinnici	enea, Italy
Bastien Chopard	University of Geneva, Switzerland
Maciej Ciesielski	University of Massachusetts, USA
Pietro Cinaglia	University of Catanzaro, Italy
Noelia Correia	Universidade do Algarve, Portugal
Adriano Cortes	University of Rio de Janeiro, Brazil
Ana Cortes	Universitat Autònoma de Barcelona, Spain
Enrique Costa-Montenegro	Universidad de Vigo, Spain
David Coster	Max Planck Institute for Plasma Physics, Germany
Carlos Cotta	University of Málaga, Spain
Peter Coveney	University College London, UK
Alex Crimi	AGH University of Krakow, Poland
Daan Crommelin	CWI Amsterdam, The Netherlands
Attila Csikasz-Nagy	King's College London, UK/Pázmány Péter Catholic University, Hungary
Javier Cuenca	University of Murcia, Spain
António Cunha	UTAD, Portugal
Pawel Czarnul	Gdansk University of Technology, Poland
Pasqua D'Ambra	IAC-CNR, Italy
Alberto D'Onofrio	University of Trieste, Italy
Lisandro Dalcin	KAUST, Saudi Arabia

Bhaskar Dasgupta	University of Illinois at Chicago, USA
Clélia de Mulatier	University of Amsterdam, The Netherlands
Ewa Deelman	University of Southern California, Marina del Rey, USA
Quanling Deng	Australian National University, Australia
Eric Dignum	University of Amsterdam, The Netherlands
Riccardo Dondi	University of Bergamo, Italy
Rafal Drezewski	AGH University of Krakow, Poland
Simon Driscoll	University of Reading, UK
Hans du Buf	University of the Algarve, Portugal
Vitor Duarte	Universidade NOVA de Lisboa, Portugal
Jacek Długopolski	AGH University of Krakow, Poland
Wouter Edeling	Vrije Universiteit Amsterdam, The Netherlands
Nahid Emad	University of Paris Saclay, France
Christian Engelmann	ORNL, USA
August Ernstsson	Linköping University, Sweden
Aniello Esposito	Hewlett Packard Enterprise, Switzerland
Roberto R. Expósito	Universidade da Coruna, Spain
Hongwei Fan	Imperial College London, UK
Tamer Fandy	University of Charleston, USA
Giuseppe Fedele	University of Calabria, Italy
Christos Filelis-Papadopoulos	Democritus University of Thrace, Greece
Alberto Freitas	University of Porto, Portugal
Ruy Freitas Reis	Universidade Federal de Juiz de Fora, Brazil
Kohei Fujita	University of Tokyo, Japan
Takeshi Fukaya	Hokkaido University, Japan
Wlodzimierz Funika	AGH University of Krakow, Poland
Takashi Furumura	University of Tokyo, Japan
Teresa Galvão	University of Porto, Portugal
Luis Garcia-Castillo	Carlos III University of Madrid, Spain
Bartłomiej Gardas	Institute of Theoretical and Applied Informatics, Polish Academy of Sciences, Poland
Victoria Garibay	University of Amsterdam, The Netherlands
Frédéric Gava	Paris-East Créteil University, France
Piotr Gawron	Nicolaus Copernicus Astronomical Centre, Polish Academy of Sciences, Poland
Bernhard Geiger	Know-Center GmbH, Austria
Pawel Gepner	Warsaw Technical University, Poland
Alex Gerbessiotis	NJIT, USA
Maziar Ghorbani	Brunel University London, UK
Konstantinos Giannoutakis	University of Macedonia, Greece
Alfonso Gijón	University of Granada, Spain

Jorge González-Domínguez	Universidade da Coruña, Spain
Alexandrino Gonçalves	CIIC – ESTG – Polytechnic University of Leiria, Portugal
Yuriy Gorbachev	Soft-Impact LLC, Russia
Pawel Gorecki	University of Warsaw, Poland
Michael Gowanlock	Northern Arizona University, USA
George Gravvanis	Democritus University of Thrace, Greece
Derek Groen	Brunel University London, UK
Loïc Guégan	UiT the Arctic University of Norway, Norway
Tobias Guggemos	University of Vienna, Austria
Serge Guillas	University College London, UK
Manish Gupta	Harish-Chandra Research Institute, India
Piotr Gurgul	SnapChat, Switzerland
Oscar Gustafsson	Linköping University, Sweden
Ahmed Abdeen Hamed	State University of New York at Binghamton, USA
Laura Harbach	Brunel University London, UK
Agus Hartoyo	TU Kaiserslautern, Germany
Ali Hashemian	Basque Center for Applied Mathematics, Spain
Mohamed Hassan	Virginia Tech, USA
Alexander Heinecke	Intel Parallel Computing Lab, USA
Jude Hemanth	Karunya University, India
Aochi Hideo	BRGM, France
Alfons Hoekstra	University of Amsterdam, The Netherlands
George Holt	UK Research and Innovation, UK
Maximilian Höb	Leibniz-Rechenzentrum der Bayerischen Akademie der Wissenschaften, Germany
Huda Ibeid	Intel Corporation, USA
Alireza Jahani	Brunel University London, UK
Jiří Jaroš	Brno University of Technology, Czechia
Qiao Jin	National Institutes of Health, USA
Zhong Jin	Computer Network Information Center, Chinese Academy of Sciences, China
David Johnson	Uppsala University, Sweden
Eleda Johnson	Imperial College London, UK
Piotr Kalita	Jagiellonian University, Poland
Drona Kandhai	University of Amsterdam, The Netherlands
Aneta Karaivanova	IICT-Bulgarian Academy of Science, Bulgaria
Sven Karbach	University of Amsterdam, The Netherlands
Takahiro Katagiri	Nagoya University, Japan
Haruo Kobayashi	Gunma University, Japan
Marcel Koch	KIT, Germany

Robin Richardson	Netherlands eScience Center, The Netherlands
Sophie Robert	University of Orléans, France
João Rodrigues	Universidade do Algarve, Portugal
Daniel Rodriguez	University of Alcalá, Spain
Marcin Rogowski	Saudi Aramco, Saudi Arabia
Sergio Rojas	Pontifical Catholic University of Valparaiso, Chile
Diego Romano	ICAR-CNR, Italy
Albert Romkes	South Dakota School of Mines and Technology, USA
Juan Ruiz	University of Buenos Aires, Argentina
Tomasz Rybotycki	IBS PAN, CAMK PAN, AGH, Poland
Katarzyna Rycerz	AGH University of Krakow, Poland
Grażyna Ślusarczyk	Jagiellonian University, Poland
Emre Sahin	Science and Technology Facilities Council, UK
Ozlem Salehi	Özyeğin University, Turkey
Ayşin Sancı	Altinay, Turkey
Christopher Savoie	Zapata Computing, USA
Ileana Scarpino	University "Magna Graecia" of Catanzaro, Italy
Robert Schaefer	AGH University of Krakow, Poland
Ulf D. Schiller	University of Delaware, USA
Bertil Schmidt	University of Mainz, Germany
Karen Scholz	Fraunhofer MEVIS, Germany
Martin Schreiber	Université Grenoble Alpes, France
Paulina Sepúlveda-Salas	Pontifical Catholic University of Valparaiso, Chile
Marzia Settino	Università Magna Graecia di Catanzaro, Italy
Mostafa Shahriari	Basque Center for Applied Mathematics, Spain
Takashi Shimokawabe	University of Tokyo, Japan
Alexander Shukhman	Orenburg State University, Russia
Marcin Sieniek	Google, USA
Joaquim Silva	Nova School of Science and Technology – NOVA LINCS, Portugal
Mateusz Sitko	AGH University of Krakow, Poland
Haozhen Situ	South China Agricultural University, China
Leszek Siwik	AGH University of Krakow, Poland
Peter Sloot	University of Amsterdam, The Netherlands
Oskar Slowik	Center for Theoretical Physics PAS, Poland
Sucha Smanchat	King Mongkut's University of Technology North Bangkok, Thailand
Alexander Smirnovsky	SPbPU, Russia
Maciej Smołka	AGH University of Krakow, Poland
Isabel Sofia	Instituto Politécnico de Beja, Portugal
Robert Staszewski	University College Dublin, Ireland

Magdalena Stobińska	University of Warsaw, Poland
Tomasz Stopa	IBM, Poland
Achim Streit	KIT, Germany
Barbara Strug	Jagiellonian University, Poland
Diana Suleimenova	Brunel University London, UK
Shuyu Sun	King Abdullah University of Science and Technology, Saudi Arabia
Martin Swain	Aberystwyth University, UK
Renata G. Słota	AGH University of Krakow, Poland
Tseden Taddese	UK Research and Innovation, UK
Ryszard Tadeusiewicz	AGH University of Krakow, Poland
Claude Tadonki	Mines ParisTech/CRI – Centre de Recherche en Informatique, France
Daisuke Takahashi	University of Tsukuba, Japan
Osamu Tatebe	University of Tsukuba, Japan
Michela Taufer	University of Tennessee, USA
Andrei Tchernykh	CICESE, Mexico
Kasim Terzic	University of St Andrews, UK
Jannis Teunissen	KU Leuven, Belgium
Sue Thorne	Hartree Centre – STFC, UK
Ed Threlfall	United Kingdom Atomic Energy Authority, UK
Vinod Tipparaju	AMD, USA
Pawel Topa	AGH University of Krakow, Poland
Paolo Trunfio	University of Calabria, Italy
Ola Tørudbakken	Meta, Norway
Carlos Uriarte	University of the Basque Country, BCAM – Basque Center for Applied Mathematics, Spain
Eirik Valseth	University of Life Sciences & Simula, Norway
Rein van den Boomgaard	University of Amsterdam, The Netherlands
Vítor V. Vasconcelos	University of Amsterdam, The Netherlands
Aleksandra Vatian	ITMO University, Russia
Francesc Verdugo	Vrije Universiteit Amsterdam, The Netherlands
Karin Verspoor	RMIT University, Australia
Salvatore Vitabile	University of Palermo, Italy
Milana Vuckovic	European Centre for Medium-Range Weather Forecasts, UK
Kun Wang	Imperial College London, UK
Peng Wang	NVIDIA, China
Rodrigo Weber dos Santos	Federal University of Juiz de Fora, Brazil
Markus Wenzel	Fraunhofer Institute for Digital Medicine MEVIS, Germany

Lars Wienbrandt	Kiel University, Germany
Wendy Winnard	UKRI STFC, UK
Maciej Woźniak	AGH University of Krakow, Poland
Xindong Wu	Hefei University of Technology, China
Dunhui Xiao	Tongji University, China
Huilin Xing	University of Queensland, Australia
Yani Xue	Brunel University, UK
Abuzer Yakaryilmaz	University of Latvia, Latvia
Xin-She Yang	Middlesex University London, UK
Dongwei Ye	University of Amsterdam, The Netherlands
Karol Życzkowski	Jagiellonian University, Poland
Gabor Závodszky	University of Amsterdam, Hungary
Sebastian Zając	SGH Warsaw School of Economics, Poland
Małgorzata Zajęcka	AGH University of Krakow, Poland
Justyna Zawalska	ACC Cyfronet AGH, Poland
Wei Zhang	Huazhong University of Science and Technology, China
Yao Zhang	Google, USA
Jinghui Zhong	South China University of Technology, China
Sotirios Ziavras	New Jersey Institute of Technology, USA
Zoltan Zimboras	Wigner Research Center, Hungary
Italo Zoppis	University of Milano-Bicocca, Italy
Chiara Zucco	University Magna Graecia of Catanzaro, Italy
Pavel Zun	ITMO University, Russia

Contents – Part V

Computational Optimization, Modelling and Simulation

Machine Learning and Data Assimilation for Dynamical Systems

Multiscale Modelling and Simulation

Computational Optimization, Modelling and Simulation

Cost-Efficient Multi-Objective Design of Miniaturized Microwave Circuits Using Machine Learning and Artificial Neural Networks

Slawomir Koziel[1,2](✉), Anna Pietrenko-Dabrowska[2], and Leifur Leifsson[3]

[1] Engineering Optimization and Modeling Center, Department of Engineering, Reykjavík University, Menntavegur 1, 102 Reykjavík, Iceland
koziel@ru.is

[2] Faculty of Electronics Telecommunications and Informatics, Gdansk University of Technology, Narutowicza 11/12, 80-233 Gdansk, Poland
anna.dabrowska@pg.edu.pl

[3] School of Aeronautics and Astronautics, Purdue University, West Lafayette, IN 47907, USA
leifur@purdue.edu

Abstract. Designing microwave components involves managing multiple objectives such as center frequencies, impedance matching, and size reduction for miniaturized structures. Traditional multi-objective optimization (MO) approaches heavily rely on computationally expensive population-based methods, especially when executed with full-wave electromagnetic (EM) analysis to guarantee reliability. This paper introduces a novel and cost-effective MO technique for microwave passive components utilizing a machine learning (ML) framework with artificial neural network (ANN) surrogates as the primary prediction tool. In this approach, multiple candidate solutions are extracted from the Pareto set via optimization using a multi-objective evolutionary algorithm (MOEA) applied to the current ANN model. These solutions expand the dataset of available (EM-simulated) parameter vectors and refine the surrogate model iteratively. To enhance computational efficiency, we employ variable-resolution EM models. Tested on two microstrip circuits, our methodology competes effectively against several surrogate-based approaches. The average computational cost of the algorithm is below three hundred EM analyses of the circuit, with the quality of generated Pareto sets surpassing those produced by the benchmark methods.

Keywords: Microwave design · multi-objective optimization · design automation · machine learning · neural networks · surrogate modeling · Pareto optimality

1 Introduction

Contemporary microwave design confronts various challenges linked to meeting diverse performance criteria [1, 2], while concurrently striving for compact circuit sizes [3–5]. The miniaturization imperative is crucial for numerous applications like the Internet of

L. Franco et al. (Eds.): ICCS 2024, LNCS 14836, pp. 3–18, 2024.
https://doi.org/10.1007/978-3-031-63775-9_1

Things, mobile healthcare, and more [6–8]. Achieving size reduction involves strategies such as multi-layer implementations or integrating geometrical modifications like line meandering, compact microstrip resonant cells, metamaterials, etc. [9–11]. Accurate assessment of intricate circuits full-wave electromagnetic (EM) analysis is indispensable [20], which proves computationally demanding, especially when repetitive simulations are required (e.g., for optimization, uncertainty quantification [12, 13]). To mitigate these high computational costs, several techniques have been introduced, including space mapping [14], response correction [15], surrogate-assisted methods [16], response feature techniques [17], machine learning [18], or restricted Jacobian updating strategies [19].

Designing microwave systems involves managing multiple criteria simultaneously, encompassing aspects like center frequency, bandwidth, return loss levels, and more. Often, these design goals conflict with each other. For instance, in the case of miniaturization, reducing size might compromise electrical performance. Achieving the best possible design trade-offs often results in a Pareto set representation [20], necessitating multi-objective optimization (MO). Many available numerical optimization methods handle only scalar objectives, requiring objective aggregation for MO [21]. However, this approach yields a single design per algorithm run. Generating the entire Pareto set requires genuine MO [20], which is primarily conducted using bio-inspired population-based algorithms [22–26]. Clearly, nature-inspired methods can produce the entire set of trade-off designs in a single run, yet, they are computationally inefficient. Particularly, direct simulation-driven MO is constrained because of excessive CPU costs.

To address the high computational costs, surrogate modeling methods [27] offer an alternative by utilizing a rapid metamodel for system evaluation instead of resource-intensive EM analysis. Surrogates can either be pre-constructed [28, 29], or iteratively refined throughout the optimization run [30, 31]. However, building reliable surrogates over larger parameter spaces poses numerical challenges, making the former approach uncommon. Consequently, most multi-objective (MO) frameworks adopt a machine learning (ML) approach, refining the metamodel iteratively using EM data gathered during the search process. The surrogate model's role lies in generating candidate designs, known as infill points [32], with a wide array of ML techniques available in literature [33, 34]. Although cost-efficient, ML tends to leave a significant portion of the search space unexplored. The search process focuses on an initially identified promising region, allocating most resources to its exploration. Moreover, the challenges in building reliable data-driven models restrict ML's applicability to simpler problems.

Addressing the challenges of EM-driven multi-objective optimization (MO) associated with high parameter space dimensionality can be achieved through domain restriction. Practically, this is done by identifying extreme non-dominated solutions (requiring extra optimization runs for individual objectives) and delineating the domain using this data [35]. The computational benefits of domain restriction lie in the smaller training datasets needed within a reduced volume, enhancing the MO process's effectiveness. Another method, performance-driven modeling [36], constructs the surrogate only within the region housing high-quality designs. Similar to domain restriction, this approach significantly improves the surrogate's predictive capability while reducing the required training data samples [37–39]. However, the said algorithms might incur high initial costs due to acquiring extreme non-dominated solutions, which could easily reach hundreds of EM simulations. Additionally, there's an assumption about the regularity of the Pareto front, posing a limitation.

The aim of this article is to introduce an innovative approach for cost-effective and reliable multi-objective optimization (MO) of compact microwave circuits. The proposed algorithm leverages machine learning (ML) with artificial neural networks (ANN). An integral part of this algorithm is a multi-objective evolutionary algorithm (MOEA) used to optimize the ANN metamodel and generate a current representation of the Pareto set. From this set, a collection of candidate designs (infill points) is derived, and their EM-simulated circuit characteristics are acquired and integrated into the ongoing dataset. The termination criterion for the algorithm is based on evaluating the similarity between the rendered Pareto sets in successive iterations. To reduce CPU cost, variable-resolution EM simulations are employed. Initial sampling starts at the lowest acceptable fidelity, gradually increasing to higher levels later on. While lower-resolution EM analyses are faster, they lack accuracy. The search process's reliability is maintained by progressively removing low-resolution points from the dataset, retaining only high-fidelity ones to form the final Pareto set. The effectiveness of this approach has been validated using two miniaturized microwave circuits, demonstrating exceptional performance. The average running cost equates to fewer than three hundred high-fidelity EM analyses of the circuit under design. This represents almost ninety percent relative acceleration over one-shot benchmark algorithms and over forty percent speedup over single-fidelity ML frameworks. Importantly, the reported savings do not affect the quality of the resulting Pareto fronts.

This work encompasses several novel contributions, notably: (i) development and deployment of a high-efficiency machine learning framework tailored for the multi-criteria design of microwave circuits; (ii) integration of variable-resolution EM simulations into a model management strategy, facilitating accelerated optimization; (iii) successful demonstration of the proposed procedure's capability in managing complex multi-objective tasks, handling parameter space dimensions of up to fifteen.

2 Multi-Objective Optimization Methodology

This section explores the details of the developed MO algorithm, whose constituent parts include a machine learning (ML) framework, ANN models, and multi-resolution EM simulations. Sections 2.1 and 2.2 delineate the formulation of the MO problem and the variable-resolution EM models, respectively. The remaining sections are devoted to providing specifics of the introduced procedure, starting from sampling and ANN model construction (Sect. 2.3), through the usage of MOEA algorithm to generate the Pareto set (Sect. 2.4), infill point rendition (Sect. 2.5), and variable-fidelity model management (Sect. 2.6). Lastly, Sects. 2.7 and 2.8 discuss the termination criteria, and reprise the entire MO framework, respectively.

2.1 MO Microwave Design Optimization

Our aim is to simultaneously minimize all considered objectives across the design space X. Hereinafter, the design objectives will be referred to as $F_k(\boldsymbol{x})$, $k = 1, \ldots, N_{obj}$, whereas the vector of objectives will be referred to as $\boldsymbol{F}(\boldsymbol{x}) = [F_1(\boldsymbol{x})\ F_2(\boldsymbol{x})\ \ldots\ F_{Nobj}(\boldsymbol{x})]^T$, with $\boldsymbol{x} = [x_1\ \ldots\ x_n]^T$ representing a designable vector (geometry parameters). The design space X is of interval-type and it is bound by lower and upper limits $\boldsymbol{l} = [l_1\ \ldots\ l_n]^T$, and $\boldsymbol{u} = [u_1\ \ldots\ u_n]^T$, respectively.

The objectives pertaining to real-world circuits are at least partially contradictory, as in the case of miniaturization which deteriorates electrical properties. When multiple objectives are present, Pareto dominance relation [20] is indispensable to compare various designs: given designs $\boldsymbol{x}$ and $\boldsymbol{y}$, $\boldsymbol{x}$ is said to dominate over y, or $\boldsymbol{F}(\boldsymbol{y}) \prec \boldsymbol{F}(\boldsymbol{x})$ if and only if $F_k(\boldsymbol{y}) \le F_k(\boldsymbol{x})$ for $k = 1, \ldots, N_{obj}$, and $F_k(\boldsymbol{y}) < F_k(\boldsymbol{x})$ for not fewer than one k. Ultimately, MO aims at finding a set of gobally non-dominated vectors within X, with $\boldsymbol{x}$ being globally non-dominated if no $\boldsymbol{y} \in X$ such at $\boldsymbol{F}(\boldsymbol{y}) \prec \boldsymbol{F}(\boldsymbol{x})$ exists.

MO procedures heavily utilize nature-inspired algorithms: tremendously expensive when applied with microwave circuit simulation models, chiefly using full-wave EM analysis. To circumvent this issue, surrogate modeling techniques are used, oftentimes, within various ML frameworks. Yet, creating dependable surrogate models remains the primary challenge, especially in highly-dimensional parameter spaces with extensive parameter ranges. Our approach alleviates the mentioned issues to some point by combining ML, multiple infill point rendition, and variable-fidelity EM analysis.

2.2 Multi-Resolution EM Models

We employ computational models in the form of full-wave EM analysis. Usually, circuit design is carried out using a model ensuring adequate accuracy $\boldsymbol{R}_f(\boldsymbol{x})$, i.e., fine (or high-fidelity) model. Unfortunately, evaluation of $\boldsymbol{R}_f(\boldsymbol{x})$ tends to be expensive, with the cost increasing with structure complexity. Lower simulation times can be obtained for models of reduced fidelity, yet, accompanied with a simultaneous degradation of model reliability. Oftentimes, two levels of EM model resolution are utilized (high-/low-fidelity), yet, the prospective benefits of integrating a spectrum of resolutions have been recently shown [40].

This research exploits multi-resolution EM simulations to speed-up the multi-objective (MO) search process. In our approach, we utilize a continuous range of fidelities $L \in [L_{\min}\ L_{\max}]$. Hereinafter, we will denote as $\boldsymbol{R}(\boldsymbol{x},L)$ the EM model of fidelity L. Moreover, $\boldsymbol{R}_f(\boldsymbol{x}) = \boldsymbol{R}(\boldsymbol{x},L_{\max})$ and $\boldsymbol{R}_c(\boldsymbol{x}) = \boldsymbol{R}(\boldsymbol{x},L_{\min})$ will be used for high- and low-fidelity models, respectively.

2.3 Sampling Procedure and ANN-Based Surrogate Modeling

Our MO framework exploits surrogates in the form of ANN regression models. The premier metamodel is built utilizing data set comprising N_{init} samples $\boldsymbol{x}_B^{(j)}, j = 1, \ldots, N_{\text{init}}$, collected over the design space X using an increased-efficacy Latin Hypercube Sampling (LHS) protocol [41]. At this step, EM simulations are carried out at the minimum level of fidelity $L_{\min}$, i.e., we have $\boldsymbol{R}(\boldsymbol{x}_B^{(j)},L_{\min}), j = 1, \ldots, N_{\text{init}}$.

As an ANN surrogate, we utilize a multi-layer perceptron [42] with the following setup: two hidden layers (each comprising 10 neurons), sigmoid activation function, network trained using Levenberg-Marquardt algorithm (maximum 1000 learning epochs, performance metric in the form of mean squared error, MSE, random split of testing/

training samples). In each iteration, the ANN model is trained anew. Using such a simple network allows for retaining short training duration (around dozen seconds for datasets comprising below two hundred samples). Thus, the cost of rendering ANNs is not taken into account in the total expenses of the MO algorithm. Figure 1 depicts the ANN-assisted rendition of frequency characteristics of a microwave coupler.

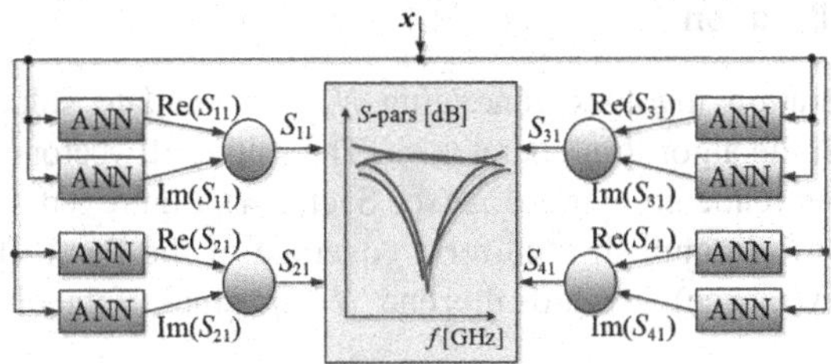

Fig. 1. Complex-valued scattering parameters of the microwave coupler represented using ANN model. ANN model renders outputs representing all pertinent device responses for specific frequencies f_1 through f_m.

Representation: Natural (floating point numbers);
Population replacement: Generational model (the new population supplants the former one); N_P stands for population size;
Selection: Pareto-ranking-based tournaments;
Other mechanisms:

- Fitness sharing with adaptively adjusted niche size;
- Multi-point elitism (non-dominated solution sets retained after each iteration);
- Termination condition based on a sufficient reduction of newly found Pareto-optimal solutions.

Recombination: A combination of an intermediate and arithmetic crossover (with equal probabilities $p_c/2$ each) [45]:

- Intermediate crossover: given $\boldsymbol{x} = [x_1 \ldots x_n]^T$ and $\boldsymbol{y} = [y_1 \ldots y_n]^T$ as the parent individuals, the offspring $\boldsymbol{z} = [z_1 \ldots z_n]^T$ is obtained so that $z_i = ax_i + (1-a)y_i$ with $0 \le a \le 1$ (a selected randomly);
- Arithmetic crossover: produces the offspring $\boldsymbol{z} = a\boldsymbol{x} + (1 - a)\boldsymbol{y}$; $0 \le a \le 1$ (a selected randomly);

Mutation: Localized random perturbations, applied with probability p_m, independently for each design variable. We have $x_i \to x_i' = x_i + \Delta x_i$, where Δx_i is a random deviation defined as [45]

$$\Delta x_i = \begin{cases} (x_{i.max} - x_i)\cdot(2(r-0.5))^\beta & \text{if } r > 0.5 \\ (x_{i.min} - x_i)\cdot(2(0.5-r))^\beta & \text{otherwise} \end{cases}$$

where $r \in [0,1]$ is a random number and $\beta = 3$

Fig. 2. The description of the search engine: multi-objective evolutionary algorithm (MOEA), which renders Pareto sets by carrying out ANN-surrogate-based multi-objective optimization.

2.4 Multi-Objective Evolutionary Algorithm (MOEA)

We employ multi-objective evolutionary algorithm (MOEA) as the main search mechanism which renders sets of Pareto-optimal designs by optimizing the ANN models. Our version of MOEA adheres to the principles included in [43]. Figure 2 illustrates the employed algorithm. The verification experiments of Sect. 3 utilize the following configuration of MOEA: $N_P = 200$ (population size), $p_{cross} = 0.1$ (crossover probability), and $p_{mut} = 0.8$ (mutation probability) [44]. The MOEA employs the rapid ANN model, rendering its CPU costs insignificant in comparison to that of an individual circuit EM simulation.

2.5 Infill Point Allocation

Each algorithm iteration involves generating N_{infill} candidate solutions, which are included into the set the algorithm operates on. The said infill vectors are derived from the latest Pareto front rendered by MOEA (see Sect. 2.4). Figure 3 depicts the selection procedure striving to accomplish a uniform coverage of the Pareto front (for a two-dimensional objective space). Towards this end, N_{infill} values of the objective F_2 are set as

$$F_j = F_{2.\min} + (F_{2.\max} - F_{2.\min})(j-1)/(N_{infill} - 1) \quad (1)$$

In (1), $F_{2.\min}$ and $F_{2.\max}$ bound the span of the Pareto set (with supplementary conditions). The choice of the infill vectors $\boldsymbol{x}_I^{(i.j)}$, $j = 1, \ldots, N_{infill}$, has to ensure that they are in close proximity of the assumed levels (here, i refers to the MO procedure iteration index).

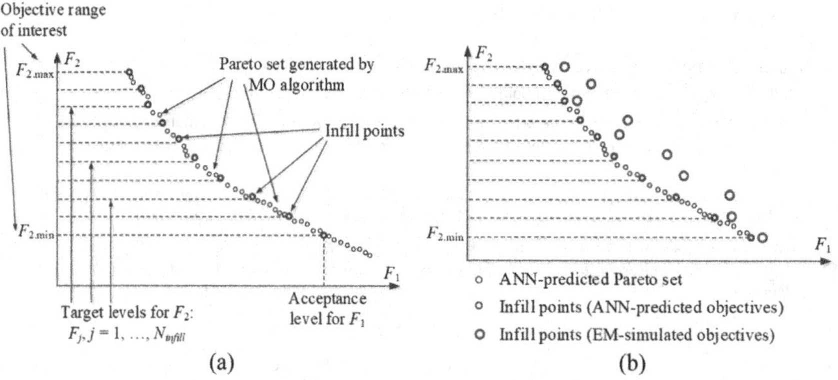

Fig. 3. (a) Infill point rendition: the chosen candidate solutions originate from the MOEA-generated Pareto front through ANN model optimization. We aim at an even distribution of the vectors concerning the objective F_2, whose range is established by the span of the Pareto set (along with optional requirements, e.g., target value of the objective F_1 as depicted above); (b) Infill points: ANN-predicted vs. EM-evaluated. The EM-predicted objectives are inferior to the ANN-rendered ones due to low surrogate accuracy. The following situation is shown: if F_2 pertains to antenna footprint, values of F_2 are identical for EM-analysis and ANN render; still, EM-predicted levels of F_1 are inferior to that ANN-predicted.

A noticeable disparity is observed among the objectives at the surrogate-predicted infill vectors and the respective values assessed based on EM simulation data (which are

Notation:

- $x_T^{(i-1.j)}$, $j = 1, \ldots, N_T$ – current dataset (available at the end of the ith iteration);
- $R(x_T^{(i-1.j)}, L^{(i-1.j)})$ – corresponding EM-simulated circuit characteristics;
- $x_I^{(i.j)}$, $j = 1, \ldots, N_{infill}$ – infill samples generated in the ith iteration;
- $R(x_I^{(i.j)}, L^{(i.j)})$ –corresponding EM-simulated circuit responses.

Rules for updating the dataset $\{x_T^{(i.j)}\}$:

- If $N_T + N_{infill} \le 2N_{init}$, then the updated dataset is obtained by concatenating $\{x_T^{(j)}\}$ and $\{x_I^{(i.j)}\}$, i.e., $\{x_T^{(i.j)}\} = \{x_T^{(i-1.j)}\} \cup \{x_I^{(i.j)}\}$, as well as the corresponding EM-simulated responses;
- If $N_T + N_{infill} > 2N_{init}$, then consider the following options:
  - If the concatenated set $\{x_T^{(i-1.j)}\} \cup \{x_I^{(i.j)}\}$ contains EM simulation data obtained at the fidelity level $L < L_{max}$, then remove $N_T + N_{infill} - 2N_{init}$ points evaluated at the (currently) lowest fidelity level;
  - If the concatenated set $\{x_T^{(i-1.j)}\} \cup \{x_I^{(i.j)}\}$ only contains EM simulation data obtained at the fidelity level $L = L_{max}$, then take $\{x_T^{(i.j)}\} = \{x_T^{(i-1.j)}\} \cup \{x_I^{(i.j)}\}$ as the updated set.

Remark: After initial sampling (before the first iteration), we have $N_T = N_{init}$, $x_T^{(0.j)} = x_B^{(j)}$, $j = 1, \ldots, N_T$, and $L^{(0.j)} = L_{min}$ for $j = 1, \ldots, N_T$ (cf. Section 2.3).

Fig. 4. The employed strategy for dataset updating.

usually worse due to low accuracy of the ANN). Still, EM-simulated objectives are of prime importance (the surrogate serves merely as a supplementary tool). The quality of the Pareto and the termination condition are evaluated using EM data.

2.6 Management Scheme of Multi-Fidelity Models. Dataset Updating

In order to lower computational expenses, the developed MO framework employs multi-fidelity EM models. In this study, the model fidelity L comes from the range between $L_{\min}$ to $L_{\max}$. The initial sampling (outlined in Sect. 2.3) is carried out using $L_{\min}$. Later, the model resolution is increased until highest fidelity $L_{\max}$ is reached. the latter is crucial for the Pareto-optimal designs to be dependable.

The strategy for managing model fidelity is straightforward and involves an affine adjustment of L. Specifically, in the ith iteration, the resolution $L^{(i)}$ is established as

$$L^{(i)} = \min\left\{L_{\max}, L_{\min} + (L_{\max} - L_{\min})\frac{i-1}{N_{transition}}\right\} \tag{2}$$

where $N_{transition}$ represents the iteration count beyond which L assumes $L_{\max}$.

The update of the dataset constitutes a crucial element of the MO procedure because of the involvement of multi-resolution EM models. When only single-resolution model is utilized, addition of the new EM data is simple, i.e., in the ith iteration, we have the complete dataset

$$\left\{x_B^{(j)}\right\}_{j=1,\ldots,N_{init}} \cup \left\{x_I^{(1.j)}\right\}_{j=1,\ldots,N_{init}} \cup \ldots \cup \left\{x_I^{(i.j)}\right\}_{j=1,\ldots,N_{init}} \tag{3}$$

whereas in the variable-fidelity approach, the samples of reduced fidelity are systematically eliminated, to ensure that high-fidelity points ultimately persist. Figure 4 illustrates the procedure for dataset updating. The said strategy keeps the dataset of the size up to $2N_{infill}$ until any samples simulated using $L < L_{\max}$ exist. Later on, the lowest-fidelity points are step-by-step ousted from $\{x_T{}^{(i.j)}\}$, until it contains solely high-fidelity. subsequently, each new ML-rendered sample is directly appended to the set.

Interpolated non-dominated Paret set: Let $\boldsymbol{F}_{nondom}{}^{(i)}$ be a $M \times 2$ matrix

$$\boldsymbol{F}^{(i)}_{nondom} = \begin{bmatrix} F^{(i)}_{I.1.1} & F^{(i)}_{I.2.1} \\ F^{(i)}_{I.1.2} & F^{(i)}_{I.2.2} \\ \vdots & \vdots \\ F^{(i)}_{I.1.M} & F^{(i)}_{I.2.M} \end{bmatrix} \tag{4}$$

where

$$F^{(i)}_{I.1.k} = F^{(i)}_{1.1} + \left[F^{(i)}_{1.K_i} - F^{(i)}_{1.1}\right]\frac{k-1}{M-1} \tag{5}$$

and

$$F^{(i)}_{I.2.k} = I\left(F^{(i)}_{I.1.k}, \left[F^{(i)}_{1.1} \ldots F^{(i)}_{1.K_i}\right], \left[F^{(i)}_{2.1} \ldots F^{(i)}_{2.K_i}\right]\right) \tag{6}$$

In (4), $I(y,[y_1 \ldots y_K],[z_1 \ldots z_K])$ is a function interpolating the value vector $[z_1 \ldots z_K]$ corresponding to the parameter vector $[y_1 \ldots y_K]$ on the input parameter y. Here, M is a large integer, e.g., 100.

Pareto-set comparison:

- Similarity metric E_i, defined for $i = 1, 2, \ldots$, as

$$E_i = \left\|\boldsymbol{F}^{(i-1)}_{nondom} - \boldsymbol{F}^{(i)}_{nondom}\right\| \tag{7}$$

where ||.|| is a Frobenius norm.

- Moving average $E_{a.i}$, defined as

$$E_{a.i} = \frac{1}{i - \max\{1, i - N_a + 1\} + 1} \sum_{k=\max\{1, i-N_a+1\}}^{i} E_k \tag{8}$$

Remark 1: The operation (2)-(4) interpolates the set $[F_{1.j}{}^{(i)}\ F_{2.j}{}^{(i)}]$, $j = 1, \ldots, K_i$ onto a matrix of a fixed size $M \times 1$ (independent of the specific number of non-dominated points K_i in iteration i), thereby enabling comparison of the Pareto sets between subsequent iterations.

Remark 2: $E_{a.i}$ of (7) smoothens out possible fluctuations of E_i being a result of the presence of stochastic components in the search process.

Fig. 5. Termination condition of the introduced ML-based MO procedure.

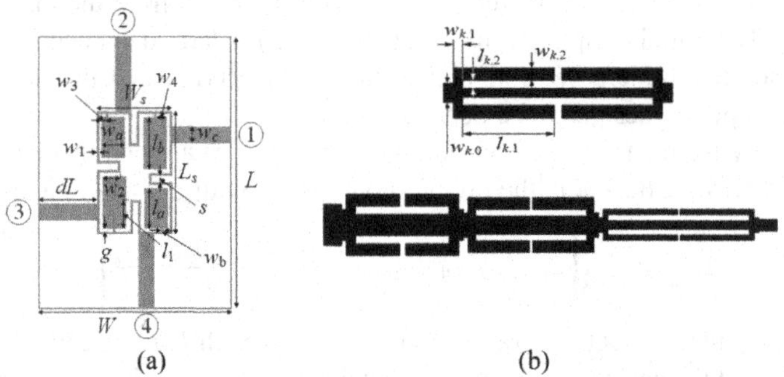

Fig. 6. Verification case studies: (a) Circuit I: a compact branch-linear coupler [45], (b) Circuit II: a miniaturized three-section 50-to-100 Ω impedance matching transformer [46]; top and bottom panels show the CMRC cell and the entire circuit, respectively.

2.7 Algorithm Termination

We employ the termination condition based on the similarity of EM-simulated design objectives among the non-dominated designs rendered by successive procedure iterations. The objective vectors are indicated by $[F^{(i)}_{1.j} \cdot F^{(i)}_{2.j}]$, $j = 1, \ldots, k_i$, and are presumed to be arranged as: $F^{(i)}_{1.j} \leq F^{(i)}_{1.k}$ for $j <$ k. The relevant notation is shown in Fig. 5. The termination condition is formulated as $E_{a.1} < \varepsilon$, with ε being the convergence threshold.

Table 1. ML-based MO algorithm: Control parameters.

Parameter	Description	Default value
N_{init}	Number of initial sample points for building the first ANN surrogate	100
N_{infill}	Number of infill points generated per MO iteration	10
$N_{transition}$	Number of iterations over which the model fidelity is changed from $L_{\min}$ to $L_{\max}$	5
N_a	The span of the moving average for Pareto front similarity metric $E_{a.i}$	5
ε	Termination threshold for the MO process	1

2.8 Algorithm Operation

Here, we summarize all the aforementioned algorithmic components. Table 1 gathers the control parameters (five in total) of the proposed ML-base MO procedure. In all numerical experiments, the default values are utilized. In particular, the number N_{init} of initial samples (we use $N_{init} = 100$) can be adjusted in order to achieve the required accuracy of the initial surrogate (e.g., around ten percent of relative RMS error). The remaining parameters are of lesser importance. For example, lower values of N_{infill} permit reducing the cost of each algorithm iteration at the expense of increasing total number of iterations necessary to yield similar Pareto set. Furthermore, higher values of N_a lead to smoothening of the moving average, which might be compensated by increasing ε.

The expenditures associated with the multi-objective (MO) process are evaluated as the total number of EM simulations performed throughout optimization run. Again, MOEA utilizes a fast ANN model, so the overhead of the Pareto set generation are negligible in comparison to the expenses associated with EM circuit simulations. The primary sampling is carried out at the minimum resolution $L_{\min}$. Later on, L is increased to $L_{\max}$ (across the first $N_{transition}$ iterations). As procedure advances, the reduced-fidelity samples are progressively eliminated from the dataset. The gradual enhancement of the distribution of infill vectors within the sub-set comprising the Pareto front leads to the improvement of ANN's accuracy in the very region. The ultimate outcome of our procedure is the high-fidelity Pareto set.

3 Verification Case Studies

Two miniaturized microstrip circuits are utilized to verify the developed MO framework: a branch-line coupler (Circuit I) and a three-section impedance matching transformer (Circuit II), which are shown in Fig. 6. Both circuits employ compact microstrip resonant cells (CMRCs). Table 2 presents all the important details on the verification structures: geometry parameters, substrate, simulation setups, and design goals. All models are simulated using transient solver of CST Microwave Studio.

For both circuits, the parameter spaces are bounded by lower and upper limits on geometry parameters (denoted as $\boldsymbol{l}$ and $\boldsymbol{u}$, respectively). The primary design goal is the same for both structures: minimization of the footprint area. Whereas the secondary goals are as follows: maximization of the 0.5-dB power division band-width (Circuit I), and minimization of the maximal value of reflection level within the assumed band (Circuit II). It is crucial to emphasize that the challenges posed by both test problems primarily arise from high dimensionality of their parameter spaces.

Table 2. Important parameters of verification circuit structures.

Circuit		I [46]			II [47]		
Geometry parameters[$] [mm]		$\boldsymbol{x} = [g\ l_{1r}\ l_a\ l_b\ w_1\ w_{2r}\ w_{3r}\ w_{4r}\ w_a\ w_b]^T$			$\boldsymbol{x} = [l_{1.1}\ l_{1.2}\ w_{1.1}\ w_{1.2}\ w_{1.0}\ l_{2.1}\ l_{2.2}\ w_{2.1}\ w_{2.2}\ w_{2.0}\ l_{3.1}\ l_{3.2}\ w_{3.1}\ w_{3.2}\ w_{3.0}]^T$		
Parameter bounds	Lower	$\boldsymbol{l} = [0.4\ 0.7\ 6.0\ 8.0\ 0.6\ 0.85\ 0.7\ 0.1\ 3.0\ 0.5]^T$			$\boldsymbol{l} = [2.5\ 0.15\ 0.75\ 0.15\ 0.3\ 2.5\ 0.15\ 0.18\ 0.24\ 1.4\ 3.0\ 0.15\ 0.2\ 0.25\ 0.55]^T$		
	Upper	$\boldsymbol{u} = [0.6\ 0.9\ 7.0\ 10\ 0.9\ 0.99\ 0.85\ 0.3\ 4\ 0.7]^T$			$\boldsymbol{u} = [3.7\ 0.3\ 0.8\ 0.55\ 0.4\ 4.5\ 0.3\ 0.52\ 0.55\ 1.8\ 4.5\ 0.5\ 0.3\ 0.3\ 1.5]^T$		
Other parameters [mm]		$L = 2dL + L_s$, $L_s = 4w_1 + 4g + s + l_a + l_b$, $W = 2dL + W_s$, $W_s = 4w_1 + 4g + s + 2w_a$, $l_1 = l_b l_{1r}$, $w_2 = w_a w_{2r}$, $w_3 = w_{3r} w_a$, and $w_4 = w_{4r} w_a$			–		
Substrate		RO4003: $\varepsilon_r = 3.38$, $h = 0.51$ mm			RF-35: $\varepsilon_r = 3.5$, $h = 0.762$ mm		
Lowest-fidelity model		LPW (L_{min})	Number of mesh cells	Simulation time [s]	LPW (L_{min})	Number of mesh cells	Simulation time [s]
		11	~22,000	100	9	~40,000	40
Highest-fidelity model		LPW (L_{max})	Number of mesh cells	Simulation time [s]	LPW (L_{max})	Number of mesh cells	Simulation time [s]
		23	~115,000	195	23	~240,000	140
Design goals[#]	F_1	Minimize footprint area [mm²]: $F_1(\boldsymbol{x}) = A(\boldsymbol{x}) = W_s L_s$					
	F_2	Center frequency: $f_0 = 1.5$ GHz Maximize the power split bandwidth, defined at the (continuous w.r.t. f_0) frequency range for which $d_S(\boldsymbol{x},f) = \lvert\,\lvert S_{21}(\boldsymbol{x},f)\rvert - \lvert S_{31}(\boldsymbol{x},f)\rvert\,\rvert \le 0.5$ dB			Minimize maximum reflection within the band from 1.75 GHz to 4.25 GHz: $F_2(\boldsymbol{x}) = \max\{f \in [1.75\ 4.25]\text{ GHz} : \lvert S_{11}(\boldsymbol{x},f)\rvert\}$ Considered are designs for which: $F_2(\boldsymbol{x}) \le -20$ dB		

[$]Parameters with subscript r are relative and unitless.
[#]Explanation of symbols: $|S_{k1}(\boldsymbol{x},f)|$ - scattering parameter S_{k1}, $k = 1, 2, 3, 4$, at design $\boldsymbol{x}$ and frequency f.

Circuits I and II were optimized using the algorithm developed in the study, as well as the control parameters and their default values (see Table 1). The ultimate result in the form of the Pareto set has been generated, where the non-dominated designs are simulated at the highest admissible fidelity. The benchmark algorithm set is comprehensive, as it comprises three surrogate-assisted MO procedures (summarized in Table 2). The two procedures (Algorithm 1 and 2) adhere to a one-shot principle, where a surrogate model is built upfront and subsequently optimized via MOEA so as to establish the Pareto set. The key distinction between the two approaches consists in the choice of metamodel: kriging interpolant (Algorithm 1), and ANN (Algorithm 2). The data sets of two cardinalities are utilized: 400 and1600 training data samples. Algorithm 3 constitutes a one-fidelity variation of the introduced framework operating solely at the highest model resolution. By comparing our approach with Algorithm 3 we are able to assess the possible computational benefits of integrating multi-fidelity EM simulations into the search process.

Table 3 presents juxtaposition of the computational expenses of the developed algorithm and benchmark techniques, assessed as the equivalent number of highest-fidelity EM simulations of the microwave structure under design. In the case of benchmark algorithms, which use solely high-fidelity EM model, the cost is calculated as the overall number of EM simulations. Figure 8 presents Pareto fronts generated using our methodology and Algorithms 1 through 3. The circuit characteristics pertinent to the designs optimal in the Pareto sense rendered by our algorithm are displayed in Figs. 9 and 10, which have been simulated using fine EM model.

Designs 1 and 3 of Circuit I (presented in Fig. 8) have been fabricated and measured. Figure 11 presents the circuit prototypes, as well as a comparison between the measured and EM-evaluated S-parameters, demonstrating their satisfactory alignment.

The developed multi-fidelity ML-based MO framework outperforms the benchmark algorithms with regard to cost-efficacy and reliability. One-shot methods (Algorithms 1 and 2) produce lesser-quality Pareto fronts as a result of the reduced accuracy of the model built in an unconstrained design space. As far as the reliability of our approach is concerned, it is superior to Algorithms 1 and 2, and similar only to the Algorithm 3. Moreover, employment of multi-fidelity EM simulations does not negatively impact the Pareto set quality.

Table 3. Benchmark algorithms.

Algorithm	General information	Surrogate model	Characterization
1	One-shot surrogate-assisted MO procedure	Kriging interpolation (Gaussian correlation functions and first-order polynomial)	• Surrogate built using N_S data samples, then optimized using MOEA; • Selected non-dominated samples are EM-simulated to yield the final result of the algorithm.
2		NN (multi-layer perceptron, two hidden layers with ten neurons each; training: Levenberg-Marquardt algorithm)	
3	Machine learning algorithm with ANN surrogates	ANN surrogates	• Initial sampling and surrogate model setup discussed in Section 2.3; • Infill point generation outlined in Section 2.4 and 2.5 (surrogate optimization using MOEA); • EM-evaluated dataset updated by adding all infill points to the existing dataset; • Solely high-fidelity EM model.

The computational efficiency of our framework is considerably better than that of both single-run and ML-based routines: the expenses of our procedure are around 260 high-fidelity EM analyses, whereas for Algorithm 3 it is close to 450 EM analyses (i.e., the savings of 42% w.r.t. single-fidelity framework have been obtained without degrading design quality). The relative savings over Algorithms 1 and 2 (using 1600 samples) are even higher (they reach up to 84%) (Table 4).

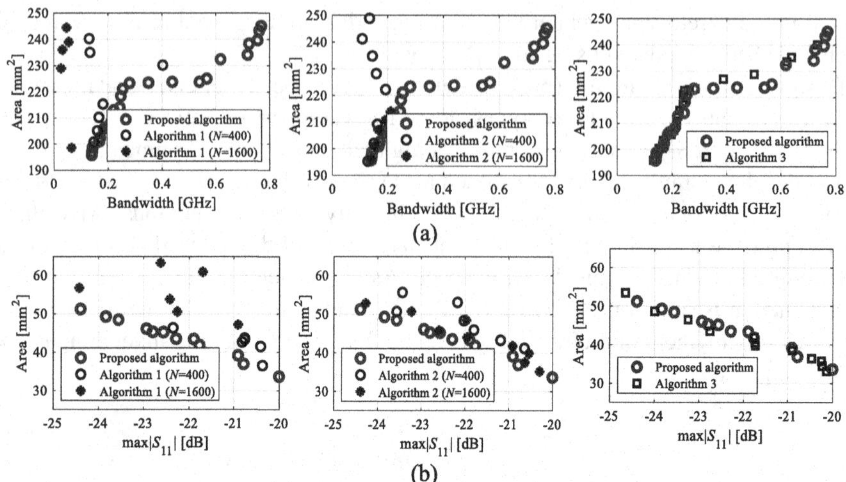

Fig. 8. Pareto fronts rendered by our and benchmark MO procedures: (a) Circuit I, (b) Circuit II.

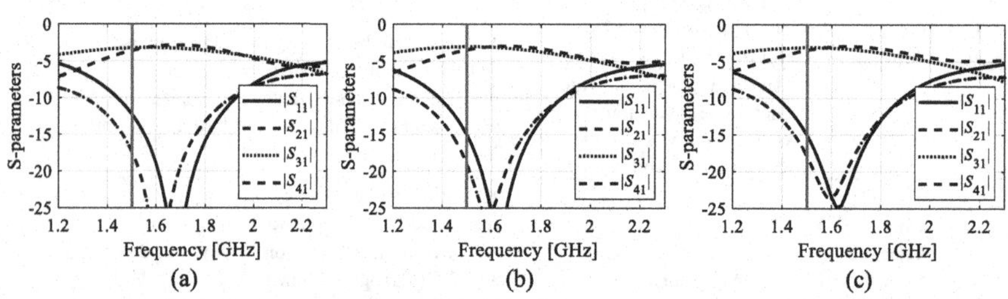

Fig. 9. Circuit I: exemplary designs from the Pareto set, S-parameters (|S_{11}| (—), |S_{21}| (····), |S_{31}| (- - -), |S_{41}| (-o-)) for: (a) BW = 770 MHz, A = 245 mm^2 (Design 1), (b) BW = 620 MHz, A = 232 mm^2 (Design 2), (c) BW = 255 MHz, A = 221 mm^2 (Design 3); BW – 0.5-dB power division bandwidth, A – circuit size. Vertical line shows the intended operating frequency.

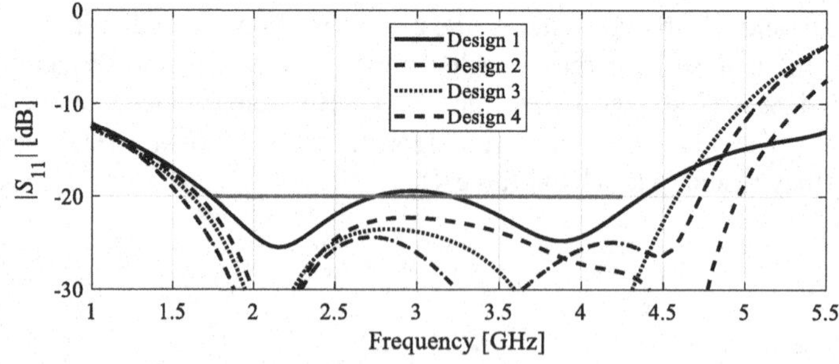

Fig. 10. Circuit II: exemplary designs from the Pareto set, reflection characteristics for: (a) A = 33.4 mm^2 (Design 1), (b) A = 41.8 mm^2 (Design 2), (c) A = 46.1 mm^2 (Design 3), (d) A = 51.2 mm^2 (Design 4). Intended operating frequency range shown by the horizontal line.

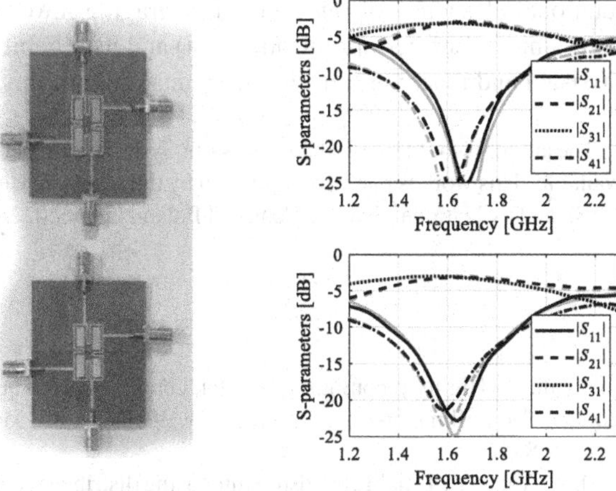

Fig. 11. Circuit I: (a) prototype photos for the Pareto-optimal points; (b) measurements (black) and EM-evaluated (gray) S-parameters. Top and bottom: Designs 1 and 3 of Fig. 8, respectively.

Table 4. Cost breakdown: the developed and benchmark procedures

Algorithm		Optimization cost#	
		Circuit I	Circuit II
This work		306.6	212.3
1	$N = 400$	400	400
	$N = 1600$	1600	1600
2	$N = 400$	400	400
	$N = 1600$	1600	1600
3		500	390

The cost of benchmark algorithms is assessed as the overall count of executed EM simulations. In the proposed approach, the cost is appraised as the equivalent number of high-fidelity EM simulations, considering the time evaluation ratio between the high- and lower-fidelity models.

4 Conclusion

This work proposed an innovative approach for accelerated multi-objective optimization of compact microwave devices. Our methodology revolves around a machine learning framework utilizing an artificial neural network (ANN) model. At each iteration of the search procedure, multiple infill points are generated and integrated into the dataset, progressively refining the surrogate model. Intermediate representations of the Pareto front are crafted by optimization of the ANN surrogate using a multi-objective evolutionary algorithm, driving designs toward the Pareto front. Additionally, the incorporation of variable-resolution EM simulations, managed by a tailored model management strategy, further enhances cost-efficiency. Rigorous numerical validation showcases the exceptional computational efficiency of our method (averaging fewer than three hundred high-fidelity EM simulations per optimization run) and its reliability compared to various surrogate-assisted and machine learning benchmark algorithms.

Acknowledgement. The authors would like to thank Dassault Systemes, France, for making CST Microwave Studio available. This work is partially supported by the Icelandic Centre for Research (RANNIS) Grant 239858 and by National Science Centre of Poland Grant 2022/47/B/ST7/00072.

References

1. Al Khanjar, K., Djerafi, T.: Highly reconfigurable patch coupler with frequency and power-dividing ratio control for millimeter-wave applications. IEEE Trans. Microwave Theory Technol. **71**, 2118–2128 (2023)
2. Li, Q., Chen, X., Ch, P., Yang, T.: Tunable bandstop filter using distributed coupling microstrip resonators with capacitive terminal. IEEE Microw. Wireless Compon. Lett. **30**, 35–38 (2020)
3. Tian, H., Dong, Y.: Wideband low-loss filter with compact size and wide stopband based on folded planar waveguide. IEEE Microw. Wirel. Technol. Lett. **33**, 651–654 (2023)
4. Arsanjani, A., Bartlett, C., Robins, L., Teschl, R., Bösch, W., Höft, M.: Metasurfaces for filter miniaturization and out-of-band rejection improvement. IEEE Microwave Wirel. Technol. Lett. **33**, 271–274 (2023)
5. Zhu, F., Wu, Y., Zhao, X., Chu, P., Luo, G.Q., Wu, K.: Compact and wide stopband bandpass filters based on dual-mode folded circular substrate integrated waveguide cavities. IEEE Trans. Microwave Theory Tech. **71**, 3102–3113 (2023)
6. Hussain, N., Kim, N.: Integrated microwave and mm-wave MIMO antenna module with 360° pattern diversity for 5G Internet of Things. IEEE Internet of Things J. **9**, 24777–24789 (2022)
7. Ilyas, S., Shoaib, N., Nikolaou, S., Cheema, H.M.: A wideband tunable power divider for SWIPT systems. IEEE Access **8**, 30675–30681 (2020)
8. Yang, L., Zhou, Y.J., Zhang, C., Yang, X.M., Yang, X., Tan, C.: Compact multiband wireless energy harvesting based battery-free body area networks sensor for mobile healthcare. IEEE J. Electromagnet. RF Microwave Med. Biol. **2**, 109–115 (2028)
9. Letavin, D.: Miniature microstrip branch line coupler with folded artificial transmission lines. Int. J. Electron. Commun. **99**, 8–13 (2019)
10. Qian, Z.Y., Chen, J.X.: Compact bandpass filter using CMRC-based dual-behavior resonator. Int. J. RF Microwave CAE **29**, e21719 (2019)
11. Chen, S., Guo, M., Xu, K., Zhao, P., Dong, L., Wang, G.: A frequency synthesizer based microwave permittivity sensor using CMRC structure. IEEE Access **6**, 8556–8563 (2018)
12. Zhu, D.Z., Werner, P.L., Werner, D.H.: Design and optimization of 3-D frequency-selective surfaces based on a multiobjective lazy ant colony optimization algorithm. IEEE Trans. Antennas Propag. **65**, 7137–7149 (2017)
13. Bandler, J.W., Rayas-Sánchez, J.E.: An early history of optimization technology for automated design of microwave circuits. IEEE J. Microwaves **3**, 319–337 (2023)
14. Zhang, W., Feng, F., Gongal-Reddy, V.W.R., Zhang, J., Yan, S., Zhang, Q.J.: Space mapping approach to electromagnetic centric multiphysics parametric modeling of microwave components. IEEE Trans. Microwave Theory Tech. **66**, 3169–3185 (2018)
15. Zhang, W., et al.: Advanced parallel space-mapping-based multiphysics optimization for high-power microwave filters. IEEE Trans. Microw. Theory Tech. **69**, 2470–2484 (2021)
16. Zhang, Z., Cheng, Q.S., Chen, H., Jiang, F.: An efficient hybrid sampling method for neural network-based microwave component modeling and optimization. IEEE Microwave Wirel. Compon. Lett. **30**, 625–628 (2020)

17. Pietrenko-Dabrowska, A., Koziel, S.: Response Feature Technology for High-Frequency Electronics. Optimization, Modeling, and Design Automation. Springer, New York (2023)
18. Wu, Q., Wang, H., Hong, W.: Multistage collaborative machine learning and its application to antenna modeling and optimization. IEEE Trans. Antennas Propag. **68**, 3397–3409 (2020)
19. Koziel, S., Pietrenko-Dabrowska, A.: Efficient gradient-based algorithm with numerical derivatives for expedited optimization of multi-parameter miniaturized impedance matching transformers. Radioengineering **28**, 572–578 (2019)
20. Deb, K.: Multi-Objective Optimization Using Evolutionary Algorithms. Wiley, New York (2001)
21. Marler, R.T., Arora, J.S.: The weighted sum method for multi-objective optimization: new insights. Struct. Multidisciplinary Optim. **41**, 853–862 (2010)
22. Ghorbaninejad, H., Heydarian, R.: New design of waveguide directional coupler using genetic algorithm. IEEE Microwave Wirel. Compon. Lett. **26**, 86–88 (2016)
23. Baumgartner, P., et al.: Multi-objective optimization of Yagi-Uda antenna applying enhanced firefly algorithm with adaptive cost function. IEEE Trans. Magn. **54**, Article no. 8000504 (2018)
24. Liu, C., Zheng, F., Kai, C.: An improved multi-objective artificial bee colony algorithm for pattern synthesis of conformal arrays. In: International Conference on Natural Computation, Fuzzy Systems and Knowledge Discovery (ICNC-FSKD), Changsha, pp. 265–270 (2016)
25. Champasak, P., Panagant, N., Pholdee, N., Bureerat, S., Yildiz, A.R.: Self-adaptive many-objective meta-heuristic based on decomposition for many-objective conceptual design of a fixed wing unmanned aerial vehicle. Aerosp. Sci. Technol. **100**, Article no. 105783 (2020)
26. Al-Tashi, Q., Abdulkadir, S.J., Rais, H.M., Mirjalili, S., Alhussian, H.: Approaches to multi-objective feature selection: a systematic literature review. IEEE Access **8**, 125076–125096 (2020)
27. Yang, W., Li, Y., Wang, H., Jiang, M., Cao, M., Liu, C.: Multi-objective optimization of high-power microwave sources based on multi-criteria decision-making. IEEE Trans. Electron Devices **70**, 3892–3898 (2023)
28. Wei, Y., Qi, G., Wang, Y., Yan, N., Zhang, Y., Feng, L.: Efficient microwave filter design by a surrogate-model-assisted decomposition-based multi-objective evolutionary algorithm. Electronics **11**, 3309 (2022)
29. De Melo, M.C., Santos, P.B., Faustino, E., Bastos-Filho, C.J.A., Sodré, A.C.: Computational intelligence-based methodology for antenna development. IEEE Trans. Antennas Propag. **10**, 1860–1870 (2022)
30. Wu, W., Wang, H., Hong, W.: Multistage collaborative machine learning and its application to antenna modelling and optimization. IEEE Trans. Antennas Propag. **68**, 3397–3409 (2020)
31. Nouri, M., Aghdam, S.A., Jafarieh, A., Mallat, N.K., Jamaluddin, M.H., Dor-Emami, M.: An optimized small compact rectangular antenna with meta-material based on fast multi-objective optimization for 5G mobile communication. J. Comput. Electron. **20**, 1532–1540 (2021)
32. Liu, B., Aliakbarian, H., Radiom, S., Vandenbosch, G.A.E., Gielen G.: Efficient multi-objective synthesis for microwave components based on computational intelligence techniques. In: Proceedings of the Design Automation Conference (DAC), San Francisco, CA, pp. 542–548 (2012)
33. An, S., Yang, S., Mohammed O.A.: A Kriging-assisted light beam search method for multi-objective electromagnetic inverse problems. IEEE Trans. Magn. **54**, paper 7001104 (2018)
34. Taran, N., Ionel, D.M., Dorrell D.G.: Two-level surrogate-assisted differential evolution multi-objective optimization of electric machines using 3-D FEA. IEEE Trans. Magn., **54**, paper 8107605 (2018)

35. Koziel, S., Pietrenko-Dabrowska, A.: Constrained multi-objective optimization of compact microwave circuits by design triangulation and Pareto front interpolation. Eur. J. Oper. Res. **299**, 302–312 (2022)
36. Koziel, S., Pietrenko-Dabrowska, A.: Performance-Driven Surrogate Modeling of High-Frequency Structures. Springer, New York (2020)
37. Pietrenko-Dabrowska, A., Koziel, S., Golunski L.: Two-stage variable-fidelity modeling of antennas with domain confinement. Sci. Rep. **12**, paper no. 17275 (2022)
38. Pietrenko-Dabrowska, A., Koziel, S.: Accelerated multi-objective design of miniaturized microwave components by means of nested kriging surrogates. Int. J. RF Microw. CAE **30**(4), paper no. e22124 (2020)
39. Koziel, S., Pietrenko-Dabrowska, A.: Fast multi-objective optimization of antenna structures by means of data-driven surrogates and dimensionality reduction. IEEE Access **8**, 183300–183311 (2020)
40. Koziel, S., Pietrenko-Dabrowska, A.: Accelerated gradient-based optimization of antenna structures using multi-fidelity simulation models. IEEE Trans. Antennas Propag. **69**(12), 8778–8789 (2021)
41. Beachkofski, B., Grandhi, R.: Improved distributed hypercube sampling. American Institute of Aeronautics and Astronautics, paper AIAA 2002–1274 (2002)
42. Vang-Mata, R. (ed.): Multilayer Perceptrons. Nova Science, Hauppauge (2020)
43. Fonseca, C.M., Fleming, P.J.: Multiobjective optimization and multiple constraint handling with evolutionary algorithms Part–I: a unified formulation. IEEE Trans. Syst. Man Cybern. Part A Syst. Hum. **28**(1), 26–37 (1998)
44. Michalewicz, Z.: Genetic Algorithms + Data Structures = Evolution Programs. Springer, New York (1996)
45. Tseng, C.H., Chang, C.L.: A rigorous design methodology for compact planar branch-line and rat-race couplers with asymmetrical T-structures. IEEE Trans. Microwave Theory Tech. **60**(7), 2085–2092 (2012)
46. Koziel, S., Pietrenko-Dabrowska, A.: Reliable data-driven modeling of high-frequency structures by means of nested kriging with enhanced design of experiments. Eng. Comput. **36**(7), 2293–2308 (2019)

Expedited Machine-Learning-Based Global Design Optimization of Antenna Systems Using Response Features and Multi-fidelity EM Analysis

Anna Pietrenko-Dabrowska[1](✉), Slawomir Koziel[1,2], and Leifur Leifsson[3]

[1] Faculty of Electronics Telecommunications and Informatics, Gdansk University of Technology, Narutowicza 11/12, 80-233 Gdansk, Poland
anna.dabrowska@pg.edu.pl, koziel@ru.is

[2] Engineering Optimization and Modeling Center, Department of Engineering, Reykjavík University, Menntavegur 1, 102 Reykjavík, Iceland

[3] School of Aeronautics and Astronautics, Purdue University, West Lafayette, IN 47907, USA
leifur@purdue.edu

Abstract. The design of antenna systems poses a significant challenge due to stringent performance requirements dictated by contemporary applications and the high computational costs associated with models, particularly full-wave electromagnetic (EM) analysis. Presently, EM simulation plays a crucial role in all design phases, encompassing topology development, parametric studies, and the final adjustment of antenna dimensions. The latter stage is especially critical as rigorous numerical optimization becomes essential for achieving optimal performance. In an increasing number of instances, global parameter tuning is necessary. Unfortunately, the use of nature-inspired algorithms, the prevalent choice for global design, is hindered by their poor computational efficiency. This article presents an innovative approach to cost-efficient global optimization of antenna input characteristics. Our methodology leverages response feature technology, ensuring inherent regularization of the optimization task by exploring the nearly-linear dependence between the coordinates of feature points and the antenna's dimensions. The optimization process is structured as a machine learning (ML) procedure, utilizing a kriging surrogate model rendering response features to generate promising candidate designs (infill points). This model is iteratively refined using accumulated EM simulation data. Further acceleration is achieved by incorporating multi-fidelity EM analysis, where initial sampling and surrogate model construction use low-fidelity EM simulations, and the ML optimization loop employs high-fidelity EM analysis. The multi-fidelity EM simulation data is blended into a single surrogate using co-kriging. Extensive verification of the presented algorithm demonstrates its remarkable computational efficiency, with an average running cost not exceeding ninety EM simulations per run and up to a seventy percent relative speedup over the single-fidelity procedure.

Keywords: Antenna design · global optimization · computer-aided design · response features · bio-inspired algorithms · variable-fidelity EM analysis

L. Franco et al. (Eds.): ICCS 2024, LNCS 14836, pp. 19–34, 2024.
https://doi.org/10.1007/978-3-031-63775-9_2

1 Introduction

The development of contemporary antennas presents considerable challenges. On one hand, stringent performance requirements, guided by both existing and emerging application areas [1–3], encompass a range of electrical and field parameters (among others, wideband and multi-band operation, high gain, circular polarization, beam scanning [4–6]), reconfigurability [7], and constraints related to small physical dimensions [8]. On the other hand, addressing the intricacies of antenna geometries crafted to meet these demands [9, 10] poses a substantial challenge on its own. Traditional parametric studies, often employed for dimension adjustments, are found lacking in the control of multiple variables, not to mention the consideration of various design objectives or constraints. As a more robust alternative, formal numerical optimization methods [11] are recommended. However, the accurate assessment of antenna responses necessitates full-wave electromagnetic (EM) analysis, a resource-intensive process. Given that EM-driven optimization often requires a large number of antenna simulations, the resulting computational expenses could be prohibitive. Even the costs associated with local parameter adjustment, whether gradient-based [12] or stencil-based [13], may be significant. Global [14] or multi-objective optimization [15], as well as uncertainty quantification [16], entail considerably higher expenses, with the usual number of objective function evaluations easily reaching several thousand.

Despite the challenges associated with global optimization, its significance in antenna design is on the rise. On one hand, there is a growing number of inherently multimodal problems, such as pattern synthesis of antenna arrays, design of frequency-selective surfaces, or the creation of metamaterials and metasurfaces [17]. On the other hand, many antenna structures incorporate various topological alterations, such as stubs, defected ground structures, and shorting pins [18], aiming to enhance antenna performance but also introducing parameter redundancy and expanding the search space [19]. Other reasons for employing global optimization methods include the need for antenna re-design across an extensive range of operating conditions (e.g., center frequencies) or the absence of an initial design of sufficient quality. Currently, global optimization is predominantly driven by bio-inspired population-based procedures [20, 21]. These algorithms process sets of candidate solutions for the given problem. Their global search capability is often attributed to information exchange among individuals in the population [22], and the utilization of partially-stochastic mechanisms such as selection and recombination [23]. Various bio-inspired algorithms, including genetic and evolutionary algorithms, particle swarm optimizers (PSO), firefly algorithm, or grey wolf optimization, are widely used [24–27]. However, new methods in this category are continually proposed [28, 29]. While population-based procedures are structurally simple and easy to handle, their computational efficiency is inferior. The typical running costs of a nature-inspired algorithm can reach several thousand objective function calls. Thus, their direct use in EM-driven design is essentially prohibitive. In practice, these algorithms are employed only when the underlying merit function is inexpensive to evaluate (especially if it is analytical or EM-based but takes less than ten seconds per simulation) or when parallelization is facilitated by available resources and licensing.

Enabling practical electromagnetic (EM)-driven optimization using bio-inspired algorithms becomes feasible through the application of surrogate modeling techniques

[30]. A common configuration for a surrogate-assisted procedure involves utilizing the metamodel as a rapid predictor to generate candidate solutions, with iterative surrogate enhancement utilizing accumulated EM simulation data [31]. These candidate solutions, known as infill points, may be generated based on different criteria aiming to enhance model reliability (exploration), identify a globally-optimum design (exploitation), or strike a balance between exploitation and exploration [32]. The role of a bio-inspired search is to globally optimize the surrogate model or identify parameter space regions with the highest expected modeling error. The computational advantages arise from the fact that the majority of operations occur at the level of the fast metamodel rather than directly using EM analysis. Commonly utilized modeling techniques encompass kriging, Gaussian Process Regression (GPR), and neural networks [33, 34]. Optimization procedures following the discussed scheme are often referred to as machine learning (ML) frameworks [35]. Despite the advantages of surrogate-assisted methods, they face challenges related to the rendition of a dependable behavioral model. The complexity arises from the prononunced nonlinearity of antenna responses and the vast size of the search space. Typically, these procedures are demonstrated using relatively simple (especially low-dimensional) test cases [36]. Mitigation methods available in the literature include performance-driven modeling [37]; however, its integration into global search procedures is not straightforward. Other techniques include multi-fidelity EM simulations [38], and response feature technology [39], which proves useful for local tuning as well as surrogate modeling. Feature-based methods capitalize on re-stating the design problem with regard to carefully appointed characteristic points of the component outputs (in the case of antennas, usually, frequency and level values of their resonances) and leveraging nearly-linear dependence between these coordinates and original designable parameters (predominantly, geometry ones). Response features have been demonstrated to efficiently regularize the design task, resulting in faster convergence, and reducing the training dataset size necessary to construct a dependable behavioral model [40, 41].

The aim of this research is to present an innovative method for surrogate-based global optimization of antenna input characteristics. Our approach integrates a machine learning (ML) framework with kriging interpolation surrogates and multi-fidelity electromagnetic (EM) simulation models. In the initial phase of the search process, encompassing parameter space sampling and initial surrogate model construction, low-fidelity EM analysis is employed. Within the primary ML loop, the high-resolution model is used, and the model (predictor) generates candidate solutions (infill points) using the particle swarm optimizer (PSO) as the underlying search engine. The infill criterion is the minimization of the predicted objective function. The model is then refined with accumulated EM simulation data, and low- and high-fidelity data samples are merged into a single metamodel using co-kriging. The entire search process is conducted by re-stating the problem in terms of the response features of the antenna under consideration. Comprehensive verification studies demonstrate the remarkable computational efficiency of the proposed technique, with an average running cost not exceeding ninety high-fidelity EM simulations. This corresponds to a significant relative acceleration of up to seventy percent over various state-of-the-art benchmark methods, including both bio-inspired and ML algorithms. Simultaneously, our algorithm exhibits excellent reliability and repeatability of results.

2 Global Optimization Using Response Features and Multi-fidelity EM Analysis

This part of the paper outlines the developed optimization framework. We start by formulating the optimization problem in Sect. 2.1 with focus on multi-band antenna input characteristics. Section 2.2 reprises the notion of response features, whereas Sect. 2.3 is devoted to the variable-resolution EM simulations. Subsequent sections delineate kriging and co-kriging modelling (Sect. 2.4), which are employed for construction of a primary low-fidelity surrogate (Sect. 2.5), and high-fidelity model refinement (Sect. 2.6). Section 2.7 offers a summary of the entire procedure.

2.1 Antenna Design Task: Formulation

As we focus on optimization of the input characteristics of multi-band antennas, our aim is to arrange antenna resonances at specific locations, i.e., intended (target) frequencies $\boldsymbol{F}_t = [f_{t.1} \ldots f_{t.K}]^T$. We also strive to enhance the impedance matching at the same locations, which corresponds to minimization of the modulus of the reflection coefficient $|S_{11}|$ at $f_{t.j}$, $j = 1, \ldots, K$. This problem utilizes a minimax objective function. Table 1 provides the basic notation, using which the design task may be expressed as

$$\boldsymbol{x}^* = \arg\min_{\boldsymbol{x}} U(\boldsymbol{x}, \boldsymbol{F}_t) \tag{1}$$

Other design problems are also conceivable, such as maximization of the impedance bandwidth or antenna gain enhancement. Yet, the particular task discussed in this study may serve as an illustrative scenario frequently encountered in practical applications.

2.2 Concept of Response Features

The main hardship in EM-driven antenna design optimization is the significant cost of multiple EM analyses, inevitably invoked by numerical search process. This expenses are troublesome for local optimization and increase radically when global algorithms are at play. In the global search, the entire design space—vast both in terms of dimensionality and span—needs to be explored, which presents a challenging task. Nonlinearity of antenna outputs, particularly for multi-band antennas, add further complexity to this undertaking. Similarly, due to the response shape, rendition of a dependable surrogate accurately representing the characteristics of interest is intricate.

The response feature approach [42] can be employed for tackling the aforementioned difficulties. The response feature technology requires that the design task is redefined with regard to characteristic points of the antenna outputs. This allows to take the advantage of the nearly-linear dependency of the feature point coordinates (typically, frequencies and levels of characteristic locations in the frequency characteristics) on antenna dimensions [39]. Applying this method permits regularization of the merit function, and speeding up the convergence of the entire search. Additionally, quasi-global search capabilities are enabled [39], whereas the data set size for creating a dependable model is largely reduced [40]. It is of paramount importance that feature points coincide with the design goals [42]. For example, in the case of impedance matching improvement of

Table 1. Multi-band antenna optimization: Basic notation.

Symbol	Meaning	Comment
$x = [x_1 \ldots x_n]^T$	Vector of antenna designable variables	Typically, antenna geometry parameters
$S_{11}(x,f)$	Antenna reflection at a given design x and frequency f	Reflection coefficient is a complex number; we handle its modulus $\|S_{11}\|$ [dB]
$F_t = [f_{t.1} \ldots f_{t.K}]^T$	Vector of target operating frequencies	Frequencies corresponding to the intended allocation of antenna resonances
$U(x,F_t)$	Objective function (to be minimized by the search procedure)	Function quantifying the design quality, we use: $U(x,F_t) = \max_x \{\|S_{11}(x,f_{t.1})\|,\ldots,\|S_{11}(x,f_{t.K})\|\}$

multi-band structures, frequency and level locations pertinent to antenna resonances are particularly useful.

Let us define a response feature vector $\boldsymbol{f}_P(\boldsymbol{x}) = [\boldsymbol{f}_f(\boldsymbol{x})^T \ \boldsymbol{f}_L(\boldsymbol{x})^T]^T$, with $\boldsymbol{f}_f(\boldsymbol{x}) = [f_{f.1}(\boldsymbol{x}) \ldots f_{f.K}(\boldsymbol{x})]^T$ and $\boldsymbol{f}_L(\boldsymbol{x}) = [f_{L.1}(\boldsymbol{x}) \ldots f_{L.K}(\boldsymbol{x})]^T$ being its horizontal and vertical coordinates, respectively. The merit function may be expressed with regard to response features as

$$U_F(\boldsymbol{x},\boldsymbol{f}_P,\boldsymbol{F}_t) = \max_{\boldsymbol{x}}\{f_{L.1}(\boldsymbol{x}), \ldots, f_{L.K}(\boldsymbol{x})\} + \beta||\boldsymbol{f}_f - \boldsymbol{F}_t||^2 \quad (2)$$

In (2), the second term $\beta \ ||\boldsymbol{f}_f - \boldsymbol{F}_t||^2$ corresponds to a regularization factor, enforcing alignment of the resonant frequencies with their required allocations. The factor β should be set so as to guarantee that the regularization term sufficiently contributes when necessary (i.e., if resonances are largely misaligned (we have $\beta = 100$). Despite the fact that (2) differs from the minimax formulation of Table 1, yet, the respective optimum solutions coincide (provided they are reachable).

2.3 Multi-resolution EM Simulations

Low-fidelity models can expedite design optimization procedures mainly by diminishing the time needed for system evaluation at the expense of deteriorating its accuracy. In antenna design, low-resolution models are realized as coarse-discretization EM models. Depending on the intricacy of the antenna topology, various levels of acceleration can be achieved. Typical speedup ranges from below three to over ten in some cases.

In this study, the role of the low-resolution surrogate, labelled as $\boldsymbol{R}_c(\boldsymbol{x})$, will be twofold: (i) rendition (random) of observable set utilized for design space pre-screening, and (ii) construction of the primary surrogate (in the form of a kriging interpolant, as explained in Sect. 2.4). As the procedure advances, co-kriging model will be employed for merging low- and high-fidelity points to establish an enhanced surrogate. We will use $\boldsymbol{R}_f(\boldsymbol{x})$ to represent the high-fidelity model.

Table 2. Basic definitions of kriging and co-kriging modeling.

Model	Component	Analytical form
Kriging	Model formulation	$s_{KR}(\boldsymbol{x}) = \boldsymbol{M}\gamma + r(\boldsymbol{x})\cdot\boldsymbol{\Psi}^{-1}\cdot(\boldsymbol{R}_c(X_{Bc}) - \boldsymbol{F}\gamma)$ where: $\boldsymbol{M}$ – $N_{Bc} \times t$ model matrix of X_{Bc}, $\boldsymbol{F}$ – $1 \times t$ vector of the evaluation point $\boldsymbol{x}$, t –number of terms used in the regression function
	Regression function coefficients	$\gamma = (X_{Bc}{}^T\boldsymbol{\Psi}^{-1}X_{Bc})^{-1}X_{Bc}\boldsymbol{\Psi}^{-1}\boldsymbol{R}_f(X_{Bc})$
	Vector of correlations between $\boldsymbol{x}$ and X_{Bc}	$r(\boldsymbol{x}) = (\psi(\boldsymbol{x}, \boldsymbol{x}_{Bc}^{(1)}), \ldots, \psi(\boldsymbol{x}, \boldsymbol{x}_{Bc}^{(N_{Bc})}))$
	Correlation matrix	$\boldsymbol{\Psi} = [\Psi_{i,j}]$ is a correlation matrix, where $\Psi_{i,j} = \psi(\boldsymbol{x}_{Bf}^{(i)}, \boldsymbol{x}_{Bf}^{(j)})$
	Correlation function	$\psi(\boldsymbol{x}, \boldsymbol{x}') = \exp\left(\sum_{k=1}^{n} -\theta_k \lvert x^k - x'^k \rvert^P\right)$
	Model identification: finding hyperparameters θ_k, $k = 1, \ldots, n$, using Maximum Likelihood Estimation (MLE)	$(\theta_1, \ldots, \theta_n) = \arg\min_{\theta_1, \ldots, \theta_n}\left[-(N_{Bf}/2)\ln(\hat{\sigma}^2) - 0.5\ln(\lvert\Psi\rvert)\right]$ where $\hat{\sigma}^2 = (\boldsymbol{R}_f(X_{Bf}) - F\alpha)^T\boldsymbol{\Psi}^{-1}(\boldsymbol{R}_f(X_{Bf}) - F\alpha)/N_{Bf}$ and $\lvert\boldsymbol{\Psi}\rvert$ is the determinant of $\boldsymbol{\Psi}$. In practice, a Gaussian correlation function ($P = 2$) is often employed, as well as $\boldsymbol{F} = [1 \ldots 1]^T$ and $\boldsymbol{M} = 1$.
Co-kriging	Model formulation	$s_{CO}(\boldsymbol{x}) = \boldsymbol{M}\gamma + r(\boldsymbol{x})\cdot\boldsymbol{\Psi}^{-1}\cdot(\boldsymbol{r} - \boldsymbol{F}\gamma)$
	Vector of correlations	$r(\boldsymbol{x}) = [\rho\cdot\sigma_c^2\cdot r_c(\boldsymbol{x}), \rho^2\cdot\sigma_c^2\cdot r_c(\boldsymbol{x}, X_{B_f}) + \sigma_d^2\cdot r_d(\boldsymbol{x})]$
	Correlation matrix	$\boldsymbol{\Psi} = \begin{bmatrix} \sigma_c^2\boldsymbol{\Psi}_c(X_{Bc}, X_{Bc}) & \rho\sigma_c^2\boldsymbol{\Psi}_c(X_{Bc}, X_{Bf}) \\ \rho\sigma_c^2\boldsymbol{\Psi}_c(X_{Bf}, X_{Bc}) & \rho^2\sigma_c^2\boldsymbol{\Psi}_c(X_{Bf}, X_{Bf}) + \sigma_d^2\boldsymbol{\Psi}_d \end{bmatrix}$ where $\boldsymbol{M} = [\rho\boldsymbol{M}_c\ \boldsymbol{M}_d]$, whereas $(\boldsymbol{F}_c, \sigma_c, \boldsymbol{\Psi}_c, \boldsymbol{M}_c)$ and $(\boldsymbol{F}_d, \sigma_d, \boldsymbol{\Psi}_d, \boldsymbol{M}_d)$ are matrices obtained from $\boldsymbol{s}_{KRc}$ and $\boldsymbol{s}_{KRf}$, respectively; parameter ρ is included in the MLE during model identification.

2.4 Kriging and Co-kriging Metamodels

Kriging and co-kriging models [38] are utilized in this study as predictors guiding optimization procedure towards global solution. We will use $\{\boldsymbol{x}_{Bc}{}^{(k)}, \boldsymbol{R}_c(\boldsymbol{x}_{Bc}{}^{(k)})\}_{k=1,\ldots,NBc}$, to denote the low-fidelity dataset comprising designs $\boldsymbol{x}_{Bc}{}^{(k)}$ with relevant responses. Let also $\{\boldsymbol{x}_{Bf}{}^{(k)}, \boldsymbol{R}_f(\boldsymbol{x}_{Bf}{}^{(k)})\}_{k=1,\ldots,NBf}$, represent the high-fidelity dataset, EM-evaluated at a high-fidelity resolution at the vectors $\boldsymbol{x}_{Bf}{}^{(k)}$.

Table 2 compiles the relevant notation as well as the details pertaining to kriging and co-kriging surrogates: $\boldsymbol{s}_{KR}(\boldsymbol{x})$ and $\boldsymbol{s}_{CO}(\boldsymbol{x})$, respectively. In general, co-kriging surrogate constitutes a composite of two models. The first one is a kriging model $\boldsymbol{s}_{KRc}$ whose training data set is of low-resolution, we have $(X_{Bc}, \boldsymbol{R}_c(X_{Bc}))$. Whereas the second model, $\boldsymbol{s}_{KRf}$, is established based on the residuals $(X_{Bf}, \boldsymbol{r})$, with $\boldsymbol{r} = \boldsymbol{R}_f(X_{Bf}) - \rho\cdot\boldsymbol{R}_c(X_{Bf})$, where ρ is derived by the Maximum Likelihood Estimation (MLE) of the second model [43]. Both $\boldsymbol{s}_{KR}(\boldsymbol{x})$ and $\boldsymbol{s}_{CO}(\boldsymbol{x})$ utilize the same correlation function, see Table 2.

2.5 Pre-screening of Design Space. Primary Surrogate Construction

In our approach, the first stage of the search process consists in gathering an ensemble of random low-fidelity parameter vectors. Using the eligible samples (i.e., such whose frequency characteristics enable feature point extraction) the primary surrogate $\boldsymbol{s}^{(0)}(\boldsymbol{x})$ is established as a kriging interpolation metamodel [38].

$$\boldsymbol{s}^{(0)}(\boldsymbol{x}) = \left[\left[s_{f.1}^{(0)}(\boldsymbol{x}) \ldots s_{f.K}^{(0)}(\boldsymbol{x})\right]^T \left[s_{L.1}^{(0)}(\boldsymbol{x}) \ldots s_{L.K}^{(0)}(\boldsymbol{x})\right]^T\right]^T \tag{3}$$

Observe that the surrogate (3) generates predictions of the feature's coordinates. The model $\boldsymbol{s}^{(0)}(\boldsymbol{x})$ is identified using the training samples comprising vectors $\boldsymbol{x}_{Bc}{}^{(j)}, j = 1, \ldots, N_{init}$, along with the corresponding features $\boldsymbol{f}_P(\boldsymbol{x}_{Bc}{}^{(j)})$, assessed based on low-resolution simulations. The generation of the data points is carried out sequentially, and only the vectors of identifiable features are incorporated (see Fig. 1). The number of the initial samples N_{init} (in practice, between 50 and 200) is selected based on the required metamodel accuracy, with the parameter $E_{\max}$ (acceptance threshold for a relative RMS error) set by user. The data generation procedure is illustrated in Fig. 2. The specific number of random points allowing to acquire N_{init} decent samples is usually twice (or three times) as large as N_{init}.

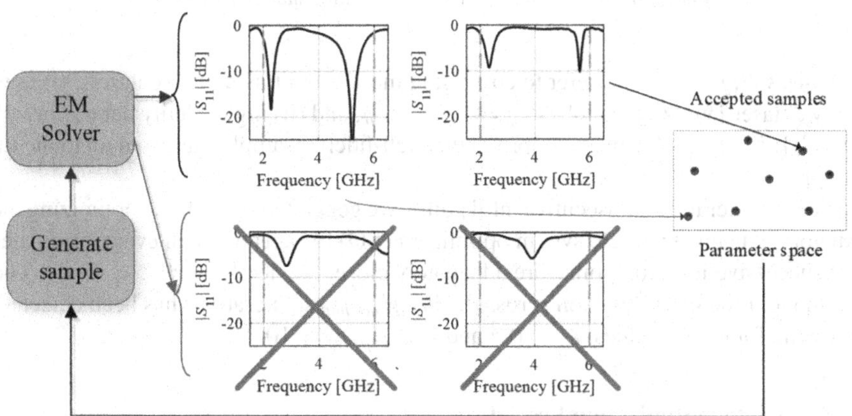

Fig. 1. Observable generation for primary model construction, which is identified using the samples whose resonances belong to the assumed target range.

2.6 PSO-Based Infill Points Generation. Co-Kriging Model

Once N_{init} low-fidelity data samples have been collected, the initial surrogate (3) is established. The main phase of the proposed procedure involves refinement of $\boldsymbol{s}^{(0)}(\boldsymbol{x})$ based on high-resolution data $\boldsymbol{x}_f{}^{(i)}$, $i = 1, 2, \ldots$, which are evaluated as

$$\boldsymbol{x}_f^{(i+1)} = \arg\min_{\boldsymbol{x}\in X} U_F(\boldsymbol{x}, \boldsymbol{s}^{(i)}(\boldsymbol{x}), \boldsymbol{F}_t) \tag{4}$$

1. Input parameters:
   - Design space X (interval [$\boldsymbol{l}$ $\boldsymbol{u}$], where $\boldsymbol{l}$ and $\boldsymbol{u}$ are lower and upper bounds for designable parameters; other constraints are possible per designer's needs);
   - Required modelling error $E_{\max}$;
2. Set $j = 0$;
3. Generate a sample $\boldsymbol{x}_{tmp} \in X$ using uniform probability distribution;
4. Evaluate antenna responses at $\boldsymbol{x}_{tmp}$ using low-resolution EM analysis;
5. Extract the feature vector $\boldsymbol{f}_P(\boldsymbol{x}_{tmp})$; assign $\boldsymbol{f}_P(\boldsymbol{x}_{tmp}) = 0$ if features are not extractable;
6. **if** $||\boldsymbol{f}_P(\boldsymbol{x}_{tmp})|| > 0$

   Set $j = j + 1$;

   Set $\boldsymbol{x}_{Bc}^{(j)} = \boldsymbol{x}_{tmp}$;

   Construct surrogate model $\boldsymbol{s}_{tmp}(\boldsymbol{x})$ using dataset $\{\boldsymbol{x}_{Bc}^{(k)},\boldsymbol{f}_P(\boldsymbol{x}_B^{(k)})\}_{k=1,\ldots,j}$;

   Estimate model error E_{tmp} using K-fold cross-validation, $K = \min\{j,10\}$;

   if $E_{tmp} < E_{\max}$

   Go to 8;

   end

   end
7. Go to 3;
8. Return $\boldsymbol{s}^{(0)}(\boldsymbol{x}) = \boldsymbol{s}_{tmp}(\boldsymbol{x})$;

Fig. 2. Primary surrogate model: Training data generation.

In (4), $\boldsymbol{s}^{(j)}, j = 1, 2, \ldots$, refer to co-kriging models built using two datasets: (i) low-fidelity dataset $\{\boldsymbol{x}_{Bc}{}^{(j)}, \boldsymbol{f}_P(\boldsymbol{x}_{Bc}{}^{(j)})\}, j = 1, \ldots, N_{init}$, and (ii) high-fidelity dataset $\{\boldsymbol{x}_f{}^{(j)}, \boldsymbol{f}_P(\boldsymbol{x}_f{}^{(j)})\}, j = 1, \ldots, I$, which comprises the high-fidelity samples gathered up to the ith iteration.

In current iteration, consecutive infill points are generated by globally optimizing the predictor $\boldsymbol{s}^{(i)}$ using a particle swarm optimizer (PSO) [44]. The search event terminates if: (i) successive iteration points are adequately close, i.e., $||\boldsymbol{x}^{(i+1)} - \boldsymbol{x}^{(i)}|| < \varepsilon$, or (ii) no betterment of the merit function across the last $N_{no\_improve}$ iterations has been detected. We use the following values: $\varepsilon = 10^{-2}$ and $N_{no\_improve} = 10$.

2.7 Complete Optimization Procedure

This section provides delineation of the complete global optimization procedure. We have only three control parameters, summarized in Table 3. The discussion on parameters ε and $N_{no\_improve}$ has been already given in Sect. 2.6. As for the last parameter $E_{\max}$, it represents an acceptance threshold for the relative RMS error of the primary model. We use $E_{\max} = 2\%$, yet, virtually any value below ten percent may be used, because the model operates at response features. This makes functional landscape rather regular, in contrast to far more rugged landscape of entire antenna responses.

Figure 3 summarizes operating steps our algorithm. First, input data is delivered (Step 1). The algorithm is launched with the pre-screening stage (Step 2; see Sect. 2.5). This is followed by building the initial model (Step 3), which later on serves as a predictor for rendition of the first high-fidelity infill point (Step 5). In Steps 6 and 8, the surrogate update is performed, and the entire infill process continues until convergence.

1. Input parameters:
   - Target operating frequencies $\boldsymbol{F}_t$ (cf. Section 2.1);
   - Definition of the response features $\boldsymbol{f}_P$ and the objective function U_F (cf. Section 2.2);
   - Design space X (interval $[\boldsymbol{l}\ \boldsymbol{u}]$, where $\boldsymbol{l}$ and $\boldsymbol{u}$ are lower and upper bounds for parameters);
   - Required modelling error E_{max};
   - Termination thresholds ε and $N_{no\_improve}$;
2. Generate the set of low-resolution initial samples $\{\boldsymbol{x}_{Bc}^{(k)}, \boldsymbol{f}_P(\boldsymbol{x}_{Bc}^{(k)})\}_{k=1,\ldots,Ninit}$, (see Section 2.5),
3. Construct the initial (kriging) surrogate model $\boldsymbol{s}^{(0)}(\boldsymbol{x})$;
4. Set $i = 0$;
5. Obtain infill point $\boldsymbol{x}_f^{(i+1)}$ by solving (4) using the PSO algorithm:
   $$\boldsymbol{x}_f^{(i+1)} = \arg\min_{\boldsymbol{x}\in X} U_F(\boldsymbol{x}, \boldsymbol{s}^{(i)}(\boldsymbol{x}), \boldsymbol{F}_t)$$
6. Update the dataset: $\{\boldsymbol{x}_{Bc}^{(k)}, \boldsymbol{f}_P(\boldsymbol{x}_{Bc}^{(k)})\}_{k=1,\ldots,Ninit} \cup \{\boldsymbol{x}_f^{(k)}, \boldsymbol{f}_P(\boldsymbol{x}_f^{(k)})\}_{k=1,\ldots,i}$
7. Set $i = i + 1$;
8. Construct the co-kriging surrogate model $\boldsymbol{s}^{(i)}(\boldsymbol{x})$ using the updated dataset;
9. **if** $||\boldsymbol{x}^{(i)} - \boldsymbol{x}^{(i-1)}|| < \varepsilon$ OR no objective function improvement for $N_{no\_improve}$ iterations
   Go to 11;
   end
10. Go to 5;
11. Return $\boldsymbol{x}^* = \boldsymbol{x}_f^{(i)}$;

Fig. 3. Pseudocode of the developed global optimization framework.

Table 3. Global multi-fidelity optimization framework: Control parameters.

Parameter	Meaning	Default value
E_{max}	Maximum value of relative RMS error of the initial surrogate model (error estimated using cross-validation), cf. Section 2.5	2%
ε	Termination threshold for convergence in argument, cf. Section 2.6	10^{-2}
$N_{no\ improve}$	Termination threshold for no objective function value improvement, cf. Section 2.6	10

3 Verification Case Studies

The efficacy of the developed global optimization procedure is verified in this section with the use of two antenna examples and comparisons with four state-of-the-art benchmark algorithms.

3.1 Test Cases and Experimental Setup

The introduced algorithm is applied to two microstrip antenna structures shown in Figs. 4 and 5 respectively. The same figures provide essential information about antenna substrate parameters and design variables. The EM models are realized using CST Microwave Studio [47]. The low-resolution models are obtained by reducing the discretization density of the structures (cf. Table 4). Our design goal is to relocate antenna resonant frequencies to the assumed target values. The target operating frequencies are $\boldsymbol{f}_t = [2.45\ 5.3]^T$ GHz (Antenna I) and $\boldsymbol{f}_t = [3.5\ 5.8\ 7.5]^T$ GHz (Antenna II). The lower and upper bounds for design variables are $\boldsymbol{l} = [15\ 3\ 0.35\ 0.2\ 1.8\ 0.5]^T$, $\boldsymbol{u} = [50\ 12\ 0.85$

1.5 4.3 2.7]^T (Antenna I), and $\boldsymbol{l}$ = [10 17 0.2 45 5 0.4 0.15 0.2 0.1 0.5 0.1]^T, $\boldsymbol{u}$ = [16 25 0.6 55 15 0.5 0.3 0.8 0.4 0.65 0.5]^T (Antenna II).

Antennas I and II were optimized using the proposed algorithm (setup: $E_{\max} = 2\%$, $\varepsilon = 10^{-2}$, $N_{no\_improve} = 10$). The benchmark algorithms have been outlined in Table 5. We have: PSO, a multiple-start gradient search, and two ML algorithms. The first one is a surrogate-based procedure with kriging metamodels, tackling complete antenna frequency characteristics. The second is akin to the feature-based algorithm proposed in Sect. 2 but only employs the high-resolution EM model at all search stages.

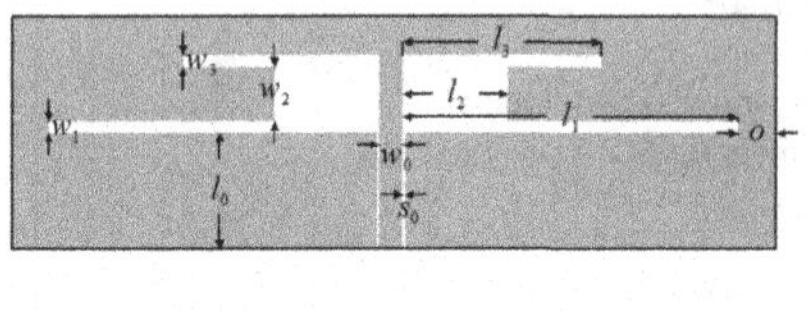

(a)

Substrate	RO4350 substrate (ε_r = 3.48, h = 0.762 mm)
Design variables	$\boldsymbol{x} = [l_1\ l_2\ l_{3r}\ w_1\ w_2\ w_3]^T$
Other variables	$l_3 = l_{3r}l_1$; l_0 = 30, w_0 = 3, s_0 = 0.15, o = 5; all dimensions except l_{3r} (which is relative) are in mm

(b)

Fig. 4. Antenna I [45]: (a) geometry, (b) essential parameters.

(a)

Substrate	ε_r = 3.2, h = 3.06 mm
Design variables	$\boldsymbol{x} = [L_1\ L_s\ L_{ur}\ W\ W_1\ dL_r\ dW_r\ g\ l_{s1r}\ l_{s2r}\ w_{ur}]^T$
Other variables	b = 1, w_f = 7.4, s = 0.5, w = 0.5, $dL_2 = L_1$, $L = L_s + g + L_1 + dL_2$, $L_u = L_{ur}W_1$, $dL = dL_rL$, $dW = dW_rW$, $l_{s1} = l_{s1r}(L - dL)$, $l_{s2} = l_{s2r}(W - dW)$, $w_u = w_{ur}(L_1 - b - s)$

(b)

Fig. 5. Antenna II [46]: (a) geometry, the light-shade grey indicates a ground-plane slot, (b) essential parameters.

Table 4. Computational models for Antennas I and II.

Antenna	EM simulation model			
	Low-fidelity $\boldsymbol{R}_c$		High-fidelity $\boldsymbol{R}_f$	
	Discretization density (# of mesh cells)	Simulation time [s]	Discretization density (# of mesh cells)	Simulation time [s]
I	~60,000	25	~410,000	92
II	~160,000	42	~800,000	165

3.2 Results and Discussion

Given the stochastic nature of the procedures under consideration, each algorithm has been executed ten times, and the resulting statistics are presented in Tables 6 and 7 for

Table 5. Benchmark algorithms.

Algorithm	Algorithm type	Setup
I	Particle swarm optimizer (PSO)	Swarm size $N = 10$, standard control parameters ($\chi = 0.73$, $c_1 = c_2 = 2.05$); number of iterations set to 50 (version I) and 100 (version II)
II	Trust-region gradient based optimizer [48]	Random initial design, response gradients estimated using finite differentiation, termination criteria based on convergence in argument and reduction of the trust region size [48]
III	ML algorithm operating on complete antenna characteristics	Algorithm highlights: • Initial surrogate set up to ensure relative RMS error not higher than 10% with the max. number of training samples equal to 400; • Optimization based on processing the antenna frequency characteristics (unlike response features in the proposed procedure); • Infill criterion: minimization of the predicted objective function [49].
IV	Feature-based ML algorithm utilizing high-fidelity EM model only	Algorithm highlights: • Surrogate model constructed at the level of response features; • Optimization process only uses high-fidelity EM simulations; • Infill criterion: minimization of the predicted objective function [49].

Antenna I and II, respectively. Figures 6 and 7 illustrate the reflection characteristics and the objective function evolution for selected runs of the algorithm. The assessment of the developed optimization procedure's performance is conducted below, taking into account factors such as the dependability of the optimization process, design quality, and cost-efficacy.

Reliability: The reliability is gauged through the success rate, representing the number of successful runs (out of ten) where the algorithm successfully positioned the antenna resonances at the intended targets. Our algorithm demonstrates a flawless success rate, comparable only to the two ML-based benchmark methods (Algorithms III and IV). Other techniques exhibit noticeable inferiority. Notably, in the case of PSO, the results highlight the significantly higher computational budget required for bio-inspired optimization.

Design Quality: Design quality is evaluated by the mean value of the cost function, which remains comparable for all algorithms. The reported differences hold minor practical importance. The seemingly poorer values displayed by Algorithms I and II result from showcasing the average performance, reduced by unsuccessful runs.

Computational Efficiency: The proposed algorithm stands out for its lower running cost compared to other global search algorithms. The average expenses, assessed as the equivalent number of high-resolution EM simulations, are only 65 and 90 for Antenna I and II, respectively. This implies a cost reduction of 30% and 74% for Antenna I and II compared to Algorithm IV (similar to the proposed one but using the high-fidelity model only). It is noteworthy that the computational benefits increase with problem complexity. The majority of antenna evaluations occur in the initial stage of the search procedure (parameter space pre-screening and initial surrogate model rendition) with the use of the low-fidelity model. Another observation is that the proposed algorithm only required 77 and 259 random observables in the first stage of the optimization process

Table 6. Optimization results for Antenna I.

Optimization algorithm		Performance figure		
		Average objective function value [dB]	Computational cost[$]	Success rate[#]
Algorithm I: PSO	50 iterations	–18.2	500	9/10
	100 iterations	–19.3	1,000	10/10
Algorithm II: Trust-region gradient-based algorithm		–13.5	84.2	6/10
Algorithm III: machine learning algorithm processing complete antenna responses		–20.7	457.8	10/10
Algorithm IV: feature-based machine learning algorithm using high-fidelity EM model		–20.3	92.3	10/10
Proposed algorithm		–23.9	65.3	10/10

[$] The cost expressed in terms of the number of EM simulations of the antenna structure under design.
[#] Number of algorithms runs at which the operating frequencies were allocated in the vicinity of the target frequencies.

Table 7. Optimization results for Antenna II.

Optimization algorithm		Performance figure		
		Average objective function value [dB]	Computational cost[$]	Success rate[#]
Algorithm I: PSO	50 iterations	–12.3	500	6/10
	100 iterations	–14.2	1,000	8/10
Algorithm II: Trust-region gradient-based algorithm		–12.1	125.4	4/10
Algorithm III: machine learning algorithm processing complete antenna responses		–14.2	473.0	7/10
Algorithm IV: feature-based machine learning algorithm using high-fidelity EM model		–17.9	347.0	10/10
Proposed algorithm		–15.3	89.8	10/10

[$] The cost expressed in terms of the number of EM simulations of the antenna structure under design.
[#] Number of algorithms runs at which the operating frequencies were allocated in the vicinity of the target frequencies.

for Antennas I and II (in both cases, achieving the modeling error E_{max}). Conversely, Algorithm III, working with complete antenna responses, could not reach the required accuracy limit, necessitating the establishment of the surrogate using 400 data samples (the allowed budget). Thus, levera:ging response features leads to a notable reduction in computational expenses during this stage of the process.

This analysis suggests that the proposed algorithm serves as a viable alternative to existing global search techniques. Although demonstrated for multi-band antennas, comparable performance is anticipated for other problem types, provided they can be reformulated using response features. The primary advantages of our technique lie in reliability, computational efficiency, and a straightforward setup: with only three control parameters, two of which relate to termination criteria.

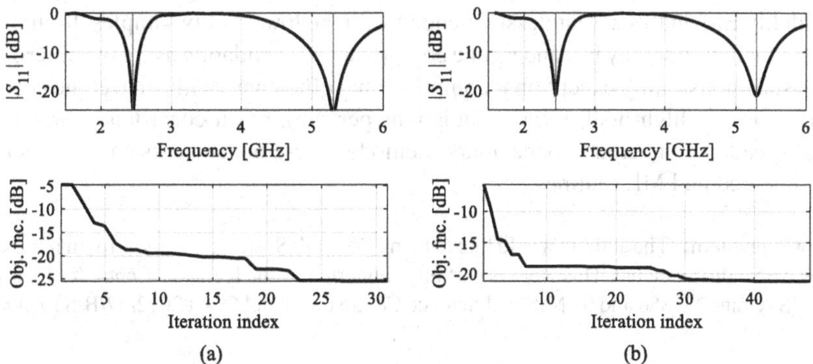

Fig. 6. $|S_{11}|$ of Antenna I evaluated for the designs generated by our algorithm (top) and evolution of the objective function value (bottom), shown for typical algorithm executions: (a) run 1, (b) run 2. The iteration counter starts upon constructing the initial surrogate model. Vertical lines represent the assumed operating frequencies, here 2.45 GHz and 5.3 GHz.

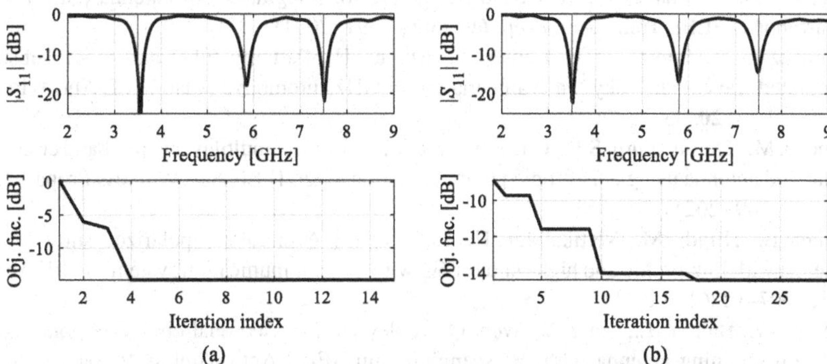

Fig. 7. $|S_{11}|$ of Antenna II evaluated for the designs generated by our algorithm (top) and evolution of the objective function value (bottom), shown for typical algorithm executions: (a) run 1, (b) run 2. The iteration counter starts after constructing the initial surrogate model. Vertical lines represent the assumed operating frequencies, here 3.5 GHz, 5.8 GHz, and 7.5 GHz.

4 Conclusion

This study has introduced an innovative approach to achieving low-cost global optimization for antenna structures. Our methodology capitalizes on the characteristics of response features, employing a feature-based reformulation to serve as a tool for regularization of the optimization task. The search process is structured within a machine learning (ML) framework, utilizing kriging interpolation surrogates as the primary predictor. This surrogate is employed to generate promising candidate solutions (infill points), with iterative refinement using accumulated electromagnetic (EM) simulation data. The involvement of multi-fidelity EM simulations is instrumental in reducing the algorithm's running cost. Specifically, initial parameter space sampling and surrogate model construction are executed using low-fidelity EM simulations, while the ML process leverages high-fidelity EM analysis, seamlessly blended with the low-fidelity samples through co-kriging. Our methodology has undergone comprehensive validation using two microstrip antennas, demonstrating exceptional computational efficiency (with an average cost not exceeding ninety high-fidelity EM simulations per run), result consistency, and a significant speedup compared to benchmark methods. These benchmarks encompass both nature-inspired and ML routines.

Acknowledgement. The authors would like to thank Dassault Systemes, France, for making CST Microwave Studio available. This work is partially supported by the Icelandic Centre for Research (RANNIS) Grant 239858 and by National Science Centre of Poland Grant 2022/47/B/ST7/00072.

References

1. Yuan, X.-T., Chen, Z., Gu, T., Yuan, T.: A wideband PIFA-pair-based MIMO antenna for 5G smartphones. IEEE Ant. Wireless Propag. Lett. **20**, 371–375 (2021)
2. Sun, L., Li, Y., Zhang, Z.: Wideband decoupling of integrated slot antenna pairs for 5G smartphones. IEEE Trans. Ant. Prop. **69**, 2386–2391 (2021)
3. Kapusuz, K.Y., Berghe, A.V., Lemey, S., Rogier, H.: Partially filled half-mode substrate integrated waveguide leaky-wave antenna for 24 GHz automotive radar. IEEE Ant. Wireless Propag. Lett. **20**, 33–37 (2021)
4. Ameen, M., Thummaluru, S.R., Chaudhary, R.K.: A compact multilayer triple-band circularly polarized antenna using anisotropic polarization converter. IEEE Ant. Wireless Propag. Lett. **20**, 145–149 (2021)
5. Aqlan, B., Himdi, M., Vettikalladi, H., Le-Coq, L.: A circularly polarized sub-terahertz antenna with low-profile and high-gain for 6G wireless communication systems. IEEE Access **9**, 122607–122617 (2021)
6. Wen, Z.-Y., Ban, Y.-L., Yang, Y., Wen, Q.: Risley-prism-based dual-circularly polarized 2-D beam scanning antenna with flat scanning gain. IEEE Ant. Wireless Propag. Lett. **20**, 2412–2416 (2021)
7. Shirazi, M., Li, T., Huang, J., Gong, X.: A reconfigurable dual-polarization slot-ring antenna element with wide bandwidth for array applications. IEEE Trans. Ant. Prop. **66**, 5943–5954 (2018)
8. Chen, C.: A compact wideband endfire filtering antenna inspired by a uniplanar microstrip antenna. IEEE Ant. Wireless Propag. Lett. **21**, 853–857 (2022)
9. Wu, Y.F., Cheng, Y.J., Zhong, Y.C., Yang, H.N.: Substrate integrated waveguide slot array antenna to generate Bessel beam with high transverse linear polarization purity. IEEE Trans. Ant. Propag. **70**, 750–755 (2022)
10. Lee, J., Kim, H., Oh, J.: Large-aperture metamaterial lens antenna for multi-layer MIMO transmission for 6G. IEEE Access **10**, 20486–20495 (2022)
11. Chen, Z., Zhang, H.C., Cheng, Q.S.: Surrogate-assisted quasi-Newton enhanced global optimization of antennas based on a heuristic hypersphere sampling. IEEE Trans. Ant. Propag. **69**, 2993–2998 (2021)
12. Nocedal, J., Wrigth, S.J.: Numerical Optimization, 2nd edn. Springer, New York (2006)
13. Conn, A.R., Scheinberg, K., Vicente, L.N.: Derivative-Free Optimization. MPS-SIAM Series on Optimization, Society for Industrial and Applied Mathematics, Philadelphia (2009
14. Xu, Y., et al.: A reinforcement learning-based multi-objective optimization in an interval and dynamic environment. Knowl.-Based Syst. **280**, paper No. 111019 (2023)

15. Zhu, D.Z., Werner, P.L., Werner, D.H.: Design and optimization of 3-D frequency-selective surfaces based on a multiobjective lazy ant colony optimization algorithm. IEEE Trans. Ant. Propag. **65**, 7137–7149 (2017)
16. Du, J., Roblin, C.: Stochastic surrogate models of deformable antennas based on vector spherical harmonics and polynomial chaos expansions: application to textile antennas. IEEE Trans. Ant. Prop. **66**, 3610–3622 (2018)
17. Blankrot, B., Heitzinger, C.: Efficient computational design and optimization of dielectric metamaterial structures. IEEE J. Multiscale Multiphys. Comp. Techn. **4**, 234–244 (2019)
18. Qian, B., Huang, X., Chen, X., Abdullah, M., Zhao, L., Kishk, A.A.: Surrogate-assisted defected ground structure design for reducing mutual coupling in 2×2 microstrip antenna array. IEEE Ant. Wireless Propag. Lett. **21**, 351–355 (2022)
19. Koziel, S., Pietrenko-Dabrowska, A.: Reliable EM-driven size reduction of antenna structures by means of adaptive penalty factors. IEEE Trans. Ant. Propag. **70**, 1389–1401 (2021)
20. Li, W., Zhang, Y., Shi, X.: Advanced fruit fly optimization algorithm and its application to irregular subarray phased array antenna synthesis. IEEE Access **7**, 165583–165596 (2019)
21. Jia, X., Lu, G.: A hybrid Taguchi binary particle swarm optimization for antenna designs. IEEE Ant. Wireless Propag. Lett. **18**, 1581–1585 (2019)
22. Ding, D., Wang, G.: Modified multiobjective evolutionary algorithm based on decomposition for antenna design. IEEE Trans. Ant. Propag. **61**, 5301–5307 (2013)
23. Zhang, H., Bai, B., Zheng, J., Zhou, Y.: Optimal design of sparse array for ultrasonic total focusing method by binary particle swarm optimization. IEEE Access **8**, 111945–111953 (2020)
24. Michalewicz, Z.: Genetic Algorithms + Data Structures = Evolution Programs. Springer, New York (1996)
25. Wang, D., Tan, D., Liu, L.: Particle swarm optimization algorithm: an overview. Soft. Comput. **22**, 387–408 (2018)
26. Jiang, Z.J., Zhao, S., Chen, Y., Cui, T.J.: Beamforming optimization for time-modulated circular-aperture grid array with DE algorithm. IEEE Ant. Wireless Propag. Lett. **17**, 2434–2438 (2018)
27. Baumgartner, P., et al.: Multi-objective optimization of Yagi-Uda antenna applying enhanced firefly algorithm with adaptive cost function. IEEE Trans. Mag. **54**, article no. 8000504 (2018)
28. Mostafa, R.R., Gaheen, M.A., El-Aziz, M.A., Al-Betar, Ewees, A.A.: An improved gorilla troops optimizer for global optimization problems A.A. Knowl.-Based Syst. **269**, paper No. 110462 (2023)
29. Ram, G., Mandal, D., Kar, R., Ghoshal, S.P.: Cat swarm optimization as applied to time-modulated concentric circular antenna array: analysis and comparison with other stochastic optimization methods. IEEE Trans. Antennas Propag. **63**, 4180–4183 (2015)
30. Easum, J.A., Nagar, J., Werner, P.L., Werner, D.H.: Efficient multi-objective antenna optimization with tolerance analysis through the use of surrogate models. IEEE Trans. Ant. Prop. **66**, 6706–6715 (2018)
31. Jones, D.R., Schonlau, M., Welch, W.J.: Efficient global optimization of expensive black-box functions. J. Global Opt. **13**, 455–492 (1998)
32. Chen, C., Liu, J., Xu P.: Comparison of infill sampling criteria based on Kriging surrogate model. Sc. Rep. **12**, Art. No. 678 (2022)
33. De Villiers, D.I.L., Couckuyt, I., Dhaene T.: Multi-objective optimization of reflector antennas using kriging and probability of improvement. In: International Symposium on Antennas and Propagation, San Diego, USA, pp. 985–986 (2017)
34. Qin, J., Dong, W., Wang, M.: Fast multi-objective optimization of multi-parameter antenna structures based on improved BPNN surrogate model. IEEE Access **7**, 77692–77701 (2019)
35. Hu, C., Zeng, S., Li C.: A framework of global exploration and local exploitation using surrogates for expensive optimization. Knowl.-Based Syst. **280**, paper No. 111018 (2023)

36. Xiao, S., et al.: Multi-objective Pareto optimization of electromagnetic devices exploiting kriging with Lipschitzian optimized expected improvement. IEEE Trans. Magn. **54**, paper No. 7001704 (2018)
37. Koziel, S., Pietrenko-Dabrowska, A.: Performance-Driven Surrogate Modeling of High-Frequency Structures. Springer, New York (2020)
38. Pietrenko-Dabrowska, A., Koziel, S.: Antenna modeling using variable-fidelity EM simulations and constrained co-kriging. IEEE Access **8**, 91048–91056 (2020)
39. Pietrenko-Dabrowska, A., Koziel, S.: Generalized formulation of response features for reliable optimization of antenna input characteristics. IEEE Trans. Ant. Propag. **70**, 3733–3748 (2021)
40. Pietrenko-Dabrowska, A., Koziel, S.: Simulation-driven antenna modeling by means of response features and confined domains of reduced dimensionality. IEEE Access **8**, 228942–228954 (2020)
41. Pietrenko-Dabrowska, A., Koziel, S., Ullah U.: Reduced-cost two-level surrogate antenna modeling using domain confinement and response features. Sc. Rep. **12**, Art. No. 4667 (2022)
42. Koziel, S.: Fast simulation-driven antenna design using response-feature surrogates. Int. J. RF & Micr. CAE **25**, 394–402 (2015)
43. Kennedy, M.C., O'Hagan, A.: Predicting the output from complex computer code when fast approximations are available. Biometrika **87**, 1–13 (2000)
44. Vinod Chandra, S.S., Anand, H.S.: Nature inspired meta heuristic algorithms for optimization problems. Computing **104**, 251–269 (2022)
45. Chen, Y.-C., Chen, S.-Y., Hsu, P.: Dual-band slot dipole antenna fed by a coplanar waveguide. In: Proceedings of IEEE Antennas and Propagation Society International Symposium, Albuquerque, NM, USA, pp. 3589–3592 (2006)
46. Consul, P.: Triple band gap coupled microstrip U-slotted patch antenna using L-slot DGS for wireless applications. In: Communication, Control and Intelligent Systems (CCIS), Mathura, India, pp. 31–34 (2015)
47. CST Microwave Studio, ver. 2021, Dassault Systemes, France (2021)
48. Conn, A.R., Gould, N.I.M., Toint, P.L.: Trust Region Methods, MPS-SIAM Series on Optimization (2000)
49. Liu, J., Han, Z., Song, W.: Comparison of infill sampling criteria in kriging-based aerodynamic optimization. In: 28thInternational Congress of the Aeronautical Sciences, Brisbane, Australia, pp. 1–10, 23–28 September 2012

Exploring Apple Silicon's Potential from Simulation and Optimization Perspective

Karol Struniawski[1], Aleksandra Konopka[1(✉)], and Ryszard Kozera[1,2]

[1] Institute of Information Technology, Warsaw University of Life Sciences - SGGW, ul. Nowoursynowska 159, 02-776 Warsaw, Poland
{karol_struniawski,aleksandra_konopka,ryszard_kozera}@sggw.edu.pl

[2] School of Physics, Mathematics and Computing, The University of Western Australia, 35 Stirling Highway, Crawley, Perth 6009, Australia

Abstract. This study explores the performance of Apple Silicon processors in real-world research tasks, with a specific focus on optimization and Machine Learning applications. Diverging from conventional benchmarks, various algorithms across fundamental datasets have been assessed using diverse hardware configurations, including Apple's M1 and M2 processors, NVIDIA RTX 3090 GPU and a mid-range laptop. The M2 demonstrates competitiveness in tasks such as *BreastCancer, liver* and *yeast* classification, establishing it as a suitable platform for practical applications. Conversely, the dedicated GPU outperformed M1 and M2 on the *eyestate1* dataset, underscoring its superiority in handling more complex tasks, albeit at the expense of substantial power consumption. With the technology advances, Apple Silicon emerges as a compelling choice for real-world applications, warranting further exploration and research in chip development. This study underscores the critical role of device specifications in evaluating Machine Learning algorithms.

Keywords: Machine Learning · Optimization · Simulation · Apple Silicon · Extreme Learning Machine

1 Introduction

In November 2020, Apple introduced a new line of processors, starting with the M1 chip that adopts a System on a Chip (SoC) design with unified memory. The M1 processor, built using 5nm process technology and containing 16 billion transistors, also integrated the task specific modules like Apple Neural Engine. Over subsequent years, Apple released upgraded versions like the M1 Pro/M1 Max in 2021, Ultra in 2022, M2 in 2022, M2 Pro/Max/Ultra in 2023, and M3, M3 Pro/Max in 2023, all promising improved performance. The Apple M1 SoC is a highly integrated processor unit that includes all of the necessary components for a fully working computer while consuming less power in general, making it available to customers in the markets without losing performance. The technological

L. Franco et al. (Eds.): ICCS 2024, LNCS 14836, pp. 35–42, 2024.
https://doi.org/10.1007/978-3-031-63775-9_3

innovations in this field have promising future and grant more investigation and research toward the development of the chips [11,14].

In addition to noteworthy features such as prolonged battery life and fanless design in MacBook Air models, contributing to their quiet and portable nature, these devices have piqued the interest of researchers due to their potential applications in scientific endeavors. More specifically, researchers can harness the computational capabilities of Apple Silicon for tasks like optimization and Machine Learning (ML) calculations [3]. This study aims to evaluate the practicality and effectiveness of Apple Silicon-powered devices in tasks commonly undertaken by researchers, with a focus on performing calculations. The primary objective in this study is to evaluate various fundamental optimization and ML algorithms across diverse datasets. The notable gap in the existing literature is detected, where performance assessments are conducted on a single dataset [10] or are limited to a single ML method. This paper is an extension of Kasperek et al. [9] suggesting the possibility to further expand their research to CUDA-enabled devices. This study employed a CUDA device - RTX 3090.

2 Apple Silicon Overview

The Apple Silicon processors, such as the M1, utilize a Unified Memory Architecture (UMA) that allows for shared memory access across different modules of the SoC [7]. This means that the RAM is a single pool of memory that all parts of the processor can access, enabling the GPU to utilize more system memory while other parts of the SoC ramp down, without the need to shuttle data between different memory spaces [10]. In contrast, traditional CPU devices have separate memory spaces for the GPU and CPU, requiring data movement between these spaces, which can be inefficient. The benefits of UMA are particularly evident in the context of ML tasks, where the Apple Silicon chip offers hardware acceleration support, making it a tempting option for researchers. Additionally, the use of UMA has been found to be beneficial when only a small random portion of data is accessed for a set of benchmarks, highlighting its efficiency [13]. In the realm of SoC architectures, many-core architectures with shared memory are preferred for flexible and programmable solutions in computationally intensive application domains, including ML and embedded processing [12].

The MacOS operating system leverages the concept of shared memory to enhance performance by expanding UMA with swap memory, albeit with a trade-off in effectiveness [6]. This approach allows a more flexible allocation of resources, particularly in the context of Apple Silicon devices, where different RAM sizes are available. Therefore, comparing the performance of devices without considering the RAM utilization may lead to incorrect conclusions.

The low-energy SoC chip offers clear advantages, notably in terms of extended battery life and optimal performance per watt power. Significantly, the operational efficiency of Apple Silicon devices remains consistent whether operating on battery power or when connected to an external power source, a capability not commonly observed in conventional computing systems.

These advantages become even more pronounced given the escalating energy prices in Europe following the aftermath of the conflict in Ukraine [2]. For instance, the M1-powered Mac mini demonstrates an average power consumption ranging from 10W to 31W [5]. In contrast, a PC-class device equipped with an AMD R9 or Intel i9 CPU and dedicated GPU like the NVIDIA RTX 3090 can consume up to 800W at peak performance (calculated based on the cumulative peak power consumption of individual PC components as per manufacturer specifications). The significance of power consumption is underscored by the current global scenario, where electricity demand is outpacing the growth of renewable sources [8]. Highlighting this, the Cinebench R23 Single Package Power Efficiency metric reflects favorably on the SoC, registering 297 points per watt. In comparison, competitors such as the Ryzen 5 5600U score 90.8 and the Intel i5-1240P scores 64 points [1]. This underlines the efficiency and energy-conscious performance of the low-energy SoC chip in a landscape where power consumption considerations are paramount.

3 Methodology

In preceding experiments that compared the NVIDIA V100 and A100 GPUs with the M1 and M1 Ultra, the obtained results were promising, showcasing the superior performance of Apple Silicon over both GPUs [10]. Despite of the aforementioned GPUs produced impressive results, they do not represent the pinnacle of current GPU capabilities, with the NVIDIA V100 providing 14.13 TFLOPS Float32 precision to the RTX 3090's 35.58. To comprehensively assess the performance of selected ML classifiers across diverse hardware platforms and data types the six benchmark datasets are employed [4], wherein the number of samples, features and classes for each task is specified (see Table 1). The objective is to measure the execution time of each classifier on three distinct hardware platforms: Apple's M1 with 8GB RAM and M2 with 16GB RAM, a high-performance NVIDIA RTX 3090 GPU with 24GB memory and a mid-range laptop configuration featuring an Intel Core i5 11500h processor and an NVIDIA RTX 3050ti graphics card. The intended experiment also aimed to compare the performance of mobile devices (M1/M2) with an i5-powered laptop. Surprisingly, the unplugged i5 showed four times longer performance on average compared to the plugged-in scenario. On the other hand, the M1/M2 devices maintained consistent computational power on battery. Regrettably, the i5's limited battery life led to the decision to forgo the experiment before completion.

This approach ensures a comprehensive evaluation that extends beyond the previously explored GPUs, providing insights into the real-world performance of the classifiers across a spectrum of hardware configurations. A variety of ML methods, including Extreme Learning Machine (ELM), k-Nearest Neighbors (kNN), Multi-Layer Perceptron (MLP), Random Forest (RF) and Support Vector Machine (SVM), are employed for the datasets (see Table 1, column labels)

Table 1. Details of datasets utilized in performance evaluations.

	BreastCancer	*eyestate1*	*liver*	*musk*	*waveform*	*yeast*
Samples	699	762	345	1682	500	150
Features	9	14	6	166	21	8
Classes	2	3	4	2	10	10

4 Experimental Results

All experiments were conducted in Python 3.11 using Tensorflow 2.15, scikit-learn 1.4.0 and numpy 1.26.3 on MacOS 14.2.1 or Windows 11. The outcomes, illustrated in Figs. 2, 3, 4, 5, and 6, represent the time taken for training and testing during 10 times repeated 10-fold cross-validation, ensuring result significance by mitigating odd observations.

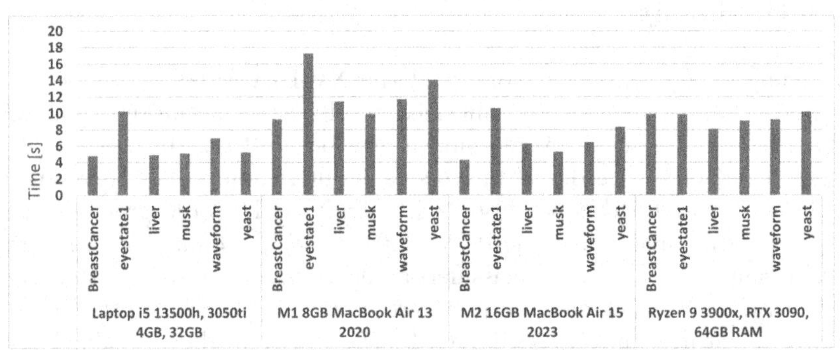

Fig. 1. Extreme Learning Machine with 100 neurons in hidden layer execution time.

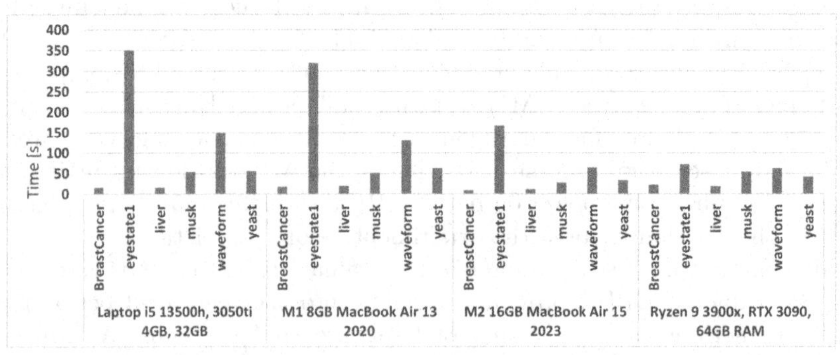

Fig. 2. Extreme Learning Machine with 1000 neurons in hidden layer execution time.

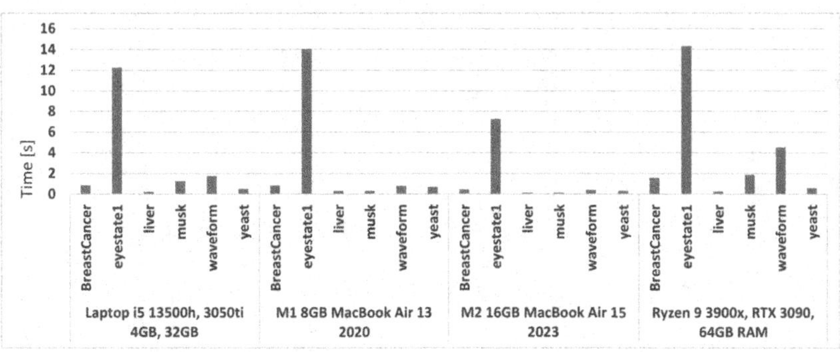

Fig. 3. K-Nearest Neighbours classifier execution time.

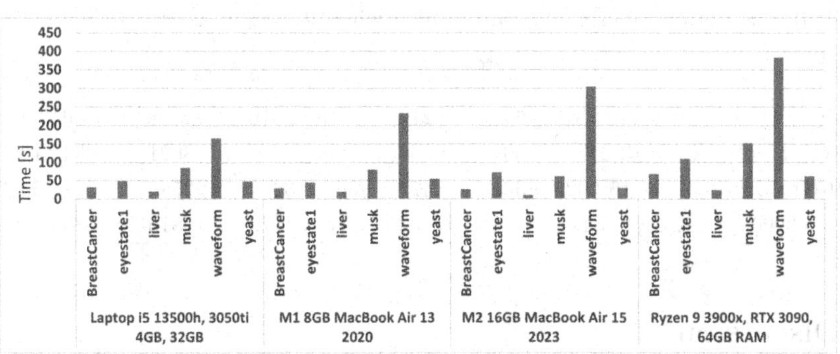

Fig. 4. Multi-Layer Perceptron execution time.

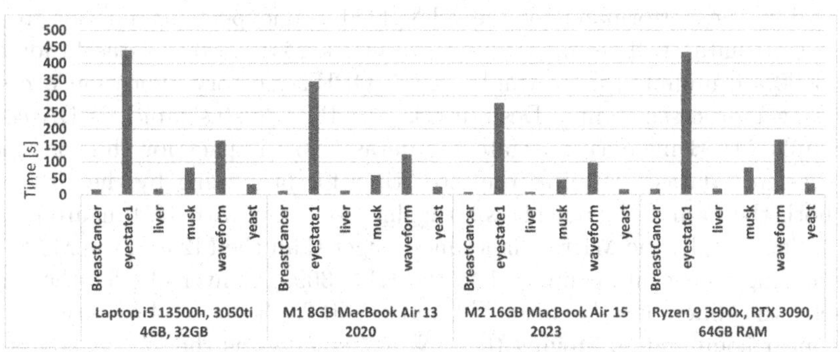

Fig. 5. Random Forest execution time.

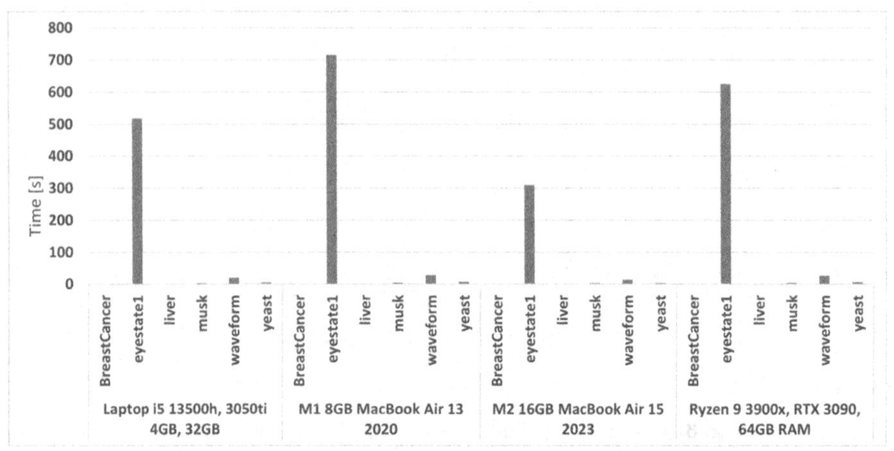

Fig. 6. Support Vector Machine classifier execution time.

Table 2. Performance evaluation results, where *bc* states as *BreastCancer*.

Device	*bc*	*eyestate1*	*liver*	*musk*	*waveform*	*yeast*	sum
Laptop i5 11500h, 3050ti, 32 GB	657	16210	670	2516	46898	3360	70311
M1 8 GB MacBook Air 13 2020	661	69000	700	17777	38621	3034	129792
M2 16 GB MacBook Air 15 2023	331	38477	342	3055	22585	1582	66372
Ryzen 9 3900x, RTX 3090, 64 GB	401	6058	376	1170	17816	1633	27454

5 Discussion

In the case of ELM with 100 hidden layer units, subtle differences emerge, with the most notable discrepancy found in the execution time on the M1 8 GB device, which is approximately twice as long as the M2 16 GB counterpart (see Fig. 1). Surprisingly, the i5 laptop yields comparable results to the M2 16 GB.

Notably, the performance of the RTX 3090 is unexpectedly inferior to the M2 on each dataset. The differences in results across various datasets for the RTX 3090 are minimal, indicating that the GPU's memory allocation necessitated longer processing time. Despite this, the RTX 3090's rapid CUDA cores and ample 24 GB memory mitigate the dataset size impact for this classifier. The *eyestate1* dataset requires the most time for processing by the classifier, with both the i5 and M1 machines struggling for ELM with 1000 neurons (see Fig. 2). Specifically, the M1 requires times longer than the M2 and the M2 takes twice as long to execute compared to the RTX 3090. Conversely, for the other datasets, the differences between M2 and the RTX 3090 are less pronounced and appear comparable. Moving to kNN, M2 emerges as the fastest across all devices, surpassing the RTX 3090 by a few times for the *waveform* dataset. Inexplicably, the RTX 3090 delivers suboptimal results despite having updated drivers and configurations, consistent across repeated experiments (see Fig. 3).

Similar patterns are observed with MLP with topology (10, 10), where the RTX 3090 consistently produces the worst results, yet for the *waveform* dataset, the i5 device is the fastest, followed by the M1 and then the M2 (see Fig. 4).

In the context of RF, the RTX 3090 once again yields subpar results, while the M2 proves to be the fastest (see Fig. 5). Similar trends persist for the SVM method, with the RTX 3090 delivering suboptimal results and the M2 demonstrating the fastest performance (see Fig. 6).

Considering the real-world scenario where all classifiers run on a given dataset, the aim is to compare the models' overall performance. Combining the total time required for a device to run all classifiers on different datasets, along with the ultimate sum of running all classifiers on all datasets, reveals interesting insights (see Table 2). The RTX 3090 emerges as the leader with a combined time of 27454 s, whereas the M2 is twice as slow. The i5 device demonstrates comparable performance to the M2, while the M1 lags as the slowest due to memory-intensive tasks. A closer examination highlights the substantial impact of the *eyestate1* dataset, where the RTX 3090 outperforms the M2 sixfold, showcasing the dedicated GPU's potential for more complex datasets. Conversely, M2 excels in tasks such as *BreastCancer*, *liver* and *yeast*, underscoring its competitive edge in certain scenarios against an 800W machine with a 30W device.

6 Conclusion

In conclusion, the results reveal nuanced variations in the performance of ML classifiers across diverse datasets and hardware configurations. The ELM with 100 hidden layer units showcases subtle differences, with notable disparities in execution time between devices. ELM with 1000 neurons introduces new dynamics, impacting performance across datasets.

In specific algorithms like kNN, MLP, RF and SVM, the Apple M2 processor consistently demonstrates promising performance compared to the Nvidia RTX 3090 GPU, highlighting the efficacy of Apple Silicon in real-world applications, especially taking into account the performance per watt power.

These findings underscore the importance of considering device specifications and configurations when assessing the practicality and effectiveness of ML algorithms. The competitive edge of Apple Silicon, particularly the M2 processor, is evident in various scenarios, showcasing its potential for tasks such as *BreastCancer*, *liver* and *yeast*, even against higher-power GPU counterparts.

References

1. Cinebench Scores. https://nanoreview.net/en/cpu-list/cinebench-scores
2. Ari, A., et al.: Surging energy prices in Europe in the aftermath of the war: how to support the vulnerable and speed up the transition away from fossil fuels. IMF Working Papers **2022**(152), A001 (2022). https://doi.org/10.5089/9798400214592.001.A001
3. Dalakoti, V., Chakraborty, D.: Apple M1 chip vs Intel (X86). EPRA Int. J. Res. Develop. **7**(5), 207–211 (2022)

4. Dheeru, D., Karra Taniskidou, E.: UCI machine learning repository (2017). http://archive.ics.uci.edu/ml
5. Frumusanu, A.: The 2020 Mac Mini unleashed: Putting Apple Silicon M1 to the test (2020). https://www.anandtech.com/show/16252/mac-mini-apple-m1-tested
6. Golden, R., Case, A.: In lieu of swap: analyzing compressed RAM in Mac OS X and Linux. Digi. Invest. **11**, S3–S12 (2014). https://doi.org/10.1016/j.diin.2014.05.011
7. Hart, C.: CDRAM in a unified memory architecture. In: Proceedings of the COMPCON, pp. 261–266 (1994). https://doi.org/10.1109/CMPCON.1994.282913
8. IEA. Electricity market report - July 2021 (2021). https://www.iea.org/reports/electricity-market-report-july-2021
9. Kasperek, D., Podpora, M., Kawala-Sterniuk, A.: Comparison of the usability of Apple M1 processors for various Machine Learning tasks. Sensors **22**(20) (2022). https://doi.org/10.3390/s22208005
10. Kenyon, C., Capano, C.: Apple silicon performance in scientific computing. arXiv (2022). https://doi.org/10.48550/arXiv.2211.00720
11. Liao, X., Li, B., Li, J.: Impacts of Apple's M1 SoC on the technology industry. In: Proceedings of the 2022 7th International Conference on Financial Innovation and Economic Development (ICFIED 2022), pp. 355–360 (2022). https://doi.org/10.2991/aebmr.k.220307.056
12. Luan, H., Gatherer, A.: Combinatorics and geometry for the many-ported, distributed and shared memory architecture. In: 2020 14th IEEE/ACM International Symposium on Networks-on-Chip (NOCS), pp. 1–6 (2020).https://doi.org/10.1109/NOCS50636.2020.9241708
13. Xu, H., Lin, P.H., Emani, M., et al.: Xunified: a framework for guiding optimal use of GPU Unified Memory. IEEE Access **10**, 82614–82625 (2022). https://doi.org/10.1109/ACCESS.2022.3196008
14. Zhang, Z.: Analysis of the advantages of the M1 CPU and its impact on the future development of Apple. In: 2021 2nd International Conference on Big Data and Artificial Intelligence and Software Engineering (ICBASE), pp. 732–735 (2021). https://doi.org/10.1109/ICBASE53849.2021.00143

Deep Neural Network for Constraint Acquisition Through Tailored Loss Function

Eduardo Vyhmeister[(✉)], Rocio Paez, and Gabriel Gonzalez-Castane

Insight Centre of Data Analytics, University College Cork, Cork, Ireland
eduardo.vyhmeister@insight-centre.org

Abstract. The importance of extracting constraints from data is emphasized by its potential practical applications in solving real-world problems. While constraints are commonly used for modeling and problem-solving, methods for learning constraints from data are still relatively scarce. Moreover, the complex nature of modeling requires expertise and is susceptible to errors, making constraint acquisition methods valuable for automating this process through learning constraints from examples or behaviours of solutions and non-solutions. This study introduces a novel approach grounded in Deep Neural Networks (DNN) based on Symbolic Regression, where suitable loss functions are used to extract constraints directly from datasets. With this approach, constraints can be directly formulated. Additionally, given the wide range of pre-developed architectures and functionalities of DNNs, potential connections and extensions with other frameworks are foreseeable.

Keywords: Constraint Acquisition · Deep Neural Network · Symbolic Regression

1 Introduction

The importance of learning constraints from data is emphasized by its potential applications in addressing real-world problem-solving scenarios. Despite the widespread use of constraints for modeling and problem-solving, the methodologies for learning constraints from data are relatively limited. Recent efforts have been directed towards surveying and contextualizing constraint learning within the broader machine-learning landscape. This involves recognizing subtle distinctions from standard function learning, identifying challenges, and exploring potential applications of constraint learning in diverse fields. These endeavours contribute to a deeper understanding of the role of constraint learning and pave the way for its integration into a wider array of problem-solving domains.

In terms of existing approaches for constraint acquisition, various systems (covered in the following section) have shown capabilities to extract systems constraints using iterative methods, oracle feedback, examples and constraints catalogues, to reduce the constraint space. Most of these deductive methods have

L. Franco et al. (Eds.): ICCS 2024, LNCS 14836, pp. 43–57, 2024.
https://doi.org/10.1007/978-3-031-63775-9_4

different limitations that restrict broad applicability. For example, the dependence on user interaction may pose challenges, especially if users are not domain experts or if providing examples requires a deep understanding of the problem. The number of queries or interactions needed to converge to an accurate model can also be a limitation. Furthermore, for some methods based on exemplification, a priori knowledge of the positive or negative nature of the examples is required to limit the search spaces, which implies the need for previous knowledge of the system behaviour. Additionally, the effectiveness of these methods relies on the expressiveness of the predefined constraint libraries or the set of possible constraints. If the library lacks certain types of constraints relevant to a specific problem domain, the models generated may be limited in their accuracy.

In the broader context of constraint acquisition, the interplay among machine learning, data mining, and constraint satisfaction has attracted considerable attention. While constraints are conventionally employed in logical and optimization problems, their application in the domains of machine learning and data mining is becoming increasingly prominent (e.g. [19]).

The term "constraint learning" is precisely defined in this context as the inductive learning of a constraint theory from examples. This definition distinguishes constraint learning from the deductive processes associated with clause learning or constraint acquisition of non-ML solver technologies. The exploration of constraints in the context of machine learning and data mining signifies a multifaceted and evolving area of research with implications for diverse problem-solving domains.

The integration of Deep Neural Networks (DNN) into constraint learning brings benefits, leveraging their ability to efficiently extract information directly from data. DNN can learn intricate representations, discover relevant features, and adapt to different data types. DNN excel at learning intricate and hierarchical representations from large data sets [8]. They can automatically discover relevant features and patterns in the input, which can be advantageous in capturing complex relationships within constraint networks. Furthermore, Deep learning models can be trained in an end-to-end fashion, allowing them to learn both the feature representation and the constraint-solving process simultaneously. This may lead to more seamless integration and optimization. Additionally, DNN are highly adaptable to different types of data and problem domains [9]. They can be trained to handle diverse constraint types and can potentially generalize well to new problem instances. DNN can automatically extract relevant features from the input data, potentially reducing the need for manual feature engineering in the modeling process. This can be particularly beneficial when dealing with complex and high-dimensional constraint spaces, nevertheless that does not imply that expert knowledge could not be integrated within DNN methodologies [11,12]. In fact, active learning can be naturally integrated into neural network-based approaches. The model can dynamically query the user or another system for specific examples, enabling more efficient and targeted learning of constraints. Deep learning models can take advantage of parallel processing capabilities, which can lead, together with increasing computational power, to faster training times [9]. This can be crucial in scenarios where quick acquisition of constraint knowledge is essential. Furthermore, pre-trained neural network models can be fine-tuned for specific constraint learning tasks [10]. This leverages knowledge learned from one context to another, potentially speeding up the learning process for similar new problems. Importantly, Deep learning techniques can be seamlessly integrated with other machine learning methods or other pre-trained DNN, providing a holistic approach to constraint learning. This can include combining neural networks with symbolic reasoning or other constraint-solving techniques.

While these benefits are promising, it is important to note that the success of neural network-driven approaches in constraint learning depends on factors such as data availability, problem complexity, and the specific characteristics of the application domain. Additionally, the interpretability and transparency of neural network models should be considered, especially in domains where understanding the reasoning behind constraint decisions is crucial.

Symbolic regression (SR) is a computational method employed to uncover mathematical equations, commonly implemented through genetic programming. This entails using evolutionary algorithms to explore and identify the most suitable equation that aligns with a given dataset. While SR has proven effective in revealing fundamental laws governing physical systems based on empirical data, its scalability is hindered by the combinatorial complexity of the underlying problem [13,14].

The synergy between SR and DNN is pivotal in the pursuit of identifying behaviours inherent in a dataset. SR operates by exploring mathematical expressions to discover the optimal model that aligns with the given data. On the other hand, Neural Networks (NN) and Machine Learning (ML) techniques excel in capturing intricate non-linear relationships between input and output variables.

For instance, Schmidt et al. [16] and Martius et al. [17] introduced a DNN approach named the Equation Learner (EQL), where traditional activation functions are replaced with primitive functions (PFs). This novel approach enables the DNN to perform SR, allowing it to learn analytical expressions and extrapolate to unseen domains. The EQL is implemented as an end-to-end differentiable feed-forward network, facilitating efficient gradient-based training. This integration of SR and DNN methodologies represents a powerful strategy for uncovering and understanding complex relationships within datasets.

To evaluate the potential of DNN for constraint acquisition, with all the possible benefits previously described, and, at the same time, facilitate the transparency of the DNN for improved user acceptance of the final results, this work proposes the use of an EQL-based architecture with tailored loss functions to extract system constraints by using tailored loss function. The manuscript is organized as follows: In Sect. 2 are presented recognized systems for constraint acquisition. In Sect. 3 an introduction to the EQL, the tailored loss function, and their combination, are described. This section also presents the experiments used for the system evaluation. Section 4 discusses the results. Finally, Sect. 5 presents the main remarks and conclusions.

2 Related Works

Constraint programming involves both modeling and solving. In modeling, problems are defined using variables with specific values and rules, while in solving, values are identified that satisfy all constraints simultaneously. Despite its potential, user-friendliness limitations pose challenges for constraint programming. The declarative nature of this approach enables the solving of problem models using standard methods, but fully leveraging this potential requires increased automation in modeling [1].

Constraint acquisition plays a crucial role in the automation of complex and error-prone tasks associated with modeling constraint programming problems. By using automated approaches constraints not previously considered or improvements on mathematical representations of those already considered could be achieved, allowing at the end to have better models to represent the system behaviour. The constraint acquisition role increase is important especially when the number of features and their trends complexity is increased. This process enables the extraction of constraints from data that represents both solutions and non-solutions. Various constraint acquisition systems have been proposed to support non-expert users in their modeling tasks [7].

As an illustrative example, Baldicenau and Simonis [4] have introduced a methodology where constraints are constructed from a catalogue as primitives. Their ModelSeeker tool leverages this approach to analyze a substantial dataset with up to 6500 variables and 7000 samples. In this methodology, examples are organized as a matrix, and the system identifies constraints in the global constraint catalogue that are satisfied by rows or columns across all examples. An integral part of their workflow includes dominance checks and trivia removal, processes that assess constraints between each other or utilize specific rules to simplify and eliminate irrelevant constraints. These last steps can contribute significantly to enhancing the efficiency and relevance of the acquired constraints but, it is noteworthy that, as the system's complexity increases, users are required to provide a larger number of examples for the target set of constraints to be effectively learned, limiting the applicability of the approach to highly complex systems and feedback from expert knowledge.

Bessiere et al. [2] propose QuAcq, an iterative method generating partial queries and utilizing oracle feedback to reduce the constraint space. The need to use Oracle feedback imposes over QuAcq (and other approaches based on this) similar limitations discussed AS for Modelseeker. Furthermore, the approach could also suffers from a high query count for convergence. MultiAcq [5] extends QuAcq, learning multiple constraints on negative examples. MQuAcq [6] combines QuAcq and MultiAcq, effectively reducing query complexity.

Other approaches, such as PREDICT&ASK introduce a distinct algorithm focused on predicting missing constraints within a partially learned network [7]. The algorithm uses the local data structure which contains all constraints that are candidates for recommendation. Through recommendation queries, the approach enhances user interaction, as demonstrated through experimental comparisons against QuAcq. Even though the recommendation approach could enhance system performance, the need for expert knowledge within the identification process forces to have similar limitations as previously discussed systems.

Hassle-sls integrates metaheuristic techniques for joint learning of hard and weighted soft constraints. Despite advancements, runtime issues arise from evaluating multiple MAX-SAT models. Hassle-gen [18] addresses this by incorporating a genetic algorithm and an efficient model evaluation procedure, contributing significantly to the state of the art.

EOCCA [20] is a fast algorithm for acquiring constraints from feasible solution examples. It preprocesses data using x-means partitioning, standardization, and principal components analysis. Then, the algorithm wraps the preprocessed data into hyper-ellipsoids using constructed constraints.

Even though different approaches explore constraint learning through decision trees, genetic programming, local search, and other optimization approaches, a considerable gap exists in using DNN directly for constraint learning, highlighting the relevance of the work presented here.

3 Methodology

The general goal of the approach is to find expressions of the form $-B+\sum_{i=1}^{F} A_i \cdot f(X_i)$ that delimit (i.e. constraints) the space in which the dataset is distributed. In this expression A and B are the corresponding terms of the constant matrices to be solved during the training process and F is the number of features in the dataset. It is important to note that while $f(X_i)$ is represented as a linear expression in this work, it has the flexibility to assume any form by setting sound primitives in the EQL. For a better understanding of the EQL and its interaction with the tailored loss functions, we first include a short description of SR.

3.1 Symbolic Regression

In the context of SR, the goal is to model a system represented by an unknown analytical function, denoted as $\phi : \mathbb{R}^n \rightarrow \mathbb{R}^m$. The observed data, denoted as $x, y = (x_1, y_1), \ldots, (x_N, y_N)$, where $x \in \mathbb{R}^n$ and $y \in \mathbb{R}^m$, is generated from $y = \phi(x) + \xi$, where ξ represents a term of additive zero-mean noise. The goal of SR methods is to construct a function $\psi : \mathbb{R}^n \rightarrow \mathbb{R}^m$ that minimizes the empirical error on the training set and generalizes well to future data.

Given that the true analytical function ϕ is unknown, the resulting mathematical expression, denoted as ψ, is designed to be interpretable and capable of extrapolating the data. This interpretability stems from the fact that the constructed model is not a black-box, making it easier to understand and analyze. The process involves considering a predefined set of "primitive functions," denoted as $\mathbf{f} = f_1, f_2, \ldots, f_n$, upon which the regression method can build. The selection of these primitive functions is crucial for accurately capturing the mathematical representation of the underlying model.

For instance, common choices for the set of primitive functions, denoted as $\mathbf{f}$, may include constants (C), linear terms (x), quadratic terms (x^2), trigonometric functions ($\sin(x)$), exponential functions (e^x), sigmoid functions ($S(x) = 1/(1 + e^{-x})$), logarithmic functions ($\ln(x)$), reciprocal functions ($1/x$), and square root functions ($\sqrt{x}$). The specific choice of primitive functions can be tailored to different applications and problem domains. In the present work to construct linear constraints, primitives were limited to linear terms and constants.

Specification of the neural architecture, together with the integration with the specific loss function used for constraint acquisition is described in Sect. 3.3

3.2 Loss Function Definition for Constraint Acquisition

ML and DNN algorithms can be fine-tuned or trained by solving a problem-structural optimization, expressed as [15]:

$$\min_f \frac{1}{N}\left[\sum_{i=1}^{N} L_\theta(f(x_i)) + \lambda R(f)\right] \tag{1}$$

In this minimization problem, a training set of size N is assessed through two main terms. The first term is the empirical risk, where $L()$ denotes the loss function, θ represents the parameter vector, and f is a functional evaluation over each element i of the set. The second term, $R(f)$, is known as the regularization term, reflecting the model's complexity. Including this second term aids in reducing the size of the final expressions. The weight parameter $\lambda \geq 0$ balances the trade-off between the empirical risk and the model complexity. Moreover, additional terms can be introduced and appropriately weighted to extend functionalities during training.

Common regression loss functions like square loss, absolute loss, Huber loss, and Log-cosh loss are designed for deriving functions in regression tasks (e.g., $f(x) = \phi(x) + \xi$ from the dataset). In constraint learning, where expressions involve both equalities and inequalities, modifying the loss function is crucial. This adaptation emphasizes adherence to specified constraints, distinct from conventional regression that mainly minimizes disparities between model predictions and observed values. The redefined loss function guides the model to respect given constraints, shifting focus from precise numerical predictions to a more comprehensive consideration of data limitations.

To address this need, it is proposed the utilization of a tailored loss function, taking into account three inter-playing notions:

1. In our approach, we track the permissible directions for adjusting the predicted function by employing two distinct expressions of error. The choice between these expressions is contingent on the type of inequality sought. For instance, if the objective is to enforce constraints of the form $f(x) \leq A$, where A is a specified parameter, the error expression $y - f(x)$ is employed. Conversely, when aiming for constraints of the form $A \leq f(x)$, the opposite expression, $f(x) - y$, is utilized. Each of these error terms is weighted to contribute to an overarching global valuation term.

2. The previous term is combined with an absolute maximum error that serves as an anchor to bound the extent of movement in the predicted equation

space. The inclusion of this anchor ensures that the optimization process does not diverge infinitely, promoting stability in the learning process.

3. Finally, an induced observation threshold derived from the spectrum of all possible errors is considered; rather than taking into account errors over the entire dataset.

This approach combining these three notions effectively constrains the model's behaviour, aligning with the overarching goal of capturing constraints inherent in the data. The graphical interpretation of these notions are shown in Fig. 1.

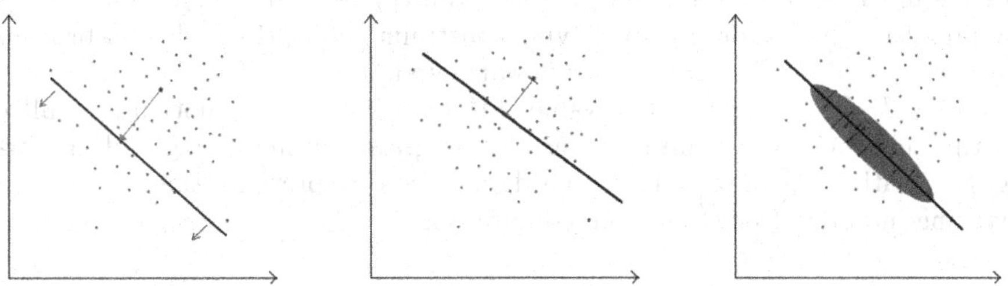

Fig. 1. Schematic representation of the different notions considered for the three terms definition of the loss functions: Notion 1 (left panel), Notion 2 (center panel), Notion 3 (right panel). See text for a full explanation

These notions are expressed within the loss function mathematical representation as three distinct and independent terms, implemented in the tuning (minimization) process. These components are carefully designed to accommodate the specific nature of the constraints being searched:

$$z = \frac{\alpha_1}{N} \sum_{i=1}^{N} L_e(y, f(x_i)) + \frac{\alpha_2}{N} \sum_{i=1}^{N} L_{P_{alpha}}(y, f(x_i)) + \alpha_3 L_{anchor}(y, f(x_i)) \tag{2}$$

$$e = \begin{cases} y - f(x), \text{ if } A \leq f(x) \\ f(x) - y, \text{ if } f(x) \leq A \end{cases} \tag{3}$$

$$L_e(y, f(x_i)) = e \tag{4}$$

$$L_{P_\gamma}(y, f(x_i)) = \begin{cases} (y_i - f(x_i))^2 \ \forall \text{ y,xi if error} \in P_\gamma \\ 0 \ \forall \text{ y,xi if error} \notin P_\gamma \end{cases} \tag{5}$$

$$L_{anchor}(y, f(x_i)) = |Max(e)| \tag{6}$$

In these equations, the minimization objective is denoted as z, and α represents the weighting factor assigned to each contribution during the training. The set P_γ comprises data points within the γ percentile.

The term L_{error} captures the directional aspect of the loss, contingent on the inequality direction (notion 1). For instance, choosing the error $e = y - f(x)$ promotes higher negative values as minimization is carried out, favouring constraints of the form $A \leq f(x)$.

On the other hand L_{P_γ} represents a quadratic loss function, with the distinctive feature that it considers only data points within the γ quartile (notion 3). This focused observation facilitates obtaining robust representations of constraints, particularly in border regions identified during the minimization process. Quantil-based analyses and loss functions have found applications in various fields such as statistics and econometrics, underscoring its versatility and effectiveness in capturing the range of potential outcomes. This adaptability makes it a promising choice for tasks involving constraint acquisition, where a broader consideration of prediction intervals is warranted.

Lastly, L_{anchor},is the absolute value of the maximum error (notion 2). Unlike L_{e}, this term acts as a counterbalance by progressively increasing and anchoring $f(x)$ within the proximity of the data points. It prevents the model from diverging infinitely from the observed data points.

3.3 EQL and Loss Function Integration

The integration of both SR and the custom loss function for constraint acquisition was done using Python as the primary programming language. TensorFlow V2 served as the framework for constructing the EQL, with activation functions explicitly specified as primitive functions defined by the user. To facilitate a direct understanding of the mathematical constraint representation and potentially offer feedback to the symbolic DNN, the *sympy* library was employed as a symbolic mathematical tool.

By leveraging Tensorflow's capabilities, the custom loss function was incorporated during the compilation process of the optimizer. The full implementation of the combined DNN architecture can be found in the associated repository (https://github.com/eduardovyhmeister/Constraint-Aquisition), including symbolic layers (tensorflow) and methods such as initialization, build, call, and configuration. These symbolic layers facilitate the integration of symbolic mathematics into the DNN, enhancing interpretability and enabling feedback mechanisms for constraint representation.

For the experiments that follow, a three-layer architecture comprising an input layer, a symbolic layer, and an output layer was used. The number of layers can be varied to serve different goals. The input layer allows the data from a varying number of features to input the system. The symbolic layer allows defining which primitive functions are active (and parametrized during training). This layer used the identity (i.e., $f(x) = x_i$) and constants (C) as primitive without any bias on the neurones, suitable for linear constraints. The output layer, which consists of a single node representing the output prediction, aggregates the evaluations on the symbolic layer using a linear activation function (that includes a bias). These layers also included L1 and L2 regularization for training considerations.

Additionally, a masking process was integrated during training in some of the experiments. If a weight parameter fell below 0.001, the connection weight value was set to 0 and could not be updated thereafter. This approach promotes the parsimony of the final equation by eliminating any contribution from connections with negligible impact.

In the Equation Learner, the values of the weights and biases in the architecture play a critical role in defining the mathematical formulation. Thus, a careful initialization of these values is essential to effectively explore the search space during the training process. For the symbolic layer, Xavier uniform initializers were used, while random uniform initializers were employed for the biases of the output layer. The random uniform initialization was confined within 0.5 times the minimum and 0.5 times the maximum of the provided data values. It is important to note that the analysis of optimal initializers for the search engine was not explicitly considered in the current work, leaving room for further investigation in future studies.

To extract the mathematical formulation from the Equation Learner (EQL), the input data should include each of the features considered in the analysis (i.e., X_i). The output was defined as an array of zeros with the same size as the instances under evaluation.

3.4 Experiments Definition and Setup

Table 1 shows the metaparameters to configure both, the EQL (first two columns) and the loss function (last two columns).

Table 1. Parameters for EQL and the loss function

Metaparameter	Value	Metaparamter	Value
Activation Functions	two $f(x) = x_i$ and two C	α_1	1.0
Epochs	400	α_2	0.5
Learning rage	1e-8	α_3	0.5
L1	0.05	γ	5.0
L2	0.05		

Testing the approach and the loss function involved creating data points randomly based on a specific set of constraints. For a two-dimensional problem defined by X_0,X_1, the experiments include data with the following configurations:

1- High Granularity Square: an area with 600 data points delimited within -5 and 25 with an additional restriction (i.e. $-5 \leq X_i \leq 25$, $i = 0, 1$; $4 < X_1 + 2X_2$)
2- Circle: a circular area of radius $\sqrt{200}$ containing 250 data points (i.e. $X_0^2 + X_1^2 \leq 200$);
3- Low Granularity Square: same area as (1) with only 100 data points.

For a three-dimensional case study (defined by X_0, X_1, X_2), 2000 data points were distributed in a cube delimited within -5 and 25 with a planar section (i.e. $-5 \leq X_i \leq 25$, $i = 0, 1, 2$; $4 < X_1 + 2X_2 - 3 * X_3$).

The data set, a combination of input (X_i) and output Y data points within the given constraints, was fed to the EQL with the tailored loss function. The training process was run 10 times using a specific error representation.

4 Results and Discussion

Figure 2 illustrates the results when two features and only one type of error was used (i.e. $y - f(x)$).

As observed in the figures, the EQL combined with tailored loss function was able to be used to extract suitable constraints for the data sets, independent of the figure shape (Square and Circle) or the granularity of the data points (600 or 100 for Square data points). As the granularity was reduced, the percentile value γ had to be reduced from 5.0 to 2.5 to improve performance. As expected, for low granularity, the points within the same percentile of error are more separated, and a γ value suitable for high granularity would imply points too distant from the boundary of the figure at low granularity. Thus, limiting the L_P by reducing the γ value contributes to a more accurate search. Thus, γ plays a crucial role depending on the system granularity.

As defined in the methodology, the directional expression of the error was specified as $y - f(x)$, implying that constraints of the form $A \leq f(x)$ should be obtained. Yet, One important outcome observed during experiments with the Circle is that constraints of the form $A \leq f(x)$ and $f(x) \leq A$ were obtained. This is probably due to the fact that no constraints on the network weights were defined and thus, possible negative values for those weights were obtained (inversion of the constraint).

Table 2 display the results of the constraints found for all the experiments performed, including the specific functionality detected in each case and the performance metric. The table includes the results for the High Granularity Square (Results from 1 to 10); Cube (results from 11 to 15); Low Granularity Square (results from 16 to 25); and Circle (Results from 26 to 35). The performance presented in the table is measured as the error quantified by the percentage of points violating the obtained constraints. This error, when approaching a value close to 0 (though not necessarily reaching 0), implies the generation of expressions closely aligned with the borders of the dataset-covered area.

As indicated in Table 2, the majority of the derived expressions demonstrate satisfactory performance in delineating boundary values, as evidenced by the minimal error committed.

It is crucial to note that the points were randomly generated over the complete surface (volume) of the 2D (3D) figures limits, thus instances laying exactly on the boundaries of the constraints may not be included in the training datasets. Consequently, the resulting mathematical expression can vary significantly based on the specific dataset used and the intricacies of the training process.

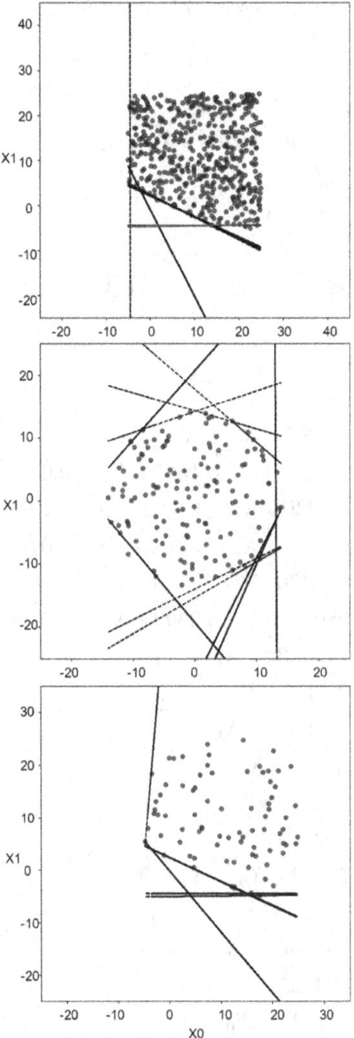

Fig. 2. Constraints obtained using $y - f(x)$ representation of error within the L_e expression: High-granularity Square example (top panel), Circle example (centre panel), Low-granularity square example (bottom panel). The red line is equivalent to Result #7 of Table 2, without masking (see text for full explanation)

The initial ten expressions listed in the table incorporate the masking process, resulting in the elimination of relatively small terms from the mathematical expressions during the training process. This elimination is enforced after the training process. For instance, Result #7 in the table represents a horizontal line, yet the result for the equivalent experiment in Fig. 2, produced with no

Table 2. Mathematical expression obtained for the 2 features problem

Result #	Expression	Error
1	$-4.402 \leq X_0$	1.45%
2	$18188.7953 \leq -4002.3703 * X_0 + X_1$	1.06%
3	$2.1469 \leq 0.4772 * X_0 + X_1$	0.91%
4	$2.1420 \leq 0.4683 * X_0 + X_1$	0.91%
5	$-4.6843 \leq -0.0151 * X0 + X_1$	4.68%
6	$10353.0651 \leq -2288.4936 * X_0 + X_1$	1.09%
7	$-4.678 \leq X_1$	0.36%
8	$-4.702 \leq X_1$	0.36%
9	$2.0978 \leq 0.4497 * X_0 + X_1$	0.91%
10	$-0.8634 \leq 1.9262 * X_0 + X_1$	1.45%
11	$0.5036 \leq 0.2023 * X_0 + 0.1505 * X_1 - 0.0439 * X_2$	1.34%
12	$-0.7109 \leq 0.2230 * X_0 + 0.0158 * X_1 + 2.6140e^{-5} * X_2$	1.65%
13	$-0.8440 \leq 0.2247 * X_0 + 0.0113 * X_1 - 0.0002 * X_2$	1.67%
14	$-0.8210 \leq -6.476e^{-5} * X_0 + 0.217 * X_1 + 7.475e^{-6} * X_2$	1.47%
15	$0.6027 \leq 0.0805 * X_0 + 0.2399 * X_1 - 4.990e^{-5} * X_2+$	1.37%
16	$-14.009 \leq -0.4832 * X_0 + X_1$	1.36%
17	$14.131 \leq -0.3313 * X_0 + X_1$	2.04%
18	$-28.594 \leq -1.9450 * X_0 + X_1$	1.36%
19	$20.61634 \leq 1.101 * X_0 + X_1$	0.68%
20	$-14.975 \leq -1.013 * X_0$	1.36%
21	$1.165 \leq 19.252 * X_0 + X_1$	0.68%
22	$14.297 \leq 0.28728 * X_0 + X_1$	1.36%
23	$2293.265 \leq -174.17 * X_0 + X_1$	2.04%
24	$-15.398 \leq -0.5665 * X_0 + X_1$	0.0%
25	$17.7929 \leq 0.84825 * X_0 + X_1$	1.36%
26	$-5.0622 \leq -0.011 * X_0 + X_1$	1.23%
27	$-0.139 \leq 1.163 * X_0 + X_1$	1.23%
28	$57.826 \leq 11.204 * X_0 + X_1$	1.23%
29	$2.353 \leq 0.459 * X_0 + X_1$	1.23%
30	$2.330 \leq 0.451 * X_0 + X_1$	1.23%
31	$2.348 \leq 0.457 * X_0 + X_1$	1.23%
32	$2.364 \leq 0.457 * X_0 + X_1$	2.47%
33	$2.3713 \leq 0.459 * X_0 + X_1$	2.47%
34	$-4.5015 \leq 0.000476 * X_0 + X_1$	1.23%
35	$2.351 \leq 0.454 * X_0 + X_1$	1.23%

masking, reveals small slopes on the lines (red line), showcasing the impact of the masking process on the final expressions.

In Results #11 to #15, which correspond to the 3D figure, no masking process was applied (only these results from Table 2). As evident in these expressions, numerous terms could be retained or eliminated based on the relative importance of different features in the final mathematical expressions. The absence of the masking approach allows for a more comprehensive inclusion of terms, potentially enhancing the richness of the expressions. However, it is essential to note that the masking process, as described earlier, contributes to the parsimony of the final mathematical expression, thereby improving readability for users. The choice between using or not using the masking approach depends on the specific goals and interpretability requirements of the application.

Despite the promising results outlined so far, several trends, other than those regularly related to the use of ML (i.e. meta parameters definitions), were observed during the evaluation of the presented approach. These trends were primarily associated with dimensionality and initialization.

In terms of dimensionality, as the number of features increased, unexpected inequalities with an unusual number of features were observed (e.g. See Result #11 in Table 2). These constraints could arise from the combination of constraints (e.g., combining $-5 \leq X_0$ and $-5 \leq X_1$ as $-10 \leq X_0 + X_1$). A refinement stage, together with an extended number of epochs during training (to improve parameter determination), could be used to secure obtaining more precise inequality expressions. As previously suggested in literature [4], the incorporation of dominance checks and trivia removal could be considered during this refinement stage. These techniques evaluate constraints between each other or apply specific rules to simplify and eliminate irrelevant constraints, contributing to a more refined and accurate set of inequalities.

The initialization stage involves specifying the weights within the Equation Learner (EQL), and thus initial constraints assumptions to be corrected during training. These initial assumptions could drive gradient values during the training process, and consequently, the direction in which mathematical expressions are explored. In the current strategy, weights were randomly initialized for testing the developed approach; however, alternative approaches could be considered to improve searches. For instance, zero initialization for certain values could be enforced, thereby guiding the search towards a reduced number of features involved in the final expression. This thoughtful initialization strategy could potentially contribute to a more effective and efficient exploration of the solution space.

5 Conclusions

In this work, a novel approach for learning constraints from data using DNN based on SR was introduced. The method demonstrated its capability to directly extract linear inequalities from datasets by settling suitable loss functions. The approach was validated on predefined datasets, revealing its effectiveness in approximating the boundaries defined by the constraints. The results, showcase the satisfactory performance of the derived expressions in delineating boundary values, with minimal error committed. The error, quantified as the percentage of points violating the original constraint areas, approached values close to 0, indicating a close alignment with the dataset-covered area borders. The outcomes support the efficacy of the proposed methodology in constraint learning.

The approach has the potential to be extended by incorporating further tailored loss functionalities terms. This extension could facilitate to address diverse challenges and achieve more robust results. Furthermore, extensions such as the use of non-linear primitives in the EQL, together with the exploration of optimal non-linear primitives and constraints, can be easily foreseen to be implemented. As expected, these considerations aim to refine the proposed approach for efficient knowledge extraction from data.

Acknowledgments. This research was partially supported by the EU's Horizon Digital, Industry, and Space program under grant agreement ID 101092989-DATAMITE. Additionally, we acknowledge Science Foundation Ireland under Grant No. 12/RC/2289 for funding the Insight Centre of Data Analytics (which is co-funded under the European Regional Development Fund).

References

1. O'Sullivan, B.: Automated modelling and solving in constraint programming. In: Twenty-Fourth AAAI Conference on Artificial Intelligence, vol. 24, no. 1, p. 2010 (2010)
2. Bessiere, C., et al.: Constraint acquisition via partial queries. In: Proceedings of the Twenty-Third International Joint Conference on Artificial Intelligence (IJCAI 2013), Beijing, pp. 475–481 (2013)
3. Daoudi, A., Mechqrane, Y., Bessiere, C., Lazaar, N., Bouyakhf, E.H.: Constraint acquisition using recommendation queriesff. In: International Joint Conference on Artificial Intelligence (IJCAI), New York, pp. 720–726 (2016)
4. Beldiceanu, N., Simonis, H.: ModelSeeker: extracting global constraint models from positive examples. In: Bessiere, C., De Raedt, L., Kotthoff, L., Nijssen, S., O'Sullivan, B., Pedreschi, D. (eds.) Data Mining and Constraint Programming. LNCS (LNAI), vol. 10101, pp. 77–95. Springer, Cham (2016). https://doi.org/10.1007/978-3-319-50137-6_4
5. Arcangioli, R., Bessiere, C., Lazaar, N.: Multiple constraint acquisition. In: Proceedings of the Twenty-Fifth International Joint Conference on Artificial Intelligence (IJCAI-16), New York (2016)
6. Tsouros, D.C., Stergiou, K.: Efficient multiple constraint acquisition. Constraints **25**, 181–225 (2020)
7. Daoudi, A., Mechqrane, Y., Bessiere, C., Lazaar, N., Bouyakhf, E.H.: Constraint acquisition using recommendation queries. In: International Joint Conference on Artificial Intelligence (IJCAI), pp. 720– 726 (2016)
8. Bau, D., Zhu, J.-Y., Strobelt, H., Lapedriza, A., Zhou, B., Torralba, A.: PNAS **117**(48), 30071–30078 (2020)
9. Donges, N.: (Updated Matthew Urwin). 4 Disadvantages of Neural Networks. (2023). https://builtin.com/data-science/disadvantages-neural-networks. Accessed 12 Jan 2024

10. Church, K.W., Chen, Z., Ma, Y.: Emerging trends: a gentle introduction to fine-tuning. Nat. Lang. Eng. **27**, 763–778 (2021)
11. Dash, T., Chitlangia, S., Ahuja, A., Srinivasan, A.: A review of some techniques for inclusion of domain-knowledge into deep neural networks. Sci. Rep. **12**, 1040 (2022)
12. Sun, L., Gao, H., Pan, S., Wang, J.X.: Surrogate modeling for fluid flows based on physics-constrained deep learning without simulation data. Comput. Methods Appl. Mech. Eng. **361**, 112732 (2020)
13. He, B., Lu, Q., Yang, Q., Luo, J., Wang, Z.: Taylor genetic programming for symbolic regression. In: Proceedings of the Genetic and Evolutionary Computation Conference, pp. 946–954 (2022)
14. La Cava, W., et al.: Contemporary symbolic regression methods and their relative performance. In: Proceedings of the Neural Information Processing Systems Track on Datasets and Benchmarks, vol. 1 (2021)
15. Wang, Q., Ma, Y., Zhao, K., Tian, Y.: A comprehensive survey of loss functions in machine learning. Annal. Data Sci. **9**, 187–212 (2022)
16. Schmidt, M., Lipson, H.: Distilling free-form natural laws from experimental data. Science **324**(5923), 81–85 (2009)
17. Martius, G., Lampert, C.H.: Extrapolation and learning equations. arXiv preprint arXiv:1610.02995 (2016)
18. Berden, S., Kumar, M., Kolb, S., Guns, T.: Learning MAX-SAT models from examples using genetic algorithms and knowledge compilation. In: 28th International Conference on Principles and Practice of Constraint Programming (CP 2022). Schloss Dagstuhl-Leibniz-Zentrum für Informatik (2022)
19. Maragno, D., Wiberg, H., Bertsimas, D. Birbil, I., den Hertog, D., Fajemisin, A.O.: Mixed-iteger optimization with constraint learning. arXiv preprint arXiv:2111.04469 (2023)
20. Pawlak, T.P., Litwiniuk, B.: Ellipsoidal one-class constraint acquisition for quadratically constrained programming. Eur. J. Oper. Res. **294**(1), 36–49 (2021)

Efficient Search Algorithms for the Restricted Longest Common Subsequence Problem

Marko Djukanović[1](✉), Aleksandar Kartelj[2], Tome Eftimov[3], Jaume Reixach[4], and Christian Blum[4]

[1] Faculty of Natural Sciences and Mathematics, University of Banja Luka, Banja Luka, Bosnia and Herzegovina
marko.djukanovic@pmf.unibl.org

[2] Faculty of Mathematics, University of Belgrade, Belgrade, Serbia
aleksandar.kartelj@matf.bg.ac.rs

[3] Computer Systems, Jožef Stefan Institute, Ljubljana, Slovenia
tome.eftimov@ijs.si

[4] Artificial Intelligence Research Institute (IIIA-CSIC), Campus of the UAB, Bellaterra, Spain
{jaume.reixach,christian.blum}@iiia.csic.es

Abstract. This paper deals with the restricted longest common subsequence (RLCS) problem, an extension of the well-studied longest common subsequence problem involving two sets of strings: the input strings and the restricted strings. This problem has applications in bioinformatics, particularly in identifying similarities and discovering mutual patterns and motifs among DNA, RNA, and protein molecules. We introduce a general search framework to tackle the RLCS problem. Based on this, we present an exact best-first search algorithm and a meta-heuristic Beam Search algorithm. To evaluate the effectiveness of these algorithms, we compare them with two exact algorithms and two approximate algorithms from the literature along with a greedy approach. Our experimental results show the superior performance of our proposed approaches. In particular, our exact approach outperforms the other exact methods in terms of significantly shorter computation times, often reaching an order of magnitude compared to the second-best approach. Moreover, it successfully solves all problem instances, which was not the case with the other approaches. In addition, Beam Search provides close-to-optimal solutions with remarkably short computation times.

Keywords: Longest Common Subsequence Problem · Beam search · A* search · Restricted Patterns

1 Introduction

A string is a finite sequence of characters from a finite alphabet Σ. Strings are often used as a data structure, for example, in programming languages. Moreover, they play an important role as a model for DNA, RNA, and protein

L. Franco et al. (Eds.): ICCS 2024, LNCS 14836, pp. 58–73, 2024.
https://doi.org/10.1007/978-3-031-63775-9_5

sequences. In the fields of stringology and bioinformatics, a pivotal task is to find meaningful and representative measures of structural similarity between molecular structures. Among several measures, one that has gathered significant attention from both practical and theoretical perspectives is the well-known *longest common subsequence* (LCS). In this context, a subsequence of a string s is a string obtained by deleting zero or more symbols from s without changing the order of the remaining symbols. Finding longest common subsequences has been a subject of study for over half a century. Given a set of input strings $S = \{s_1, \ldots, s_m\}$, the LCS problem aims to identify a common subsequence concerning all strings in S of maximal length [4]. Apart from bioinformatics applications, this problem has shown to be useful in various fields, such as data compression and text processing [21].

From the very beginning, scientists have been concerned with the development of efficient algorithms for the LCS problem, especially for the case $m = 2$. Notable examples include algorithms based on the dynamic programming (DP) paradigm, such as the Hirschenberg algorithm, the Hunt-Szymanski algorithm, and the Apostolico-Crochemore algorithm; see [3,16]. If m is a fixed value, the LCS problem becomes polynomially solvable by DP, with a time complexity of $O(n^m)$, where n is the length of the longest string in S. For arbitrarily large sets S, however, the problem is known to be $\mathcal{NP}$-hard [18]. Moreover, it was found that a time complexity of $O(n^m)$ is likely the tightest unless $\mathcal{P} = \mathcal{NP}$. Consequently, the existence of an efficient algorithm for the general LCS problem scenario seems unlikely. As a result, various heuristic and approximation algorithms have been proposed. In particular, beam-search-based approaches [11] and hybrid anytime algorithms [12] have proven to be very efficient. In parallel with the development of methods for solving the LCS problem, several practical variants of this problem have been introduced. These include the longest arc-preserving common subsequence problem [5,17], the constrained LCS problem [10,22], and the shortest common supersequence problem [19], among others.

In this study, we deal with the *restricted longest common subsequence* (RLCS) *problem*, originally introduced by Gotthilf et al. [14]. In addition to considering an arbitrary set of input strings S, the problem involves a set of restricted pattern strings $R = \{r_1, \ldots, r_k\}$. The objective is to find a longest common subsequence s such that none of the restricted patterns $r_i \in R$ is contained as a subsequence of s. In their work, the authors show that the RLCS problem is $\mathcal{NP}$-hard even in the case of two input strings and an arbitrary number of restricted patterns. Moreover, they develop a DP approach for general values of m and k. In this scenario, they find that RLCS is in FPT (Fixed-Parameter Tractable) when parameterized by the total length of the restricted patterns. In addition, the authors propose two approximation algorithms. The first ensures an approximation ratio of $1/|\Sigma|$, while the second one guarantees a ratio of $(k_{\min} - 1)/n_{\min}$, where $k_{\min}$ and $n_{\min}$ represent the lengths of the shortest strings in R and S, respectively.

Independently of Gotthilf et al. [14], Chen and Chao [8] proposed a DP approach specifically for the RLCS problem with $m = 2$ and $k = 1$, achieving a time complexity of $O(|s_1| \cdot |s_2| \cdot |r_1|)$. For the same special case of the RLCS problem, Deorowicz and Grabowski [9] introduced two asymptotically faster algorithms than the conventional dynamic approach, with subcubic time complexities of $O(|s_1| \cdot |s_2| \cdot |r_1| / \log(|s_1|))$ and $O(|s_1| \cdot |s_2| \cdot |r_1| / \log^{\frac{3}{2}}(|s_1|))$ by utilizing well-designed internal data structures. Farhana and Rahman [13] proposed a finite automata-based approach to solve the general RLCS problem with a time complexity of $O(|\Sigma|(\mathcal{R} + m) + nm + |\Sigma|\mathcal{R}n^k)$, where $\mathcal{R} = O(n^m)$ denotes the size of the resulting automaton. The experimental results presented in that paper emphasize the superiority of the automata approach over the classical DP approach. The contributions of this paper are as follows:

1. *An error-free DP approach.* We present a DP approach to the RLCS problem that handles an arbitrary number of input strings and restricted pattern strings. In particular, our approach avoids significant flaws identified in the DP approach from [14].
2. *General search framework.* We design a general search framework for solving the RLCS problem, which serves as the core of an exact A* algorithm and a Beam Search approach. The search process in these methods is guided by utilizing the tightest known upper bounds for the classical LCS problem with arbitrary input strings.
3. *In-depth comparative analysis.* We perform a thorough comparison of all seven approaches (three proposed in this paper and four from the literature) for the RLCS problem, using a comprehensive set of available instances for evaluation.

Our A* search shows a clear superiority over the other two exact approaches from the literature. It excels in terms of the number of optimally solved instances, while it provides significantly shorter runtimes, often by an order of magnitude, compared to the best exact approaches from the literature.

1.1 Preliminaries

The length of a string s is denoted by $|s|$, whereas $s[i], 1 \leq i \leq |s|$, stands for its i-th character. It should be noted that—in this paper—the position of the leading character is indexed with 1. For two integers $i, j \leq |s|$, $s[i, j]$ denotes a continuous part of the string s that begins with the character at position i and ends with the character at position j. If $i = j$, the single-character string $s[i] = s[i, i]$ is given, or if $i > j$, the empty string ε is assigned.

For a left position vector $p^L = (p^L_1, \ldots, p^L_m)$, $1 \leq p^L_i \leq |s_i|$, $i = 1, \ldots, m$, we denote by $S[p^L]$ the set of suffix input strings associated with the respective coordinates of this vector, i.e., $S[p^L] := \{s_i[p^L_i, |s_i|] \mid i = 1, \ldots, m\}$. Finally, we define $p^L - \mathbf{1} := (p^L_i - 1, \ldots, p^L_m - 1)$ or, more generally, for two vectors $\mathbf{p}, \mathbf{q} \in \mathbb{N}^m$, $\mathbf{p} - \mathbf{q} := (p_1 - q_1, \ldots, p_m - q_m)$.

A complete RLCS problem instance is denoted as a pair (S, R) of two sets of strings, where S contains the input strings and R the restricted pattern strings. For two integer vectors $\mathbf{p} \in \mathbb{N}^m$ and $\mathbf{q} \in \mathbb{N}^k$, a sub-problem (sub-instance) of the initial problem instance with respect to these two (left) positional vectors is denoted by $(S[\mathbf{p}], R[\mathbf{q}])$.

The remaining sections of the work are organized as follows. Section 2 presents a DP approach for solving the RLCS problem. In particular, this section addresses and corrects the shortcomings of DP proposed in [14]. Section 3 presents a simple and naive greedy algorithm as an alternative method for solving the RLCS problem. Section 4 builds on the DP approach and derives a general search framework. In particular, we propose efficient A* and Beam Search algorithms. The practical comparison between our approaches and those in the literature is detailed in Sect. 5 through a thorough experimental evaluation. The paper concludes in Sect. 6, with possible directions for future research.

2 The DP Approach for the RLCS Problem

When considering DP, it is crucial to determine whether the problem under study has the optimal substructure property. The question is if the problem can be broken down into smaller parts, such that solving these smaller subproblems leads to an optimal solution to the overall problem. Based on this concept and using the well-known DP algorithm for the LCS problem with an arbitrary number of input strings, we derive the DP approach for the RLCS problem as follows. Let (S, R) be a RLCS problem instance, $\mathbf{p} \in \mathbb{N}^m$ and $\mathbf{l} \in \mathbb{N}^k$ with $1 \leq p_i \leq |s_i|$ for $i \in \{1, \ldots, m\}$ and $1 \leq l_j \leq |r_j|$ for $j \in \{1, \ldots, k\}$. We denote by RLCS$[\mathbf{p}; \mathbf{l}]$, the length of a longest common subsequence of $S[\mathbf{p}] = \{s_1[1, p_i], \ldots, s_m[1, p_m]\}$ with no string from $R_l = \{r_1[1, l_1], \ldots, s_k[1, l_k]\}$ as a subsequence. We distinguish the following non-trivial cases:

Case 1: $s_i[p_i] = \sigma \in \Sigma$ for every $i \in \{1, \ldots, m\}$. Let us denote $J := \{j \in \{1, \ldots, k\} \mid r_j[l_j] = \sigma\}$. In the event that letter σ does not contribute to the optimal solution of this subproblem, the relevant smaller subproblem is $(S_{\mathbf{p}-\mathbf{1}}, R_{\mathbf{l}})$ or $(S_{\mathbf{p}-\mathbf{1}}, R_{\mathbf{l}^*})$ if it does contribute, where $l^*_j = l_j - 1$ for $j \in J$ and $l^*_j = l_j$ otherwise. There are two sub-cases for this case:

(a) RLCS$[\mathbf{p}; \mathbf{l}] =$ RLCS$[\mathbf{p} - \mathbf{1}; \mathbf{l}]$ if there is an index $j \in J$ such that $l_j = 1$;

(b) RLCS$[\mathbf{p}; \mathbf{l}] = \max\{$RLCS$[\mathbf{p} - \mathbf{1}; \mathbf{l}]$, RLCS$[\mathbf{p} - \mathbf{1}; \mathbf{l}^*] + 1\}$, otherwise.

Case 2: $s_{i_1}[p_{i_1}] \neq s_{i_2}[p_{i_2}]$ for some $i_1, i_2 \in \{1, \ldots, m\}$. The same is done as in the recursion for the LCS problem. That is, the recursion is given by

$$\text{RLCS}[\mathbf{p}; \mathbf{l}] = \max\{\text{RLCS}[\mathbf{p} - \mathbf{e}_i; \mathbf{l}] \mid i = 1, \ldots, m\}.$$

In [14], the authors did not provide a correct derivation for Case 1 (see Sect. 4 of the aforementioned paper). In particular, they failed to include the +1 term ("plus one") for the Case 1b and did not distinguish between the Cases 1a and 1b. This omission led to incorrect calculations in the experimental evaluations of subsequent papers from the literature, e.g. [13].

3 Greedy Algorithm

The greedy algorithm for the RLCS problem uses a constructive approach that employs the best-next heuristic. It is an extension of the greedy approach used for the LCS problem [6]. At each step, it consists of appending the feasible letter with the best greedy value to the current partial solution. The algorithm starts with an empty solution $s^p = \varepsilon$ and processes the input strings from the far left (pointing to the characters at position 1 for strings from both sets, S and R) towards their right endpoints. The pointers for this process are denoted as $\mathbf{p}$ and $\mathbf{l}$ for each set, respectively. Next, the set Σ^{cand} of those letters that occur in all suffix strings $s_i[p_i, |s_i|]$ is considered. From this set, the algorithm retains only those letters which, when appended to s^p, do not cause a violation of the restriction that the whole r_i becomes a subsequence of the extended s^p. We denote the filtered set with Σ^{feas}. The algorithm selects a letter $a \in \Sigma^{feas}$ with the smallest value calculated by:

$$g(\mathbf{p}, \mathbf{l}, a) = \sum_{i=1}^{m} \frac{Succ[p_i]_{i,a} - p_i}{|s_i| - p_i + 1} + \sum_{j=1}^{k} \frac{|\Sigma|}{|r_j| - l_j - I_{r_i[l_i]=a} + 1} \tag{1}$$

where $Succ[p_i]_{i,a}$ denotes the smallest position in s_i greater or equal to p_i, at which letter a appears (in this way we eliminate suboptimality at the local level of decisions) and I represents the indicator function. Let us denote the best letter according to these g-values by a^*. Then, the following updates are performed: $s^p = s^p \cdot a^*$, $p_i = Succ[p_i]_{i,a^*} + 1$ and $l_j = l_j + I_{r_j[\mathbf{l}_j]=a^*}$ for every $i \in \{1, \ldots, m\}$ and $j \in \{1, \ldots, k\}$ respectively. The following iterations proceed in the same way until $\Sigma^{feas} = \emptyset$. Finally, the constructed solution is returned.

Note that, at each iteration, the greedy criterion (1) favors the feasible letters whose selection excludes smaller parts of the input strings from being considered, and who make the partial solution less close to having one of the restricted strings as a subsequence.

4 The Proposed Efficient Search Methods

In this section, we first describe a general search framework for the RLCS problem based on the definition of a state graph. Afterwards, we derive an exact and a heuristic search algorithm, both based on this framework.

4.1 The General Search Framework

The state graph is the environment of our proposed algorithms. Its inner nodes represent partial solutions while its sink nodes represent complete solutions. Moreover, edges between nodes represent extensions of partial solutions. The state graph $G = (V, E)$ of an RLCS problem instance (S, R) is defined as follows.

We say that a partial solution—that is, a common subsequence s^v of the strings in S that does not contain any string from R as as subsequence—*induces* a node $v = (p^{L,v}, l^v, u^v) \in V$ if:

- $|s^v| = u^v$
- s^v is a subsequence of all $s_i[1, p_i^{L,v} - 1], i \in \{1, \ldots, m\}$ and $p_i^{L,v} - 1$ is the smallest index that satisfies this property.
- s^v contains none of the prefix strings $r_j[1, l_j^v]$ as its subsequence whereas $r_j[1, l_j^v - 1], j \in \{1, \ldots, k\}$ are all included.

Additionally, there is an edge between nodes $v_1 = (p^{L,v_1}, l^{v_1}, u^{v_1})$ and $v_2 = (p^{L,v_2}, l^{v_2}, u^{v_2})$ labelled with a letter $a \in \Sigma$, denoted by $t(v_1, v_2) = a$, if:

- $u^{v_1} + 1 = u^{v_2}$
- The partial solution inducing node v_2 is obtained by appending the letter a to the partial solution inducing node v_1.

Each edge of the state graph of an RLCS problem instance has weight one and (as mentioned above) a label denoting the letter used for the extension.

To extend a node v and determine its successor nodes (children), it is necessary to identify the letters that can feasibly extend the partial solution s^v represented by v. First, all letters occurring in each string from the set $S[p^{L,v}]$ are identified. Then, the letters that cause a violation of the restrictions are removed, i.e., letters that cause one of the restricted patterns $r_i \in R$ to be a subsequence of the partial solution obtained by extending s^v with this letter. In addition, dominated letters are also omitted. A letter a is said to *dominate* the letter b (i.e., b is dominated by a) if $Succ[p^{L,v}i]_{i,a} \leq Succ[p^{L,v}i]_{i,b}$ for every $i \in \{1, \ldots, m\}$ and $r_j[l_j^v] \notin \{a, b\}$ for all $j \in \{1, \ldots, k\}$. We denote the set of non-dominated feasible letters to extend the partial solution of a node v by Σ_v^{nd}.

For a letter $a \in \Sigma_v^{nd}$, the corresponding successor node $w = (p^{L,w}, l^w, u^w)$ of v is constructed as follows.

- $u^w = u^v + 1$, i.e., the partial solution of node v derives the partial solution of node w by appending the letter a to it: $s^w = s^v \cdot a$.
- $l_j^w = l_j^v + 1$ if $r_j[l^v] = a$ or $l_j^w = l_j^v$ otherwise.
- For the (left) position vectors, $p_i^{L,w} = Succ[p_i^{L,v}]_{i,a} + 1$.

Notably, the aforementioned data structure *Succ* can be preprocessed before the construction of an RLCS state graph is started. In this way, finding suitable position vectors of a child node is addressed in time $O(m)$.

The *root* (initial) node $r = ((1, \ldots, 1), (1, \ldots, 1), 0)$ corresponds to the empty solution $s^r = \varepsilon$, which is trivially feasible and induces the complete problem instance (S, R).

We say that a node v is *complete* if $\Sigma_v^{nd} = \emptyset$. These are the nodes that have no child nodes (successors). Note that (partial) solutions induced by complete nodes are candidates for optimal solutions. In this context, note that optimal solutions are end-points of the longest paths from the root node r to complete nodes. Since the RLCS problem is $\mathcal{NP}$-hard, generating the entire state graph is generally infeasible as its size grows exponentially with the instance size. Consequently, our algorithm proposals generate and visit nodes on the fly, making intelligent

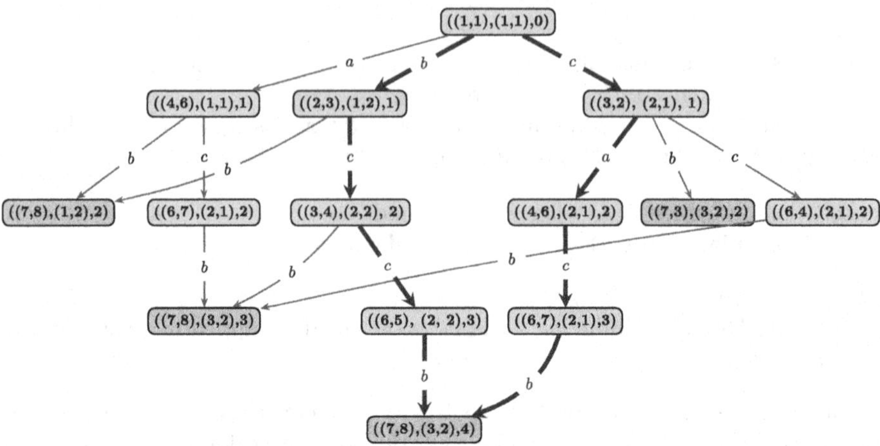

Fig. 1. Example of the full state graph in the form of a directed acyclic graph for the problem instance ($S = \{s_1 = \texttt{bcaacbb}, s_2 = \texttt{cbccacb}\}$, $R = \{\texttt{cbb}, \texttt{ba}\}$). It contains four complete nodes (light grey background). The two paths from $((7,8),(3,2),4)$ to the root node (in blue) are the longest paths in the graph. Hence, they represent two optimal solutions for this problem instance, `bccb` and `cacb` respectively

decisions to prioritize the exploration of more promising nodes, as explained in the following sections. This section concludes by showing the complete state graph of an instance in Fig. 1.

4.2 A* Search Algorithm

A* search [15] is an exact, informed search algorithm that follows the best-first search strategy for path-finding. It is the most widely used path-finding algorithm, being of high relevance in many fields e.g., in video games, in string matching and parsing, and others. The idea of the algorithm is to always expand the most promising nodes first. To rank the quality of nodes, a scoring function $f(v) = g(v) + h(v)$ is used. If the goal is to find longest paths, as in our case, functions $g()$ and $h()$ are defined as follows:

- $g(v)$ is the length of the longest path currently known from the root r to v.
- $h(v)$ is a heuristic function for estimating the length of the longest path from v to a complete (goal) node.

Note that A* works on a dynamically generated directed acyclic graph and in practice rarely examines all nodes. It has the advantageous ability to merge multiple nodes into one, which, as explained below, leads to considerable memory savings. To set up an efficient A* search for the RLCS problem, two important data structures are used:

- A hash map N with keys of the form $(p^{L,v}, l^v)$, where the corresponding value is the longest partial solution that induces a node with these vectors,

indicating the sub-instance $(S[p^{L,v}], R[l^v])$ to be solved. This data structure efficiently checks whether or not a node with the same key values has already been visited.
- A priority queue Q that contains open (not yet expanded) nodes that are prioritized based on their f-values. This structure facilitates the efficient retrieval of the most promising node.

Next, the heuristic function $h()$ must be defined. For this purpose, we opt for using the tightest known upper bound for the LCS problem. Note that any upper bound for an LCS problem instance is also an upper bound for a corresponding RLCS problem instance obtained by adding a set R of restricted strings. The upper bound we used is the minimum of two known upper bounds, denoted as UB_1 and UB_2, that is, $\mathrm{UB} = \min\{\mathrm{UB}_1, \mathrm{UB}_2\}$. For detailed information, we refer to [6] and [23]. The upper bound UB_1 determines for every letter an upper bound on the number of times this letter is potentially included in an optimal solution and then returns the sum of these values. On the other hand, UB_2 repeatedly applies dynamic programming to the LCS problem of two input strings for constructing the upper bound. It is important to note that there may be multiple nodes with the same f-value. In such cases, ties are resolved by favoring those with a higher u^v value.

The algorithm starts by initializing the root node r, which is then added to both N and Q. At the same time, the best solution s^{best} is initialized to the empty string ε. In each iteration, the most promising node v from the beginning of the priority queue Q is selected. If $f(v)$ is less than or equal to $|s^{best}|$, the search is terminated and the proven optimal solution s^{best} is returned. If the node v is complete, it is checked whether u^v is greater than $|s^{best}|$. If this is the case, s^{best} is chosen for inducing node v, which is reconstructed by traversing from v back to the root node r and reading the letters along the transitions. The algorithm then proceeds to the next iteration. If none of the above conditions are met, the node v is expanded in all possible ways by creating its children. For each child w, it is checked whether $(p^{L,w}, l^w)$ is already contained in N. If not, w is added to both N and Q. Otherwise, it is checked whether a new best path from the root node r to any node associated with $(p^{L,w}, l^w)$ has been found. In the case of a positive answer, the information in N is updated and the priority of this node is changed in Q. In the case of a negative answer, w is declared irrelevant and thus skipped in the next iteration of the algorithm.

4.3 Beam Search Algorithm

Beam Search (BS) [1] is a heuristic search algorithm that works in a "breadth-first-search" (BFS) manner, expanding nodes at each level, with a limitation on the number of nodes to be expanded. More precisely, up to $\beta > 0$ of the most promising nodes at each level are selected to generate the nodes of the next level. Parameter β ensures that the size of the BS tree remains polynomial with respect to the size of the problem instance, which makes this method applicable to various complex problems. BS is widely used in fields such as packing [1], scheduling [20], and bioinformatics [7], among others.

In addition to parameter $\beta > 0$, the effectiveness of BS strongly depends on a heuristic function $h()$ used to evaluate the "quality" of each node. The choice of $h()$ is typically a problem-specific decision. For our purpose, we opt for the tightest upper bound (UB) for the LCS problem already introduced in the previous section. The BS approach for the RLCS problem works as follows. The root node r is first generated and included in the beam B (i.e. $B = \{r\}$) and l_{best} is initialized to 0. Then, the main loop is entered. All nodes in the beam B are expanded in all possible ways. The resulting child nodes are stored in V_{ext}, while l_{best} is increased by one. The nodes from V_{ext} are then sorted in descending order according to their $h()$-values. B is emptied and the best $\beta > 0$ nodes are then added to it for the subsequent level. These steps are repeated as long as the beam B is not empty, in which case, the algorithm is stopped, and the best RLCS solution s_{best} of length l_{best} is returned.

5 Experimental Evaluation

This section presents a comprehensive experimental evaluation comparing three exact competitors: the A* search proposed in Sect. 4.2 and two existing approaches from the literature, namely, the automaton approach presented in [13] (denoted as Automaton) and the corrected version of DP initially proposed in [14], as provided in Sect. 2 (denoted as Dp). Besides the exact approaches, we also compare four heuristic methods: the Beam Search (BS) proposed in Sect. 4.3, the greedy approach from Sect. 3 (denoted as Greedy) and two approximation algorithms (denoted as Approx1 and Approx2) from the literature, proposed in [14]. As mentioned above, the source code for the Automaton approach in its original form was obtained directly from the authors of [13], along with the provided problem instances. The remaining six approaches were implemented in C++ under Ubuntu 20.0 and compiled with gcc 13.1.0 with optimization level *Ofast*. All experiments were conducted in single-threaded mode on an Intel Xeon E5-2640 with 2.40GHz and 16 GB of memory.

The experimental evaluation employs three sets of benchmark instances denoted as Random, Real and Substr-Ec. In the benchmark set Random, instances are divided into six groups based on the values of m and k, which denote the amount of input and restricted strings respectively. Five groups comprise 10 instances each, while one group comprises a single instance, as provided by the original authors. Thus, there are a total of 51 randomly generated problem instances in this set.

For the benchmark set Real, we use four real-world instances presented in [13]. Finally, to extend the scope of our experiments, we use randomly generated instances designed for a variant of the RLCS problem, namely the substring-exclusion constrained LCS problem [2]. This set consists of two sets of 10 randomly generated instances each, resulting in a total of 20 instances. We refer to this benchmark set as Substr-Ec.

Parameter Setting. Only Beam search (BS) requires the setting of one of its parameters, namely the beam width $\beta > 0$. After a preliminary experimental evaluation, we found that setting $\beta = 100$ leads to a favorable compromise between the quality of the final solutions and the required computation times.

Description of the Results. Table 1 shows the numerical results for the three exact algorithms applied to the benchmark set RANDOM. Every row presents the average results for one of the instance groups. The first four columns describe the properties of each group, specifying the number of input strings (m), the number of pattern strings (k), the length of all input strings ($|s_0|$) and the length of all pattern strings ($|r_0|$). Moreover, $\overline{|s|}$ denotes the average length of the optimal solutions for every instance group.

For each algorithm and instance group, we present the amount (*#opt*) of instances that were solved to optimality and the average runtime required for obtaining these optimal solutions. Symbol "–" denotes that the respective algorithm was not able to provide any optimal solution due to problems during runtime, such as reaching time or memory limits.

The following conclusions can be drawn from Table 1.

1. A* search and DP can find a provably optimal solution for all (51) problem instances within the given time and memory constraints. For the AUTOMATON approach, this was possible for 32 problem instances, facing memory limitations for the remaining ones.
2. In terms of runtime, the A* search emerges as the clear winner, as it shows a significantly faster performance in comparison to DP and AUTOMATON, often outperforming them by an order of magnitude. It achieves consistently short average runtimes, all below one second.
3. The runtimes of DP increase rapidly with increasing m, while it seems to be difficult for the AUTOMATON approach to handle instances with larger k values (memory problems are notable for $k \geq 3$).

Table 2 shows the averaged numerical results for the four heuristic approaches applied to the benchmark set RANDOM. The values are averages over the instances within each group (rows). As in Table 1, the first four columns correspond to the instance groups. The table is then divided into four blocks, each consisting of two columns. These blocks correspond to the approaches BS, GREEDY, APPROX1 and APPROX2 respectively. Two values are provided for each algorithm: the average quality of the best-found solutions ($\overline{|s|}$) and the average runtime for obtaining these best solutions ($\bar{t}[s]$).

The following conclusions can be drawn from Table 2.

- The most effective heuristic approach, both in terms of solution quality and time efficiency, is BS. It outperforms the second-best approach, GREEDY, by a significant margin. With runtimes of around 0.1 s on average, BS achieves a high solution quality and remains within 7% of the optimal results.
- In contrast, the two approximation algorithms APPROX1 and APPROX2 are significantly behind the rest. In particular, the computation times of

Table 1. Comparisons between the exact approaches on benchmark set RANDOM. All pattern strings in the instances of the same group are of an equal length $|r_0|$. The same also holds for the length of the input strings, which are of length $|s_0|$

m	k	$\lvert s_0\rvert$	$\lvert r_0\rvert$	$\overline{\lvert s\rvert}$	A*		DP		AUTOMATON	
					#opt	$\bar{t}[s]$	#opt	$\bar{t}[s]$	#opt	$\bar{t}[s]$
2	1	200	3	65.0	1	0.0	1	0.0	1	0.3
2	3	250	8	88.2	10	0.0	10	6.8	1	2.1
2	4	250	6	87.1	10	0.0	10	18.5	0	–
3	1	200	10	46.0	10	0.6	10	35.3	10	1.2
3	2	200	3	43.8	10	1.0	10	44.9	10	1.9
4	1	75	3	12.9	10	0.0	10	92.4	10	0.2

Table 2. Comparisons between the heuristic approaches on benchmark set RANDOM

m	k	$\lvert s_0\rvert$	$\lvert r_0\rvert$	$\overline{\lvert s\rvert}$	BS		GREEDY		APPROX1		APPROX2	
					$\overline{\lvert s\rvert}$	$\bar{t}[s]$	$\overline{\lvert s\rvert}$	$\bar{t}[s]$	$\overline{\lvert s\rvert}$	$\bar{t}[s]$	$\overline{\lvert s\rvert}$	$\bar{t}[s]$
2	1	200	3	65.0	64.0	0.1	62.0	0.0	13.0	0.0	2.0	0.0
2	3	250	8	88.2	88.2	0.1	86.2	0.0	15.9	0.0	7.0	0.0
2	4	250	6	87.1	87.1	0.1	85.0	0.0	16.4	0.0	5.1	0.0
3	1	200	10	46.0	44.9	0.1	38.2	0.0	11.2	0.0	9.0	7.0
3	2	200	3	43.8	41.0	0.1	36.6	0.0	11.5	0.0	2.0	7.0
4	1	75	3	12.9	12.9	0.0	11.5	0.0	4.2	0.0	2.0	20.9

APPROX2 grow rapidly with increasing values of m, which indicates that using DP for the LCS problem is slow for larger instances.

Table 3 shows the numerical results for all seven approaches applied to four real cases. This table is structured as follows. The first four columns show the characteristics of the problem instances, including the number of input strings (m), the number of restricted strings (k), the length pairs for the shortest and longest input strings (n), and the length pairs for the shortest and longest restricted pattern strings (p). The next seven blocks, each consisting of two columns, show the results for the aforementioned approaches. In particular, the two columns show the delivered solution quality ($|s|$) and the corresponding runtime ($t[s]$).

The following conclusions can be drawn from the numerical results in Table 3.

- A* search and AUTOMATON provide optimal solutions for all four instances. A* outperforms AUTOMATON by being approx. one order of magnitude faster. Conversely, the DP approach only achieved optimal solutions for two instances and reached memory limits for the remaining two cases.
- Among the heuristic approaches, BS stands out as the superior choice, providing optimal solutions for two out of four instances. It is closely followed by

Table 3. Comparison between all seven approaches for benchmark set REAL

m	k	n	p	A*		AUTOMATON		DP		APPROX1		APPROX2		BS		GREEDY	
				$\overline{\lvert s\rvert}$	$\overline{t}[s]$	$\overline{\lvert s\rvert}$	$\overline{t}[s]$	$\overline{\lvert s\rvert}$	$\overline{t}[s]$	$\overline{\lvert s\rvert}$	$\overline{t}[s]$	$\overline{\lvert s\rvert}$	$\overline{t}[s]$	$\overline{\lvert s\rvert}$	$\overline{t}[s]$	$\overline{\lvert s\rvert}$	$\overline{t}[s]$
4	2	(141, 146)	(1, 2)	48	0.1	48	3.4	–	–	15	0	–	–	48	0.1	27	0
3	2	(255, 293)	(1, 4)	74	0.4	74	8.7	74.0	69.7	29	0	0.0	26.9	72	0.1	57	0
5	4	(98, 123)	(1, 4)	16	0.1	16	2.0	–	–	9	0	–	–	16	0.0	15	0
3	2	(124, 185)	(4, 5)	37	0.1	37	0.5	37.0	19.3	10	0	3.0	2.8	32	0.0	30	0

GREEDY, which provides solutions of reasonable quality with exceptionally short computation times. On the other hand, both approximation algorithms, APPROX1 and APPROX2, provide impractical results. In particular, APPROX2 does not provide any results for instances with $m \geq 4$.

Table 4 reports the numerical results for all seven approaches on the benchmark set SUBSTR-EC. This table is structured similarly to Table 2 with a small difference in the instance description. In particular, in addition to m and k, instances are also described by their *index* number (there are 10 (indexed) instances in each of the two groups).

The following conclusions can be derived from Table 4.

- A* and DP achieve optimal solutions for all 20 instances. However, A* shows again a notable advantage in terms of computation times (a difference of two orders of magnitude). The AUTOMATON approach encounters difficulties in terms of memory usage for the instances with $k = 3$, as already observed for the RANDOM benchmark set.
- Regarding the heuristic approaches, BS proves to be an outstanding performer as it provides optimal solutions for all 20 problem instances with remarkably short computation times (at most 0.1 s). GREEDY is able to derive optimal solutions for all (10) instances with $k = 2$. However, in case of the instances with $k = 3$, the performance of GREEDY deteriorates as an optimal solution is only produced in 3 (out of 10) cases.
- In contrast, the other two algorithms provide extremely fast solutions, but they deviate significantly from the known optimal solutions.

5.1 Statistical Analysis

To determine statistical differences between the results of the seven competing approaches, we conducted a pairwise statistical analysis using a one-sided Wilcoxon rank-sum test, as shown in Fig. 2. The null hypothesis, asserting that the first algorithm yields superior (larger) results compared to the second algorithm, was assessed at a significance level (α) of 0.05. For instance, at the point

Table 4. Comparison between the approaches on benchmark set SUBSTR-EC. The instances comprised in the group $m = 2, k = 1$ have both input strings of length 100 and pattern strings of length 40. The instances comprised in the group $m = 2, k = 3$ have both input strings of length 250 and all pattern strings of length 8

m	k	*ind.*	A*		AUTOMATON		DP		APPROX1		APPROX2		BS		GREEDY	
			$\overline{s}$	$\overline{t}[s]$	$\overline{s}$	$\overline{t}[s]$	$\overline{s}$	$\overline{t}[s]$	$\overline{s}$	$\overline{t}[s]$	$\overline{s}$	$\overline{t}[s]$	$\overline{s}$	$\overline{t}[s]$	$\overline{s}$	$\overline{t}[s]$
2	1	0	47	0.0	47.0	0.2	47	0.1	7	0	39	0.0	47	0.1	47	0.0
		1	42	0.0	42.0	0.2	42	0.1	6	0	39	0.0	42	0.0	42	0.0
		2	44	0.0	44.0	0.2	44	0.1	6	0	39	0.0	44	0.1	44	0.0
		3	44	0.0	44.0	0.2	44	0.1	8	0	39	0.0	44	0.0	44	0.0
		4	46	0.0	46.0	0.2	46	0.1	7	0	39	0.0	46	0.0	46	0.0
		5	47	0.0	47.0	0.2	47	0.1	7	0	39	0.0	47	0.1	47	0.0
		6	45	0.0	45.0	0.2	45	0.1	7	0	39	0.0	45	0.0	45	0.0
		7	48	0.0	48.0	0.2	48	0.1	7	0	39	0.0	48	0.1	48	0.0
		8	43	0.0	43.0	0.2	43	0.1	8	0	39	0.0	43	0.0	43	0.0
		9	45	0.0	45.0	0.2	45	0.1	7	0	39	0.0	45	0.1	45	0.0
2	3	0	90	0.0	–	–	90	7.0	16	0	7	0.0	90	0.1	87	0.0
		1	84	0.0	–	–	84	8.1	14	0	7	0.0	84	0.1	83	0.0
		2	87	0.0	–	–	87	6.5	16	0	7	0.0	87	0.1	87	0.0
		3	91	0.0	–	–	91	5.6	16	0	7	0.0	91	0.1	90	0.0
		4	89	0.0	–	–	89	7.4	16	0	7	0.0	89	0.1	85	0.0
		5	87	0.0	–	–	87	7.5	15	0	7	0.0	87	0.1	83	0.0
		6	88	0.0	–	–	88	6.4	17	0	7	0.0	88	0.1	87	0.0
		7	91	0.0	–	–	91	9.5	17	0	7	0.0	91	0.1	91	0.0
		8	89	0.0	–	–	89	5.9	15	0	7	0.0	89	0.1	83	0.0
		9	86	0.0	–	–	86	7.0	17	0	7	0.0	86	0.1	86	0.0

of intersection of A* search (on the x-axis) and the GREEDY approach (on the y-axis), a p-value resulting from a one-sided Wilcoxon rank-sum test between the outcomes of these two algorithms is provided (with a value of 1.0). This indicates substantial evidence in favor of retaining the null hypothesis over the alternative. Hence, we infer that the A* search results statistically outperform those of the GREEDY approach. The solution quality achieved by A* is superior across all 75 instances compared to other competitors. In addition, remember that the running times of A* are significantly shorter than those of DP and AUTOMATON approaches.

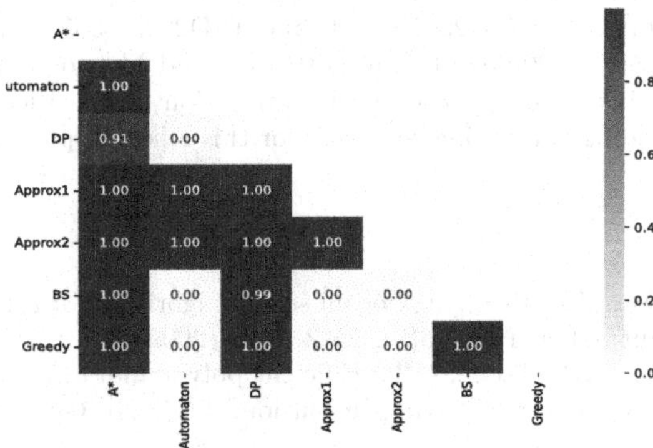

Fig. 2. Post-hoc pairwise statistical comparison between the seven competing approaches on all problem instances using the (one-sided) Wilcoxon rank-sum test

6 Conclusions and Future Work

This work has dealt with the RLCS problem, an extension of the well-known LCS problem. First, we corrected a previously proposed dynamic programming approach for the RLCS problem. Then, a comprehensive search framework based on the concept of a state graph was introduced. Using this framework, we developed both an exact A* search algorithm and a heuristic Beam Search. Our results, validated on 75 problem instances from the literature, obtained from the authors of [13], emphasize the effectiveness of the proposed methods. The exact approach showed strong performance by being the only algorithm to provide provably optimal solutions for all instances, with computation times being an order of magnitude shorter than those of the best exact competitor from the literature. In addition, the Beam Search showed promising results by providing optimal solutions for many instances employing remarkably short computation times.

For future research, it would be interesting to explore the limits of our proposed exact approach by testing it on much larger instances. Moreover, the known scalability and robustness of Beam Search present considerable potential and also deserve further investigation. Improving the heuristic guidance for Beam Search is another research direction; possible ideas include developing a probabilistic search model or incorporating machine learning techniques.

Acknowledgments. The research of M. Djukanović is partially supported by the Ministry for Scientific and Technological Development and Higher Education of the Republic of Srpska, B&H in the course of the bilateral research project between B&H and Slovenia entitled "Theoretical and computational aspects of some graph problems with the application to graph network information spreading" and the COST Action ROAR-NET under no. CA22137. A. Kartelj was supported by grant 451-03-47/2023-01/200104 funded by the Ministry of Science Technological Development and Innovations of the Republic of Serbia. J. Reixach and C. Blum are supported by grants TED2021-129319B-I00 and PID2022-136787NB-I00 funded by MCIN/AEI/10.13039/501100011033. The authors would like to thank the Compute Cluster Unit of the Institute of Logic and Computation at the Vienna University of Technology for providing computing resources for this research project.

References

1. Akeb, H., Hifi, M., M'Hallah, R.: A beam search algorithm for the circular packing problem. Comput. Oper. Res. **36**(5), 1513–1528 (2009)
2. Ann, H.Y., Yang, C.B., Tseng, C.T.: Efficient polynomial-time algorithms for the constrained lcs problem with strings exclusion. J. Comb. Optim. **28**(4), 800–813 (2014)
3. Apostolico, A., Guerra, C.: The longest common subsequence problem revisited. Algorithmica **2**, 315–336 (1987)
4. Bergroth, L., Hakonen, H., Raita, T.: A survey of longest common subsequence algorithms. In: Proceedings Seventh International Symposium on String Processing and Information Retrieval. SPIRE 2000, pp. 39–48. IEEE (2000)
5. Blum, C., Blesa, M.J.: A hybrid evolutionary algorithm based on solution merging for the longest arc-preserving common subsequence problem. In: 2017 IEEE Congress on Evolutionary Computation (CEC), pp. 129–136. IEEE (2017)
6. Blum, C., Blesa, M.J., Lopez-Ibanez, M.: Beam search for the longest common subsequence problem. Comput. Oper. Res. **36**(12), 3178–3186 (2009)
7. Carlson, J.M., Chakravarty, A., Gross, R.H.: Beam: a beam search algorithm for the identification of cis-regulatory elements in groups of genes. J. Comput. Biol. **13**(3), 686–701 (2006)
8. Chen, Y.C., Chao, K.M.: On the generalized constrained longest common subsequence problems. J. Comb. Optim. **21**(3), 383–392 (2011)
9. Deorowicz, S., Grabowski, S.: Subcubic algorithms for the sequence excluded LCS problem. In: Gruca, A.., Czachórski, T., Kozielski, S. (eds.) Man-Machine Interactions 3. AISC, vol. 242, pp. 503–510. Springer, Cham (2014). https://doi.org/10.1007/978-3-319-02309-0_55
10. Djukanovic, M., Berger, C., Raidl, G.R., Blum, C.: On solving a generalized constrained longest common subsequence problem. In: Olenev, N., Evtushenko, Y., Khachay, M., Malkova, V. (eds.) OPTIMA 2020. LNCS, vol. 12422, pp. 55–70. Springer, Cham (2020). https://doi.org/10.1007/978-3-030-62867-3_5
11. Djukanovic, M., Raidl, G.R., Blum, C.: A beam search for the longest common subsequence problem guided by a novel approximate expected length calculation. In: Nicosia, G., Pardalos, P., Umeton, R., Giuffrida, G., Sciacca, V. (eds.) LOD 2019. LNCS, vol. 11943, pp. 154–167. Springer, Cham (2019). https://doi.org/10.1007/978-3-030-37599-7_14
12. Djukanovic, M., Raidl, G.R., Blum, C.: Finding longest common subsequences: new anytime A* search results. Appl. Soft Comput. **95**, 106499 (2020)
13. Farhana, E., Rahman, M.S.: Constrained sequence analysis algorithms in computational biology. Inf. Sci. **295**, 247–257 (2015)
14. Gotthilf, Z., Hermelin, D., Landau, G.M., Lewenstein, M.: Restricted LCS. In: Chavez, E., Lonardi, S. (eds.) SPIRE 2010. LNCS, vol. 6393, pp. 250–257. Springer, Heidelberg (2010). https://doi.org/10.1007/978-3-642-16321-0_26
15. Hart, P.E., Nilsson, N.J., Raphael, B.: A formal basis for the heuristic determination of minimum cost paths. IEEE Trans. Syst. Sci. Cybernet. **4**(2), 100–107 (1968)

16. Hirschberg, D.S.: Algorithms for the longest common subsequence problem. J. ACM **24**(4), 664–675 (1977)
17. Lin, G., Chen, Z.Z., Jiang, T., Wen, J.: The longest common subsequence problem for sequences with nested arc annotations. J. Comput. Syst. Sci. **65**(3), 465–480 (2002)
18. Maier, D.: The Complexity of Some Problems on Sequences. Princeton University (1978)
19. Mousavi, S.R., Bahri, F., Tabataba, F.S.: An enhanced beam search algorithm for the shortest common supersequence problem. Eng. Appl. Artif. Intell. **25**(3), 457–467 (2012)
20. Sabuncuoglu, I., Bayiz, M.: Job shop scheduling with beam search. Eur. J. Oper. Res. **118**(2), 390–412 (1999)
21. Storer, J.A.: Data Compression: Methods and Theory. Computer Science Press, Inc. (1987)
22. Tsai, Y.T.: The constrained longest common subsequence problem. Inf. Process. Lett. **88**(4), 173–176 (2003)
23. Wang, Q., Pan, M., Shang, Y., Korkin, D.: A fast heuristic search algorithm for finding the longest common subsequence of multiple strings. Proc. AAAI Conf. Artif. Intell. **24**, 1287–1292 (2010)

Adaptive Hyperparameter Tuning Within Neural Network-Based Efficient Global Optimization

Taeho Jeong[1], Pavankumar Koratikere[1], Leifur Leifsson[1(✉)], Slawomir Koziel[2,3], and Anna Pietrenko-Dabrowska[3]

[1] School of Aeronautics and Astronautics, Purdue University, West Lafayette, IN 47907, USA
{jeong183,pkoratik,leifur}@purdue.edu

[2] Engineering Optimization and Modeling Center, Department of Engineering, Reykjavík University, Menntavegur 1, 102 Reykjavík, Iceland
koziel@ru.is

[3] Faculty of Electronics Telecommunications and Informatics, Gdansk University of Technology, Narutowicza 11/12, 80-233 Gdansk, Poland
anna.dabrowska@pg.edu.pl

Abstract. In this paper, adaptive hyperparameter optimization (HPO) strategies within the efficient global optimization (EGO) with neural network (NN)-based prediction and uncertainty (EGONN) algorithm are proposed. These strategies utilize Bayesian optimization and multi-armed bandit optimization to tune HPs during the sequential sampling process either every iteration (HPO-1itr) or every five iterations (HPO-5itr). Through experiments using the three-dimensional Hartmann function and evaluating both full and partial sets of HPs, adaptive HPOs are compared to traditional static HPO (HPO-static) that keep HPs constant. The results reveal that adaptive HPO strategies outperform HPO-static, and the frequency of tuning and number of tuning HPs impact both the optimization accuracy and computational efficiency. Specifically, adaptive HPOs demonstrate rapid convergence rates (HPO-1itr at 28 iterations, HPO-5itr at 26 for full HPs; HPO-1itr at 13, HPO-5itr at 28 iterations for selected HPs), while HPO-static fails to approximate the minimum within the allocated 45 iterations for both scenarios. Mainly, HPO-5itr is the most balanced approach, found to require 21% of the time taken by HPO-1itr for tuning full HPs and 29% for tuning a subset of HPs. This work demonstrates the importance of adaptive HPO and sets the stage for future research.

Keywords: Neural networks · surrogate-based optimization · hyperparameter optimization · sequential sampling

1 Introduction

The engineering design optimization process merges computational simulations with optimization methods to determine optimal design under specific

L. Franco et al. (Eds.): ICCS 2024, LNCS 14836, pp. 74–89, 2024.
https://doi.org/10.1007/978-3-031-63775-9_6

constraints. This process is typically iterative, which increases the computational cost. Surrogate-based optimization (SBO) presents an effective solution by employing data-driven surrogate models instead of repetitive simulations, and can save computational cost while yielding optimum results [14].

The efficient global optimization (EGO) algorithm [8] is a widely used SBO technique that typically employs kriging [3] as its surrogate model. This algorithm adopts a sequential sampling method, starting with an initial kriging model. Each iteration involves selecting a new sample point that maximizes the expected improvement (EI) criterion, balancing exploration and exploitation. The new samples are then integrated into the kriging model for subsequent iterations, progressively enhancing the performance [16]. However, the computational cost associated with kriging increases quickly with more samples [12].

Neural networks (NNs) are effective surrogate models for managing large, nonlinear data sets. They learn complex input-output relationships, enabling accurate predictions of new data outcomes [14]. Their performance largely depends on the data's quality and quantity, as NNs do not typically employ sequential sampling, a process often referred to as one-shot optimization or sampling [7]. This is primarily due to their inability to estimate uncertainty, a requirement for exploration. Sequential sampling could lead to convergence at local minima without exploration and limiting further refinement of the surrogate model. Therefore, optimizing an initial data set size is important when using NNs [17].

Koratikere et al. [10] recently introduced the efficient global optimization using neural networks (EGONN) algorithm, which combines sequential sampling with NN-based predictions. This method uniquely employs a secondary NN to model uncertainty, allowing EGO to combine with NNs' predictive abilities for sequential sampling. A significant advantage of this approach is simplifying the preliminary setup and enhancing the efficiency of NNs in global modeling and optimization tasks [9].

The structure of NNs is defined using hyperparameters (HPs), including structural parameters and learning algorithms, which need to be set before training. Consequently, HP optimization (HPO) is important for constructing an optimally performing NN [15,17]. In one-shot optimization, the optimal HP configuration is determined for the initial data set and kept static throughout the algorithm run [7]. However, HPs require occasional tuning as the optimal NN structure and algorithms may vary with the number of training samples for sequential sampling.

This work proposes an adaptive HPO for sequential sampling within the EGONN algorithm. This adaptive HPO addresses the limitations associated with static HP, offering a more dynamic and responsive optimization strategy. It combines Bayesian optimization (BO) [15] and multi-armed bandit (MAB) optimization [5] to optimize continuous and discrete HPs, respectively. These strategies balance exploration and exploitation, enabling efficient identification of the HP configuration. This integrated approach facilitates finding the best HP configuration with fewer design iterations and reduced time [1,4].

Additionally, this paper conducts a comparative analysis of different HPO strategies within the EGONN framework, examining static HP tuning, tuning HPs every five iterations, and every iteration. These strategies are evaluated based on metrics such as the number of infill points required for convergence and the time cost, using the EGO as a benchmark. The analysis utilizes the three-dimensional Hartmann function by tuning a complete set and a subset of HPs for numerical experiments.

The remainder of the paper is structured as follows. Section 2 provides an overview of the employed SBO methods and HPO. Section 3 describes numerical experiments for evaluating the proposed methods. Finally, Sect. 4 summarizes the essential findings and identifies potential future research directions.

2 Methods

This section introduces the basics of SBO and specific sequential sampling algorithms for optimization. It then describes the HPO methods and adaptive HPO strategies employed in this work.

2.1 Surrogate-Based Optimization

SBO is effective for design optimization problems that minimize an objective function $f(\mathbf{x}_d)$, subject to constraints $g_i(\mathbf{x}_d) \leq 0$ and $h_j(\mathbf{x}_d) = 0$, which can be written as:

$$\begin{aligned} \min_{\mathbf{x}_d} \quad & f(\mathbf{x}_d) \\ \text{subject to} \quad & g_i(\mathbf{x}_d) \leq 0,\ i = 1, \dots, q, \\ & h_j(\mathbf{x}_d) = 0,\ j = 1, \dots, r, \end{aligned} \tag{1}$$

where $\mathbf{x}_d = (x_{d_1}, x_{d_2}, ..., x_{d_n})$ is the vector of design variables, q is the number of the inequality constraints, and r is the number of the equality constraints. These problems often require extensive evaluation using simulations, making SBO a valuable tool for efficiently finding solutions using a data-driven surrogate model instead of directly evaluating the objective function or constraints [3]. The effectiveness of SBO, however, is contingent upon the accuracy of the surrogate model. Inaccurate approximations can result in sub-optimal optimization outcomes. Furthermore, constructing a reliable surrogate model requires substantial initial data. SBO iteratively refines the surrogate model by adding a sample point, resulting in progressively more accurate approximations of the objective function. [16].

Figure 1 presents a flowchart outlining the SBO process with sequential sampling utilized in this research. The process is initiated by generating an initial set of samples through a deterministic Halton sequence sampling [2]. These samples are used to acquire observations for training the surrogate models, such as NN or kriging. Then, infill criteria, such as the expected improvement, are maximized to explore and exploit the design space using differential evolution [11]. The new infill point is added to the training data for the next iteration, continuing until the allocated budget is exhausted or a convergence criterion is satisfied.

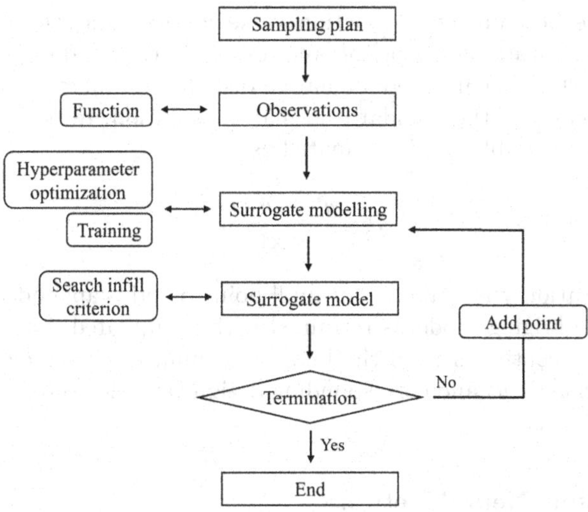

Fig. 1. A flowchart illustrating SBO with sequential sampling and adaptive HPO

2.2 Efficient Global Optimization

EGO is a widely used SBO method to optimize a black box function that uses a kriging model for prediction and uncertainty estimation [8]. As described in Algorithm 1, the process begins by generating an initial sample plan $\mathbf{X}$ and observing the corresponding outputs $\mathbf{Y}$. In each iteration, a kriging model is created using the available data set. Then, EI [3] is maximized to find the next infill point $\mathbf{p}$, which balances the exploration of areas characterized by high uncertainty against the exploitation of areas with low estimated function values. The EI function is formulated as

$$EI(\mathbf{x}) = \left[f(\mathbf{x}^*) - \hat{f}(\mathbf{x})\right]\Phi(Z) + \hat{s}(\mathbf{x})\phi\left(\frac{f(\mathbf{x}^*) - \hat{f}(\mathbf{x})}{\hat{s}(\mathbf{x})}\right), \quad (2)$$

Algorithm 1. Efficient global optimization [8].

Create initial sampling plan $\mathbf{X}$
Compute objective function value $\mathbf{Y}$ at initial sampling $\mathbf{X}$
while infill budget lasts **do**
    Create kriging model using $(\mathbf{X}, \mathbf{Y})$
    Find the best sample y^* in $\mathbf{Y}$
    Maximize $EI(\mathbf{x})$ to find next infill point $\mathbf{p}$
    Compute objective function value $\mathbf{y}$ at $\mathbf{p}$
    Append $(\mathbf{X}, \mathbf{Y})$ with $(\mathbf{p},y)$
end while
Return best sample $(\mathbf{x}^*,y^*)$ in appended data set

where y^* is the best objective function value in the data set, $\hat{y}(\mathbf{x})$ and $\hat{s}(\mathbf{x})$ represent the surrogate model prediction and the corresponding uncertainty in the prediction. Φ and ϕ are the cumulative distribution function and probability density function of the standard normal distribution, respectively, and the standard normal variable (Z) is evaluated as:

$$Z = \frac{y^* - \hat{y}(\mathbf{x})}{\hat{s}(\mathbf{x})}. \tag{3}$$

The new observation $\mathbf{y}$ is evaluated at infill point $\mathbf{p}$ and is appended to the data set $(\mathbf{X}, \mathbf{Y})$. The kriging model is retrained with an updated data set, and this iterative cycle is continued until the budget for infill is exhausted. Finally, the best-observed sample y^* and corresponding design variable vector $\mathbf{x}^*$ is returned as the optimum design.

2.3 EGO Using Neural Networks

The recently developed EGONN algorithm enhances the EGO process by incorporating two separate NNs for modeling the objective function and uncertainty [10]. The first NN (NN_y) models the objective function and the second NN (NN_u) models uncertainty. A third NN (NN_g) is added if there are constraints in the problem [9]. As described in Algorithm 2, the EGONN process is similar to EGO but there are two major differences. The process starts with generating two different sampling plans, $\mathbf{X}$ and $\mathbf{X}_u$, with corresponding function values Y and Y_u. Within the iterative cycle, a NN model for $\hat{y}$ is created using ($\mathbf{X}$, $\mathbf{Y}$). Then, $\hat{y}$ model is used for predicting the values at $\mathbf{X}$ and $\mathbf{X}_u$. The prediction error can be computed since the true values are available. In the next step, the $\hat{s}$ model is created using $\mathbf{X}$ and $\mathbf{X}_u$, and the corresponding prediction error. In this way, EGONN creates two NN models, one for prediction and the other for uncertainty estimation.

The remaining part of the cycle is the same as EGO. The next step is to maximize EI($\mathbf{x}$) using the prediction and uncertainty models to find the next

Algorithm 2. Efficient global optimization using neural networks [10].

Create initial sampling plan $\mathbf{X}$ and $\mathbf{X}_u$
Compute objective function value $\mathbf{Y}$ and $\mathbf{Y}_u$ for initial samples
while infill budget lasts **do**
    Create NN model for $\hat{y}$ using $(\mathbf{X}, \mathbf{Y})$
    Create NN model for $\hat{s}$ using prediction error of $\hat{y}$ for $\mathbf{X}$ and $\mathbf{X}_u$
    Find the best sample y^* in $\mathbf{Y}$
    Maximize $EI(\mathbf{x})$ to find next infill point $\mathbf{p}$
    Compute objective function value $\mathbf{y}$ at $\mathbf{p}$
    Append $(\mathbf{X}, \mathbf{Y})$ with $(\mathbf{p}, y)$
end while
Return best sample $(\mathbf{x}^*, y^*)$ in appended data set

Table 1. A table of the HPs with their respective types and ranges or options

Name	Type	Bounds/options
Learning rate (x_1)	$\mathbb{R}^+$	$10^{-4} \leq x_1 \leq 10^{-1}$
Epoch number (x_2)	$\mathbb{Z}^+$	$10^3 \leq x_2 \leq 5 \cdot 10^3$
Activation function (x_3)	[-]	(relu, elu, tanh, sigmoid)
Optimizer (x_4)	[-]	(SGD, Adam, RMSprop, Adagrad)
Hidden layer number (x_5)	$\mathbb{Z}^+$	$1 \leq x_5 \leq 4$
Initial neuron number (x_6)	$\mathbb{Z}^+$	$4 \leq x_6 \leq 16$

infill point. The function value at the infill point is evaluated, and the (**X**, **Y**) data set is updated. This cycle is repeated until the infill budget is exhausted. Recently, a constrained version of EGONN was proposed, which uses a third NN model for representing the constraints [9].

2.4 Hyperparameter Optimization

A critical step while creating an NN model involves configuring its structure and the training algorithm, which are problem-dependent and must be tuned for good performance [7]. Table 1 illustrates the HPs used in this study with their respective ranges and details. The number of hidden layers and neurons defines the NN's depth and pattern recognition capacity of input data features. The activation function introduces non-linearity in the NN, which is essential to recognize complex patterns. The learning rate controls the optimization pace to ensure efficient learning while the optimizer guides the learning through weight updates. Finally, the epoch size indicates the number of iterations of the optimization algorithm [18].

The HPO problem involves minimizing the root mean squared error (RMSE) of NN prediction on the validation set with respect to $\mathbf{x} = (x_1, x_2, \ldots, x_6)$, where each x_i denotes a distinct tunable HP within its respective bounds. The HPO process is executed during steps 4 and 5 of Algorithm 2 for all the NNs, which is written as

$$\mathbf{x}^* = \min_{\mathbf{x} \in \chi} \text{RMSE}(\mathbf{x}), \tag{4}$$

where $\mathbf{x}^*$ is the optimal HP configuration, χ is the HP domain, and RMSE is the discrepancy between a model's predictions $\hat{f}(\mathbf{x}, \mathbf{x}_d)$ and actual observation $f(\mathbf{x}_d)$ for the number of validation data points n_v, which is formulated as

$$\text{RMSE} = \sqrt{\frac{\sum_{i=1}^{n_v} [f(\mathbf{x}_d^{(i)}) - \hat{f}(\mathbf{x}, \mathbf{x}_d^{(i)})]^2}{n_v}}. \tag{5}$$

The optimization process concludes upon meeting a predefined convergence criterion, typically defined as either reaching a maximum number of HP sets or attaining a target loss function value.

This paper evaluates the impact of different HPO frequencies on NN performance and computational efficiency. It contrasts static HPO with tuning the HPs every five iterations and every iteration, focusing on the optimal balance between their time efficiency and effectiveness in optimizing the function.

BO [15] offers a solution for continuous HPs with infinite choices. It adopts an adaptive strategy, using observations from previous evaluations to decide subsequent sets of the HPs to evaluate. BO begins by constructing a smooth surrogate model using a Gaussian process (GP), characterized by a mean function, which offers the best prediction based on existing observations, and a kernel function, which measures prediction uncertainty [12,17]. The radial basis function (RBF) kernel [15], also known as the Gaussian kernel, is used in GP regressions, quantifying similarities between data points. The RBF kernel includes HPs such as the length scale and the variance, which are internally tuned in BoTorch [1].

The predictions and uncertainty estimates are combined to form an acquisition function. In this study, the EI in (2) is taken to be the acquisition function, which is maximized to select the most promising HP configuration for the next observation. After each observation, the surrogate model is updated, and this iterative process continues until it converges with the optimal HP configuration.

MAB optimization [5], in contrast to BO, is particularly suited for discrete HPs with a finite set of choices. In this work, MAB optimization utilizes the Thompson sampling (TS) algorithm [4] for sequentially selecting discrete HP sets. TS bases its choice on the probability of each HP set being the most effective, balancing between exploring lesser-known options and exploiting those that have yielded promising results.

TS begins by assigning a prior distribution to each HP configuration's reward probability. For every iteration, it samples a reward value from the distribution of each HP configuration, selecting the one with the highest sampled value. The observed reward from the chosen configuration updates its distribution. This iterative process continues until the HP configuration with the highest reward probability distribution is identified, which is the optimal HP configuration [4].

3 Numerical Experiments

This section presents the results from employing HPO strategies for SBO with sequential sampling, applied to a three-dimensional Hartmann function. The investigation explores the impact of varying the number of HPs optimized in each scenario.

3.1 Problem Formulation of the Hartmann Function

The three-dimensional Hartmann function minimization problem [13], which has four local minima and one global minimum, is defined as:

$$\min_{\mathbf{x}_d} f(\mathbf{x}_d) = -\sum_{i=1}^{4} \alpha_i \exp\left(-\sum_{j=1}^{3} A_{ij}(x_{dj} - P_{ij})^2\right), \tag{6}$$

where $\alpha = [1.0, 1.2, 3.0, 3.2]^T$,

$$\mathbf{A} = \begin{bmatrix} 3.0 & 10 & 30 \\ 0.1 & 10 & 35 \\ 3.0 & 10 & 30 \\ 0.1 & 10 & 35 \end{bmatrix},$$

and

$$\mathbf{P} = 10^{-4} \begin{bmatrix} 3689 & 1170 & 2673 \\ 4699 & 4387 & 7470 \\ 1091 & 8732 & 5547 \\ 381 & 5743 & 8828 \end{bmatrix}.$$

The lower and upper bounds are $\mathbf{x}_{d_L} = (0, 0, 0)$ and $\mathbf{x}_{d_U} = (1, 1, 1)$, respectively, with the global minimum $f(\mathbf{x}_{\mathbf{d}}^*)$ of -3.86 located at $\mathbf{x}_d^* = (0.11,\ 0.56,\ 0.85)$.

3.2 Strategies and Algorithm Setup

HPO strategies within the EGONN framework include static HPO (HPO-static), tuning HPs every five iterations (HPO-5itr), and tuning every iteration (HPO-1itr). The HPs are tuned using the open-source tool Ax [1] built on PyTorch [6], which integrates BO and MAB optimization. Ax constructs a combined search space for the continuous and discrete HPs where each type of HP is optimized by its respective algorithm simultaneously.

The minimization problem (6) is tackled using the SBO methodology depicted in Fig. 1 and EGONN algorithm (cf. Algorithm 2). The initial sample size is 10 with 45 infill points. Deterministic Halton sequence sampling [2] ensures consistency in the starting points across all methods. This study evaluates the efficacy of various HPO strategies by comparing their convergence performance and time efficiency against the traditional EGO method (cf. Algorithm 1) as a benchmark. Initially, a full set of the HPs, as detailed in Table 1, is optimized, followed by the optimization of a subset of the HPs based on preliminary results.

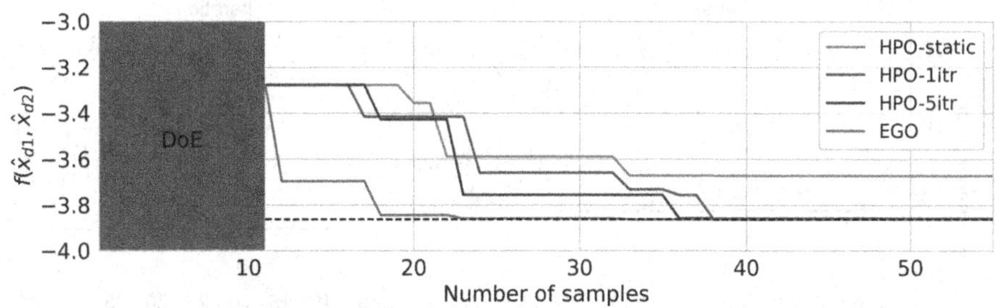

Fig. 2. Optimization of the three-dimensional Hartmann function utilizing EGO and EGONN with various HPO strategies that tune a full set of HPs

3.3 Optimizing the Full Set of HPs

The minimization of the three-dimensional Hartmann function is explored through EGONN, employing NNs optimized with a set of six tunable HPs, and EGO. Figure 2 illustrates varied convergence rates towards the minimum with increasing sample numbers. While all methods gradually approach the minimum, they require different numbers of samples for accurate approximation. The results indicate that while EGO outperforms EGONN, the two adaptive HPO strategies show comparable performance. EGO achieves convergence in 12 infills, whereas EGONN with HPO-5itr and HPO-1itr take 26 and 28 infills, respectively. HPO-static shows no significant improvement after 22 infills.

Figure 3 presents the history of six HPs across various HPO strategies. While the HPO-static approach uses a fixed HP configuration, including the ReLU activation function, Adagrad optimizer, two hidden layers, 15 neurons, a learn-

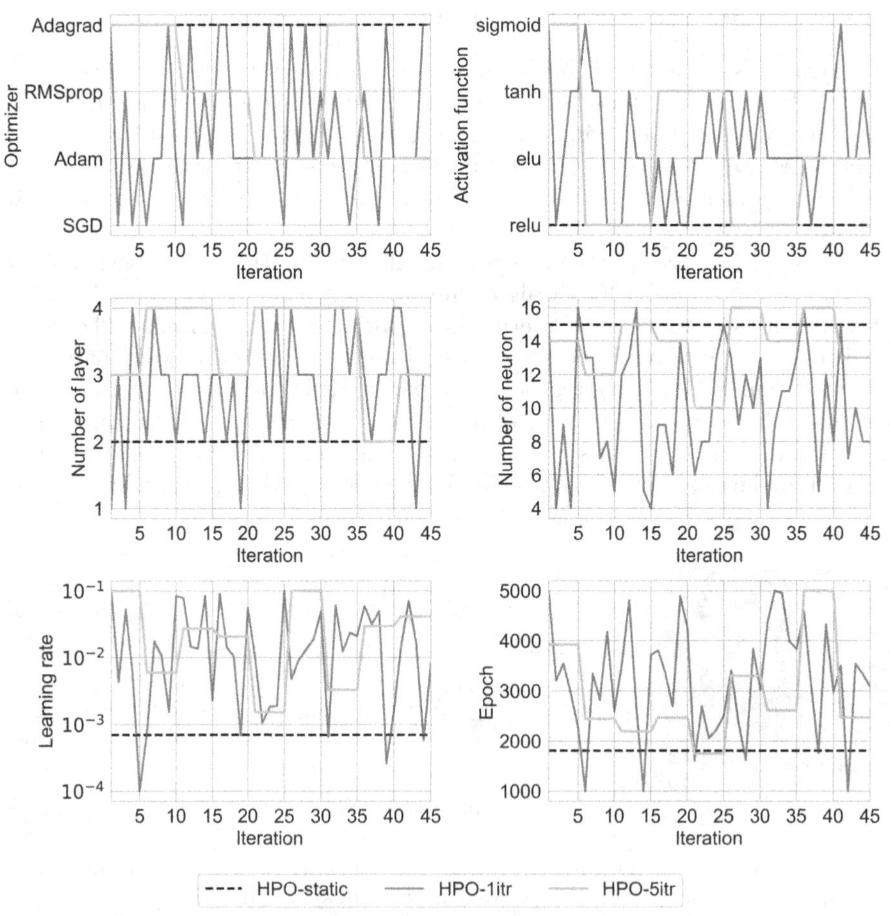

Fig. 3. Convergence analysis of 6 HPs across iterations of different HPO strategies within the EGONN algorithm for optimizing a three-dimensional Hartmann function

Table 2. Percentage of each HP calculated for each method using a full set of HPs

Hyperparameter		Methods		
		HPO-1itr	HPO-5itr	HPO-static
Activation	relu	20.0%	**37.5%**	**100%**
	elu	**44.4%**	25.0%	0.0%
	tanh	28.9%	25.0%	0.0%
	sigmoid	6.7%	12.5%	0.0%
Optimizer	SGD	15.6%	0.0%	0.0%
	Adam	**48.9%**	**44.4%**	0.0%
	RMSprop	11.1%	22.2%	0.0%
	Adagrad	24.4%	33.3%	**100%**
Layer	1	8.9%	0.0%	0.0%
	2	20.0%	11.1%	**100%**
	3	**46.7%**	33.3%	0.0%
	4	24.4%	**55.6%**	0.0%
Neuron	$4 \sim 8$	**37.8%**	0.0%	0.0%
	$8 \sim 12$	31.1%	11.1%	0.0%
	$12 \sim 16$	31.1%	**88.9%**	**100%**
Epoch	$1 \sim 2\ (10^3)$	13.3%	11.1%	**100%**
	$2 \sim 3\ (10^3)$	26.7%	**55.6%**	0.0%
	$3 \sim 4\ (10^3)$	**37.8%**	22.2%	0.0%
	$4 \sim 5\ (10^3)$	22.2%	11.1%	0.0%
LR	$10^{-3} \sim 10^{-4}$	11.1%	0.0%	**100%**
	$10^{-2} \sim 10^{-3}$	26.7%	33.3%	0.0%
	$10^{-1} \sim 10^{-2}$	**60.0%**	**66.7%**	0.0%

ing rate of 0.0007, and 1805 epochs for NN_y, adaptive HPO strategies show enhanced flexibility and effectiveness by adjusting HPs to fit better the evolving data set size. The trends, however, indicate that the HP convergence is not directly correlated with increased sample size. This is attributed to the nonlinear and multimodal relationship between HPs and optimization performance, indicating multiple HP configurations can yield a similar optimized NN performance. This suggests continuous HP tuning with each added infill point may be unnecessary, yet adaptive HPO is critical for identifying the most effective HPs for NNs across different data set sizes.

Table 2 outlines the percentage distribution of HP selections, with bolded most frequently chosen HPs. The data reveals that the choice of activation functions is nearly uniformly distributed for both the adaptive HPO strategies, indicating no clear preference. However, a noticeable trend is the selection of a higher number of layers, suggesting a move towards more complex NN architectures as sample sizes increase. Additionally, the Adam optimizer, alongside higher learning rates and epoch sizes, is consistently preferred in the

Table 3. Time cost for iterations of HPO of a full set of HPs and surrogate training in the three-dimensional Hartmann function optimization

Time cost (hr)		Methods			
		HPO-1itr	HPO-5itr	HPO-static	EGO
HPO	Mean	0.21	0.19	-	-
	SD	0.09	0.08	-	-
	Total time	9.40	1.75	0.09	-
Training	Mean	$8.9{\cdot}10^{-3}$	$9.1{\cdot}10^{-3}$	$6.9{\cdot}10^{-3}$	$1.6{\cdot}10^{-4}$
	SD	$3.1{\cdot}10^{-3}$	$3.6{\cdot}10^{-3}$	$1.7{\cdot}10^{-3}$	$1.1{\cdot}10^{-5}$
	Total time	0.36	0.35	0.32	$7.5{\cdot}10^{-3}$
Cumulative time		9.76	2.10	0.41	$7.5{\cdot}10^{-3}$

adaptive HPO strategies, indicating the critical HPs for achieving convergence. The preference for HPs in the adaptive HPO strategies explains the limitations of the static HPO approach. As sample sizes increase, the necessity for complex HP configurations grows, emphasizing the effectiveness of adaptive HPO strategies.

A comprehensive analysis of the time costs throughout all 45 iterations is presented in Table 3. It specifies each method's mean and standard deviation (SD) of HPO and surrogate training per iteration along with the total time cost, demonstrating significant variations in time cost. This variability is mainly attributed to the growing training data sets as iterations progress. Despite these variations in HPO time costs, the time required to train NNs shows minimal differences across strategies. The finding reveals that the time cost of HPO-1itr is 465% of HPO-5itr's time cost, primarily due to the frequency of HPO. Consequently, HPO-5itr stands out as the most efficient EGONN approach, balancing the convergence speed and the overall computational time.

3.4 Optimizing a Subset of the HPs

Analyzing the optimization convergence history in Fig. 2 and the HP distribution in Table 2 suggests that a strategic HP selection could enhance the efficiency

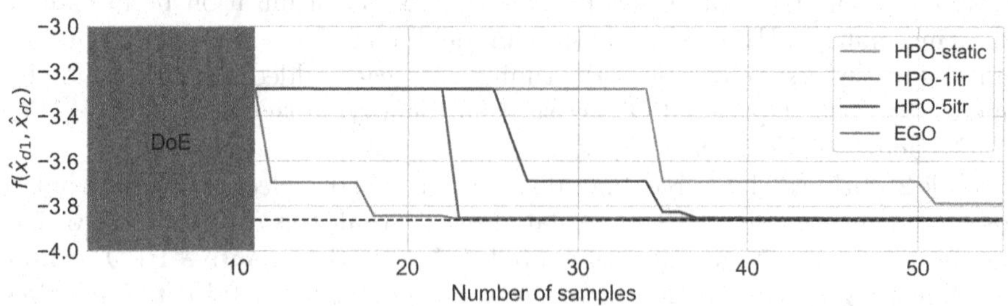

Fig. 4. Optimization of the three-dimensional Hartmann function utilizing EGO and EGONN with HPO-static, HPO-5itr, and HPO-1itr strategies for a subset of HPs

of HPO. The activation function, number of epochs, and learning rate are identified as critical HPs for tuning. Although the optimizer choice also influences performance, the consistent selection of the Adam optimizer from the previous evaluation reflects its effectiveness for this test case. The preference for a high learning rate necessitates further optimization because it is a continuous parameter offering a broad spectrum of possibilities. Based on the superior performance of HPO-5itr in the previous evaluation, the mean values for the number of neurons and layers are adopted, which are 13 and 3, respectively. The options for the activation function and the ranges of the epoch size and the learning rate are maintained as detailed in Table 1.

Figure 4 displays the convergence performance of each HPO strategy on the subset of the HPs. Despite a similar trend to the previous test, noticeable improvements are observed in the HPO-static and HPO-1itr strategies. HPO-1itr achieves early convergence at 13 iterations, while HPO-static approaches closer to the minimum from −3.67 to −3.79. In contrast, HPO-5itr shows limited improvement, converging at 25 iterations.

The convergence history and HP distribution, as shown in Fig. 5 and Table 4, indicate the differences in HP selection across the strategies. The improved performance of the HPO-static is attributed to its initial choice of a high learning rate. However, the HPO-static strategy chose a relatively low epoch size, contrasting with the higher epoch sizes preferred by adaptive HPO strategies.

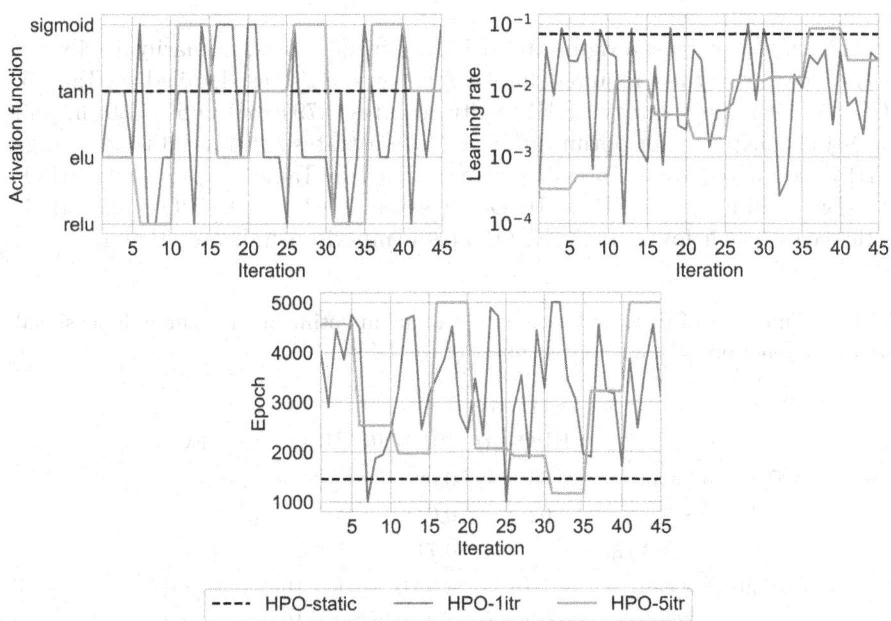

Fig. 5. Convergence analysis of 3 HPs with various HPO strategies in EGONN for the three-dimensional Hartmann function optimization

Table 4. Percentage distribution of each HP for the test using a subset of the HPs

Hyperparameter		Methods		
		HPO-1itr	HPO-5itr	HPO-static
Activation	relu	15.6%	22.2	0.0%
	elu	**35.6%**	22.2%	0.0%
	tanh	24.4%	22.2%	**100%**
	sigmoid	24.4%	**33.3%**	0.0%
Epoch	$1 \sim 2\ (10^3)$	17.8%	22.2%	**100%**
	$2 \sim 3\ (10^3)$	17.8%	22.22%	0.0%
	$3 \sim 4\ (10^3)$	**33.3%**	11.1%	0.0%
	$4 \sim 5\ (10^3)$	31.1%	**44.4%**	0.0%
LR	$10^{-3} \sim 10^{-4}$	13.3%	22.2%	0.0%
	$10^{-2} \sim 10^{-3}$	28.9%	22.2%	0.0%
	$10^{-1} \sim 10^{-2}$	**55.6%**	**55.6%**	**100%**

Despite a similar HP distribution to the previous test for HPO-1itr, retaining the optimizer and the number of neurons and layers enhanced convergence performance. Additionally, the adaptation of activation functions to sigmoid for HPO-5itr and tanh for HPO-static indicates strategic adjustments to improve optimization outcomes.

The analysis reveals a significant decrease in time costs, primarily attributed to the decreased time required for tuning fewer HPs, as detailed in Table 5. EGONN with the strategy of HPO-1itr requires 3.79 h, HPO-5itr 1.01 h, and HPO-static 0.40 h to for optimization. It demonstrates substantial time savings in HPO compared to the previous case outlined in Table 3. Specifically, HPO-1itr saves 5.97 h for 45 HPOs, HPO-5itr saves 1.04 h for 9 HPOs, and HPO-static saves 0.04 h for a single HPO. This comparison indicates that reducing

Table 5. Time cost of HPO and surrogate training in optimizing the three-dimensional Hartmann function with a selected subset of the HPs

Time cost (s)		Methods			
		HPO-1itr	HPO-5itr	HPO-static	EGO
HPO	Mean	0.08	0.08	-	-
	SD	0.02	0.01	-	-
	Total time	3.43	0.71	0.05	-
Training	Mean	$8.7 \cdot 10^{-3}$	$8.6 \cdot 10^{-3}$	$6.7 \cdot 10^{-3}$	$1.6 \cdot 10^{-4}$
	SD	$2.9 \cdot 10^{-3}$	$3.2 \cdot 10^{-3}$	$1.6 \cdot 10^{-3}$	$1.1 \cdot 10^{-5}$
	Total time	0.33	0.30	0.36	$7.5 \cdot 10^{-3}$
Cumulative time		3.79	1.01	0.41	$7.5 \cdot 10^{-3}$

the number of HPOs directly correlates with increased time efficiency. It notably reduces the gap between HPO-1itr and HPO-5itr from 7.66 h to 2.75 h, stressing the importance of the streamlined HPO processes.

4 Conclusion

This paper proposes the use of adaptive hyperparameter optimization (HPO) strategies within the efficient global optimization (EGO) with neural network (NN)-based prediction and uncertainty (EGONN). Numerical experiments with the three-dimensional Hartmann function revealed that while traditional EGO is effective, integrating NNs through EGONN with HPO strategies enhances performance for optimization problems.

Moreover, this study demonstrates the importance of appropriate HPO in the success of SBO for their adaptive tuning as the complexity of the data set increases. It shows that a static HPO strategy could not adapt to the sequential sampling process, leading to sub-optimal performance. In contrast, the adaptive HPO strategies demonstrated their ability to converge with fewer iterations by effectively reaching the optimum. Tuning HPs every five iterations emerged as the most efficient approach, balancing optimization performance and computational burden. Additionally, the optimization of a subset of the HPs reveals that the most effective adaptive HPO strategies can vary as the number of tunable HPs changes.

Future research directions include establishing a predefined tolerance threshold based on numerical metrics, such as the loss function values, and identifying a subset of the HPs for iterative tuning guided by global sensitivity analysis. Such investigation could yield more reliable NN performance throughout the SBO iterations while saving computational resources. Moreover, expanding the scope of test cases to more complex and higher-dimensional problems, such as airfoil shape optimization in transonic viscous flow, will provide further insights into the efficiency of EGONN with adaptive HPO versus EGO.

Acknowledgments. This work was supported in part by the U.S. National Science Foundation (NSF) award number 2223732 and by the Icelandic Centre for Research (RANNIS) award number 239858.

References

1. Baird, S.G., Liu, M., Sparks, T.D.: High-dimensional Bayesian optimization of 23 hyperparameters over 100 iterations for an attention-based network to predict materials property: a case study on CrabNet using ax platform and SAASBO. Comput. Mater. Sci. **211**, 111505 (2022). https://doi.org/10.1016/j.commatsci.2022.111505
2. Faure, H., Lemieux, C.: Generalized halton sequences in 2008: a comparative study. ACM Trans. Model. Comput. Simul. (TOMACS) **19**(4), 1–31 (2009). https://doi.org/10.1145/1596519.1596520

3. Forrester, A.I., Keane, A.J.: Recent advances in surrogate-based optimization. Prog. Aerosp. Sci. **45**(1), 50–79 (2009). https://doi.org/10.1016/j.paerosci.2008.11.001
4. Gupta, N., Granmo, O.C., Agrawala, A.: Thompson sampling for dynamic multi-armed bandits. In: 2011 10th International Conference on Machine Learning and Applications and Workshops, vol. 1, pp. 484–489 (Honolulu, Hawaii, 18–21 December 2011). https://doi.org/10.1109/ICMLA.2011.144
5. Henri, S., Vlachou, C., Thiran, P.: Multi-armed bandit in action: optimizing performance in dynamic hybrid networks. IEEE/ACM Trans. Netw. **26**(4), 1879–1892 (2018). https://doi.org/10.1109/TNET.2018.2856302
6. Imambi, S., Prakash, K.B., Kanagachidambaresan, G.R.: PyTorch. In: Prakash, K.B., Kanagachidambaresan, G.R. (eds.) Programming with TensorFlow. EICC, pp. 87–104. Springer, Cham (2021). https://doi.org/10.1007/978-3-030-57077-4_10
7. Jeong, T., Koratikere, P., Leifsson, L.T.: Automated hyperparameter tuning for airfoil shape optimization with neural network models. In: AIAA SCITECH 2024 Forum, p. 2671 (Orlando, FL, USA, 8–12 January 2024). https://doi.org/10.2514/6.2024-2671
8. Jones, D.R., Schonlau, M., Welch, W.J.: Efficient global optimization of expensive black-box functions. J. Global Optim. **13**(4), 455–492 (1998). https://doi.org/10.1023/A:1008306431147
9. Koratikere, P., Leifsson, L., Koziel, S., Pietrenko-Dabrowska, A.: Constrained aerodynamic shape optimization using neural networks and sequential sampling. In: Mikyška, J., de Mulatier, C., Paszynski, M., Krzhizhanovskaya, V.V., Dongarra, J.J., Sloot, P.M. (eds.) ICCS 2023. LNCS, vol. 10475, pp. 425–438. Springer, Cham (2023). https://doi.org/10.1007/978-3-031-36024-4_33
10. Koratikere, P., Leifsson, L.T., Barnet, L., Bryden, K.: Efficient global optimization algorithm using neural network-based prediction and uncertainty. In: AIAA SCITECH 2023 Forum. AIAA (National Harbor, MD, USA, 23–27 January 2023). https://doi.org/10.2514/6.2023-2683
11. Lampinen, J.: A constraint handling approach for the differential evolution algorithm. In: Proceedings of the 2002 Congress on Evolutionary Computation, vol. 2, pp. 1468–1473. IEEE (2002). https://doi.org/10.1109/CEC.2002.1004459
12. Lim, Y.F., Ng, C.K., Vaitesswar, U., Hippalgaonkar, K.: Extrapolative Bayesian optimization with gaussian process and neural network ensemble surrogate models. Adv. Intell. Syst. **3**(11), 2100101 (2021). https://doi.org/10.1002/aisy.202100101
13. Mehdad, E., Kleijnen, J.: Classic kriging versus kriging with bootstrapping or conditional simulation. J. Oper. Res. Soc. **66** (2015). https://doi.org/10.1057/jors.2014.126
14. Queipo, N.V., Haftka, R.T., Shyy, W., Goel, T., Vaidyanathan, R., Tucker, P.K.: Surrogate-based analysis and optimization. Prog. Aerosp. Sci. **41**(1), 1–28 (2005). https://doi.org/10.1016/j.paerosci.2005.02.001
15. Snoek, J., Larochelle, H., Adams, R.P.: Practical Bayesian optimization of machine learning algorithms. In: Advances in Neural Information Processing Systems, vol. 25 (2012)
16. Viana, F.A., Haftka, R.T., Watson, L.T.: Efficient global optimization algorithm assisted by multiple surrogate techniques. J. Global Optim. **56**, 669–689 (2013). https://doi.org/10.1007/s10898-012-9892-5

17. Wu, J., Chen, X.Y., Zhang, H., Xiong, L.D., Lei, H., Deng, S.H.: Hyperparameter optimization for machine learning models based on Bayesian optimization. J. Electron. Sci. Technol. **17**(1), 26–40 (2019). https://doi.org/10.11989/JEST.1674-862X.80904120
18. Yang, L., Shami, A.: On hyperparameter optimization of machine learning algorithms: theory and practice. Neurocomputing **415**, 295–316 (2020). https://doi.org/10.1016/j.neucom.2020.07.061

Hypergraph Clustering with Path-Length Awareness

Julien Rodriguez[1,2(✉)], François Galea[3], François Pellegrini[2], and Lilia Zaourar[3]

[1] University of Montpellier, LIRMM, CNRS, Montpellier, France
julien.rodriguez@lirmm.fr

[2] Université de Bordeaux & INRIA, 33405 Talence, France
francois.pellegrini@u-bordeaux.fr

[3] Université Paris-Saclay, CEA List, 91120 Palaiseau, France
{francois.galea,lilia.zaourar}@cea.fr

Abstract. Electronic design automation toolchains require solving various circuit manipulation problems, such as floor planning, placement and routing. These circuits may be implemented using either Very Large-Scale Integration (VLSI) or Field Programmable Gate Arrays (FPGAs). However, with the ever-increasing size of circuits, now up to billions of gates, straightforward approaches to these problems do not scale well. A possible approach to reduce circuit complexity is to cluster circuits. In this work, we consider the problem of clustering combinatorial circuits, without cell replication. We propose a dedicated clustering algorithm based on binary search and study and improve the existing parameterized approximation ratio from $M^2 + M$ [8] (with M being the maximum size of each cluster) to M under specific hypothesis. We present an extension of the weighting schemes introduced in [19] to model path length more accurately. This weighting scheme is combined with clustering methods based on a recursive matching algorithm. We evaluate and compare our approximation algorithm and recursive matching on several circuit instances and we obtain better results for a large number of instances with our algorithm than recursive matching.

Keywords: Clustering · Hypergraph · Digital electronic circuit

1 Introduction

Our research interest concerns circuit prototyping on multi-FPGA platforms, to map efficiently circuits that are too big to fit into a single FPGA. In this case, it is necessary to partition the circuit into several parts that have to be placed on the different components (*i.e.*, FPGAs) of the platform. Traditional partitioning tools use a classic multilevel scheme consisting of three phases: *coarsening*, *initial partitioning*, and *refinement* [11]. The coarsening phase recursively uses a clustering algorithm to reduce the problem size. During the second phase, an initial partitioning is computed on the smallest coarsened hypergraph. Finally,

L. Franco et al. (Eds.): ICCS 2024, LNCS 14836, pp. 90–104, 2024.
https://doi.org/10.1007/978-3-031-63775-9_7

in the third phase, for each coarsening level, the solution for the coarser level is extended to the finer level, and then refined using a local refinement algorithm. The clustering algorithms presented in this paper concern the first step of the multilevel framework described above.

Several such clustering algorithms already exist in the literature; Z. Donovan [6] defines two classes to categorize them: CA and CN. CA algorithms aim at finding clustering of circuits that minimize signal propagation delay, while allowing logic replication (see e.g. [14–16,18]). On the opposite, CN algorithms compute clustering that minimize circuit delay without cell replication (see e.g. [8,10]).

An optimal solution for the CA problem can be computed in polynomial time. However, unbounded replication can yield very large circuits [14,18]. In the context of circuit placement on multi-FPGA platforms, the number of resources is limited. Thus, it is necessary to perform either a disjoint clustering, or to bound the number of replications. In this work, we focus on the CN problem.

The remainder of the paper is organized as follows. Section 2 presents the notations, definitions, and previous works on CN. We introduce our weighting scheme for clustering in Sects. 3. In Sect. 4, we introduce our clustering algorithms and complexity results. Our experiments are outlined in Sect. 5. We conclude and give perspectives in Sect. 6.

2 Preliminaries

Combinatorial circuits are often modeled as directed hypergraphs, *i.e.*, a generalization of directed graphs in which the notion of arc is extended to that of hyperarc. A hyperarc can connect one or more source vertices to one or more sink vertices. In a combinatorial circuit, a net (or wire) can connect more than two gates, and there typically is one single signal source per net. Hence, we consider only hyperarcs that comprise a single source vertex. Several works rely on a graph model to represent combinatorial circuits, such as: [8–10,15,16,18]. While this model is relevant to represent dependencies between outputs and inputs of gates, it is not adequate to evaluate the number of cut wires and does not model critical path [19]. Consequently, in this work, we will represent circuits using a Directed Acyclic Hypergraph (DAH) model [19] to measure and control the size of the cut, while also relying on an underlying graph model to compute clustering scores between gates and solve the CN problem without considering the number of cut hyperarcs.

2.1 Notations and Definitions

Let $H \stackrel{\text{def}}{=} (V, A, W_V, W_a)$ be a directed hypergraph, defined by a set of vertices V and a set of hyperarcs A, with a vertex weight function $W_V : V \rightarrow \mathbb{R}^+$ and a hyperarc weight function $W_a : A \rightarrow \mathbb{R}^+$. Every hyperarc $a \in A$ is a subset of vertex set V: $a \subseteq V$. Let $s^-(a)$ be the source vertex set of hyperarc a, and $s^+(a)$ its sink (destination) vertex set. We consider each hyperarc has a single source, so $\forall a, |s^-(a)| = 1$. As hyperarcs connect vertices, let $\Gamma(v)$ be the set of neighbor vertices of vertex v, and $\Gamma^-(v) \subseteq \Gamma(v)$ and $\Gamma^+(v) \subseteq \Gamma(v)$ the sets of its inbound and outbound neighbors, respectively.

In the model we propose, hypergraphs that model circuits will be represented as sets of interconnected DAHs, according to a red-black vertex coloring scheme. Red vertices correspond to I/O (Inputs/Outputs) ports and registers, and black vertices to combinatorial circuit components. Let $V^R \subset V$ and $V^B \subset V$ be the red and black vertex subsets of V, such that $V^R \cap V^B = \emptyset$ and $V^R \cup V^B = V$. A hypergraph or sub-hypergraph H is a DAH iff its red vertices $v_R \in V^R$ are either only sources or sinks (*i.e.*, $\Gamma^-(v_R) = \emptyset$ or $\Gamma^+(v_R) = \emptyset$), and no cycle path connects a vertex to itself.

Using this definition, we can represent circuit hypergraphs as *red-black hypergraphs*, *i.e.*, set of DAHs that share some of their red vertices. Let $\mathbf{H}(\mathbf{V}, \mathbf{A}) \stackrel{\text{def}}{=} \{H_i, i \in \{1 \ldots n\}\}$ be a red-black hypergraph, such that every H_i is a DAH and an edge-induced sub-hypergraph of $\mathbf{H}$. Consequently, $\mathbf{V} = \bigcup_i V_i$, $\mathbf{A} = \bigcup_i A_i$, $\mathbf{V}^R = \bigcup_i V_i^R$, and $\mathbf{V}^B = \bigcup_i V_i^B$. Moreover, $\forall i, j$ with $i \neq j$, if $V_{i,j} = V_i \cap V_j \neq \emptyset$, then H_i and H_j share source and/or sink vertices, *i.e.*, $V_{i,j} \subset \mathbf{V}^R$.

In this model, the paths in $\mathbf{H}$ to consider when addressing the objective of minimizing path-cost degradation during partitioning are only the paths interconnecting red vertices, as these red-red paths represent register-to-register paths in digital electronic circuits. Since only red vertices are shared between DAHs in $\mathbf{H}$, red-red paths only exist within a single DAH and can never span across several DAHs.

Let us define $\mathbf{P}$ as the set of red-red paths in $\mathbf{H}$, such that $\mathbf{P} \stackrel{\text{def}}{=} \{p | p \text{ is a path in } H \in \mathbf{H}\}$. From these paths and a function $d_{\max}(u, v)$ which computes the maximum distance between vertices u and v of some DAH H, we can define the longest path distance for H as: $d_{\max}(H) \stackrel{\text{def}}{=} \max(d_{\max}(u, v) | u, v \in H)$ and, by extension, for $\mathbf{H}$, as: $d_{\max}(\mathbf{H}) \stackrel{\text{def}}{=} \max(d_{\max}(H) | H \in \mathbf{H})$.

A clustering $\mathcal{C}$ of $\mathbf{H}$ is a splitting of $\mathbf{V}$ into vertex subsets $\mathcal{C}_i$, called clusters, such that: (i) all clusters $\mathcal{C}_i$, given a capacity bound M, respect the capacity constraint: $\sum_{v \in \mathcal{C}_i} W_V(v) \leq M$; (ii) all clusters are pairwise disjoint: $\forall i \neq j, \mathcal{C}_i \cap \mathcal{C}_j = \emptyset$; and (iii) the union of all clusters is equal to $\mathbf{V}$: $\bigcup_i \mathcal{C}_i = \mathbf{V}$.

2.2 Related Works

A. A. Diwan *et al.* [5] address a similar problem, consisting in placing nodes of a memory access structure on disk pages such that a path through several nodes traverses as few disks as possible. Their data structure is a DAG, and their objective is to cluster the DAG such that the number of shared edges per cluster along a path is minimized. The problem is similar to the unweighted case of the CN problem. The authors also present a polynomial-time algorithm for trees, and show that the problem is NP-hard for unweighted DAGs.

More recently, Z. Donovan *et al.* [7–9] have studied the combinatorial circuit clustering problem, with and without vertex replication. They propose several algorithms to solve this problem. The authors present NP-hardness proofs for the

DAG circuit clustering problem with minimization of critical path degradation during the clustering step, e.g., minimization of the number of cut penalties along the critical paths. They propose exact exponential algorithms and approximation algorithms parameterized by the size of the cluster. Further details of this work can be found in Z. Donovan's thesis [6]. Other work on combinatorial circuit clustering to minimize critical path degradation by placing neighboring vertex pairs in different clusters are available [3,16].

3 Model and Weighting Schemes

Criticality is a metric used in [1,2,19], to classify the cells of a circuit according to the cost of the combinatorial path they traverse. The *criticality* of a vertex v is equal to the length of the longest path traversing v, $d_{\max}(v)$. In this section, we present the various state-of-the-art weighting schemes used to measure vertex criticality. We propose a new weighting scheme that models more finely the criticality per vertex pair which we use to cluster red-black hypergraphs.

In the CN problem, an additional constant cost D is added between two neighboring vertices placed in different clusters. Consequently, in our model, the distance between any two vertices u and v (*i.e.*, path cost) may increase during clustering, due to the additional cost that paths have to incur across clusters. Let us recall that the distance function between two vertices in a red-black hypergraph is defined by $d_{\max}(u,v)$ which is equal to the longest path between u and v. Let D be the penalty associated with the distance between two vertices u and v placed in different clusters; let us recall that the distance function for some clustering $\mathcal{C}$, is thus: $d^{\mathcal{C}}_{\max}(u,v) \geq d_{\max}(u,v)+D$. The objective function f_p is defined as the minimization of the longest path of H subject to clustering $\mathcal{C}$: $f_p = \min d_{\max}(H^{\mathcal{C}})$. We extend the definition of the $CN{<}w, M, \Delta{>}$ problem defined by Z. Donovan *et al.* [6] to red-black hypergraphs as follows:

> *Given a red-black Hypergraph* $H = (V, A)$, *with a vertex-weight function* $w : V \to \mathbb{R}^+$, *delay function* $d : V \to \mathbb{R}^+$, *maximum degree* Δ, *constant* D, *and a cluster capacity* M, *the goal is to partition* V *into clusters such that: (i) the weight of each cluster is bounded by* M; *and (ii) the maximum delay-length of any red-red path of* H *is minimized.*

To be consistent with previous definitions of the CN problem, we will keep the Δ parameter, even though we will not use it in the following.

As we exposed previously, the criticality of a vertex v measures the value of the longest path through v. Consequently, criticality seems to be an interesting weighting scheme for measuring the attractiveness between two connected vertices.

Delay Propagation. Several previous works have proposed metrics for clustering, with the objective of path minimization [1,2]. For example, C. Ababei *et al.* [1] presented a weighting scheme based on delay propagation to drive *min-cut* tools, *i.e.*, the weight between two vertices u and v is equal to the longest path from a red source vertex to vertices u and v. This method calculates local weights along subpaths from red source vertices to any vertex. Thus, within each DAH, $H = (V, A)$ of H:

$$l(u) = \begin{cases} d(u) & \text{if } \Gamma^-(u) = \emptyset\,, \\ d(u) + \max\limits_{v \in \Gamma^-(u)} l(v) & \text{otherwise}\,. \end{cases} \tag{1}$$

For any vertex $u \in V$, the value $l(u)$ corresponds to the maximum path cost from any source vertex to u. Therefore, the maximum path cost within some DAH will be found at the level of its sink vertices. A calculation on the subpath does not indicate whether their subpath is on the critical path. Cutting anywhere along a path has the same detrimental effect as adding a penalty to the total path cost. It is to alleviate these issues that the next metric have been made.

Delay Retro-Propagation. As critical vertices must be labeled with the same weight, the delay propagation scheme is not adequate. Hence, we have first devised a new weighting scheme based on the back-propagation of path cost:

$$r(u) = \begin{cases} l(u) & \text{if } \Gamma^+(u) = \emptyset\,, \\ \max\limits_{v \in \Gamma^+(u)} r(v) & \text{otherwise}\,. \end{cases} \tag{2}$$

For any $u \in V$, the value $r(u)$ represents an upper bound for the path cost of the longest red-red path traversing u. If u belongs to a path of maximum path cost, then $r(u)$ is equal to that path cost.

This weighting scheme accounts better for the overall impact of the cut along a path because, unlike the previous method, the information is back-propagated to all predecessors. However, it may include heavy vertices that do not belong to a longest red-red path, as shown in Fig. 1. To overcome this problem, we need to define the value of the local critical path through each pair of vertices. For this reason, we have proposed a second weighting system in the next subsection.

Refined Delay Retro-Propagation. In this subsection, we present a weighting scheme based on the cost of the local critical path. This scheme retro-propagates critical information throughout the red-black hypergraph and avoids non-critical heavy vertices. The l, r, and r^* metrics are used as weighting schemes, as represented in Fig. 1.

Let $r^*(u, v)$ be the criticality value between connected vertices u and v, defined as follows:

$$r^*(u, v) = \begin{cases} l(u) & \text{if } u = v\,, \\ r(v) - \left(\max\limits_{u' \in \Gamma^-(v)} l(u') - l(u) \right) & \text{otherwise}\,. \end{cases} \tag{3}$$

$\max_{u' \in \Gamma^-(v)} l(u')$ represents the value of the arcs along the local critical path, which is the longest red-red path traversing v such that, for every other

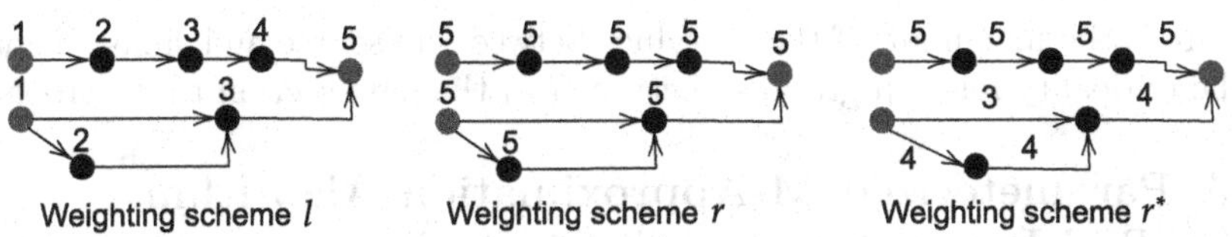

Fig. 1. An example of the three weighting schemes: l [1], r, and r^*. We consider a unit delay for each vertex and a delay equal to zero for each arc. In this example, we can clearly see that scheme l does not effectively weight critical vertices. Scheme r weights critical vertices correctly, but considers non-critical vertices. Scheme r^* is more relevant in its weighting, with respect to our objective.

$l(u) < \max_{\forall u' \in \Gamma^-(v)} l(u')$, arcs (u, v) are not in the local critical path. It is a more accurate metric for improving the behavior of clustering algorithms because, in the context of circuit clustering, the aim is to group critical vertices together. If the relationships between vertices reflect correctly criticality, then the clustering algorithm can take advantage of this. An example of the computation of r^* is represented in Fig. 2.

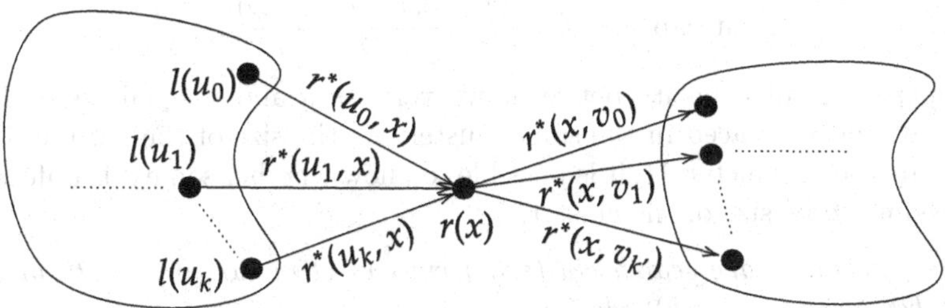

Fig. 2. This figure exhibits an example of schemes and how r^* is computed. $r^*(u_i, x)$ and $r^*(x, v_i)$ are the values of the local critical path between pairs of vertices (u_i, x) and (v_i, x) in this subgraph. There is a maximum value for each $l(u_i)$, $\overline{w}$. For each u_i, $\overline{w} - l(u_i)$ represents the contribution of u_i to the local critical path value $r(x) = \max_{v_i \in \Gamma^+(x)} r^*(x, v_i)$.

For each combinatorial sub-circuit modeled with a DAH, the r^* vertex-vertex criticality relation defines a graph G such that G is a DAG and is simple. Every hyperarc in the DAH defines a group of arcs in the DAG, in which each arc connecting the source vertex to a sink vertex. The cut weight of arcs correspond to the r^* value between source and sink in arcs. Hence, the cut weight of this hyperarc is the maximum of the r^* values between its source and sinks. We will use the criticality relation graph structure G in the next section to do proofs.

4 A Parameterized M-Approximation Algorithm for Red-Black Hypergraph Clustering

Since the clustering problem is NP-hard and there is no approximation algorithm with a constant factor in the general case, approximation algorithms have been proposed to provide acceptable solutions in reasonable time, such as the parameterized $M^2 + M$ approximation algorithm presented by Z. Donovan *et al.* [9]. We propose an improved approximation ratio under delay hypothesis and a direct clustering algorithm based on binary search.

4.1 Binary Search Clustering (BSC)

Let $H = (V, A)$ be a DAH, and $p_{\max}$ its critical path. Let ϕ be a feasible minimum cost, $\phi \in [|p_{\max}| \times d, |A| \times D]$, with D the inter-cluster delay and d the intra-cluster delay. Given a fixed value ϕ, we can define a cut capacity for each pair of vertices (u, v) as:

$$\text{cut\_cap}(u,v) \;=\; \frac{\max(0, r^*(u,v) - \phi)}{D}. \tag{4}$$

Suppose the cut capacity between two vertices u and v equals zero. Then, u and v should be placed in the same cluster. As the size of the cluster is constrained by the parameter M, it is possible to know whether some ϕ is unfeasible, by exceeding some size of the cluster.

Lemma 1. *The binary search clustering runs in $O(m \cdot \log(m))$, with m being the number of arcs ($r^*(u,v)$ relations).*

Proof. Algorithm 1 contains a `while` loop that will perform at most $\log(m)$ iterations. Lines 1 and 2 of the algorithm define the lower and upper bounds of the binary search. Even if the hypergraph is a path, *i.e.*, if the lower bound is equal to m and the upper bound is equal to m^2, the number of iterations of the `while` loop will be in $O(\log(m^2))$, which does not change the order of complexity.

Line 6 calls a procedure that works in $O(m)$. Indeed, the procedure computes the cut capacity of every arc and merges every pair of vertices with a cut capacity equal to zero. In line 10, to cluster the remaining unclustered vertices connected by arcs with non-negative cutting capacity, BSC calls a $O(m)$ procedure which loops over hyperarcs and works as follow: for each hyperarc, try to cluster yet unclustered vertices with other vertices in the hyperarc. Hence, the complexity of this algorithm is in $O(m \cdot \log(m))$.

The algorithm presented by Z. Donovan [6] has a complexity in $O(2^{\Delta \cdot M} + |V|^{O(1)})$ time. For a sufficiently large M, this algorithm can become impractical. The BSC algorithm has a complexity in $O(m \cdot \log_2(m))$, which is more attractive in practice. Also, circuit instances are relatively sparse, that is, m is not higher than the number of vertices.

Algorithm 1. Binary Search Clustering

Require: H, M, D, d
Ensure: $\mathcal{C}$ a clustering of H
$\overline{\phi} \leftarrow |A| \times D$
$\underline{\phi} \leftarrow |p_{\max}| \times d,\ p_{\max} \in H$
while $\overline{\phi} > \underline{\phi}$ **do**
  $\phi_{target} \leftarrow \frac{\underline{\phi}+\overline{\phi}}{2}$
▷ Compute the cut capacity for every pair $(uv) = \frac{\max(0, r^*(u,v) - \phi_{\text{target}})}{D}$ and for all pair with cut capacity equal to zero, place u and v into the same cluster.
  $\mathcal{C} \leftarrow$fusion_cut_cap(H, ϕ_{target}, max_size)
  if $\max_{c \in \mathcal{C}} |c| \leq M$ **then**
    $\overline{\phi} \leftarrow \phi_{\text{target}}$
  else
    $\underline{\phi} \leftarrow \phi_{\text{target}}$
  end if
end while
`Try to cluster yet unclustered vertices by looping over hyperarcs.`
return $\mathcal{C}$

Theorem 1. *The binary search clustering is an M-approximation algorithm for $CN{<}[w], M, \Delta{>}$ when $|p_{\max}| \times d > D$, $D \gg d$ and $\frac{D}{d} \leq M$, with $p_{\max}$ the critical path, d an intra-cluster delay, D an inter-cluster delay and M the maximum size of clusters.*

Proof. Let $H = (V, A)$ be a DAH, and $G = (V, A)$ be its corresponding r^*-weighted DAG. Let $|p_{\max}|$ be the longest path in H. As each vertex have a weight w, we will consider that $w = 1$. Let $\text{Sol}^*(H)$ be the optimal solution for a vertex-set clustering of H, an intra-cluster delay d, and an inter-cluster delay D, such that $D \gg d$ and $\frac{D}{d} \leq M$.

$$\text{Sol}^*(H) \geq \left(\left\lceil \frac{|p_{\max}|}{M} \right\rceil - 1\right) \times D + \left(|p_{\max}| - 1 - \left(\left\lceil \frac{|p_{\max}|}{M} \right\rceil - 1\right)\right) \times d\,. \quad (5)$$

Let $p_{\max}$ the critical path; we suppose $|p_{\max}| \times d > D$. Hence, we obtain:

$$|p_{\max}| \times d - D > 0\,. \quad (6)$$

In many cases, the propagation time of a circuit's critical path is longer than the time it takes to transfer a signal from one FPGA to another. However, there are circuits for which this is not true, although they are very few. Therefore, this proof applies only to circuits that satisfy the Eq. 6.

The BSC algorithm groups vertices using a direct approach based on cut capacity. This makes it more practical than a recursive coupling approach. Let $\text{Sol}_{\text{bsc}}(H)$ be the solution produced by our algorithm BSC, presented as Algorithm 1. It can be bounded by the worst solution. A worst-case solution is one in which each vertex forms a cluster. Hence, we have:

$$\text{Sol}_{\text{bsc}} \leq (|p_{\max}| - 1) \times D\,. \quad (7)$$

Then, the approximation ratio is defined by:

$$\frac{\mathrm{Sol}_{\mathrm{bsc}}(H)}{\mathrm{Sol}^*(H)} \leq \frac{(|p_{\max}|-1)\times D}{\left(\left\lceil\frac{|p_{\max}|}{M}\right\rceil-1\right)\times D+\left(|p_{\max}|-1-\left(\left\lceil\frac{|p_{\max}|}{M}\right\rceil-1\right)\right)\times d}. \quad (8)$$

Let us calculate the approximation ratio for $|p_{\max}| > M$ and $|p_{\max}| \leq M$. In the case when $|p_{\max}| > M$:

$$\left\lceil\frac{|p_{\max}|}{M}\right\rceil = \frac{|p_{\max}|+(M+r)}{M}. \quad (9)$$

By applying Eq. 9, we obtain:

$$\frac{\mathrm{Sol}_{\mathrm{bsc}}(H)}{\mathrm{Sol}^*(H)} \leq \frac{(|p_{\max}|-1)\times D}{(|p_{\max}|-r)\times D+(M-1)\times|p_{\max}|\times d-(M-r)\times d}.$$

By applying Eq. 6, we obtain:

$$\frac{\mathrm{Sol}_{\mathrm{bsc}}(H)}{\mathrm{Sol}^*(H)} \leq M\frac{(|p_{\max}|-1)\times D}{(|p_{\max}|-r+M-1)\times D-(M-r)\times d}.$$

Let us study the positivity of the expression $DM-Dr-(M-r)d$, we obtain:

$$DM-Dr > (M-r)d = DM-Dr-(M-r)d > 0. \quad (10)$$

By applying Eq. 10, we obtain:

$$\frac{\mathrm{Sol}_{\mathrm{bsc}}(H)}{\mathrm{Sol}^*(H)} \leq M.$$

In the case when $|p_{\max}| \leq M$, we have:

$$\left\lceil\frac{|p_{\max}|}{M}\right\rceil = 1. \quad (11)$$

By applying Eq. 11 to Eq. 8, we obtain:

$$\frac{(|p_{\max}|-1)\times D}{(|p_{\max}|-1)\times d}.$$

Since we have $\frac{D}{d} \leq M$, we obtain:

$$\frac{(|p_{\max}|-1)\times D}{(|p_{\max}|-1)\times d} = \frac{D}{d} \leq M.$$

Hence, parameterized approximation ratio is M for $CN{<}[w], M, \Delta{>}$ under the condition specified in the Theorem 1. In the general case, the ratio remain M^2+M.

4.2 Heavy-Edge Matching

The heavy-edge matching (HEM) approach for graph coarsening, presented by G. Karypis [12], is widely used in (hyper)graph partitioning tools [13,17] and yields efficient results in many cases. The advantage of this algorithm is that, in the unconstrained case, it almost halves the size of the instance during each of the first stages of the multilevel framework, which makes its complexity more interesting than that of our Algorithm 1. However, we will show in this subsection that HEM and other algorithms dedicated to 2-matching introduced by Z. Donovan *et al.* [6,8], do not capture path topology adequately. An example is presented in Fig. 3 for a clustering with $M > 2$. We will also show that HEM, applied to the DAG weighted with the r^* scheme, yields an approximation ratio of 2 for the $CN<[1], 2, \Delta>$ problem. This algorithm differs from the two algorithms presented by Z. Donovan *et al.* [6,8]: one of them looks for a dominant matching, and otherwise returns an arbitrary clustering, while the other is based on a linear programming rounding algorithm.

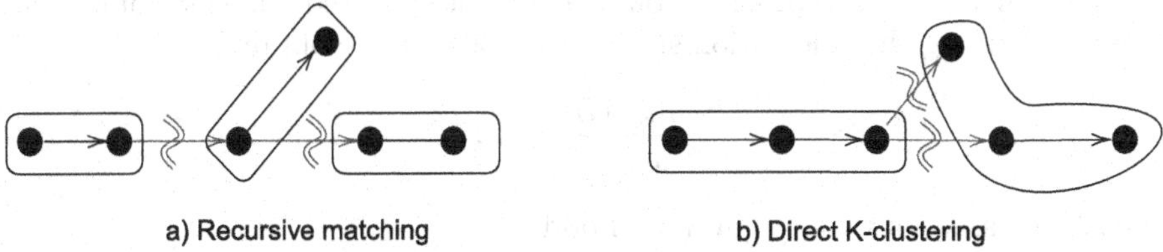

Fig. 3. This figure presents the effects of recursive matching vs. direct k-way clustering. On the left is a solution produced by a recursive matching algorithm for clustering with $M = 3$. On the right is the result of a direct clustering. As we can see, direct clustering produces less cut and clusters than recursive matching approach.

In the example shown in Fig. 3, the recursive methods will match vertices only once and cannot match them at the next level, because new vertices have a weight equal to 2. A direct clustering algorithm like our Algorithm 1 will produce in this case a result as good as recursive matching methods. This suggests that a direct clustering algorithm will be more interesting than a recursive coupling algorithm when M is large.

Theorem 2. *Let H be a DAH and $G = (V, A)$ be its corresponding r^*-weighted DAG. The HEM algorithm applied to the DAG for $CN<[1], 2, \Delta>$ is a 2-approximation algorithm.*

Proof. Let d be the intra-cluster delay and D be the inter-cluster delay, such that $D \gg d$. Let $p_{\max}$ be the critical path, with $|p_{\max}| \times d > D$. Hence:

$$|p_{\max}| \times d - D > 0\,. \tag{12}$$

Let $\mathrm{Sol}^*(H)$ be the optimal solution for a vertex set clustering of H. In the best case, for a size of cluster bounded by 2, the critical path will be coupled $\frac{|p_{\max}|}{2}$ times, which will yield the following lower bound for $\mathrm{Sol}j(H)$:

$$\mathrm{Sol}^*(H) \geq \left(\left\lceil\frac{|p_{\max}|}{2}\right\rceil - 1\right) \times D + \left(|p_{\max}| - 1 - \left(\left\lceil\frac{|p_{\max}|}{2}\right\rceil - 1\right)\right) \times d\,. \quad (13)$$

Let $\mathrm{Sol}_{\mathrm{HEM}}(H)$ be the solution produced by the HEM scheme on our proposed DAG model. It can be bounded by the worst possible solution, in which every vertex forms a cluster. Hence:

$$\mathrm{Sol}_{\mathrm{HEM}}(H) \leq (|p_{\max}| - 1) \times D\,. \quad (14)$$

Then, the approximation ratio is defined by:

$$\frac{\mathrm{Sol}_{\mathrm{HEM}}(H)}{\mathrm{Sol}^*(H)} \leq \frac{(|p_{\max}| - 1) \times D}{\left(\left\lceil\frac{|p_{\max}|}{2}\right\rceil - 1\right) \times D + \left(|p_{\max}| - 1 - \left(\left\lceil\frac{|p_{\max}|}{2}\right\rceil - 1\right)\right) \times d}\,. \quad (15)$$

Let us calculate the approximation ratio for the even and odd cases of $|p_{\max}|$. By performing the calculation similar to proof 4.1, we obtain:

$$\frac{\mathrm{Sol}_{\mathrm{HEM}}(H)}{\mathrm{Sol}^*(H)} = 2\,, \quad (16)$$

for both cases, when $|p_{\max}|$ is even and odd.

5 Experimental Results

To validate our models and algorithms, we have performed experiments on benchmarks of 19 logic circuits (B01-14 and B17-22) presented in F. Corno *et al.* [4]. These circuits consist of acyclic combinatorial blocks, bounded by their input and output registers. Every combinatorial block can therefore be modeled as a DAH. Their computation time is conditioned by their critical path, defined as the longest path between two registers (*i.e.*, two red vertices). These circuits have a number of cells from 51 (B01), to 233685 (B19).

Remember that we want to minimize the number of cuts on the critical path. In fact, in our problem, a cut on a path means an additional delay in the path cost. Thus, the compared algorithms aim to group the red-black hypergraphs by minimizing the delay path-length, *i.e.*, the maximum path cost $p_{\max}$. Since the execution time and the number of clusters are important parameters, we measure and compare them. Recall that clustering minimizing the number of clusters refers to the bin-packing problem, which is known to be NP-hard.

To compare them, we measured the degradation of the critical path produced by algorithm $\mathcal{A}$ for each instance I, calculated by: $(\mathrm{Sol}_{\mathcal{A}}(I) - p^I_{\max})/p^I_{\max}$. BSC and HEM algorithm were run 10 times for each circuit and for each size of cluster. The average of these 10 runs was used to calculate the path-cost averages for all instances per size of clusters.

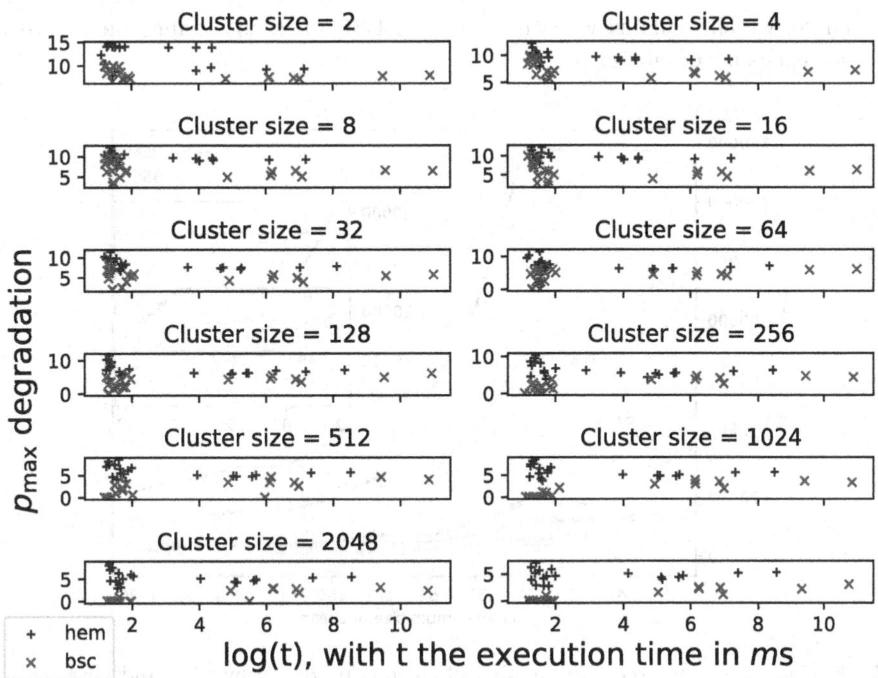

Fig. 4. Results of BSC and HEM on each circuit (point) for M values ranging from 2 to 4096. Each labeled point "+" is a clustering result calculated by the HEM algorithm and each labeled point "×" is for BSC. Each point is defined by the degradation of the critical path (ordinate) as a function of its logarithmic execution time (abscissa). As shown in all sub-figures, the "×" points are positioned below the "+" points, which are based on a lower critical path degradation for BSC. In addition, each point positioned to the left is based on a lower execution time. For two circuits, BSC take more execution time than HEM.

The results in Fig. 4 show that our BSC clustering algorithm, applied to circuit hypergraph, outperforms the HEM algorithm for critical path degradation. It can be shown that HEM points are more on the left side than BSC points. This rely on the fact that HEM takes less execution time than BSC. However, the execution time of HEM increase according to the size of cluster, that is, some HEM points moves from left to right. Indeed, as we increase the size of the clusters, we notice that HEM makes more recursive calls. Even if these recursive calls are executed on reduced hypergraphs, this increases the runtime. As a result, the complexity of HEM can be described by an additional factor of $\log_2(M)$, while the BSC algorithm admits a time complexity that depends only on the number of hyperedges. In practice, however, we find that the execution time of the BSC algorithm varies slightly as a function of M during the grouping phase, since this phase differs for each M. For BSC, however, these variations remain negligible, which explains why the points of BSC does not change on abscissa of 4 for each size of cluster M.

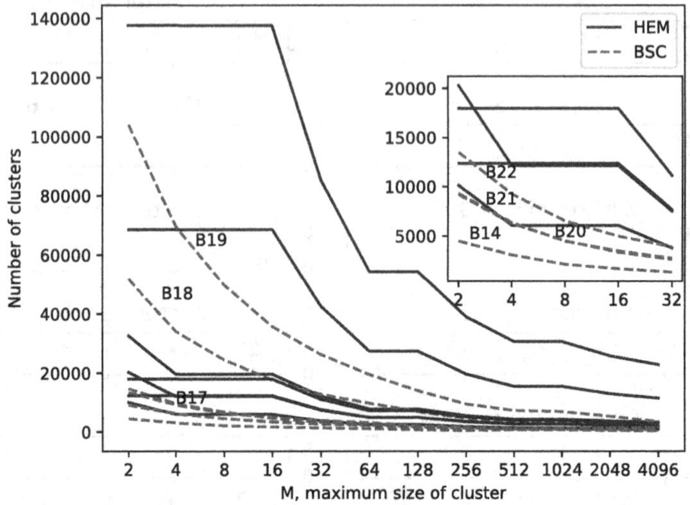

Fig. 5. Comparison between the number of clusters produced by BSC and HEM on a subset of highest circuits of vertices size for M values ranging from 2 to 4096. Each plain lines corresponds to HEM number of clusters and hatched lines to BSC number of clusters. Results show a less number of clusters for BSC than HEM. Subfigure shows us a zoom of B14, B20, B21, and B22 number of clusters for M values ranging from 2 to 32.

The results in Fig. 5 show that each BSC curves are under HEM curves, that is, our BSC clustering algorithm produces less number of clusters compared to the HEM algorithm. This can be explained by the fact that BSC directly groups a set of related vertices and applies a second refinement step that tends to reduce the number of clusters. In contrast, the HEM algorithm recursively groups vertices in pairs, which can more easily lead to situations where there are several adjacent clusters of size $M/2 + 1$ that cannot be merged.

Nevertheless, in the context of a multilevel scheme, both algorithms can be used, but BSC clusters vertices directly and does not create clustering levels as HEM does. An adaptation is necessary for such use.

6 Conclusion

In this work, we studied the combinatorial circuit clustering problem for delay minimization (CN) and presented a brief state-of-the-art in Sect. 2.2.

The aims of clustering algorithms is to select vertices to merge. Hence, the key is to define an attractiveness between the vertices that models the objective.

In Sect. 3, we presented existing weighting schemes l, r for attractiveness between pairs of vertices with our r^* weighting scheme. We showed that our r^* weighting scheme appears to be a better model to cluster critical vertices than the l and r.

In Sect. 4, we demonstrated that the approximation ratio parameterized by the size of cluster is in M if $|p_{\max}| \times d > D$, $D \gg d$ and $\frac{D}{d} \leq M$. This result improves the existing $M^2 + M$-approximation ratio under conditions mentioned above. We will continue to explore different hypotheses to find better bounds for this problem.

In Sect. 5, experimental results show that BSC produces less critical path degradation results for a majority of circuit instances. Moreover, BSC produces less number of clusters than HEM. Future works will investigate the efficiency of the BSC algorithm in a multilevel scheme. Along this work, a study should be conducted to determine which size of cluster is better, as the size of the BSC cluster must be adjusted to create multiple levels.

References

1. Ababei, C., Selvakkumaran, N., Bazargan, K., Karypis, G.: Multi-objective circuit partitioning for cutsize and path-based delay minimization. In: Proceedings of the 2002 IEEE/ACM International Conference on Computer-Aided Design, ICCAD 2002, pp. 181–185. Association for Computing Machinery, New York (2002). https://doi.org/10.1145/774572.774599
2. Çatalyürek, Ü.V., et al.: More Recent Advances in (Hyper)Graph Partitioning. Technical report arXiv:2205.13202, arXiv (2022). http://arxiv.org/abs/2205.13202
3. Cong, J., Wu, C.: Global clustering-based performance-driven circuit partitioning. In: Proceedings of the 2002 International Symposium on Physical Design, ISPD 2002, pp. 149–154. Association for Computing Machinery, New York (2002). https://doi.org/10.1145/505388.505424
4. Corno, F., Reorda, M.S., Squillero, G.: RT-level ITC'99 benchmarks and first ATPG results. J. IEEE Des. Test Comput. **17**(3), 44–53 (2000)
5. Diwan, A.A., Rane, S., Seshadri, S., Sudarshan, S.: Clustering techniques for minimizing external path length. In: Proceedings of the 22th International Conference on Very Large Data Bases, pp. 342–353 (1996)
6. Donovan, Z.N.: Algorithmic Issues in some Disjoint Clustering Problems in Combinatorial Circuits. Thesis dissertation, West Virginia University Libraries (2018). https://doi.org/10.33915/etd.3721. https://researchrepository.wvu.edu/etd/3721
7. Donovan, Z.N., Mkrtchyan, V., Subramani, K.: Complexity issues in some clustering problems in combinatorial circuits. arXiv:1412.4051 (2017). http://arxiv.org/abs/1412.4051
8. Donovan, Z., Subramani, K., Mkrtchyan, V.: Disjoint clustering in combinatorial circuits. In: Colbourn, C.J., Grossi, R., Pisanti, N. (eds.) IWOCA 2019. LNCS, vol. 11638, pp. 201–213. Springer, Cham (2019). https://doi.org/10.1007/978-3-030-25005-8_17
9. Donovan, Z.N., Subramani, K., Mkrtchyan, V.: Analyzing clustering and partitioning problems in selected VLSI models. Theory Comput. Syst. **64**(7), 1242–1272 (2020)
10. Kagaris, D.: On minimum delay clustering without replication. Integration **36**(1), 27–39 (2003)

11. Karypis, G., Aggarwal, R., Kumar, V., Shekhar, S.: Multilevel hypergraph partitioning: applications in VLSI domain. IEEE Trans. Very Large Scale Integr. (VLSI) Syst. **7**(1), 69–79 (1999). https://doi.org/10.1109/92.748202
12. Karypis, G., Kumar, V.: Analysis of multilevel graph partitioning. In: Proceedings of the 1995 ACM/IEEE Conference on Supercomputing, pp. 29–es (1995)
13. Karypis, G., Kumar, V.: Hmetis: a hypergraph partitioning package. ACM Trans. Architect. Code Optim. (1998)
14. Lawler, E.L., Levitt, K.N., Turner, J.: Module clustering to minimize delay in digital networks. IEEE Trans. Comput. **100**(1), 47–57 (1969)
15. Murgai, R., Brayton, R.K., Sangiovanni-Vincentelli, A.: On clustering for minimum delay/ara, pp. 6–9. IEEE Computer Society (1991). https://doi.org/10.1109/ICCAD.1991.185176. https://www.computer.org/csdl/proceedings-article/iccad/1991/00185176/12OmNASILTx
16. Pan, P., Karandikar, A.K., Liu, C.L.: Optimal clock period clustering for sequential circuits with retiming. IEEE Trans. Comput. Aided Des. Integr. Circuits Syst. **17**(6), 489–498 (1998)
17. Pellegrini, F.: Scotch and PT-scotch graph partitioning software: an overview. In: Uwe Naumann, O.S. (ed.) Combinatorial Scientific Computing, pp. 373–406. Chapman and Hall/CRC (2012). https://doi.org/10.1201/b11644-15. https://inria.hal.science/hal-00770422
18. Rajaraman, R., Wong, M.D.F.: Optimal clustering for delay minimization. In: Proceedings of the 30th International Design Automation Conference, DAC 1993, pp. 309–314. Association for Computing Machinery, New York (1993). https://doi.org/10.1145/157485.164907. https://doi.org/10.1145/157485.164907
19. Rodriguez, J., Galea, F., Pellegrini, F., Zaourar, L.: A hypergraph model and associated optimization strategies for path length-driven netlist partitioning. In: Mikyška, J., de Mulatier, C., Paszynski, M., Krzhizhanovskaya, V.V., Dongarra, J.J., Sloot, P.M. (eds.) ICCS 2023. LNCS, vol. 10475, pp. 652–660. Springer, Cham (2023). https://doi.org/10.1007/978-3-031-36024-4_50

Adaptive Sampling for Non-intrusive Reduced Order Models Using Multi-task Variance

Abhijnan Dikshit[1(✉)], Leifur Leifsson[1], Slawomir Koziel[2,3], and Anna Pietrenko-Dabrowska[3]

[1] School of Aeronautics and Astronautics, Purdue University, West Lafayette, IN 47907, USA
{adikshit,leifur}@purdue.edu

[2] Department of Engineering, Engineering Optimization and Modeling Center, Reykjavík University, Menntavegur 1, 102 Reykjavík, Iceland
koziel@ru.is

[3] Faculty of Electronics Telecommunications and Informatics, Gdansk University of Technology, Narutowicza 11/12, 80-233 Gdansk, Poland
anna.dabrowska@pg.edu.pl

Abstract. Non-intrusive reduced order modeling methods (ROMs) have become increasingly popular for science and engineering applications such as predicting the field-based solutions for aerodynamic flows. A large sample size is, however, required to train the models for global accuracy. In this paper, a novel adaptive sampling strategy is introduced for these models that uses field-based uncertainty as a sampling metric. The strategy uses Monte Carlo simulations to propagate the uncertainty in the prediction of the latent space of the ROM obtained using a multi-task Gaussian process to the high-dimensional solution of the ROM. The high-dimensional uncertainty is used to discover new sampling locations to improve the global accuracy of the ROM with fewer samples. The performance of the proposed method is demonstrated on the environment model function and compared to one-shot sampling strategies. The results indicate that the proposed adaptive sampling strategies can reduce the mean relative error of the ROM to the order of 8×10^{-4} which is a 20% and 27% improvement over the Latin hypercube and Halton sequence sampling strategies, respectively at the same number of samples.

Keywords: Adaptive sampling · Field-based uncertainty · Reduced order modeling · Multi-task Gaussian process · Monte Carlo simulation

1 Introduction

The need to reduce the computational cost of high-fidelity simulations has been of great interest in the fields of science and engineering. A computationally cheaper alternative to high-fidelity simulations has been identified in the form of reduced order modeling (ROM) strategies. ROMs can be intrusive [2], i.e., they alter the governing equations of a problem and offer a simpler alternative that

L. Franco et al. (Eds.): ICCS 2024, LNCS 14836, pp. 105–119, 2024.
https://doi.org/10.1007/978-3-031-63775-9_8

is faster to solve, or they can be non-intrusive [24], i.e., they can be data-driven models that use high-fidelity data and machine learning techniques to create an approximation of simulation output. Intrusive ROMs typically require access to the simulation source code which may be difficult to acquire. Implementing intrusive ROMs also requires extensive knowledge and effort to manipulate the governing equations and simulation source code. These challenges can be overcome by using non-intrusive ROMs that only require access to the data generated by a high-fidelity simulation. Non-intrusive ROMs have gained popularity and have been extensively applied in the domain of fluid mechanics to predict high-dimensional field-based variables [6,12,20].

A common issue with the creation of ROMs is that they often require a large number of samples to achieve the desired approximation quality of high-dimensional field-based variables. These samples are usually generated in one-shot using, for example, Latin hypercube sampling (LHS) [22] or low discrepancy sequences, such as Halton sequences [17]. These methods are usually space-filling design of experiments, however, these methods do not generate sampling plans that are tailored to the approximation quality of non-intrusive ROMs. As a result, a higher number of samples are required to lower the prediction error of the ROM leading to a large number of high-fidelity simulations.

To overcome this disadvantage, adaptive sampling strategies can be used to choose samples within a parameter space that directly target the prediction quality of the ROM. An extensive overview of adaptive sampling and different types of sampling strategies is given in [15] and [9]. The initial work on adaptive sampling of ROMs was focused on the sampling of POD-based ROMs. Braconnier et al. [5] and Guenot' et al. [11] developed sampling strategies based on a measure of quality of POD basis generation and the prediction quality of the low-dimensional space interpolation model of the ROM. Wang et al. [7] extended the work of Guenot et al. [11] by developing a fully-automated approach that utilizes the proposed adaptive sampling metrics in a combined manner.

Yang and Xiao [23] proposed an adaptive sampling algorithm for a POD-based ROM that used the variance estimate and gradient information of the low-dimensional space modeled by a Gaussian process regression model. They employ a weighting strategy for the variance of the prediction of each modal coefficient and the magnitudes of the gradients. The weights are based on the singular values of the modal coefficients. This weighting strategy restricts the use of this method to only POD-based ROMs. The work of Karcher and Franz [14] and Franz et al. [8] developed adaptive sampling strategies that can be implemented for both POD-based and manifold learning ROMs. Some of the sampling metrics used in these studies include manifold-filling metrics, residual of governing equations, and error estimations, such as leave-one-out cross-validation error and mean squared error of the interpolation model of the low-dimensional space.

Most of the previous work on adaptive sampling strategies for ROMs has typically utilized the approximation quality of the low-dimensional space directly. This approach does not account for the approximation quality in the high-dimensional space which is the actual quantity of interest in a ROM prediction. The current work will quantify the variance in the high-dimensional solution of the ROM by the use of forward propagation through the backmapping procedure of the ROM. The high-dimensional variance estimate will then be utilized as a sampling objective in an adaptive sampling algorithm. It is hypothesized that the introduction of high-dimensional variance information will improve the performance of the adaptive sampling.

In this work, a novel adaptive sampling strategy is proposed for ROMs based on the multi-task variance of the latent space obtained by the free-form parameterization multi-task Gaussian process (MTGP) formulation [4]. In this case, the latent space is modeled using a MTGP model to predict all the latent space coefficients simultaneously. Monte Carlo sampling is used to propagate the multi-task variance of the latent space to the high-dimensional solution space to estimate the high-dimensional variance and employ it to discover new sampling locations to improve the global accuracy of the non-intrusive ROM. The proposed adaptive sampling strategy is a novel approach as it uses high-dimensional variance information as an infill criteria to search the design space for new samples rather than using only the latent space variance. This strategy will introduce more information in the form of field-based variance information into the search for new samples that could lead to a better search and possibly higher accuracy with fewer samples. The proposed method is also agnostic to the dimensionality reduction method of the ROM. To illustrate the possible benefits, the proposed algorithm is benchmarked against traditional one-shot sampling strategies, in particular, LHS and Halton sequences.

The next section will describe the novel adaptive sampling algorithm proposed in this paper. The section following that will describe the application of the adaptive sampling algorithm to a 2D analytical test function and a performance comparison with benchmarking methods. The final section of the paper will present the main conclusions and possible future work.

2 Field-Based Adaptive Sampling

This section introduces the high-dimensional field-based adaptive sampling strategy for ROMs proposed in this work.

2.1 Proposed Adaptive Sampling Algorithm

The general process of adaptive sampling is driven by an optimization problem that chooses a new sampling point by maximizing a sampling objective. This can be mathematically represented as

$$\mathbf{x}_{new} = \text{argmax}\, f(\mathbf{x}), \tag{1}$$

where $\mathbf{x} = (x_1, x_2, ...x_n)$ is a vector of model parameters, $f(\mathbf{x})$ denotes the sampling objective and $\mathbf{x}_{new}$ is the new sampling point. The sampling objective itself is based on a measure of the prediction uncertainty of the ROM. In the case of the field-based adaptive sampling proposed here, each point in the field will be assigned a prediction uncertainty rather than an aggregated prediction uncertainty being used for the entire field prediction. However, to facilitate the calculation of a sampling objective the prediction uncertainty from all points in the field must be aggregated using mathematical operations such as averaging. This means that in the case of field-based adaptive sampling proposed here, the sampling problem should be represented as

$$\max_{\mathbf{x}} f = g(\hat{\sigma}(\mathbf{x})), \tag{2}$$

where $\hat{\sigma}(\mathbf{x})$ is the prediction uncertainty field for a given set of parameters and $g(\hat{\sigma}(\mathbf{x}))$ is an aggregation function to aggregate the prediction uncertainty in the field in a meaningful way.

The main contribution of this work is to introduce high-dimensional information in the adaptive sampling algorithm. To do this, the variance estimate of the latent space prediction provided by an MTGP model will be propagated to the high-dimensional solution to attain a high-dimensional variance estimate. The proposed propagated uncertainty sampling algorithm is shown in Algorithm 1. The main features of the algorithm are discussed in detail in the following subsections. The algorithm described in these subsections will be repeated until a fixed computational budget is reached. Here, the computational budget refers to the number of samples that can be obtained within the limits of computational cost.

2.2 Dimensionality Reduction

To start the algorithm, an initial sampling plan $\mathbf{X} = \{\mathbf{x}^{(1)}, ..., \mathbf{x}^{(n)}\}$ is generated using a one-shot sampling strategy. In this work, Halton sequences are used to generate the initial sampling plan that contains only a few number of samples. This is used as the base sampling plan to create the initial ROM and calculate variance estimates of the ROM. The high-dimensional field-based

Algorithm 1. Adaptive sampling using field-based variance estimate.

Require: initial data sets $(\mathbf{X}, \mathbf{Y})$, computational budget N_{max}
  repeat
    obtain POD modes $\mathbf{\Phi}$ through SVD
    project data to low dimensional space, $\mathbf{z} \leftarrow \mathbf{\Phi}^T\mathbf{Y}$
    fit $MTGP_z$ to the data set $(\mathbf{X}, \mathbf{z})$
    $\mathbf{x}_{new} \leftarrow \text{argmax } g(\hat{\sigma}(\mathbf{x}))$
    $\mathbf{X} \leftarrow \mathbf{X} \cup \mathbf{x}_{new}$
    get field-based data $(\mathbf{y}_{new})$ at $\mathbf{x}_{new}$
    $\mathbf{Y} \leftarrow \mathbf{Y} \cup \mathbf{y}_{new}$
  until computational budget expires
  return final sampling plan, $\mathbf{X}$

solutions $\mathbf{Y} = \{\mathbf{y}^{(1)}, ..., \mathbf{y}^{(n)}\} \in \mathbb{R}^{m \times n}$ are calculated for each sample in $\mathbf{X}$ to create the data for the ROM. $\mathbf{Y}$ is a matrix containing n fields of dimensionality m. To focus on the adaptive sampling algorithm itself, the simplest possible dimensionality reduction method is used to create the ROM in this work. The proposed algorithm, however, is agnostic of the dimensionality reduction method and any dimensionality reduction method could be applied along with it. POD is applied to the high-dimensional solutions $\mathbf{Y}$ to generate the low-dimensional latent space for the data. The modes of the high-dimensional space, $\boldsymbol{\Phi} = [\phi^{(1)}, ..., \phi^{(d)}] \in \mathbb{R}^{m \times d}$, are generated using singular value decomposition (SVD) and the truncation of the modes is performed according to the relative information criterion [19]. $\boldsymbol{\Phi}$ represents the matrix of POD modes with $d << n$ and $d << m$ as the dimensionality of the latent space coefficients will be much lower than the original dimensionality. The optimal number of modes for the POD algorithm is chosen according to a relative information content of 0.9999.

2.3 Multi-task Gaussian Process Models

After POD is applied to the high-dimensional solutions, a low-dimensional space is generated for the data. A parametric mapping is created between the input parameters and the low-dimensional space using a MTGP model that can simultaneously predict multiple latent space coefficients. This model is denoted as $MTGP_z$ from here onwards. MTGPs are chosen as they will provide a variance estimate of the low-dimensional space that can be propagated to the high-dimensional solution using the truncated POD modes and Monte Carlo sampling.

MTGPs are multi-task learning models that are applied to Gaussian processes [4]. They aim at learning multiple correlated tasks simultaneously and boost the prediction capability of the model by modelling the correlations among the different outputs that are trying to be predicted. The MTGP models used in this work are created using GPyTorch [10], a Python implementation of Gaussian process models built on the framework of PyTorch [18]. The MTGP models were trained using a radial basis function kernel wrapped with a multi-task kernel and the hyperparameters were optimized using the maximum log-likelihood criterion. These models provide an individual variance estimate of the prediction of each output.

2.4 Uncertainty Propagation

The main feature of the proposed adaptive sampling algorithm is the propagation of the variance estimate provided by $MTGP_z$ from the low-dimensional space to the high-dimensional solution space. The uncertainty propagation algorithm used in this work is given in Algorithm 2. One of the most well-known ways of propagating uncertainty is to use Monte Carlo sampling [16]. Monte Carlo sampling uses a large number of samples drawn from the distribution of uncertain variables and uses these samples to estimate the statistics of an output that is dependent on the uncertain variables. In this work, a collection of m samples, $z^{(m)}(\mathbf{x}) \sim \mathcal{N}(\boldsymbol{\mu}, \boldsymbol{\Sigma})$, is drawn from the posterior distribution of the $MTGP_z$

Algorithm 2. Propagating variance estimate to high-dimensional space.

Require: point in parameter space $\mathbf{x}$, fitted $MTGP_z$ model, number of Monte-Carlo samples M

for $m = 1, ..., M$ **do**

draw sample from $MTGP_z$ posterior of low-dimensional space at point of interest $z^{(m)}(\mathbf{x}) \sim \mathcal{N}(\boldsymbol{\mu}, \boldsymbol{\Sigma})$

predict high-dimensional solution, $\hat{\mathbf{y}}(\mathbf{x}) = \boldsymbol{\Phi} z^{(m)}(\mathbf{x})^T$

end for

estimate $\hat{\sigma}(\mathbf{x})$ from data of Monte Carlo samples

model. Each sample is then projected to the high-dimensional space to obtain the predicted high-dimensional field $\hat{\mathbf{y}}(\mathbf{x})$. Once the high-dimensional solutions of all the samples have been found, the standard deviation, $\hat{\sigma}(\mathbf{x})$, of the prediction of the ROM at each point of the solution domain will be estimated. Essentially, each point of the solution domain is treated as having a distribution with a mean value and standard deviation. The resulting standard deviation field will be used as an estimate of the uncertainty in the high-dimensional field-solution predicted by the ROM.

2.5 Scalarized Sampling Objective

The propagated uncertainty will provide a measure of variance or standard deviation for every grid point in the high-dimensional solution. A scalarization or aggregation method is required to enable the use of an optimization algorithm to find the next infill point for the algorithm. In this work, two different scalarization methods are employed. One of them involves using the average of the variance while the other involves using the maximum value of the variance obtained. The average variance is calculated as

$$\hat{\sigma}_{avg}(\mathbf{x}) = \frac{1}{m}\sum_{i=1}^{m} \hat{\sigma}_i(\mathbf{x}), \tag{3}$$

where $\hat{\sigma}_i(\mathbf{x})$ is the standard deviation of the ith grid point in the high-dimensional field and m is the total number of grid points in the field. Similarly, the maximum value of the variance can be determined as

$$\hat{\sigma}_{max}(\mathbf{x}) = \max_{i=1,2,\ldots,m} \hat{\sigma}_i(\mathbf{x}), \tag{4}$$

Once the standard deviation estimate has been calculated, the sampling objective can be defined. To promote the exploration of the parameter space by the adaptive sampling algorithm, the minimum Euclidean distance between sampling points will be included in the sampling objective [1,13]. A further consideration in the design of the sampling objective is the fact that non-intrusive ROMs are regression models and therefore, the standard deviation may not be zero at an already sampled point as is the case for interpolation models. To prevent placement of new infill points near existing sampling points, a threshold distance is defined below which the standard deviation is artificially set to zero. This will prevent an optimization algorithm from looking in regions close to existing samples while maximizing the sampling objective. After including all of these considerations, the sampling objective for the algorithm is defined as

$$f(\mathbf{x}) = \begin{cases} d(\mathbf{x}, \mathbf{X})\hat{\sigma}_{scalar}(\mathbf{x}), & d(\mathbf{x}, \mathbf{X}) \geq \alpha D_{min}^{max} \\ 0, & \text{otherwise} \end{cases}, \tag{5}$$

where $\hat{\sigma}_{scalar}(\mathbf{x})$ is the scalarized field standard deviation from either 3 or 4, $d(\mathbf{x}, \mathbf{X})$ is the minimum Euclidean distance between the point of interest and existing sampling points and αD_{min}^{max} is the threshold distance. D_{min}^{max} represents the maximum of the minimum Euclidean distance between any two existing sampling points and α is a parameter used to tune the value of the threshold distance. This sampling objective will be maximized to locate the new infill location, $\mathbf{x}_{new}$, for the adaptive sampling algorithm.

3 Numerical Experiments

This section discusses the test case used to assess the proposed method and the results obtained.

3.1 2D Environment Model Function

The analytical environment model function (EMF) [3] will be used to test the proposed adaptive sampling algorithm. EMF aims to model the spill of a pollutant caused by a chemical accident over a spatial and temporal coordinate grid. Since this function is a multi-output function, it can provide a good test bed for understanding the behavior of adaptive sampling algorithms for ROMs without the need to run high-fidelity simulations. ROMs will be trained to predict the outputs of this function on a 10×10 grid leading to the requirement of modelling 100 outputs in total. The EMF is defined as

$$h(s, t|\mathbf{x}) = \frac{M}{\sqrt{4\pi D t}} \exp(\frac{-s^2}{4Dt}) + \frac{1_{t>\tau} M}{\sqrt{4\pi D(t-\tau)}} \exp(\frac{-(s-L)^2}{4D(t-\tau)}), \tag{6}$$

where $\mathbf{x} = (M, D, L, \tau)$ are the model parameters, and s and t represent the locations of a grid point in the 10×10 grid. The original function has four input parameters, however, in this work, a 2D version of the problem is considered which fixes two parameters of the function. A 2D version is considered as it is easy to study and visualize different aspects of the proposed algorithm in 2D. In this 2D version, the values of L and τ are fixed at their calibrated values of 1.505 and 30.125, respectively [3]. The parameters $M \in [7, 13]$ and $D \in [0.02, 0.12]$ are considered to be the variables in the parameter space.

3.2 Comparison Methods and Metric

The results obtained from the adaptive sampling algorithm are compared to Halton sequences [17] and the mean performance of 10 LHS sampling plans [22] at various numbers of samples. The mean relative error in the parameter space is used as a measure of the overall prediction quality of a ROM built by using different types of sampling plans. The mean relative error is calculated as

$$e_{rel} = \frac{1}{N}\sum_{i=1}^{N}\frac{||\mathbf{W}^{pred} - \mathbf{W}^{*}||_2}{||\mathbf{W}^{*}||_2}, \tag{7}$$

where $\mathbf{W}^{pred}$ is the prediction of the field solution from the ROM, $\mathbf{W}^{*}$ is the true field solution and N is the number of testing samples for the ROM.

3.3 Setup of the Sampling Algorithm

The initial sampling plans, created using Halton sequences, consist of 10 and 20 samples. The testing data to assess the prediction quality of the ROM built from different sampling plans is an LHS sampling plan of 100 samples and common to each of the methods of generating samples. The initial sample sizes were chosen based on heuristics that have been suggested based on the dimensionality of the parameter space and computational budget of the algorithm [15]. These samples are used to train the initial ROM and propagate the uncertainty from the low-dimensional space to the high-dimensional solution space. The Monte Carlo sampling for uncertainty propagation uses 1000 samples drawn from the posterior of the MTGP model to estimate the field standard deviation of the prediction of the pollutant spill multiplied by the minimum Euclidean distance between samples as described in 5. The value of α is set to 0.10 for all numerical experiments. A fixed budget of 40 samples is chosen as the stopping criteria. Both the mean standard deviation and the maximum standard deviation in the field prediction of the pollutant spill were used as the sampling objectives of the problem. The sampling objective is maximized using differential evolution [21] with a population size of 200, a recombination constant of 0.7 and a mutation rate of 0.9.

3.4 Results

Figure 1 shows the variation of the mean relative error of the prediction of the ROM starting from 10 and 20 initial samples for a fixed budget of 40 samples. It can be seen that the adaptive sampling strategies quickly reduce the prediction error of the ROM as compared to the LHS and Halton sequence sampling plans, especially when using 20 initial samples.

With 10 initial samples, the sampling plans driven by $\hat{\sigma}_{avg}(\mathbf{x})$ and $\hat{\sigma}_{max}(\mathbf{x})$ achieve a final mean relative error of $9.892 \cdot 10^{-4}$ and $1.042 \cdot 10^{-3}$, respectively. This shows that both the sampling strategies achieved a similar performance with 10 initial samples. The LHS sampling and Halton sequence sampling strategy are

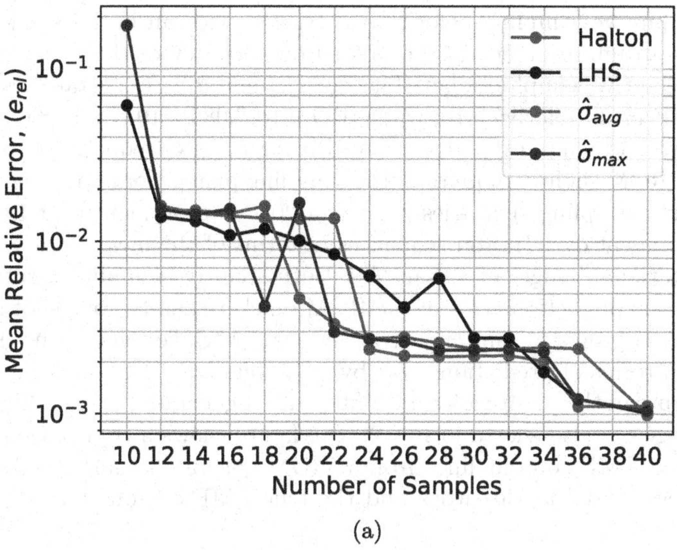

(a)

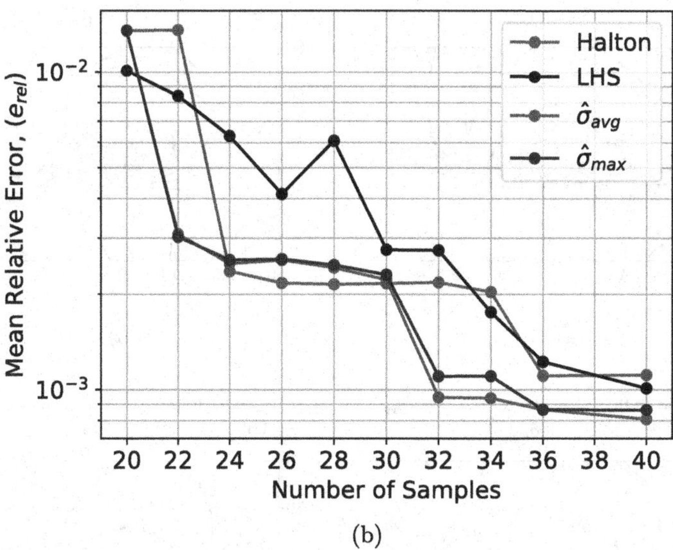

(b)

Fig. 1. Mean relative error of the prediction of the ROM constructed using the proposed sampling methods and benchmark one-shot sampling methods for (a) 10 initial samples and (b) 20 initial samples.

able to achieve a mean relative error of 1.010×10^{-3} and 1.111×10^{-3}, respectively. This means that in this case the adaptive sampling strategies perform on par with the one-shot sampling strategies.

If the adaptive sampling algorithm is started with an initial sample size of 20, there is an improvement in the prediction accuracy of the ROM generated using the adaptive sampling plan. The $\hat{\sigma}_{avg}(\mathbf{x})$ and $\hat{\sigma}_{max}(\mathbf{x})$ strategies are able to reduce the mean relative error of the ROM to 8.086×10^{-4} and 8.638×10^{-4}, respectively. The sampling plan produced using $\hat{\sigma}_{avg}(\mathbf{x})$ reduces the prediction error of the ROM slightly more than the sampling plan generated using $\hat{\sigma}_{max}(\mathbf{x})$. The adaptive sampling strategies achieve a 20% improvement over LHS and a 27% improvement over Halton sequences in terms of the mean relative error at the computational budget of 40 samples. Both adaptive sampling strategies can also reduce the prediction error of the ROM quicker than the one-shot sampling strategies. The proposed sampling strategies can overtake the one-shot sampling strategies in terms of prediction error by a sample size of 32 samples.

The nature of the prediction error of the ROM generated using different sampling strategies is also shown in Fig. 2. The figure shows a comparison between the predicted EMF contour lines from a ROM generated using all of the sampling strategies used in this work and the true EMF contour lines. The figure

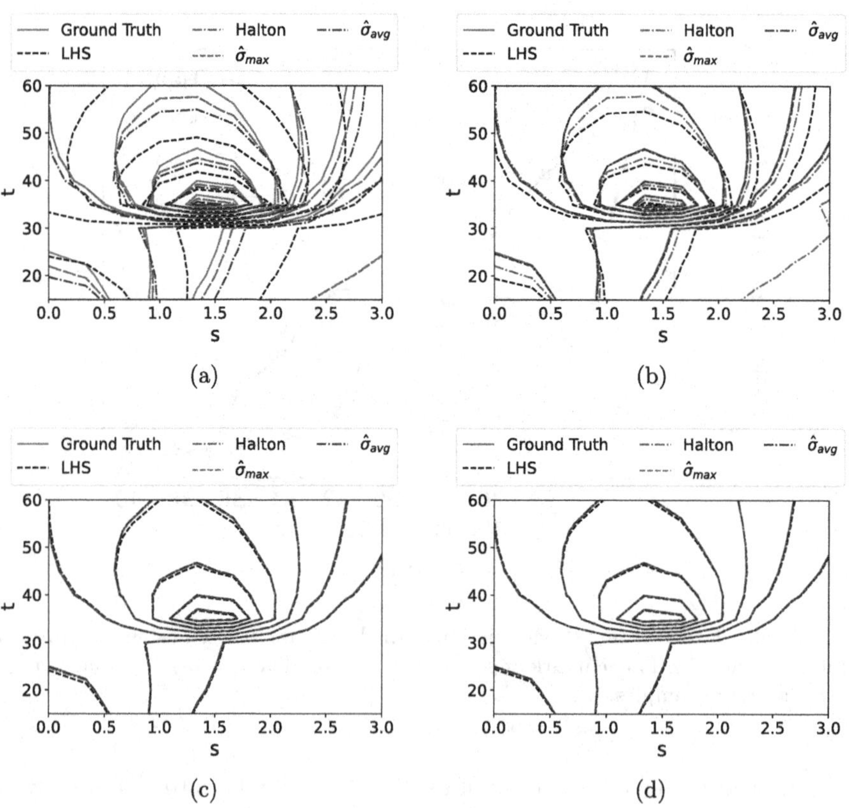

Fig. 2. True and predicted contour lines of EMF for $M = 11.3830$ and $D = 0.0204$ with (a) 10 initial samples, (b) 20 samples, (c) 30 samples and (d) 40 samples.

illustrates that the adaptive sampling strategies can produce a ROM that predicts the EMF more accurately with a fewer number of samples than a ROM generated using one-shot sampling strategies. This is because the contour lines predicted using the adaptive sampling strategies are closer to the true contours as compared to the contour lines predicted using one-shot sampling strategies.

Figure 3 shows the sampling plans that are generated using the adaptive sampling strategies. For comparison, Fig. 4 presents sampling plans of 40 samples that were generated using the benchmark one-shot sampling strategies. It can be seen that the adaptive sampling strategies place many samples on the edge of the parameter space in addition to the samples placed in the central region. There may be an influence of the distance criteria in placing the samples near the edge of the sample space as this will tend to maximize the minimum distance between samples. This is, however, also a beneficial aspect of the sampling strategy as

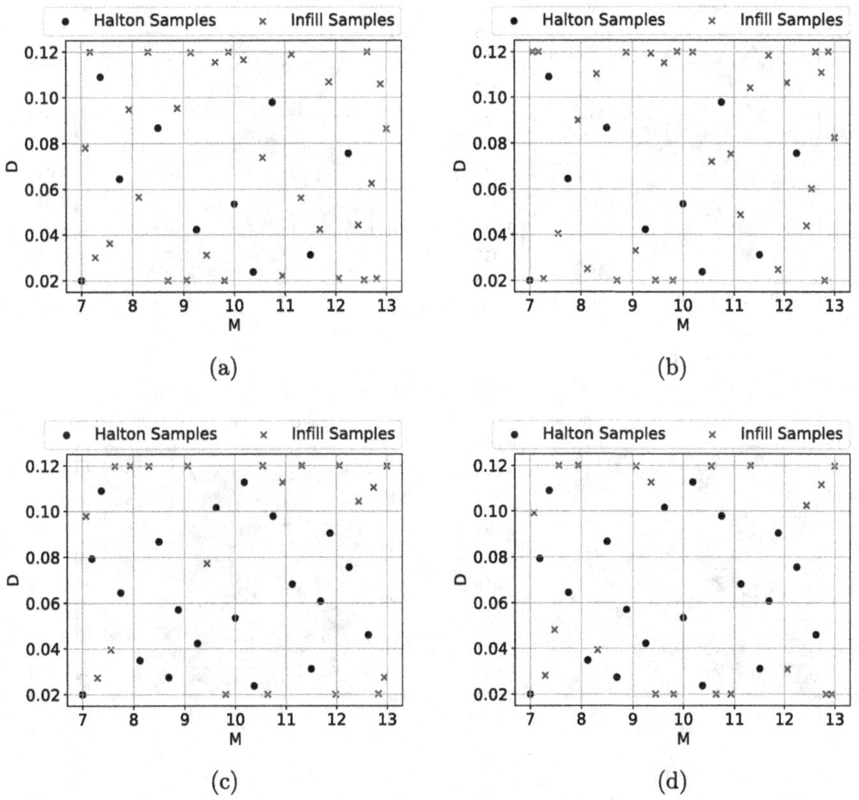

Fig. 3. Sampling plans of the 2D parameter space for the environment model function generated using adaptive sampling strategies using (a) mean propagated variance and 10 initial samples, (b) maximum propagated variance and 10 initial samples, (c) mean propagated variance and 20 initial samples, and (d) maximum propagated variance and 20 initial samples.

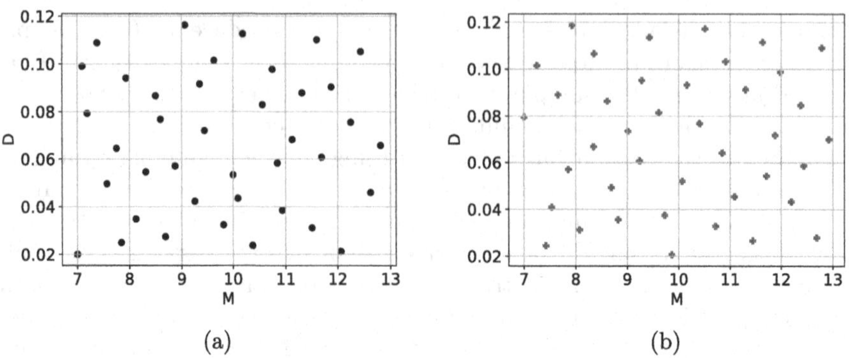

Fig. 4. Benchmark one-shot sampling plans generated using (a) Halton sequences and (b) LHS.

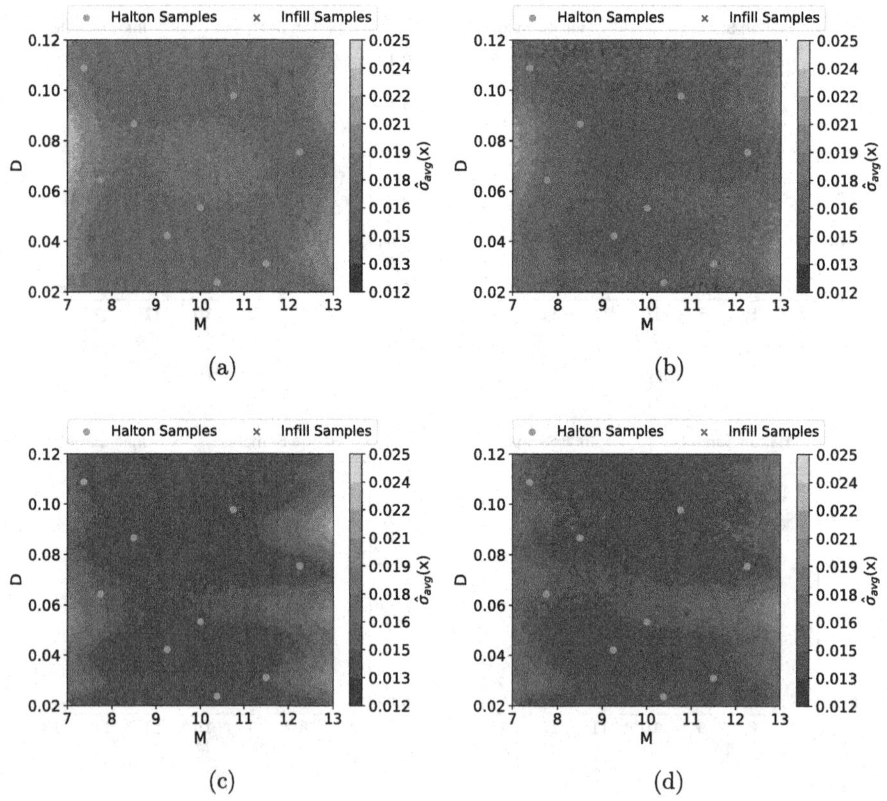

Fig. 5. Contours of $\hat{\sigma}_{avg}(\mathbf{x})$ plotted for (a) 25 samples, (b) 30 samples, (c) 35 samples, and (d) 40 samples starting with 10 initial samples.

it tends to explore the edge of the parameter space more than typical one-shot sampling strategies.

The evolution of the contours of $\hat{\sigma}_{avg}(\mathbf{x})$ with increasing number of samples is shown in Fig. 5. The contours show that higher values of $\hat{\sigma}_{avg}(\mathbf{x})$ occur near the edges of the parameter space rather than the central region. This is also an influencing factor in the placement of infill samples near the edge of the parameter space. It is also evident that as the number of samples increases the value of $\hat{\sigma}_{avg}(\mathbf{x})$ decreases across the parameter space. This illustrates that the average field uncertainty of the ROM decreases as the number of samples increases in the sampling algorithm.

4 Conclusion

This paper proposes an adaptive sampling strategy for non-intrusive ROMs based on forward propagation of the multi-task variance information of the latent space. The posterior distribution of the multi-task Gaussian process model of the latent space is sampled using Monte Carlo sampling and the standard deviation of the high-dimensional field solution is calculated. The overall field standard deviation is scalarized using the average value or the maximum value. A sampling objective is designed that maximizes the product of the variance estimate and the minimum distance between the samples. The adaptive sampling process is carried out for a fixed computational budget.

The proposed adaptive sampling algorithm is demonstrated on the environment model function. The proposed sampling strategies outperform both of the one-shot strategies especially with an initial sample size of 20 samples. In the case of the EMF, using the average value of the field uncertainty yields slightly lower prediction errors at the termination of the adaptive sampling algorithm as compared to using the maximum value of the field uncertainty. Both adaptive sampling strategies tend to place more samples near the edge of the parameter space as compared to the one-shot sampling strategies. This is due to the Euclidean distance criterion included in the sampling objective and larger field uncertainty in locations near the edges of the parameter space.

Future work will aim at applying the adaptive sampling strategies to higher dimensional and complex engineering problems, such as modeling flow around an airfoil or wing. The sampling objective of the strategy is a very important aspect of the method that will be given more attention in future work. Multiple different scalarization methods and objective formulations will be investigated to improve the performance of the algorithm.

Acknowledgments. This work was supported in part by the U.S. National Science Foundation (NSF) award number 2223732 and by the Icelandic Centre for Research (RANNIS) award number 239858.

References

1. Aute, V., Saleh, K., Abdelaziz, O., Azarm, S., Radermacher, R.: Cross-validation based single response adaptive design of experiments for Kriging metamodeling of deterministic computer simulations. Struct. Multidiscip. Optim. **48**, 581–605 (2013). https://doi.org/10.1007/s00158-013-0918-5
2. Benner, P., Gugercin, S., Willcox, K.: A survey of projection-based model reduction methods for parametric dynamical systems. SIAM Rev. **57**, 483–531 (2015). https://doi.org/10.1137/130932715
3. Bliznyuk, N., Ruppert, D., Shoemaker, C., Regis, R., Wild, S., Mugunthan, P.: Bayesian calibration and uncertainty analysis for computationally expensive models using optimization and radial basis function approximation. J. Comput. Graph. Stat. **17**, 270–294 (2008). https://doi.org/10.1198/106186008X320681
4. Bonilla, E.V., Chai, K., Williams, C.: Multi-task Gaussian process prediction. In: Advances in Neural Information Processing Systems, vol. 20, Vancouver, Canada, 3–6 December 2007
5. Braconnier, T., Ferrier, M., Jouhaud, J.C., Montagnac, M., Sagaut, P.: Towards an adaptive POD/SVD surrogate model for aeronautic design. Comput. Fluids **40**, 195–209 (2011). https://doi.org/10.1016/j.compfluid.2010.09.002
6. Decker, K., Iyengar, N., Rajaram, D., Perron, C., Mavris, D.: Manifold alignment-based nonintrusive and nonlinear multifidelity reduced-order modeling. AIAA J. **61**, 454–474 (2023). https://doi.org/10.2514/1.J061720
7. Du, X., Wang, J., Martins, J.R.: A fully automated adaptive sampling strategy for reduced-order modeling of flow fields. In: AIAA SciTech 2023 Forum. National Harbor, MD, 23–27 January 2023. https://doi.org/10.2514/6.2023-0534
8. Franz, T., Zimmermann, R., Görtz, S.: Adaptive sampling for nonlinear dimensionality reduction based on manifold learning. Model. Simul. Appl. **17**, 255–269 (2017). https://doi.org/10.1007/978-3-319-58786-8_16
9. Fuhg, J.N., Fau, A., Nackenhorst, U.: State-of-the-art and comparative review of adaptive sampling methods for Kriging. Arch. Comput. Methods Eng. **28**, 2689–2747 (2021). https://doi.org/10.1007/s11831-020-09474-6
10. Gardner, J.R., Pleiss, G., Bindel, D., Weinberger, K.Q., Wilson, A.G.: GPyTorch: blackbox matrix-matrix Gaussian process inference with GPU acceleration. In: Proceedings of the 32nd International Conference on Neural Information Processing Systems, pp. 7587–7597, Montreal, Canada, 3–8 December 2018
11. Guénot, M., Lepot, I., Sainvitu, C., Goblet, J., Coelho, R.F.: Adaptive sampling strategies for non-intrusive POD-based surrogates. Eng. Comput. (Swansea, Wales) **30**, 521–547 (2013). https://doi.org/10.1108/02644401311329352
12. Halder, R., Fidkowski, K.J., Maki, K.J.: Non-intrusive reduced-order modeling using convolutional autoencoders. Int. J. Numer. Meth. Eng. **123**, 5369–5390 (2022). https://doi.org/10.1002/nme.7072
13. Jiang, P., Shu, L., Zhou, Q., Zhou, H., Shao, X., Xu, J.: A novel sequential exploration-exploitation sampling strategy for global metamodeling. IFAC-PapersOnLine **48**, 532–537 (2015). https://doi.org/10.1016/j.ifacol.2015.12.183
14. Karcher, N., Franz, T.: Adaptive sampling strategies for reduced-order modeling. CEAS Aeronaut. J. **13**, 487–502 (2022). https://doi.org/10.1007/s13272-022-00574-6
15. Liu, H., Ong, Y.S., Cai, J.: A survey of adaptive sampling for global metamodeling in support of simulation-based complex engineering design. Struct. Multidiscip. Optim. **57**, 393–416 (2018). https://doi.org/10.1007/s00158-017-1739-8

16. Metropolis, N., Rosenbluth, A.W., Rosenbluth, M.N., Teller, A.H., Teller, E.: Equation of state calculations by fast computing machines. J. Chem. Phys. **21**, 1087–1092 (1953). https://doi.org/10.1063/1.1699114
17. Niederreiter, H.: Random number generation and Quasi-Monte Carlo methods. SIAM (1992). https://doi.org/10.1137/1.9781611970081.fm
18. Paszke, A., et al.: PyTorch: an imperative style, high-performance deep learning library. In: 33rd Conference on Neural Information Processing Systems (NeurIPS 2019), Vancouver, Canada, 8–14 December 2019
19. Rajaram, D., Perron, C., Puranik, T.G., Mavris, D.N.: Randomized algorithms for non-intrusive parametric reduced order modeling. AIAA J. **58**, 5389–5407 (2020). https://doi.org/10.2514/1.J059616
20. Sabater, C., Stürmer, P., Bekemeyer, P.: Fast predictions of aircraft aerodynamics using deep-learning techniques. AIAA J. **60**, 5249–5261 (2022). https://doi.org/10.2514/1.J061234
21. Storn, R., Price, K.: Differential evolution-a simple and efficient heuristic for global optimization over continuous spaces. J. Global Optim. **11**, 341–359 (1997). https://doi.org/10.1023/A:1008202821328
22. Tang, B.: Orthogonal array-based Latin hypercubes. J. Am. Stat. Assoc. **88**(424), 1392–1397 (1993). https://doi.org/10.1080/01621459.1993.10476423
23. Yang, M., Xiao, Z.: POD-based surrogate modeling of transitional flows using an adaptive sampling in Gaussian process. Int. J. Heat Fluid Flow **84**, 108596 (2020). https://doi.org/10.1016/j.ijheatfluidflow.2020.108596
24. Yu, J., Yan, C., Guo, M.: Non-intrusive reduced-order modeling for fluid problems: a brief review. Proc. Inst. Mech. Eng. Part G: J. Aerosp. Eng. **233**, 5896–5912 (2019). https://doi.org/10.1177/0954410019890721

GraphMesh: Geometrically Generalized Mesh Refinement Using GNNs

Ainulla Khan(✉), Moyuru Yamada, Abhishek Chikane, and Manohar Kaul

Fujitsu Research of India, Bangalore, India
{ainulla.khan,yamada.moyuru,abhishek.chikane,manohar.kaul}@fujitsu.com

Abstract. Optimal mesh refinement is important for finite element simulations, facilitating the generation of non-uniform meshes. While existing neural network-based approaches have successfully generated high quality meshes, they can only handle a fixed number of vertices seen during training. We introduce GraphMesh, a novel mesh refinement method designed for geometric generalization across meshes with varying vertex counts. Our method employs a two-step process, initially learning a unified embedding for each node within an input coarse mesh, and subsequently propagating this embedding based on mesh connectivity to predict error distributions. By learning a node-wise embedding, our method achieves superior simulation accuracy with reduced computational costs compared to current state-of-the-art methods. Through experimentation and comparisons, we showcase the effectiveness of our approach across various scenarios, including geometries with different vertex counts. We validated our approach by predicting the local error estimates for the solution of Poisson's equation.

Keywords: Finite Element Methods · Mesh Refinement · Graph Neural Networks

1 Introduction

In the fields of structural mechanics [2], aerospace [3], geophysics [4], and acoustics [5] physics simulation is a crucial tool, allowing for the exploration of phenomena that are otherwise challenging to study directly. For these applications, simulations rely on mathematical models represented mainly using partial differential equations (PDEs) [1]. These PDEs are solved on a specific geometric domain using Finite Element Method (FEM) [18]. Meshing is the initial step in FEM, wherein a physical structure is discretized into a mesh of finite elements. Each element represents a segment of the overall behavior of the structure. The method solves PDEs to provide the Finite Element (FE) solution by utilizing mesh of the structure. The effectiveness of FEM highly depends on the mesh quality. The mesh needs to be sufficiently detailed to capture critical structural behaviors, however finer meshes yield longer computations. Therefore, mesh refinement methods focus on generating an optimal mesh, where only

L. Franco et al. (Eds.): ICCS 2024, LNCS 14836, pp. 120–134, 2024.
https://doi.org/10.1007/978-3-031-63775-9_9

specific regions with simulation errors are refined for improving the simulation accuracy. The traditional mesh refinement method involves a multi-step process: initially generating coarse meshes for the geometry, followed by computing the FE solution on these initial meshes. Subsequently, a posteriori error estimation is conducted to refine the mesh elements in high-error regions. These steps repeat until the mesh meets the user's specified tolerance. Despite its effectiveness in generating non-uniform meshes, this method is computationally intensive because of the repeated execution of complex error estimation step [7]. Significant advancements have been witnessed through introduction of deep learning techniques in the area of mesh refinement to overcome the limitations of the traditional methods.

MeshingNet [9] is a significant work in the domain of mesh refinement using deep learning techniques [8]. It introduces a novel approach to mesh generation, employing deep neural networks to automate the mesh refinement. The key innovation in MeshingNet is its ability to generate high-quality meshes for geometries by learning from existing mesh datasets. This approach significantly reduces the reliance on manual meshing and expert knowledge. This approach uses Triangle software [6] for dataset generation, which involves Delaunay-based mesh refinement. The reported results outperform a non deep learning based method, ZZ estimation [11]. However, a crucial limitation of MeshingNet is the lack of the generalization capability to handle a number of edges that are different from those in training data, further the method accommodates only a single dimension for learning the boundary condition with other limited features.

There are other works that use graph neural networks for learning mesh-based simulations. But, they focus on simultaneously approximating physical quantities for time-dependent PDEs and perform Adaptive Mesh Refinement (AMR) on geometries across multiple time steps [12]. The method particularly excels in handling External Dynamic Lagrangian Systems, where mesh refinement occurs progressively across different time steps adapting progressively to external (time dependent) conditions. Our method is different from this work as our method can be used for various PDEs (e.g. Poisson's Equation, Linear Elasticity) while researchers in [12] consider only time-dependent PDEs. On the other-hand, GraphMesh deals with automatic mesh refinement on domains where the approach of AMR may not be applicable.

A recent work known as GMR-Net [13] has introduced an effective mesh refinement approach using Graph Convolutional Networks (GCNs) [17]. The above work leverages supervised learning specifically for elliptic PDEs. In this method, the model exhibits the capability to predict local error density at each nodal region of the coarse mesh with advanced input features based on polygon vertices to enhance prediction accuracy. However, a major limitation of this method lies in its inability to generalize over geometries characterized by a number of vertices beyond a predefined range. This limitation occurs due the use of a fixed input embedding size for features, with padding employed for accommodating a restricted range of polygonal geometries based on the number of vertices.

The current landscape of mesh refinement methodologies exhibits limitations in their performance, particularly when faced with tested geometries that deviate from the training set. Addressing this challenge necessitates the development of a model capable of accommodating input polygonal geometries with an arbitrary number of vertices. Such a model should not only demonstrate adaptability to diverse geometries but also excel in delivering accurate refinements while simultaneously optimizing simulation time.

In response to the existing limitations in mesh refinement methods, we present GraphMesh, a geometrically generalized mesh refinement model to handle polygonal geometries with any given numbers of vertices. Enclosed within a generalized Embedding-Decoder based two-stage framework GraphMesh is designed to refine meshes in two sequential steps, leading to the generation of high-quality optimal meshes. In the first step, GCN-based network is employed to embed the node-wise input features to a latent embedding. Subsequently, a second GCN employs a residual connections based architecture on the obtained node-wise latent embeddings to predict simulation errors for the solution of Poisson's equation. This two-stage process enhances the adaptability of our framework, ensuring effective performance across any shaped geometries.

Leveraging the capabilities of Graph Neural Networks, our proposed framework handles polygonal structures with varying numbers of edges, addressing a crucial limitation observed in existing methodologies. The essence of this approach is not only the model's adaptability to diverse polygonal configurations but also its significant impact on enhancing simulation accuracy while concurrently reducing simulation time. Our work distinguishes itself through the following contributions:

1. We propose a novel architecture that performs mesh refinement on any given geometric shape irrespective of its number of vertices, without re-training.
2. Our method shows superior results in the trade-off between simulation accuracy and simulation cost over the state of the art existing method.

The following section provides background details, followed by Sect. 3, which discusses our methodology. Section 4 outlines the experimental setup, while Sect. 5 delves into the results. The main findings from our experiments are then discussed in Sect. 6, and the conclusion, along with potential future applications of our research, is presented in Sect. 7.

2 Background

2.1 Mesh Generation

Mesh representations of shapes involve discretization of the domain leading to the generation of both CoM and FiM of the given geometry. In this work, we generated meshes by initially creating diverse input geometries through the methodology employed in [13] and subsequently utilizing the Gmsh software on the generated geometries. The geometry generation process first, entails the use of two concentric circles with proper spacing. Then the internal area formed by these circles is subsequently divided into N sections, where N represents the number of vertices required for the specific geometry under generation. Subsequently, within each section, an interior point is randomly generated which forms the polygon vertex, lastly joining of the points in each section leads to the formation of a unique geometric shape.

2.2 FE Solution and Simulation Error

The generated meshes are used for simulations, executed with the aid of an open-source FE solver FreeFEM [15]. The problem we address in this paper involves the solution of Poisson's equation $\nabla u + \varphi = 0$ within generated meshes with the boundary condition of $u = \alpha$ imposed on the polygon boundaries. Here, φ and α represents the PDE parameter and boundary value of the input meshes respectively. Following the methodology outlined by Minseong et al. [13], we estimated the simulation error values e at each nodal location of input CoM. The error estimation process utilizes the simulation results u_{CoM} and u_{FiM} corresponding to both CoM and FiM, respectively.

2.3 Characteristic Length

We facilitated the mesh refinement using refinement module of Gmsh. This refiner operates on the basis of an element's l, which can be dynamically adjusted to enhance accuracy. Specifically, the refinement process focuses on regions where the e around a given node is notably large. As the characteristic length decreases, meshes are locally refined in the vicinity of nodal region, optimizing the resolution in areas with significant error deviations. The optimal length around a node, is precisely defined by a user-defined parameter β as:

$$l = \frac{\beta}{e}. \tag{1}$$

Here, β plays a role in governing the refinement density of the meshes, offering users the flexibility to tailor the refinement process based on specific considerations and desired levels of accuracy.

3 Methodology

3.1 Overview

The core of our approach involves training a GNN model to learn a mapping between parameters, including geometry G, boundary conditions (BC), and the corresponding region-specific simulation error e. To enhance geometric generalization, we leverage the inherent properties of GNNs, which are independent of the number of nodes within a given graph. Specifically, we employ a two-step approach, consisting of the Vertex-Graph G^{nv} for each node n in the input CoM and the Main-Graph G^m representing the input CoM itself (Fig. 1). The graph G^{nv} captures embeddings for each CoM node n using geometric features with respect to polygon vertices and boundary, PDE features, regardless of the number of edges or vertices. The nodal features between the vertex x and node n of G^{nv} is represented by f_{nx}. Consequently, the graph G^m predicts e at each nodal region of the input CoM. The predicted simulation errors (denoted as $\hat{e}$) are then subjected to an established mathematical function (Eq. (1)), yielding an optimal length l. These optimal lengths are subsequently employed to guide the mesh refinement process.

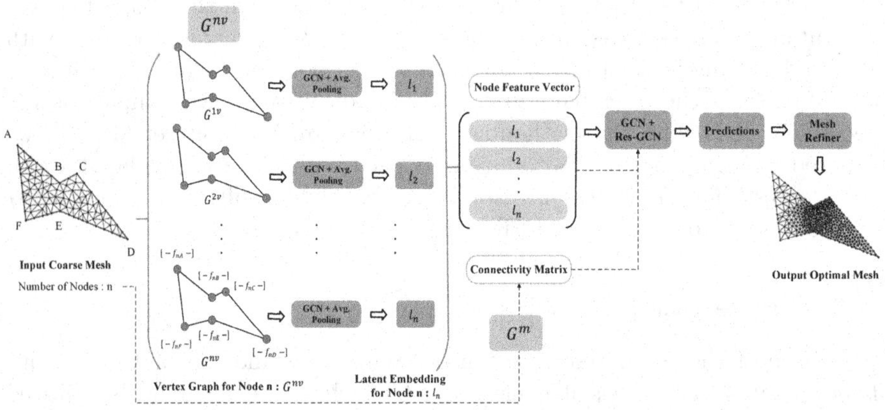

Fig. 1. Architectural Framework with the Vertex-Graph G^{nv} containing node features f_{nx} and Main-Graph G^m, demonstrated through an example input mesh.

By predicting e our proposed method provides control of mesh refinement through adjustments in mesh refinement parameters in Eq. 1. This presents a versatile and adaptable strategy for refining meshes at the user's discretion (by the choice of β), without necessitating re-training for each choice of l (see Eq. 1). Moreover, this approach also facilitates a rigorous evaluation methodology that enables the observation of the relationship between simulation accuracy and simulation time across varying refinement levels during inference. This not only provides insights into the trade-offs between accuracy and computational efficiency but also contributes to a more better understanding of the model's performance under different refinement configurations.

We employ a non-neural network based conventional method using the Gmsh software [14] for the generation of ground-truths, encompassing an input geometry, CoM and FiM of the geometry. Simulations are conducted on both the coarse and fine meshes, generating region-wise error distributions based on the L1-norm of simulation results. Subsequently, acknowledging the time-consuming and resource-intensive nature of fine mesh generation, the automation strategy is to use GNN-based supervised learning to predict the region-wise simulation errors.

3.2 Data Preparation

For training we use the geometric and PDE features extracted from the CoM. The geometric features, for defining node attributes, are mean value coordinates (MVC), vertex distance, and depth. These features contribute to a representation of nodal attributes, for extraction of the mesh characteristics during the error estimation process.

MVC. The MVC feature computes the weight of a polygon vertex p for the node n in the given input CoM. Considering an input CoM with k vertices the weight η_v for vertex v is computed as [16]:

$$\begin{cases} \eta_v = \frac{q_v}{\sum_{j=1}^{k} q_j} \\ q_v = \frac{tan(\theta_{v-1}/2)+tan(\theta_v/2)}{\|p_v - v\|} \end{cases}, \tag{2}$$

with θ_v = angle $\angle p_v v p_{v+1}$.

This feature, employed in learning the node's position relative to polygon vertices, demonstrates robustness against variations in polygon geometry [16].

Vertex Distance and Depth. The Vertex Distance feature provides insights into the number of hops separating each polygon vertex p from the node n in the given input CoM, offering information on spatial information. The Depth feature gives the closest polygon vertex p to the node n within the CoM [13].

PDE and Boundary Condition. The PDE features relates to the PDE parameter φ of the Poisson's equation and the boundary values α on all the boundary edges. For the boundary condition corresponding α value is used for the polygon vertex p.

Our model takes e, as the ground truth feature in our dataset. This unique approach aims to make our model independent of the refinement parameter β.

3.3 Architectural Details

3.3.1 Vertex Graph for Nodal Representations

The formation of the vertex graph G^{nv} is central to our methodology, aiming to obtain a node-wise representation of the overall polygon shape. The graph for a node n of input CoM is given by $G^{nv} = (V^{nv}, E^{nv})$ where G^{nv}represent the external geometry of the polygon. Here, vertices of the polygon represents nodes V^{nv} (A,B,C,D,E and F as given in Fig. 1), and polygon edges represent the links E^{nv} (Fig. 1).

The input node features f_{nx} for a given node n depends on polygon vertex x of the given input CoM. (see Sect. 3.2). Previous state-of-the-art method (GMR-Net), resorted to padding to handle these features only up to polygons with pre-defined 9 polygon vertices. Each node n in the graph used in GMR-Net comprised a 29-dimensional node feature vector, with 9, 9 and 9 dimensions allocated to MVC, Vertex Distance, and BC, respectively. Since for the node n these features are different with respect to each polygon vertex p. The remaining two dimensions were dedicated to depth and φ. Notably, a padding mechanism using -1 was employed to address polygons with fewer than 9 edges [13].

Our model introduces a novel concept by constructing a vertex graph for each node n of input CoM, aiming to accommodate features for any number of polygon vertices. This results in a fixed dimensional feature vector for the nodes of the vertex graph. In this configuration, the features f_{nx} of a node n concerning the polygon vertex x are specifically assigned to the corresponding node in the vertex graph. Consequently, a given input CoM consists of sub graphs equivalent to the number of nodes in the mesh. Furthermore, each sub graph contains nodes equivalent to the number of polygon vertices with a fixed dimensional node feature vector irrespective of the polygon geometry. This arrangement makes the model adaptable to diverse polygonal shapes, independent of number of edges without re-training. After the training phase, the latent embedding of the nodes in the input coarse mesh stores information corresponding to the polygon corners.

The vertex graph G^{nv} is designed to encode the polygon's distinct geometry into a representative embedding. Within this graph, the message-passing step within the GCN [17] gives an updated node feature in G^{nv} vector $x_i^{'}$ using the function given by:

$$x_i^{'} = f^v \bigoplus_{k \in N(i)} (x_k W). \tag{3}$$

where f^v denotes the implementation of MLP and $\bigoplus_{k \in N(i)}$ denote averaging over the one-hop neighbours $N(i)$ of node feature vector x_i with learnable parameter W. This function aggregates the neighborhood's information for each node via its connected edges. Further, to obtain a graph level latent representation for the nodes in main graph G^m, a subsequent average pooling layer is introduced over the nodes of the corresponding vertex graph G^{nv}. For information aggregation, we experimentally opted for a single-layered GCN with 128 hidden-dimensions on the vertex graph, hence integrating one-hop neighborhood-based information.

3.3.2 Main Graph for Prediction of Simulation Error

The latent representations of nodes within the CoM, learned from the vertex graph, act as the node features for the main graph G^m and the edge information is obtained using the linkages between nodes in CoM. Subsequently, the graph G^m is passed through a GCN architecture based on Residual Network [13]. The architecture employs information aggregation over a 6-hop neighborhood, featuring skip-connections every two layers of GCN, and a hidden dimension size of 128. This ResNets-based GCN enables the extraction of node-level e within the CoM. The proposed two-step process enables an initial compression of polygon vertex-based information into a fixed-sized latent embedding followed by leveraging this embedding for predicting e values. This introduces a generalization ability to our approach, making it universally applicable to various polygon geometries without any limitations concerning the number of polygon vertices.

3.4 Training and Evaluation

During the training phase, both of our graphs are trained using a supervised approach, with ground truth error values associated with each node in G_m. We train GraphMesh with the L1 loss function between the ground truth e and model predicted $\hat{e}$ across N nodes, as outlined in Eq. 4. To evaluate the model's performance, we calculate the average simulation error estimates A_e across nodes in the refined mesh. This involves, first obtaining node-wise l values using the model-predicted $\hat{e}$ with the help of Eq. 1. The Gmsh mesh refiner is then employed to generate refined meshes based on these l values. Subsequently simulations are conducted on these refined meshes, to compute A_e by comparing the simulation results of the model-predicted refined mesh with the corresponding FiM (Sect. 2.2) followed by node-wise averaging.

Furthermore, we assess methods for achieving optimal simulation accuracy within minimal time. In this evaluation, we generate multiple refined meshes with varying refinement levels by adjusting β values based on a single model-predicted $\hat{e}$. For each refined mesh, we compute A_e values while recording the simulation time. The results are visually represented, illustrating the variation in A_e and simulation time under this evaluation.

$$L = \frac{1}{N}\sum_{i=1}^{N} |\hat{e}_i - e_i|. \tag{4}$$

4 Experimental Setup

4.1 Dataset

The CoM in our study contained elements within the range of 250–550, with corresponding fine meshes FiM containing 15–20 times the number of elements present in their coarse mesh counterparts. The positional coordinates of each node in both the CoM and FiM were constrained within the 0 to 1 range. For GraphMesh, we generated train-validation dataset (Sect. 2.2) featuring 8000 training CoM and FiM, alongside 1500 validation CoM and FiM.

4.2 Experiment Details

To assess the performance of our proposed method, our experiments were performed using one Nvidia A30 24 GB GPU. We implemented our neural network and training scripts using PyTorch [19] and PyTorch Geometric [20]. The batch size was set to 32 and Adam optimizer [21] with a learning rate of 0.0005 was used for optimization. The training process spanned 100 epochs, and the corresponding loss curve is depicted in Fig. 2.

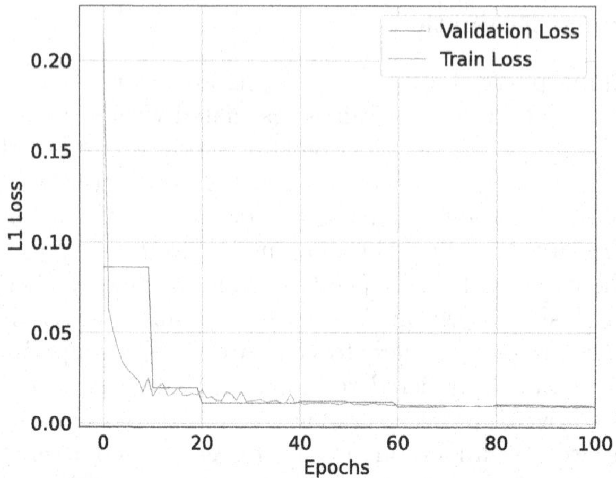

Fig. 2. Training and validation loss curves for GraphMesh

4.3 Governing Equation and Boundary Condition

The displacement field u represented the FE solution. The PDE parameter φ was randomly assigned values within the range of 0 to 1. Additionally, each boundary (α) in the mesh was subject to a random displacement ranging from 1 to 100. Further, the boundary values were normalized between 0 and 1 during training.

4.4 Evaluation Details

We conduct evaluation of our model across two distinct data settings. (1) In-distribution data setting, where both the training and validation sets comprise meshes with polygons having 6, 7, 8, and 9 vertices (E). (2) Out-of-distribution data setting, where the training set includes meshes of polygons with E equal to 6, 7, 8, and 9, and the validation set introduces meshes of polygons with E equal to 10 and 12. We generated refined meshes with model predicted $\hat{e}$ using Gmsh's l, with β set to $0.0084 * max(\hat{e})$, where $max(\hat{e})$ is the maximum of predicted simulation error values among the nodes of the corresponding mesh. The authors of [13] assessed their technique, GMR-Net, across various methods and proved its superiority. Therefore, we conduct a performance comparison between our model and the current state-of-the-art method, GMR-Net, both quantitatively and qualitatively, to encompass all existing methods. Qualitative assessments involve comparing mesh refinements, while quantitative evaluations employ simulations on output refined meshes from GraphMesh to obtain A_e across nodes of the refined mesh. Notably, while GraphMesh inherently handles out-of-distribution data settings, GMR-Net required feature down-sampling for inferencing. In the case of GMR-Net, features corresponding to vertices exceeding the 9-vertex limit are averaged out and equally distributed among the nine vertex features to facilitate fair comparisons without re-training.

Leveraging our method's independence from the refinement parameter β, we also conducted evaluations on a case with varying refinement levels. We set 10 values of β between $0.0084 * max(e)$ and $0.01 * max(e)$. The corresponding simulation error versus simulation time plots are observed across varying refinement level-based meshes.

5 Results

5.1 Mesh Comparison

In the evaluation of refinement results, we conducted a comparison using the data from both the in-distribution (Fig. 3) and out-of-distribution (Fig. 4) datasets, considering polygons with all polygon edges subjected to displacements as boundary conditions during training. The figures present refined meshes from GraphMesh and GMR-Net, along with an error distribution plot depicting the distribution of e on the evaluated geometries. The error distribution plot is obtained based on the comparison of simulation results between the CoM and FiM, as explained in Sect. 2.2. Ideally, the model should refine regions with simulation errors. Figures 3 and 4 highlight GraphMesh's accurate capture of refinement regions, resulting in the generation of improved non-uniform meshes.

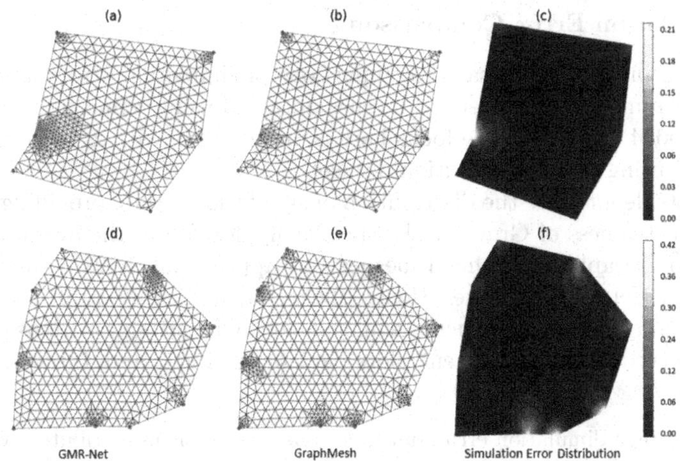

Fig. 3. Output refined meshes obtained for in-distribution dataset ($E = 6$ in (a),(b) and (c); $E = 9$ in (d),(e) and (f)) using the methods GMR-Net ((a) and (d)) and GraphMesh ((b) and (e)) with error distribution of the corresponding CoM ((c) and (f))

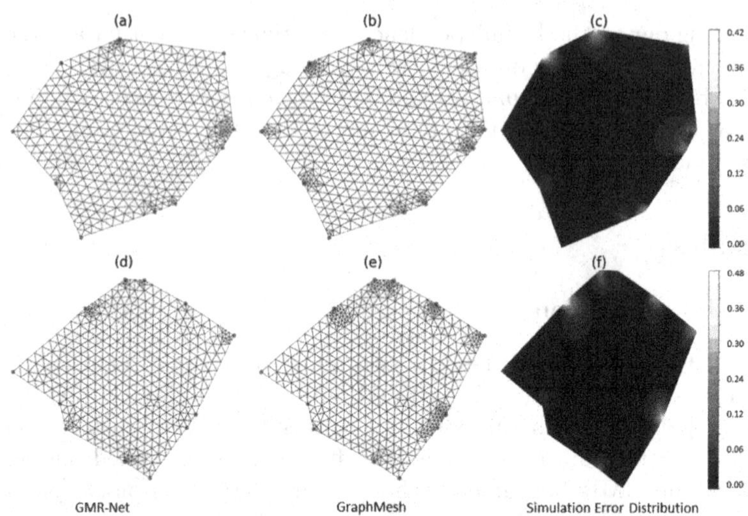

Fig. 4. Output refined meshes obtained for out-of-distribution dataset (E = 10 in (a),(b) and (c); E = 12 in (d),(e) and (f)) using the methods GMR-Net ((a) and (d)) and GraphMesh ((b) and (e)) with error distribution of the corresponding CoM ((c) and (f))

5.2 Simulation Error Comparison

In Fig. 5, we present quantitative comparisons of simulation errors across a set of 100 unseen meshes, with refined meshes obtained using refinement parameter equal to $0.0084 * max(e)$. The focus here is on polygons with E ranging from 6 to 9, constituting an in-distribution dataset.

The plots deomstrate the distribution of simulation errors, providing insights into the effectiveness of GraphMesh. Specifically, for $E = 6$ the average simulation error on GraphMesh refined meshes was approximately 21% smaller than that of GMR-Net refined meshes. Hence, the results highlight GraphMesh's capability to yield accurate refinements while efficiently identifying regions of errors, as also observed in the refinement plots depicted in Fig. 3 and Table 1.

Table 1. Average Simulation error and time comparison for in-distribution validation meshes

Metric	Method	E = 6	E = 7	E = 8	E = 9
Average Simulation Error	GMR-Net	0.0063	0.0051	0.0055	0.0062
	GraphMesh	**0.005**	**0.0044**	**0.0046**	**0.0051**
Average Simulation Time (s)	GMR-Net	0.0065	0.0073	0.0075	0.0076
	GraphMesh	**0.0047**	**0.0055**	**0.0059**	**0.0060**

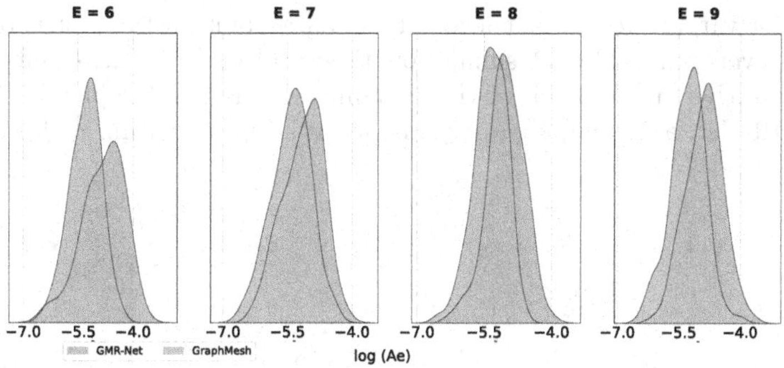

Fig. 5. Comparison of simulation errors for in-distribution dataset

In Fig. 6, we present quantitative results for the out-of-distribution dataset, involving polygons with 10 and 12 edges during inference. These results, obtained using the GMR-Net method (as shown in Fig. 6), were acquired through a down-sampling technique explained in Sect. 4.4. Figure 4 displays the qualitative evaluations, which further support the findings in Fig. 6. The proposed method accurately captures refinement regions, whereas GMR-Net fails to refine certain regions, leading to outcomes closer to a uniformly coarse mesh.

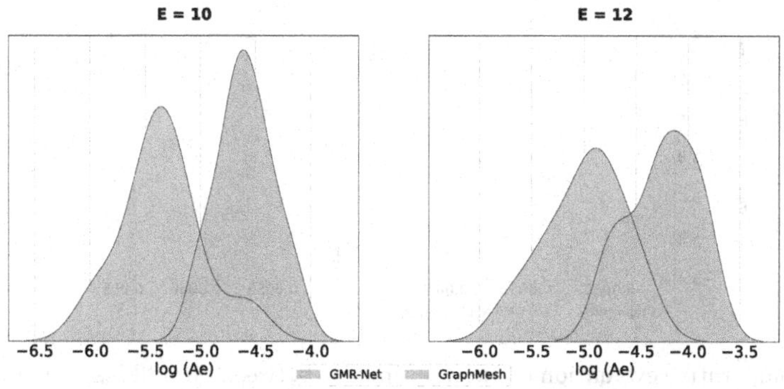

Fig. 6. Comparison of simulation errors for out-of-distribution dataset

5.3 Comparison for Various Simulation Times

Since mesh size depends on the refinement level parameter (β), the simulation time will also depend on β. Therefore, to plot errors at different times, we vary β between 0.0084 * max(e) and 0.01 * max(e) and record the corresponding simulation errors. Figures 7 and 8 compare GraphMesh and GMR-Net for both in-distribution and out-of-distribution datasets, in terms of a specific simulation accuracy for varying refinement levels. Each data point on the plot is obtained using the averaged values of simulation time and errors. The results clearly demonstrate that, under each of the 10 refinement levels, GraphMesh outperforms GMR-Net, achieving superior simulation accuracy within a shorter time frame.

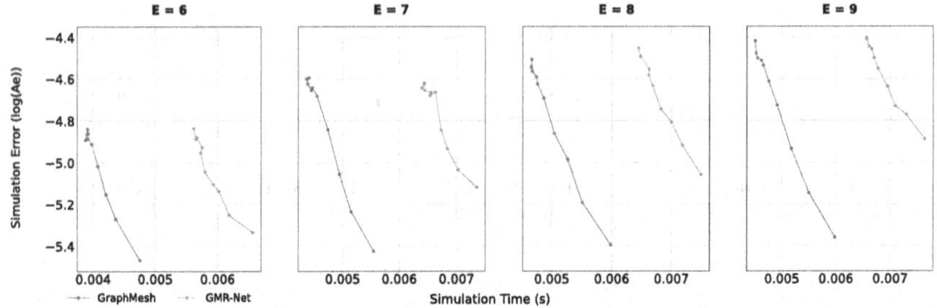

Fig. 7. Comparative evaluation of mesh optimality between GraphMesh and GMR-Net for in-distribution dataset

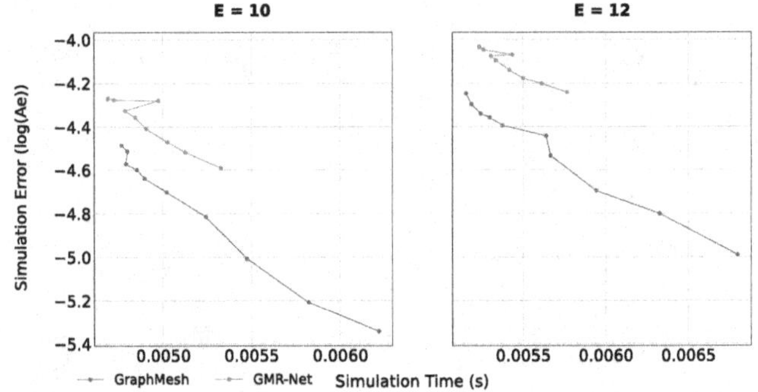

Fig. 8. Comparative evaluation of mesh optimality between GraphMesh and GMR-Net for out-of-distribution dataset

6 Discussion

We focus on several key aspects of GraphMesh that addresses limitations encountered in previous approaches. GMR-Net's ablation study had emphasized the significance of geometric node features based on input polygon vertices, restricting its adaptability to polygons within a fixed range of vertex counts. GraphMesh addresses this limitation by leveraging GNNs. The introduction of the vertex graph G^{nv} in GraphMesh allows for handling features from any input geometry, proving especially advantageous for intricate shapes where determining the precise number of edges might be challenging. Beyond generalization, the vertex graph favours learning of the inherent topology of a given polygon. The latent embeddings for each node, obtained through learnable aggregations, provide information about the node's positioning relative to polygon corners, boundaries, and PDE-defined conditions. This information is then used in the main graph, which utilizes the mesh connectivity information to predict node-wise simulation errors effectively. The generalization capabilities of GraphMesh across various datasets, both in-distribution and out-of-distribution, highlight the GraphMesh's improvement compared to GMR-Net under the considered scenarios.

7 Conclusion

This paper introduces GraphMesh, a novel methodology designed to enhance geometric generalization in the domain of automatic mesh refinement, specifically emphasizing its adaptability to out-of-distribution geometries. Our approach was tested through the solution of Poisson's equation across diverse scenarios, with comparative assessments against GMR-Net's output refined meshes. We presented comparisons that demonstrated the generalization ability and effectiveness of GraphMesh. Approximately, GraphMesh achieved an improvement in simulation accuracy of 21% with a 27% reduction in simulation time on comparison with existing state of the art method. Potential avenues for improvement include the incorporation of advanced topological features to address complexities in geometries and extensions to accommodate additional boundary conditions and diverse governing equations in various domains.

Disclosure of Interests. The authors do not have any competing interests that are applicable to the content presented in this article.

References

1. Zienkiewicz, O.C., Taylor, R.L., David, F.: The Finite Element Method for Solid and Structural Mechanics, 7th edn. Elsevier, New York (2014)
2. Panthi, S.K., Ramakrishnan, N., Pathak, K.K., Chouhan, J.S.: An analysis of springback in sheet metal bending using finite element method (FEM). J. Mater. Process. Technol. **186**, 120–124 (2007)
3. Economon, T.D., Palacios, F., Copeland, S.R., Lukaczyk, T.W., Alonso, J.J.: SU2: an open-source suite for multiphysics simulation and design. AIAA J. **54**, 828–846 (2016)
4. Zhengyong, R., Jingtian, T.: 3D direct current resistivity modeling with unstructured mesh by adaptive finite-element method. Geophysics **75**, H7–H17 (2010)
5. Steffen, M., Bodo, N.: Computational Acoustics of Noise Propagation in Fluids: Finite and Boundary Element Methods, vol. 578. Springer, Cham (2008). https://doi.org/10.1007/978-3-540-77448-8

6. Shewchuk, J.R.: Delaunay refinement algorithms for triangular mesh generation. Comput. Geom. **22**(1–3), 21–74 (2002)
7. Ainsworth, M., Oden, J.T.: A posteriori error estimation in finite element analysis. Comput. Methods Appl. Mech. Eng. **142**, 1–88 (1997)
8. Bank, R.E., Weiser, A.: Some a posteriori error estimators for elliptic partial differential equations. Math. Comput. **44**, 283–301 (1985)
9. Zhang, Z., Wang, Y., Jimack, P.K., Wang, H.: MeshingNet: a new mesh generation method based on deep learning. In: Krzhizhanovskaya, V.V., et al. (eds.) ICCS 2020. LNCS, vol. 12139, pp. 186–198. Springer, Cham (2020). https://doi.org/10.1007/978-3-030-50420-5_14
10. Zhang, Z., Jimack, P.K., Wang, H.: MeshingNet3D: efficient generation of adapted tetrahedral meshes for computational mechanics. Adv. Eng. Softw. **157**, 103021 (2021)
11. Zienkiewicz, O., Zhu, J.: Adaptivity and mesh generation. Int. J. Numer. Meth. Eng. **32**, 783–810 (1991)
12. Pfaff, T., Fortunatoet, M., Sanchez-Gonzalez, A., Battaglia, P.: Learning mesh-based simulation with graph networks. In: International Conference on Learning Representations, Vienna (2020)
13. Minseong, K., Jaeseung, L., Jibum, K.: GMR-Net: GCN-based mesh refinement framework for elliptic PDE problems. Eng. Comput. **39**, 3721–3737 (2023)
14. Geuzaine, C., Remacle, F.: Gmsh: a three dimensional finite element mesh generator with built-in pre- and post-processing facilities. Int. J. Numer. Meth. Eng. **79**, 1309–1331 (2009)
15. Hecht, F.: New development in FreeFem++. J. Numer. Math. **20**, 251–266 (2012)
16. Floater, M.S.: Mean value coordinates. Comput. Aided Geom. Des. **20**, 19–27 (2003)
17. Kipf, T.N., Welling, M.: Semi-supervised classification with graph convolutional networks. In: International Conference on Learning Representations (2017)
18. Abdelaziz, Y., Nabbou, A., Hamouine, A.: A state-of-the art review of the X-FEM for computational fracture mechanics. Appl. Math. Model. **33**, 4269–4282 (2009)
19. Paszke, A., et al.: PyTorch: an imperative style, high performance deep learning library. In: Advances in Neural Information Processing Systems (2019)
20. Fey, M., Lenssen, J.E.: Fast graph representation learning with PyTorch. In: ICLR Workshop on Representation Learning on Graphs and Manifolds (2019)
21. Kingma, D.P., Ba, J.: Adam: a method for stochastic optimization. In: International Conference on Learning Representations (2015)

Gradient Method for Solving Singular Optimal Control Problems

Mariusz Bodzioch(✉)

Faculty of Mathematics and Computer Science, University of Warmia and Mazury in Olsztyn, Olsztyn, Poland
mariusz.bodzioch@matman.uwm.edu.pl

Abstract. Solving an optimal control problem consists in finding a control structure and corresponding switching times. Unlike in a bang-bang case, switching to a singular control perturbs the control structure. The perturbation of one of the switching times affects any subsequent singular intervals in the control, as the trajectories move along different singular arcs with different values of singular controls. It makes the problem of finding optimal solutions extremely difficult. In this paper, we discuss a gradient method for solving optimal control problems, when singular intervals are present in the optimal structure. The method is based on applying the necessary conditions of optimality given by the Pontryagin Maximum Principle, where the control variable enters the Hamiltonian linearly. To demonstrate the method, we formulate a nonlinear optimal control problem and then, using the proposed algorithm, we solve the problem and find the optimal control structure and corresponding switching times. Lastly, we compare the results with results obtained using three popular optimisation modelling languages: Pyomo, AMPL and JuMP. These languages serve as interfaces for solving the optimal control problem with the non-linear optimisation algorithm Ipopt. Our case study shows that the presented method not only computes the switching times accurately, but also moves precisely along the singular arc.

Keywords: Gradient method · Optimal control · Singular control · Mathematical modelling

1 Introduction

In general, optimal control deals with the problem of finding a control that achieves a certain optimality criterion. Indeed, the need to control the dynamics of various objects arises in engineering, biological, ecological, and medical applications. In population control problems, for example, the aim is to minimise (or maximise) the size of one of the considered populations, however, other control objectives can also be considered. Solving an optimal control problem consists in finding a control structure and corresponding switching times. When singular controls do not appear in the optimal structure, computing bang-bang controls is quite simple. However, the appearance of singular controls makes the problem of finding optimal solutions extremely difficult.

L. Franco et al. (Eds.): ICCS 2024, LNCS 14836, pp. 135–149, 2024.
https://doi.org/10.1007/978-3-031-63775-9_10

Numerical methods for optimal control can be categorised into two main types: indirect and direct methods. Indirect methods are based on the Pontryagin Maximum Principle, where the optimal control problem is transcribed to a Hamiltonian boundary-value problem and then solved numerically using a differential-algebraic equation solver. On the other hand, direct methods involve approximating the problem and transforming it into a nonlinear programming problem, which is subsequently solved using well-developed software such as Ipopt [13]. For an overview and comparison of direct and indirect methods, we refer to [1,7,8,10] and references therein.

Note that, even if the method accurately computes the switching times, the obtained control may exhibit oscillations within the singular region, i.e. the control oscillates infinitely many times between the bounds [14]. This typically occurs when the optimal control consists of singular and bang-bang intervals. Thus, when applying a computational method without a priori information about the optimal control structure, both direct and indirect methods may produce incorrect results. In this paper, we present a gradient method for solving singular control problems in the presence of singular intervals in the optimal structure.

As an example, we formulate an optimal control problem for a heterogeneous population dynamics model with a non-standard objective functional introduced by us in [2]. This functional additionally penalises the size of the control-resistant population. From a biological standpoint, this is related to control-resistant subpopulations, such as chemotherapy-resistant cancer cells, virus-resistant, antibiotic-resistant strains of bacteria, or any pesticide-resistant insect populations. In our previous papers we showed that the resistance penalty gives rise to locally-optimal singular controls, see e.g. [3–5], as well as [9]. From a practical perspective, singular controls correspond to low-control time-varying control schedules. We analyse a simple model that allows us to use the gradient algorithm with the minimum required mathematical complexity. However, the algorithm can be applied to more detailed and complex models, where the control variable enters the Hamiltonian linearly.

This paper is organised as follows. Firstly, we formulate an optimal control problem and recall some formulas, concepts and notions, related to singular control. Next, we discuss gradient methods for finding optimal controls in the bang-bang form and derive a modified gradient method for solving optimal control problems when singular intervals are present in the optimal structure. We illustrate the applicability of the method using a simple mathematical model. Finally, we numerically solve a nonlinear optimal control problem, using both direct methods and the derived indirect method. Conclusions are drawn based on the performed numerical simulations.

2 Optimal Control Problem

A dynamic optimisation problem, in which the state $\mathbf{n} = \mathbf{n}(t)$ is linked in time to a control function $u = u(t)$, is called an optimal control problem. More precisely, the solution to ordinary differential equations $\dot{\mathbf{n}} = F(\mathbf{n}, u(t))$ is shaped by the control with an objective $J = J(u)$ that is optimised over all possible responses subject to external controls. Let us now formulate the problem to be considered.

The dynamics is defined by a continuously differentiable vector field $\mathbf{f}$ and control vector field $\mathbf{g}$ as

$$\dot{\mathbf{n}} = \mathbf{f}(\mathbf{n}) + \mathbf{g}(\mathbf{n})u(t), \tag{1}$$

$\mathbf{n} \in \mathbb{R}^n$, with initial condition $\mathbf{n}(0) = \mathbf{n}_0$. The vector field $\mathbf{f}$ represents the uncontrolled dynamics, while the vector field $\mathbf{g}$ represents the influence of the control on the system. We define the objective (or cost) functional in so-called Bolza form as the integral of a Lagrangian L and a penalty term M,

$$J(u(\cdot)) = M(T, \mathbf{n}(T)) + \int_0^T L(\mathbf{n}(t), u(t))dt, \tag{2}$$

where T is a fixed terminal time. Then, the optimal control problem is as follows: minimise the objective (2) over all admissible controls $u : [0, T] \to [0, u_{\max}]$ subject to dynamics (1) over an interval $[0, T]$.

Let $\mathbf{p} : [0, T] \to (\mathbb{R}^n)^*$. We define the so-called Hamiltonian function

$$H(\mathbf{p}, \mathbf{n}, u) = \mathbf{p}^T\left(\mathbf{f}(\mathbf{n}) + \mathbf{g}(\mathbf{n})u\right) + L(\mathbf{n}, u). \tag{3}$$

The Pontryagin Maximum Principle provides the first-order necessary conditions for optimality for our optimal control problem.

Theorem 1. *If u^* is an optimal control with corresponding trajectory $\mathbf{n}^*$, then there exists a co-state vector $\mathbf{p} : [0, T] \to (\mathbb{R}^n)^*$, which satisfies the adjoint equation*

$$\dot{\mathbf{p}} = -\frac{\partial H}{\partial \mathbf{n}}, \tag{4}$$

with terminal condition $\mathbf{p}(T) = \frac{\partial M}{\partial \mathbf{n}}(T, \mathbf{n}^*)$, such that the Hamiltonian H is minimised a.e. on $[0, T]$ by u^* along the optimal trajectory with a constant minimum value c, i.e.

$$H(\mathbf{p}(t), \mathbf{n}^*(t), u^*(t)) = \min_{0 \le v(t) \le u_{\max}} H(\mathbf{p}(t), \mathbf{n}^*(t), v(t)) \equiv c.$$

The adjoint equation, together with the model dynamics, forms a two-point boundary value problem that is related to the optimal control through the minimising condition. It should be pointed out that the system may have multiple solutions. To determine the globally optimal control, we need to find all of them.

If the Hamiltonian is linear in control, its minimising property motivates the definition of the switching function

$$\Phi(t) = \frac{\partial H}{\partial u}. \tag{5}$$

If u^* is an optimal control, then $u^*(t) = \begin{cases} 0 & \text{if } \Phi(t) > 0, \\ u_{\max} & \text{if } \Phi(t) < 0. \end{cases}$ Thus, the condition $\Phi(t) = 0$ is the first-order necessary condition for the Hamiltonian to be minimal.

If $\Phi(\tau) = 0$ for some $\tau \in [0, T]$, but Φ is non-zero in some neighbourhood of τ, then the function Φ changes sign at time τ and the optimal control switches between 0 and $u_{\max}$: from 0 to $u_{\max}$ if $\dot{\Phi}(\tau) < 0$ and from $u_{\max}$ to 0 if $\dot{\Phi}(\tau) > 0$. This type of control is called bang-bang control. If Φ and all its derivatives vanish identically on some interval, then the control admits intermediate values between 0 and $u_{\max}$ and we say that the control is singular. In general, determining all points where $\Phi(t) = 0$, $t \in [0, T]$, is extremely complicated.

In many practical problems, optimal controls consist of finite concatenations of bang-bang and singular controls. Generally, the standard procedure for determining the structure of the optimal control is to analyse the switching function and its derivatives. A convenient tool in this analysis is the Lie bracket. We have

$$\begin{aligned}\dot{\Phi} =&\mathbf{p}^T[\mathbf{f},\mathbf{g}] - \frac{\partial L}{\partial \mathbf{n}}^T \mathbf{g}, \\ \ddot{\Phi} =&\mathbf{p}^T\left[\mathbf{f},[\mathbf{f},\mathbf{g}]\right] - \frac{\partial L}{\partial \mathbf{n}}^T\left([\mathbf{f},\mathbf{g}] + \frac{\partial \mathbf{g}}{\partial \mathbf{n}}\mathbf{f}\right) - \mathbf{f}^T\frac{\partial^2 L}{\partial \mathbf{n}^2}\mathbf{g} \\ &+\left(\mathbf{p}^T\left[\mathbf{g},[\mathbf{f},\mathbf{g}]\right] - \frac{\partial L}{\partial \mathbf{n}}^T\frac{\partial \mathbf{g}}{\partial \mathbf{n}}\mathbf{g} - \mathbf{g}^T\frac{\partial^2 L}{\partial \mathbf{n}^2}\mathbf{g}\right)u.\end{aligned} \tag{6}$$

The Lie bracket of two differentiable vector fields $\mathbf{f}$ and $\mathbf{g}$ can be defined as $[\mathbf{f},\mathbf{g}](\mathbf{n}) = \frac{\partial \mathbf{g}}{\partial \mathbf{n}}\mathbf{f}(\mathbf{n}) - \frac{\partial \mathbf{f}}{\partial \mathbf{n}}\mathbf{g}(\mathbf{n})$. For the derivative of these formulas we refer to [9].

If a control is singular on some interval $I \subset [0, T]$, then $\Phi = \dot{\Phi} = \ddot{\Phi} = ... = 0$ identically on I. If we solve these equations for the co-state variable $\mathbf{p}$ and the term multiplying u is nonzero, the equation $\ddot{\Phi} = 0$ can be solved for u, determining the singular control

$$\begin{aligned}-\mathbf{p}^T\left[\mathbf{f},[\mathbf{f},\mathbf{g}]\right] + \frac{\partial L}{\partial \mathbf{n}}^T\left([\mathbf{f},\mathbf{g}] + \frac{\partial \mathbf{g}}{\partial \mathbf{n}}\mathbf{f}\right) + \mathbf{f}^T\frac{\partial^2 L}{\partial \mathbf{n}^2}\mathbf{g}& \\ = \left(\mathbf{p}^T\left[\mathbf{g},[\mathbf{f},\mathbf{g}]\right] - \frac{\partial L}{\partial \mathbf{n}}^T\frac{\partial \mathbf{g}}{\partial \mathbf{n}}\mathbf{g} - \mathbf{g}^T\frac{\partial^2 L}{\partial \mathbf{n}^2}\mathbf{g}\right)&u_{\text{sing}},\end{aligned} \tag{7}$$

as a feedback formula that only depend on the state variables $\mathbf{n}$ of the system (1), and does not depend on the co-state variables $\mathbf{p}$ of the system (4).

Singular control, however, is not necessarily minimising, but it can also be maximising. In that case, instead of being the optimal control, it would represent the worst possible option (see [9]). The strengthened Legendre-Clebsch condition provides a high-order necessary condition for the optimality of singular controls, allowing the distinction between these two classes. If a minimising control u is singular of order 1 on an open interval $I \subset [0, T]$, then $\frac{\partial}{\partial u}\frac{d^2}{dt^2}\frac{\partial H}{\partial u} < 0$ for all $t \in I$.

Three main types of solutions to single-input optimal control problems can be distinguished [6]: (i) singular controls do not exist, (ii) singular controls exist and are locally minimising (the strengthened Legendre-Clebsch condition is satisfied, $\frac{\partial}{\partial u}\frac{d^2}{dt^2}\frac{\partial H}{\partial u} < 0$), and (iii) singular controls exist, but are locally maximising (the strengthened Legendre-Clebsch condition is not satisfied, $\frac{\partial}{\partial u}\frac{d^2}{dt^2}\frac{\partial H}{\partial u} > 0$).

In case (i), the optimal control is bang-bang with a small number of switchings that can be easily established. In case (ii), the optimal control consists of both bang-bang and singular intervals, where a bang-bang control with a larger number of switchings is not optimal. In case (iii), the optimal control consists of a potentially very large number of switchings. The singular arc is the limit of bang-bang trajectories where the number of switchings increases and tends to infinity. In this limit, the singular arc corresponds to scenario (ii), which is not optimal. Thus, in this case, more switchings do not improve the result and typically the solution is bang-bang with a small number of switchings. This scenario is by far the most difficult of the three scenarios.

3 Gradient Methods for Finding Optimal Control

The computation of optimal controls using numerical methods is often inefficient and may fail when optimisation is performed without prior knowledge of their structure. When singular controls are not present in the optimal structure, computing bang-bang controls is quite simple. Following the ideas presented in [12] (cf. [9]), optimal switching times for bang-bang control problems with a specified number of switches can be found using iterative methods. In these methods, the gradient of the objective functional with respect to the switching times is computed in each iteration. The number of switches is taken arbitrarily at the beginning of the algorithm. Results obtained for different numbers of switches are then compared to identify the optimal solution. Indeed, defining an upper limit for the number of switches is always possible since too many switches are not applicable. In brief, we can describe this approach as follows. Arbitrarily select switching times $0 = t_0 < t_1 < ... < t_k < t_{k+1} = T$. Let u_i denote the value of the bang-bang control on the interval $[t_i, t_{i+1}]$ for $i = 0, ..., k$. Note that the control's value on the first interval determines the sequence $u = (u_1, ..., u_k)$. Solve the state equations (1) for a given control u. Using the transversality condition, $\mathbf{p}(T) = [\omega_1, \omega_2]^T$, the adjoint co-state variables $\mathbf{p}$ can be computed by integrating the adjoint equation (4) backward in time (since the equations depend only on the state variables $\mathbf{n}$). Subsequently, using formulas (5) and (6), the switching function Φ and its derivative $\dot{\Phi}$ can be easily evaluated at the switching times t_i. By introducing small changes, iteratively update the switching times until they agree with the zeros of the computed switching function. Formally, when a control u is perturbed by δu, the first variation δJ can be found as

$$\delta J = J(u + \delta u) - J(u) \approx \int_0^T \frac{\partial H}{\partial u} \delta u dt.$$

Note that the derivative with respect to δu is actually given by the switching function $\Phi = \frac{\partial H}{\partial u}$. Depending on the value of the switching function at the switching time t_i for the control used, increase or decrease the lengths of the intervals $[t_{i-1}, t_i]$, where the increment is simply taken as

$$\delta t_i = (-1)^i \alpha \frac{\partial H}{\partial u}\bigg|_{t=t_i} = (-1)^i \alpha \Phi(t_i).$$

Here, the parameter α represents an adaptive step-size parameter (learning rate) ensuring the convergence of the procedure.

However, the method described above cannot be directly applied if locally optimal singular controls exist. The presence of any subsequent singular interval in the control will perturb the control structure and switching times. Furthermore, switching to singular control will cause the trajectory to follow a different singular arc with different values of the singular control.

Therefore, we propose a modified version of the aforementioned algorithm (cf. [2]). Suppose that the structure of a control is fixed, meaning that the optimal control consists of a finite sequence with elements from 0, 1, and S, corresponding to no control, full control and singular control, respectively. However, the structure of the optimal control does not provide any information about the switching times. Treating the objective functional as a function of switching times, with a slight abuse of notation, an approximate gradient of the objective functional with respect to the switching times can be computed using finite differences

$$\frac{\partial J}{\partial t_i} \approx \frac{J(t_1, ..., t_i + \Delta, ..., t_k) - J(t_1, ..., t_i, ..., t_k)}{\Delta} \tag{8}$$

for some positive constant $\Delta \ll 1$. Algorithm 1 describes the procedure.

Algorithm 1. Gradient method.

1. Assume k, structure of optimal control, initial switching times $\mathbf{t} = (t_1, ..., t_k)$ and initial value of learning rate α.
2. Solve the state equations (1) for a given u (defined by the assumed structure of optimal control), taking into account the singular control form, defined by (7).
3. Compute the co-state variables $\mathbf{p}$ by integrating the adjoint equation (4) backward in time using the terminal condition.
4. Evaluate the Hamiltonian H given by (3) along the controlled trajectory.
5. Compute the gradient $\nabla J(\mathbf{t})$, according to formula (8).
6. Adapt the learning rate α to ensure the gradient descent.
7. Compute new switching times $\mathbf{t} - \alpha \nabla J(\mathbf{t})$.
8. Repeat steps 2–7 until a prescribed tolerance has been reached.

To ensure gradient descent, it is beneficial to reduce the learning rate as the training progresses. This can be done by implementing predefined learning rate schedules or employing adaptive learning rate algorithms. In our approach, we utilise the Adaptive Moment Estimation algorithm (Adam) and, if necessary, its modifications. However, other methods for adjusting the learning rate can also be considered (see e.g. [11]). It is important to note that, for the gradient to be well-defined, we typically require that $t_{i+1} - t_i > \Delta$ for all i. However, during iterations, the switching times change and potentially cross, i.e. $t_{i+1} - t_i$ may fall below Δ for some i. In this case the corresponding interval is removed from the control structure, resulting in trajectories with a reduced number of switchings. In general, gradient methods are more suitable for finding local rather than global minima. Thus, to ensure optimality, the method should be run with different starting control structures and several random initial switching times for each control structure.

4 Numerical Computations of Optimal Controls

4.1 Mathematical Model

As an example, we consider a simple mathematical model describing the coexistence of control-sensitive and control-resistant subpopulations of the same species. The growth of the first one follows the exponential growth function, while the second population growths according to the logistic law, sharing resources with the first population. We assume that they differ in their reaction to control. In other words, we consider a scenario in which the growth of the control-sensitive subpopulation limits the growth of the control-resistant subpopulation. Intentionally, we choose as simple as possible mathematical model, because our aim is to investigate how the proposed objective functional affects the structure of the optimal control. The non-dimensional model considered in this study is as follows

$$\begin{aligned} \dot{n}_1 &= \lambda_1 n_1 - n_1 u, \\ \dot{n}_2 &= \lambda_2 n_2 (1 - n_2 - n_1), \end{aligned} \tag{9}$$

where n_1 and n_2 are the non-dimensional sizes of control-sensitive and control-resistant subpopulations, respectively; $u = u(t)$, $u : [0, T] \to [0, 1]$ is the non-dimensional control; T is the time horizon; λ_1 and λ_2 are growth rates.

4.2 Nonlinear Objective Functional

Mathematical models typically represent some underlying processes arising from medical, biomedical, physical, economical, or engineering problems. However, the form of objective functional is artificially imposed from the outside and usually there exist several options that can be used in a particular situation. The form of the objective depends on whether the system response and properties are satisfactory with respect to other criteria that were not included in the model dynamics. Let us now define the following objective functional

$$\begin{aligned} J(u(\cdot)) &= M\left(T, n_1(T), n_2(T)\right) + \int_0^T L\left(n_1(t), n_2(t), u(t)\right) dt \\ &= \omega_1 n_1(T) + \omega_2 n_2(T) \\ &\quad + \int_0^T \left(\eta_1 n_1(t) + \eta_2 n_2(t) + \xi G\left(\tfrac{n_2(t) - n_1(t)}{\epsilon}\right) + \theta u(t)\right) dt, \end{aligned} \tag{10}$$

where ω_1, ω_2, η_1, η_2, ξ, θ are non-negative parameters (weights), $\epsilon > 0$, and $G : \mathbb{R} \to (0, 1)$ is a twice continuously differentiable function. Terms $\omega_1 n_1(T)$, $\omega_2 n_2(T)$ penalise the size of the entire population at the end of the assumed fixed control interval $[0, T]$, while $\eta_1 n_1(t)$, $\eta_2 n_2(t)$ penalise the size of the entire population during the control. The linear term $\theta u(t)$ represents the overall amount of control given and penalises side-effects (cumulative negative effects, toxicity or costs) related to their usage. It should be pointed out that the linear term containing the control u has a clear biological meaning, but makes the mathematical analysis difficult. The non-standard term $\xi G\left(\frac{n_2(t)-n_1(t)}{\epsilon}\right)$ is an activation function introduced to penalise time periods during which the population is control-resistant ($n_2 > n_1$). We require the following properties

1. $G(x) \to 0$ as $x \to -\infty$; $G(x) \to 1$ as $x \to \infty$;
2. $G'(x) > 0$ for all x; $xG''(x) < 0$ for $x \neq 0$;
3. $G(0) = \frac{1}{2}$ and $G'(0) = \frac{1}{2}$.

Notice that $G(x) \approx 1$ whenever $x \gg 0$ and $G(x) \approx 0$ when $x \ll 0$. Thus, the term increases whenever $n_2 > n_1$ and stays roughly constant if $n_1 < n_2$. The function G can be thought of as a smoothed version of the Heaviside function, where the parameter ϵ controls the steepness of the slope, i.e. it determines how close the control-resistant population needs to become to the control-sensitive one for the penalty to be applied. The concept of this function was introduced by us in [2]. For numerical purposes, we can take $\frac{1}{2}\left(1 + \tanh \frac{x}{\epsilon}\right)$ as the function of G, as shown in Fig. 1. However, other types of activation functions can also be applied.

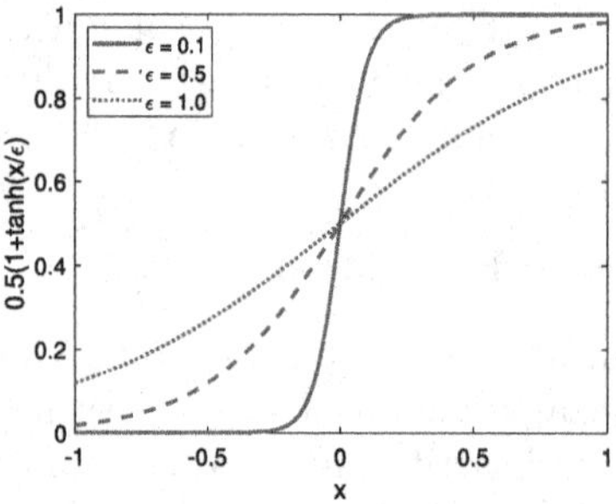

Fig. 1. Activation function $G(x) = \frac{1}{2}\left(1 + \tanh \frac{x}{\epsilon}\right)$

4.3 Optimal Control Problem

As it was mentioned earlier, our goal is to control population dynamics (growth). The optimal control problem may be formulated as follows: for a fixed terminal time T find a measurable function $u : [0, T] \to [0, 1]$ minimising the objective functional (10) subject to dynamics (9).

Let us introduce the following notations

$$\mathbf{n} = \begin{bmatrix} n_1 \\ n_2 \end{bmatrix}, \quad \mathbf{f}(\mathbf{n}) = \begin{bmatrix} \lambda_1 n_1 \\ \lambda_2 n_2 (1 - n_2 - n_1) \end{bmatrix}, \quad \mathbf{g}(\mathbf{n}) = \begin{bmatrix} -n_1 \\ 0 \end{bmatrix}, \quad \mathbf{p} = \begin{bmatrix} p_1 \\ p_2 \end{bmatrix}.$$

Following the theory described in Sect. 2, by elementary but tedious calculations, we get

$$H = p_1(\lambda_1 - u)n_1 + p_2\lambda_2 n_2(1 - n_2 + n_1) + \eta_1 n_1 + \eta_2 n_2 + \tfrac{\xi}{\epsilon} G'\left(\tfrac{n_2 - n_1}{\epsilon}\right) + \theta u,$$

where p_1 and p_2 satisfy

$$\begin{aligned} \dot{p}_1 &= -p_1(\lambda_1 - u) + p_2\lambda_2 n_2 - \eta_1 + \tfrac{\xi}{\epsilon} G'\left(\tfrac{n_2 - n_1}{\epsilon}\right), \\ \dot{p}_2 &= -p_2\lambda_2(1 - 2n_2 - n_1) - \eta_2 - \tfrac{\xi}{\epsilon} G'\left(\tfrac{n_2 - n_1}{\epsilon}\right), \end{aligned} \tag{11}$$

with terminal conditions $p_1(T) = \omega_1$, $p_2(T) = \omega_2$. From the definition of the switching function Φ and relations for its derivatives, we can state that the singular control has the following form

$$u_{\text{sing}} = \lambda_1 - \frac{\lambda_2 n_2(1 - n_2 - n_1)}{n_1} + \frac{\epsilon^2\lambda_2(\eta_2 - \eta_1)n_2 + 2\epsilon\xi\lambda_2 n_2 G'\left(\frac{n_2 - n_1}{\epsilon}\right)}{\xi n_1 G''\left(\frac{n_2 - n_1}{\epsilon}\right)} \tag{12}$$

and the corresponding singular trajectory lies in the following singular arc

$$\begin{aligned} 0 = F_{\text{arc}}(n_1, n_2; c) := \epsilon\,(\eta_1 + \lambda_1\theta - c + (\eta_2 - \eta_1)n_2) \\ + \epsilon\xi G\left(\tfrac{n_2 - n_1}{\epsilon}\right) - \xi(1 - n_1 - n_2)G'\left(\tfrac{n_2 - n_1}{\epsilon}\right). \end{aligned} \tag{13}$$

This is a consequence of the Pontryagin Maximum Principle and the condition of constancy of the Hamiltonian, $H \equiv c = const.$ Note that the above formulas are obtained as feedback formulas that only depend on the state variables $\mathbf{n}$ of the system (9), but do not depend on the co-state variables $\mathbf{p}$.

As it was mentioned earlier, $\frac{\partial H}{\partial u} = \Phi$, thus to verify if the Legendre-Clebsch condition is fulfilled along the singular arc, we need to determine the coefficient next to the control u in the expression for $\ddot{\Phi}$, i.e. $\frac{\partial}{\partial u}\frac{d^2\Phi}{dt^2} = -\frac{n_1^2}{\epsilon^2}\xi G''\left(\frac{n_2 - n_1}{\epsilon}\right)$. Using the assumptions about G, we see that the Legendre-Clebsch condition is satisfied only if $n_1 > n_2$.

Regarding the problem under consideration, we can have any of these three cases: (i) if the term G that penalises the resistance is omitted from the objective functional ($\xi = 0$), the singular control does not exist and the optimal control is bang-bang, (ii) if $n_1 > n_2$, then the Legendre-Clebsch condition is satisfied and the optimal control consists of bang-bang and singular intervals, and (iii) if $n_1 < n_2$, then the Legendre-Clebsch condition is not satisfied and the singular control is not optimal.

4.4 Numerical Optimisation

In this section we solve the optimal control problem using numerical computations. The goal is to minimise the objective functional under the model dynamics with the terminal non-dimensional time T chosen to be 14. To solve the optimal control problem and compute the optimal solution, we use the gradient method described in the previous section. Possible optimal control structures for different numbers of switches k are listed in Table 1.

Table 1. Possible structures of optimal controls for different numbers of switches k. Characters 0, 1, S denote no control, full control and singular control

k	possible structures											
0	0				1				S			
1	01		$0S$		10		$1S$		$S0$		$S1$	
2	010	$01S$	$0S0$	$0S1$	101	$10S$	$1S0$	$1S1$	$S01$	$S0S$	$S10$	$S1S$
3	0101	$01S0$	$0S01$	$0S10$	1010	$10S0$	$1S01$	$1S10$	$S010$	$S0S0$	$S101$	$S1S0$
	$010S$	$01S1$	$0S0S$	$0S1S$	$101S$	$10S1$	$1S0S$	$1S1S$	$S01S$	$S0S1$	$S10S$	$S1S1$

The gradient method was run for each of the control structures with several random initial switching times. An adaptive learning rate algorithm, specifically the Adam Optimisation Algorithm, was employed to ensure effective gradient descent. The initial value of α was empirically set in the range of $0.0001-0.001$, while the exponential decay rates for the first and second moments were set within the range of $0.9-0.9999$. However, alternative methods for adjusting the learning rate could also be considered. To solve equations and find switching times numerically, the standard MATLAB solver ode45 and the MATLAB event model were used with an error tolerance set to 10^{-9}. Each simulation used the following parameter values: $\lambda_1 = 0.2$, $\lambda_2 = 0.1$, $\omega_1 = 5$, $\omega_2 = 10$, $\eta_1 = 2$, $\eta_2 = 3$, $\theta = 0.5$, $\epsilon = 0.01$, $\xi = 1$.

For the initial condition $[0.44, 0.05]$, representing an initially control-sensitive population, it was found that the minimal value of the objective functional, 11.8659, is achieved for a control of type $1S1$ with two switching times: 2.0822 and 13.8674. By evaluating the Hamiltonian H along the obtained controlled trajectory and solving the adjoint equations (11) backwards in time using the terminal condition, we verified that the Hamiltonian is indeed constant ($H \equiv c = 1.0112$, constant up to the method order). Additionally, we confirmed that the switching function has appropriate signs. The optimal solution with corresponding control, trajectory and switching function are shown in Fig. 2. The optimal control starts with a full control interval as it penalises the control-sensitive subpopulation. Then, the control switches to singular part, defined by (12). As depicted in Figs. 2a and 2c, a singular control maintains the size of the control-sensitive subpopulation just above the size of the control-resistant one. It follows directly from the penalty-activation function G included in the objective functional. The singular interval in the middle is crucial for preserving the sensitive subpopulation. The control ends with a full dose interval to penalise the population size at the terminal time. The red dotted curve in Fig. 2c represents the singular arc, given by (13), where the right part minimises, while the left part – maximises the objective. By computing the derivative of the switching function on co-state and state variables, we determine that in the region between the two parts of the singular arc, the control can only switch from 0 to 1, while in the other regions (outside), it may switch only from 1 to 0.

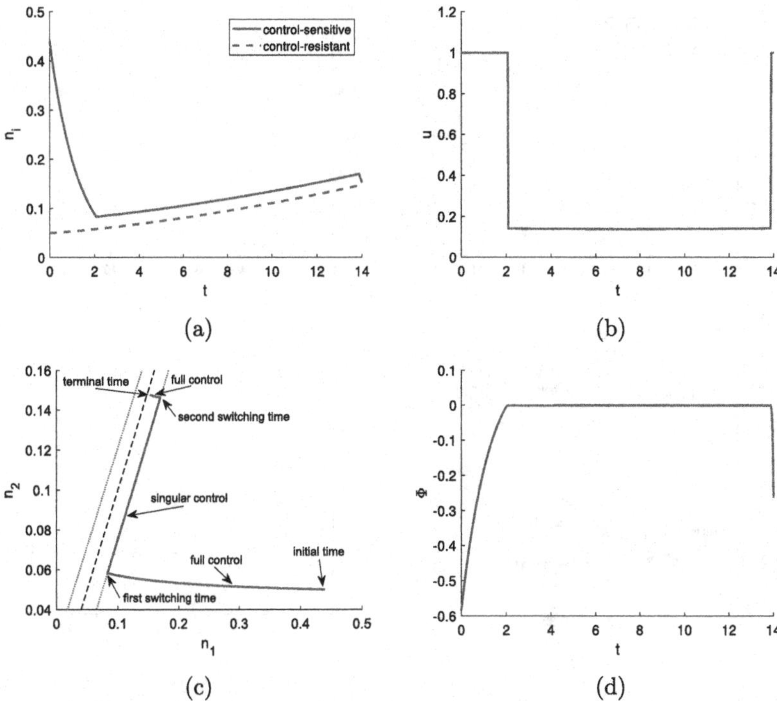

Fig. 2. Optimal solution (a), together with the corresponding optimal control (b), trajectory (c) and the switching function (d). Red dotted curve in (c) depicts the singular arc. Initial condition was chosen to be $n_1(0) = 0.44$, $n_2(0) = 0.05$ (Color figure online)

The initial condition [0.24, 0.25] was chosen to represent an initially control-resistant population. In this case, the Legendre-Clebsch condition is not satisfied (for $n_2 > n_1$), indicating that the optimal control starting from the singular interval cannot be optimal. This reduces the number of possible control structures. In this scenario, the optimal trajectory was numerically found to be $0S1$. The minimal value of the objective cost is 30.9052, while switching times are 1.0180 and 13.5811. The optimal solution, corresponding control, trajectory and switching function are shown in Fig. 3 (the constant value of $H \equiv c = 2.2095$). The control begins with a no-control interval, and when the population becomes sufficiently control-sensitive, a singular control is used.

4.5 Comparative Analysis

Now, we utilise three popular languages of optimisation modelling:

- Pyomo: A Python-based open-source optimisation modelling language with its differential algebraic equation extension Pyomo.DAE.

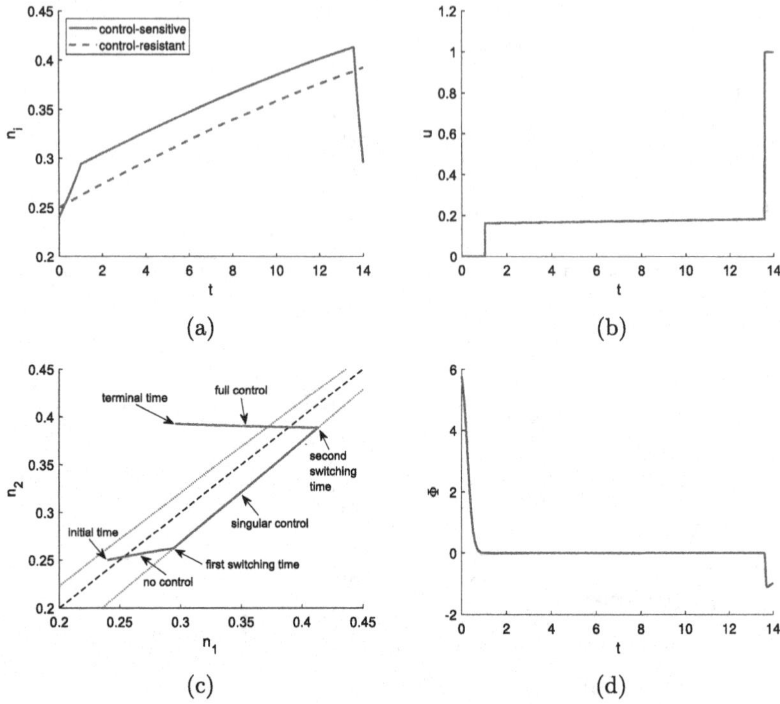

Fig. 3. Optimal solution (a), together with the corresponding optimal control (b), trajectory (c) and the switching function (d). Red dotted curve in (c) depicts the singular arc. Initial condition was chosen to be $n_1(0) = 0.24$, $n_2(0) = 0.25$ (Color figure online)

- AMPL: A Mathematical Programming Language, specifically designed for expressing and solving mathematical programming problems.
- JuMP: An algebraic modelling language that is a collection of supporting packages for mathematical optimisation embedded in the Julia language.

These modelling languages serve as interfaces for solving the optimal control problem using the non-linear optimisation algorithm Ipopt, with a regularly spaced grid of 1000 points and the forward Euler method. Ipopt is an implementation of an interior point method and is capable of solving constrained non-linear programming problems (see [13] for details).

Figures 4a–4b depict the results obtained using the Pyomo modelling language with the Ipopt solver, for initial conditions $[0.44, 0.05]$ and $[0.24, 0.25]$, respectively. When penalising the side effects (control costs, $\theta = 0.5 > 0$), a control with a large number of "fast" switchings is obtained. It was verified that increasing the number of discretisation points (grid points) increases the number of switching points and has virtually no effect on the solution. The optimal singular trajectory is the limit of bang-bang trajectories with an increasing number of switchings tending to infinity. It is important to note that using the approx-

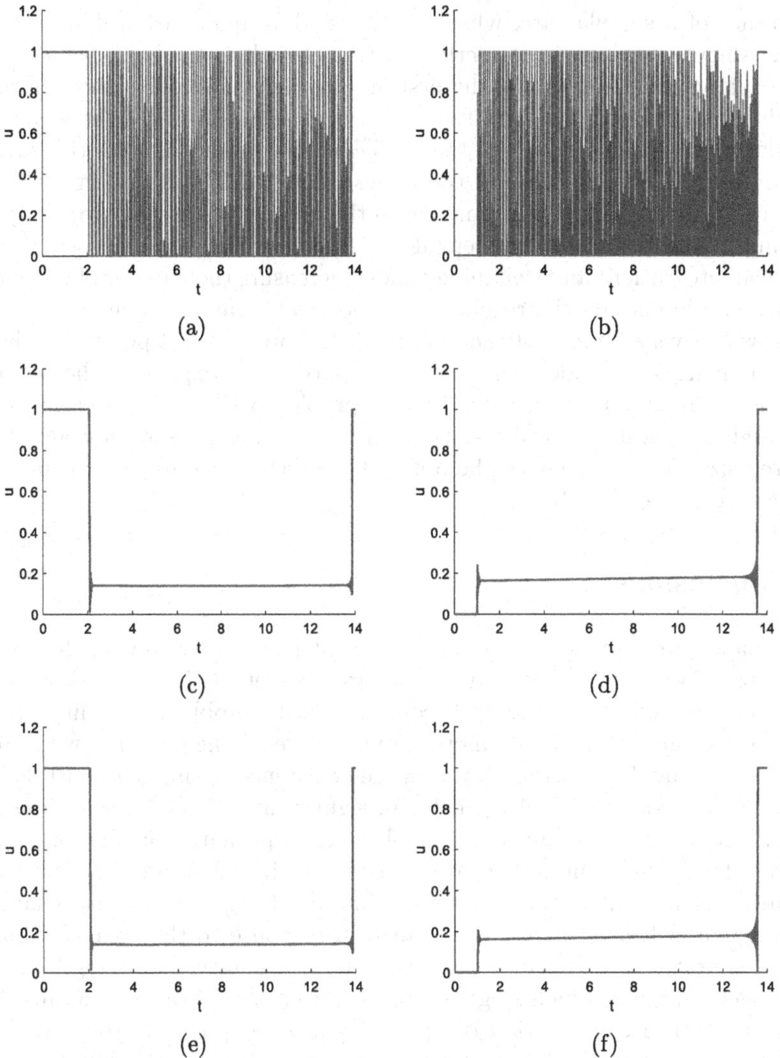

Fig. 4. Optimal controls of the model (9)–(10) obtained by Pyomo (a)–(b), AMPL (c)–(d) and JuMP (e)–(f). The initial condition in (a), (c) and (e) was set to be [0.44, 0.05], while in (b), (d) and (f) was set to be [0.24, 0.25]

imation method to determine a singular arc based on oscillations is generally not feasible. This phenomenon supports the concept of "chattering" (see [14]), where the control oscillates many times between the bounds. Figures 4c–4d show the results obtained using the AMPL language, while Figs. 4e–4f showcase the result obtained using the JuMP language. These results closely align with those obtained using the gradient method, but numerical artifacts on the singular interval can still be observed. This should be treated as artifacts caused by the appearance of a singular arc, where the method is unable to switch on to the optimal singular trajectory correctly and then switch to the bang-bang control.

In each optimisation result, the first and last switches follow those obtained from the gradient method, up to the grid-step size. Additionally, the mean optimal control values differ by less than 0.5% across methods, with the smallest value obtained using Pyomo and the largest using AMPL. The number of grid points typically has a negligible impact on the qualitative results, primarily contributing to a more accurate numerical calculation of the objective cost (through numerical integration) and switching times. Increasing the number of grid points may have an impact on the results, but a noticeable improvement in the result occurs with a very significant increase in the number of grid points and is sensitive to changes in model parameters, in particular weights in the objective functional. For example, when solving the problem with $\theta = 0$, each modelling method gives almost identical results, with minor differences attributable to the grid step size. In this scenario, the optimal trajectories closely approximate the singular arc defined by (13).

5 Conclusions

In this paper, we discussed a modified gradient method for solving the optimal control problems when singular intervals are present in the optimal structure. The appearance of singular controls complicates the problem of finding optimal solutions, because trajectories move along different singular arcs with different values of singular controls. However, the existence of singular controls leads to intermediate-value controls, which can significantly impact the optimisation results. The structure of optimal control depends primarily on the form of the objective functional, but also changes based on initial conditions and model dynamics. Even if the optimal control is initially bang-bang, it may change to bang-singular, when the dynamics change in response to the control. Singular controls are often more natural candidates for optimality than bang-bang controls, which arise only when singular controls do not exist or are inadmissible.

The numerical computation of optimal strategies poses a problem of high complexity. Direct methods can successfully run if all optimal control variables are bang-bang. Many different approaches can be applied after discretising the optimal control problem, even when the exact number of switching times is not known a priori. However, numerical procedures for solving singular control problems are usually problematic. Direct forward-backward sweep methods are not advisable. Indeed, some methods can detect the structure of the optimal solution and accurately compute the switching times without prior information about optimal control. As we showed, it is possible that the discretisation of an optimal control problem generates numerical artifacts and determining the exact optimal trajectory becomes impossible. In contrast to other numerical algorithms and modelling interfaces, the presented algorithm, after switching to the singular interval, moves along the particular singular arc that can be derived analytically. This approach provides the best possible approximation of optimal controls. A priori assumption that the optimal control structure is fixed effectively eliminates non-optimal controls, which could only be numerical artifacts.

References

1. Atkins, S., Aghaee, M., Martcheva, M., Hager, W.: Computational and Mathematical Population Dynamics. World Scientific (2023). https://doi.org/10.1142/13045
2. Bajger, P., Bodzioch, M., Foryś, U.: Singularity of controls in a simple model of acquired chemotherapy resistance. Discrete Contin. Dyn. Syst. Ser. B **24**(5), 2039–2052 (2019). https://doi.org/10.3934/dcdsb.2019083
3. Bajger, P., Bodzioch, M., Foryś, U.: Numerical optimisation of chemotherapy dosage under antiangiogenic treatment in the presence of drug resistance. Math. Methods Appl. Sci. **43**, 10671–10689 (2020). https://doi.org/10.1002/mma.6958
4. Bodzioch, M., Bajger, P., Foryś, U.: Angiogenesis and chemotherapy resistance: optimizing chemotherapy scheduling using mathematical modeling. J. Cancer Res. Clin. Oncol. **147**(8), 2281–2299 (2021). https://doi.org/10.1007/s00432-021-03657-9
5. Bodzioch, M., Bajger, P., Foryś, U.: Competition between populations: preventing domination of resistant population using optimal control. Appl. Math. Model. **114**, 697–693 (2023). https://doi.org/10.1016/j.apm.2022.10.016
6. Bonnard, B., Chyba, M.: Singular Trajectories and their Role in Control Theory, Mathematics and Applications, vol. 40. Springer, Cham (2003)
7. Falcone, M., Ferretti, R., Grüne, L., McEneaney, W.M. (eds.): Numerical Methods for Optimal Control Problems. Springer, Cham (2019). https://doi.org/10.1007/978-3-030-01959-4
8. Foroozandeh, Z., do Rosário de Pinho, M., Shamsi, M.: On numerical methods for singular optimal control problems: an application to an AUV problem. Discrete Contin. Dyn. Syst. Ser. B **24**(5), 2219–2235 (2019). https://doi.org/10.3934/dcdsb.2019092
9. Schättler, H., Ledzewicz, U.: Optimal Control for Mathematical Models of Cancer Therapies. IAM, vol. 42. Springer, New York (2015). https://doi.org/10.1007/978-1-4939-2972-6
10. Pager, E.R., Rao, A.V.: Method for solving bang-bang and singular optimal control problems using adaptive Radau collocation. Comput. Optim. Appl. **81**, 857–887 (2022). https://doi.org/10.1007/s10589-022-00350-6
11. Ruder, S.: An overview of gradient descent optimization algorithms. CoRR abs/1609.04747 (2016). http://arxiv.org/abs/1609.04747
12. Śmieja, J., Świerniak, A., Duda, Z.: Gradient method for finding optimal scheduling in infinite dimensional models of chemotherapy. J. Theoret. Med. **3**, 25–36 (2000). https://doi.org/10.1080/10273660008833062
13. Wächter, A., Biegler, L.: On the implementation of an interior-point filter line-search algorithm for large-scale nonlinear programming. Math. Program. **106**, 25–57 (2006). https://doi.org/10.1007/s10107-004-0559-y
14. Zelikin, M., Borisov, V.: Theory of chattering control: with applications to astronautics, robotics, economics, and engineering. Birkhäuser (1994). https://doi.org/10.1007/978-1-4612-2702-1

Multiobjective Optimization of Complete Coverage and Path Planning for Emergency Response by UAVs in Disaster Areas

Krzysztof Trojanowski[1], Artur Mikitiuk[1(✉)], Jakub Grzeszczak[1], and Frédéric Guinand[1,2]

[1] Cardinal Stefan Wyszyński University in Warsaw, Warsaw, Poland
{k.trojanowski,a.mikitiuk,jakub.grzeszczak}@uksw.edu.pl
[2] Normandy University of Le Havre, Le Havre, France
frederic.guinand@univ-lehavre.fr

Abstract. Complete Coverage and Path Planning methods operate on many models depending on initial constraints and user demands. In this case, we optimize paths for a set of UAVs in the disaster area divided into rectangular regions of different sizes and priorities representing the expected number of victims. Paths maximize the number of victims localized in the first minutes of the UAVs' operation and minimize the entire operation makespan. The problem belongs to the domain of multiobjective optimization; therefore, we apply the Strength Pareto Evolutionary Algorithm 2, which is equipped with several problem-specific perturbation operators. In the experimental part, we use SPEA2 to four selected test cases from a TCG-CCPP generator powered by actual data on residents in selected regions in Poland published by Statistics Poland.

Keywords: Multiobjective Optimization · Coverage Path Planning · Unmanned Aerial Vehicles · Terrain Coverage · Heuristic Optimization

1 Introduction

In recent times, UAVs have found an increasing number of applications, both civil, like agricultural, scientific, or emergency, and military, like surveillance or battlefield activity. In the presented case, we develop applications for the first reconnaissance in the disaster areas caused by floods or earthquakes. When base transceiver stations (BTSs) are out of order, and ground communication tracks and roads have been flooded or destroyed, victims cannot communicate about their location, state, and needs. In this case, UAVs represent the first, immediate response to the problem of reconnaissance and communication delivery. The UAV application has many aspects that must be subject to analysis, like communication, path planning, or emergency management. UAVs' mobility features and availability or access to additional information to predict the localization of victims in the path-planning process significantly impact the rescue action's plan and effectiveness.

L. Franco et al. (Eds.): ICCS 2024, LNCS 14836, pp. 150–165, 2024.
https://doi.org/10.1007/978-3-031-63775-9_11

In this research, we develop a path-planning procedure for a team of simultaneously operating UAVs. The paths are optimized to satisfy the two objectives: to get information about the condition and location of all persons in the disaster area and to minimize the time to find the victims demanding immediate help. UAVs are equipped with Mobile Base Stations (MBSs). Turned-on mobile phones of the victims help UAVs discover phones' precise GPS coordinates and owners' identity through phone or IMEI numbers. Additionally, we consider the terrain, which is not uniform in terms of population; selected regions, like villages and cities, have more chances to have people present than others, like farmlands or wastelands. Therefore, the shape of UAVs' paths matters for the average waiting time for help. Time is a critical factor in helping victims. The first minutes following an accident are the most important in emergency management.

We continue our earlier research described in [6,13]. In those publications, we formulated a model of the disaster area and proposed a Test Case Generator and a path-planning method for a team of UAVs. The method implements the Local Search approach but operates on a new model-specific solution representation and uses a number of new perturbation operators. We formulated a function expressing the number of victims localized in the first minutes of the UAVs' work and maximized it in the experimental research. In this paper the main contribution is a new two-stage optimization and computer simulations:

1. In the first stage, we generate an initial population of solutions. We make them randomly or employ the earlier proposed Local Search. In this method, a newly generated solution is better than the current one when its overall operation makespan is shorter. When they have equal makespans, the one with fewer victims missed in the first minutes of the UAVs' operation is better. This second criterion is a reformulated function calculating the number of victims localized in the first minutes of the UAVs' work.
2. In the second stage, we apply one of the multiobjective optimization heuristics, The Strength Pareto Evolutionary Algorithm 2 (SPEA2), using the initial population from the previous stage. Two criteria are minimized: the overall operation makespan and the number of victims missed in the first minutes of the UAVs' operation. Both stages use the same previously proposed representation of solutions, but in the latter one, we propose new perturbation operators.
3. In the experimental part, we used an earlier proposed benchmark from the Test Case Generator. The nondominated sets from SPEA2 vary depending on the initial population and settings of evolutionary parameters. We find their most beneficial settings and indicate the pros and cons of multi-criteria optimization compared to the single-criteria approach.

The paper consists of seven sections. Section 2 presents the model of the disaster area. Section 3 discusses the representation of the set of paths for a team of UAVs. Section 4 describes two solution evaluation criteria defining the objective space. Section 5 presents SPEA2 and problem-specific operators. The experimental part of the research is described in Sect. 6. Finally, Sect. 7 concludes the paper.

2 Representation of a Disaster Area Map

A team of UAVs operates over a rectangle disaster area. We aim to generate UAV paths that cover the entire area and can cross but should not overlap. The precise location of the disaster victims is unknown. However, some estimations about their density in selected places exist. Therefore, we divide the area into a set of non-overlapping regions covering the entire area and having priorities representing chances for the presence of victims. The regions are rectangles of different sizes; every place within a region has the same population density. In the presented research, the region's priority depends on the population density in the region; however, in general, the priority may also originate from natural disaster circumstances like the epicenter's location and the strength of tremors in the case of quakes or hydrographical conditions in the case of floods. Every region must belong to one of the paths because we have to localize all victims.

The sub-areas with a similar population density are approximated by rectangle regions since path generation algorithms satisfying UAV mobility features exist for such types of regions [2,4,8]. Models dividing the area into regions were proposed in numerous approaches to the Coverage Path Planning problem [1,7,9–12]. However, none have regions with assigned priorities for the coverage order. For the experimental research, we used the Test Case Generator for Problems of Complete Coverage and Path Planning (TCG-CCPP) [6], which proceeds data on residents in a 1-kilometer grid from the "2021 Population and Housing Census in Poland" published by Statistics Poland[1]. The generator transforms an input grid of size n by n squares with population density estimated from the census into an area divided into rectangular regions with different population density levels.

A region has two attributes: a surface area and an expected number of citizens/victims (ENV). Still, when we divide the region's ENV by the area, we derive the third attribute—population density. ENV for a path equals the sum of ENVs for the regions in this path, and ENV for the disaster area equals the sum of ENVs for all paths.

As mentioned earlier, regions from TCG-CCPP are rectangular. Hence, they are easy to cover by UAV flight trajectories. Therefore, we do not build precise paths within these regions. We rather optimize the order of visiting regions, where neighbor regions in the path do not necessarily have to be adjacent on the terrain map. Thus, the given path-planning problem is a combinatorial one.

3 The Problem Solution—UAVs' Paths

When a disaster occurs, all paths for u UAVs must be fully defined before the UAVs start operating over the disaster area. Hence, the operational area model must already exist and be available for computations. Briefly, the area must be divided into a set of convex regions concerning their priorities, and we need to know the number of available UAVs, their mobility features, and starting

[1] Geostatistics Portal, https://geo.stat.gov.pl, the date of access: Dec 28, 2022.

locations. A central computational unit generates paths when the rescue teams with UAVs have already been dispatched and are on their way, or even earlier, during periodic safety inspections and tests.

We need paths for all UAVs, so a single solution consists of u paths containing lists of regions' IDs. In the sum of paths, no ID can be omitted to ensure the finding of all victims. Moreover, the sum must contain no duplicated IDs to avoid redundant workloads during the paths' execution. The solution execution is called a round. Managing distances between UAVs is unnecessary since the paths are generated offline, so no interactive navigation occurs.

A region's coverage time for a UAV equals a surface unit's coverage time multiplied by the surface area. The traversing time equals the distance to traverse divided by the UAV's cruising velocity. The surface unit's coverage time and the UAV's cruising velocity are the problem's parameters that are constant over the entire round. Each path execution time considers the sum of the overall time of regions' coverage and the time necessary to traverse between them. For any two consecutive regions in the path being adjacent to each other in the terrain, traverse time equals zero. Otherwise, the traverse time equals the execution time for a UAV's direct flight in a straight line between the two closest points of the region's boundaries. It is important to stress that the location of the region's entry and exit points does not change the number of found victims over time.

4 Evaluation of UAVs' Paths

We use a model where the area has the size of $n \times n$ one-kilometer base units and is fully covered by rectangle regions whose side lengths are also expressed in base units. UAV paths consist of two types of tasks: (1) a region covering and (2) traversing from one region to another. Execution of the region coverage takes time proportional to the region's area in square kilometers multiplied by a unit coverage time T_{sc}. During the region's coverage, the number of localized victims grows linearly. When UAV traverses from one region to another, no victims are found. The traverse path is a straight line, and its execution takes time inversely proportional to the UAV's cruising velocity V_{cr}.

A model of an example problem and its example solution S are presented in Fig. 1. The left part of the figure shows the disaster area divided into regions, whereas the right part shows example paths for a team of two UAVs. Both UAVs start from the position s and execute paths $s246s$ and $s153s$, respectively, where digits represent the region's IDs. Figure 2 shows charts of the number of already found victims over time for the two parts of S, that is S_1 executing $s246s$ and S_2 — $s153s$, and example values of $T_{\text{sc}} = 2$ and $V_{\text{cr}} = 1$. The chart of the number of already localized victims over time for S is the sum of charts calculated for each UAV. Detailed discussion on the charts can be found in [13].

We use two criteria to evaluate solutions. The first one equals the makespan of the round, that is, the path execution time of the UAV that finished last:

$$f_1(\mathbf{x}) = \max(T_i) \quad \text{where } i \in \{1, \ldots, u\} \tag{1}$$

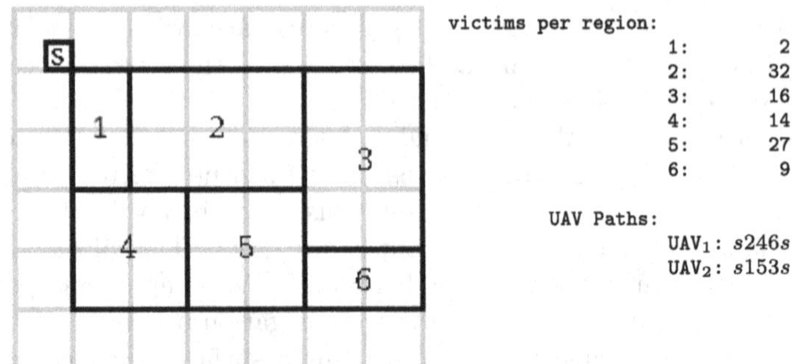

Fig. 1. An example test case and its solution: layout of an area divided into regions (on the left) and paths for two working UAVs having a starting point s in the corner of the area (on the right); extracted from [13]

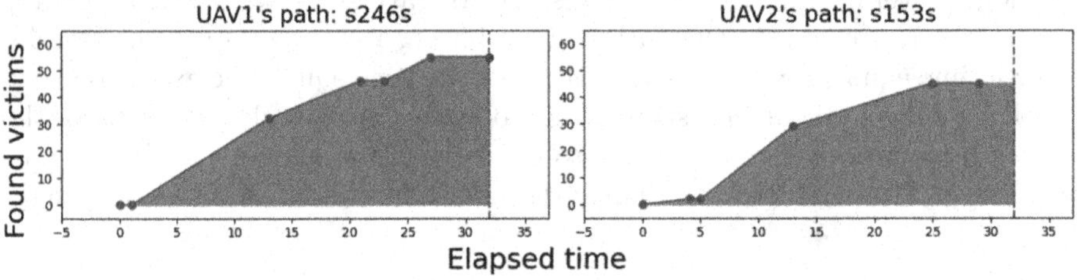

Fig. 2. Graphs of the number of detected victims over time; extracted from [13]

The second one is the sum of the areas over the curves describing the number of victims found over time, called S_{AOC}. Due to the different execution times of the UAVs' paths, the curves of the UAVs that finished earlier than the last one are extended to its length, called $T_{\max}$.

In Fig. 2, $S_{\mathrm{AOC}}(i)$ equals the sum of differences between the curve's bounding rectangle area of size $T_{\max} \cdot ENV(S_i)$ and the gray area under the curve. For an ideal solution, S_{AOC} equals 0. However, it is an unrealistic scenario. The second optimization function f_2 looks as follows:

$$f_2(\mathbf{x}) = \frac{S_{\mathrm{AOC}}}{f_1(\mathbf{x}) \times v_{\mathrm{exp}}} \qquad \text{where } S_{\mathrm{AOC}} = \sum_{i=1}^{u} ext(S_{\mathrm{AOC}}(i), T_{\max}) \tag{2}$$

$S_{\mathrm{AOC}}(i)$ is an area over the curve defined by the i-th UAV's path. The parameter v_{exp} defines the ENV of the entire disaster area, and $ext(S_{\mathrm{AOC}}(i), T_{\max})$ is a function that returns the area over the i-th curve extended to $T_{\max}$. The area of the sum of rectangle boundings for S_{AOC} equals $f_1(\mathbf{x}) \cdot v_{\mathrm{exp}}$.

Eventually, the optimization goal is: $\min[f_1, f_2]$.

5 The Optimization Method

In the experimental part, we use one of Pareto-based evolutionary multiobjective optimization algorithms [3], namely Strength Pareto Evolutionary Algorithm Version 2 (SPEA2) [14]. Main steps of SPEA2 are presented in Algorithm 1.

Algorithm 1. SPEA2

```
1: Initialization P                    ▷ generate an initial population of solutions
2: Evaluation(P)                       ▷ evaluate two objective values for each solution
3: Evaluation_SF(P)                    ▷ evaluate SPEA2 fitness
4: E ← Update(P)                       ▷ initialize archive (external set)
5: repeat
6:     Evaluation_SF(E)                ▷ evaluate SPEA2 fitness
7:     P ← Tournament(E)               ▷ perform binary tournament selection
8:     P′ ← Variation(P)               ▷ apply recombination and mutation
9:     Evaluation(P′)                  ▷ evaluate two objective values for each solution
10:    Evaluation_SF(P′ ∪ E)           ▷ evaluate SPEA2 fitness
11:    E ← Update(P′ ∪ E)
12: until termination condition met
```

The algorithm manages two populations of solutions: a reference set P and an archive E. In the first step of the algorithm, we generate an initial population of solutions for P. For each of them, we evaluate the two objective functions (Step #2). Henceforth, the position of solutions in the objective space is known, and one can assess the SPEA2 fitness of solutions (Step #3). Eventually, we copied the entire reference set to the archive since we assumed that the archive size equals the reference set size (Step #4). Then, the main loop starts.

In the first step of the main loop (Step #6), we evaluate SPEA2 fitnesses of solutions in the archive E, which is redundant in the first execution of the main loop but not in the next ones. Then, the binary tournament selection on the archive's population generates clones to save to the current reference set P (Step #7). Variation operators perturb solutions in P and make its new representation, P'. There are two types of variation operators: binary (crossover) and unary (mutation). Solutions represent a set of u paths for all UAVs. Therefore, the operators have to consider the complex structure of the solution representation.

For pairs of solutions randomly selected from P, we perform two variation operators: crossover and mutation (Step #8). Crossover implements the PMX operator [5]. PMX operates on two paths: permutations of elements from n unique IDs. In our case, we have a team of UAVs, and a solution consists of a set of their paths being sequences of regions' IDs. For the aim of the representation adaptation to PMX, we concatenate the UAVs' paths into one list of regions' IDs and remember the paths' lengths, that is, concatenation points. This way, we always obtain lists of the same constant length regardless of the UAVs' path lengths and know how to reverse the process of merging paths into one list. The lists' lengths equal the total number of regions. Then, the PMX operator is executed in a typical manner. First, for the two parents randomly selected from P, it selects uniformly two cut points along the list at random. Then, the substrings between the two cut points are copied into offsprings. Next, the remaining values are copied from respective parents, following the rules defined in PMX, to guarantee the feasibility of newly created paths. In the last step, we restore the division into UAVs' paths concerning the remembered concatenation points.

Then, we perform the mutation operator. Four mutation operators adjusted to the current problem-specific representation have been proposed in [13]. In this research, two new operators are proposed. Hence, we have six operators in hand.

All the operators apply three modification procedures, each based on two evaluation criteria. Modification procedures are as follows:

1. **M**ove—in a randomly selected UAV's path, we insert a randomly selected sector in front of another randomly selected sector,
2. **E**xchange—any two randomly selected sectors in two randomly selected UAV's paths are inserted in front of randomly selected sectors in the opposite paths,
3. **H**op—we insert a randomly selected sector from a randomly selected UAV's path in front of another randomly selected sector of another randomly selected UAV's path.

In the three procedures, neither the UAVs' paths nor the new locations for sectors in these paths are selected uniformly at random. Chances for a path to be selected may depend on one of two criteria: its total execution time (the longer, the more chances — criterion *pcT*) or the total expected number of victims in all sectors in the path (the more victims found, the more chances — criterion *pcP*). Chances for selecting a location in a path for a sector s may also depend on one of two criteria. One of them is the traverse time (criterion *scT*). That is, for each pair of sectors s_i and s_j, we calculate traverse times from s_i to s and s to s_j, and the chances are inversely proportional to the sum of the two (the shorter traverse time, the more chances). Chances for selecting a location in a path for a sector s may also depend on the difference between the population density in s and in sector s_i in front of which s will be located (criterion *scD*). In this case, for each sector s_i in the path, we calculate the difference between its population density and the population density in s, namely $\mathrm{diff}(s_i, s)$. When all the differences are greater than zero, the chances are proportional to them. Otherwise, the chances are proportional to: $1+\mathrm{diff}(s_i, s)-\min(\mathrm{diff}(s_j, s))$ $j \in \{1, \ldots, \text{the path's length}\}$.

Eventually, we obtain six mutation operators labeled with the first letter of the modification procedure and the selection criteria symbols: M[*pcT*, *scT*], M[*pcP*, *scD*], E[*pcT*, *scT*], E[*pcP*, *scD*], H[*pcT*, *scT*], H[*pcP*, *scD*]. The operators originate from the four problem-specific perturbation operators presented in [13]. These four operators have symbols LD:GD, LD:GT, LT:GD, and LT:GT. They can be mapped on the current six ones as follows: operator LD:GD works the same as application of the two: M[*pcP*, *scD*] and E[*pcP*, *scD*], LD:GT — M[*pcP*, *scD*] and E[*pcT*, *scT*], LT:GD — M[*pcT*, *scT*] and E[*pcP*, *scD*], LT:GT — M[*pcT*, *scT*] and E[*pcT*, *scT*].

Since we did not know which mutation would be the most beneficial, we applied a simple reward system. We use a table T_{mut} containing six cells where initial values in all cells equal 1/6. They represent the probability of the mutation operator selection for the six operators. The mutation step begins with a random selection of the operator. Probability selection depends on the factors in T_{mut}, which add up to one. The operator receives a prize when the mutated offspring is better than its parent, that is, the respective cell in T_{mut} grows by δ_{rew}. Then, values in cells are normalized to sum up to one again.

The remaining steps of the main loop represent generic procedures of SPEA2. In Step #9, we evaluate the two objective functions for all new solutions stored in P' and in Step #10 — the SPEA2 fitness of solutions. Then, a new archive E content is selected from the sum: P' and E (Step #11). It is a ranking selection based on the solution's SPEA2 fitness values. The main loop is executed for a limited number of the SPEA2 fitness evaluations, and this number is one of the algorithm's parameters.

The following parameters control the algorithm: the reference set size $|P|$, the archive size $|E|$, crossover probability p_{cross}, mutation probability p_{mut}, the table T_{mut} storing coefficients representing mutation operators' chances for application, prize δ_{rew} for mutation operators for improving the solution, and a stopping condition which is the maximum number of SPEA2 fitness function calls max_{nffc}.

We have five ways of the initial population generation: random initialization and using the outcome of four versions of Local Search with four problem-specific perturbation operators: LD:GD, LD:GT, LT:GD, and LT:GT [13]. LS optimizes just the first objective function, f_1. In the case of a tie, the second objective function, f_2 defined in Eq. (2), decides. SPEA2 is executed five times for every test case, once using a random population and four times — populations generated by LS.

6 Experimental Research

6.1 Plan of Experiments

In the experimental part, we did tests with the SPEA2 algorithm. A solution represents a set of paths for $u = 10$ UAVs, and the total expected number of victims $v_{\text{exp}} = 100\%$ of the population in the test area, hence each region must appear just once in one and only one path. Therefore, the sum of the paths' lengths in all solutions is the same and equals the total number of regions. The algorithm parameters are as follows: population size equals 30, $p_{\text{cross}} \in \{0.25, 0.5, 0.75, 1\}$, $p_{\text{mut}} \in \{0.25, 0.5, 0.75, 1\}$ and $\delta_{\text{rew}} \in \{0.001, 0.005, 0.01, 0.05\}$.

The size of the disaster area's side $n = 30$, the unit coverage time $T_{\text{sc}} = 10$, the UAV's cruise velocity during traverse from one region to another nonadjacent one $V_{\text{cr}} = 1$, and stopping condition, that is, the maximum number of SPEA2 fitness function calls $max_{\text{nffc}} = 500000$. We have four testing areas from TCG-CCPP covering: Gliwice, Lodz, Warsaw (SW), and Gdansk Bay.

We did five groups of experiments. They differ in the origin of the SPEA2 initial population. In four groups, the initial population is the outcome of the Local Search application using one of the four versions of perturbation: LD:GD, LD:GT, LT:GD, or LT:GT [13]. Each version of LS was called 30 times to generate a population of 30 solutions. In the fifth group, the initial population was generated randomly.

Objective space consists of two criteria: f_1 as defined in Eq. (1) and f_2 as defined in Eq. (2). Both criteria are minimized. Experiments aim at verification of the ability of multiobjective optimization algorithms to find valuable sets of nondominated solutions in the given objective space. Additionally, we investigate the role of initial populations for SPEA2 in searching for the most prominent sets and the influence of the initial population on the tuning of SPEA2 evolutionary parameters.

6.2 Results of Experiments

Each of the Figs. 3, 4, 5, and 6 presents the five most promising sets of solutions found by SPEA2 using initial populations generated by four LS operators for the four testing areas. Figure 7 shows the five most promising sets of results obtained using random initial populations for the four areas.

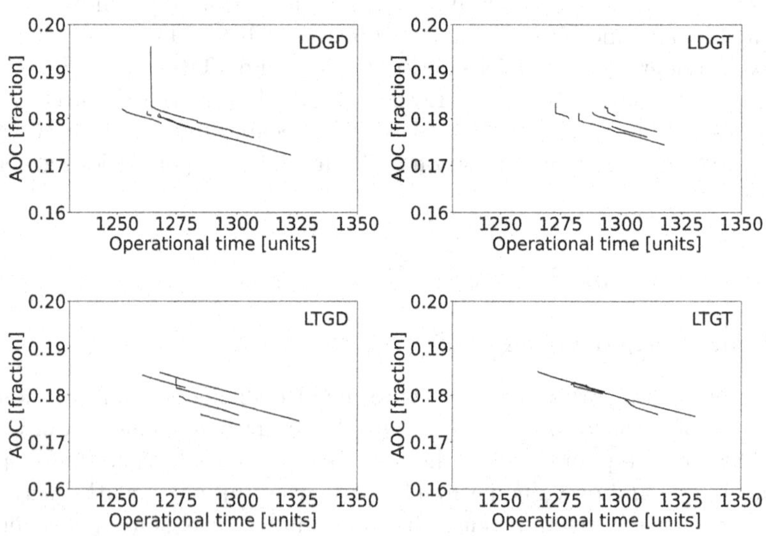

Fig. 3. The five most promising sets of solutions found by SPEA2 using initial populations generated by four LS operators for the Gliwice area

On most diagrams, there is no set containing both the best solution in terms of makespan and the best solution in terms of the area over the curve. Only the results obtained from the initial population generated by LT:GT for the Warsaw

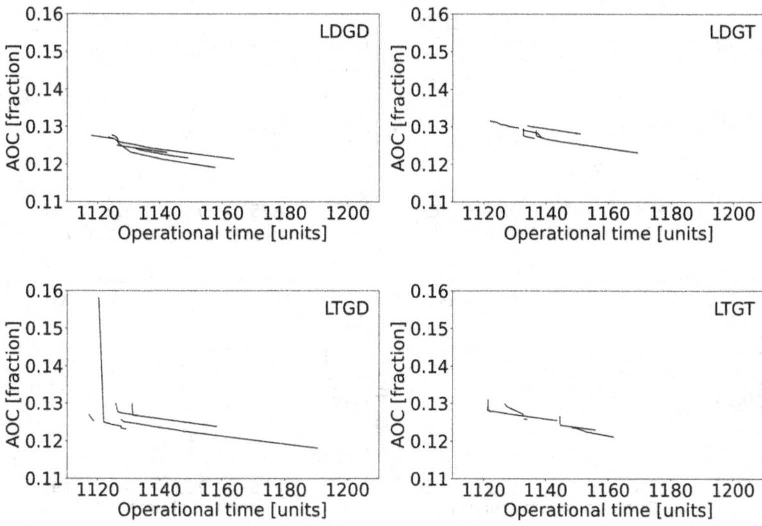

Fig. 4. The five most promising sets of solutions found by SPEA2 using initial populations generated by four LS operators for the Lodz area

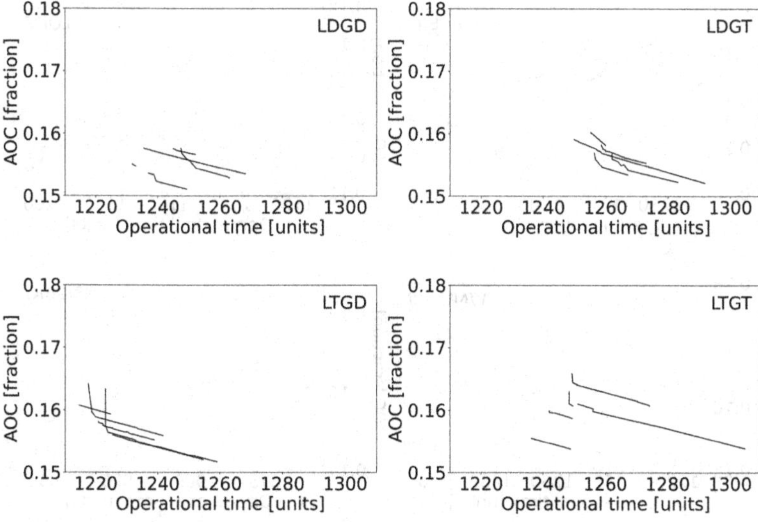

Fig. 5. The five most promising sets of solutions found by SPEA2 using initial populations generated by four LS operators for the Warsaw (SW) area

area and from the random initial populations for the Warsaw and the Gdansk Bay areas contain a set with the best solutions in both terms. On the other hand, there are diagrams with some sets that contain only solutions worse than those of the remaining sets.

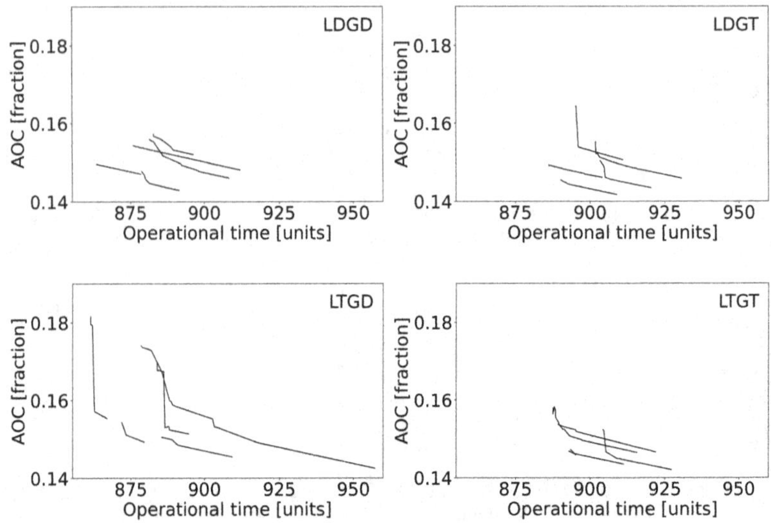

Fig. 6. The five most promising sets of solutions found by SPEA2 using initial populations generated by four LS operators for the Gdansk Bay area

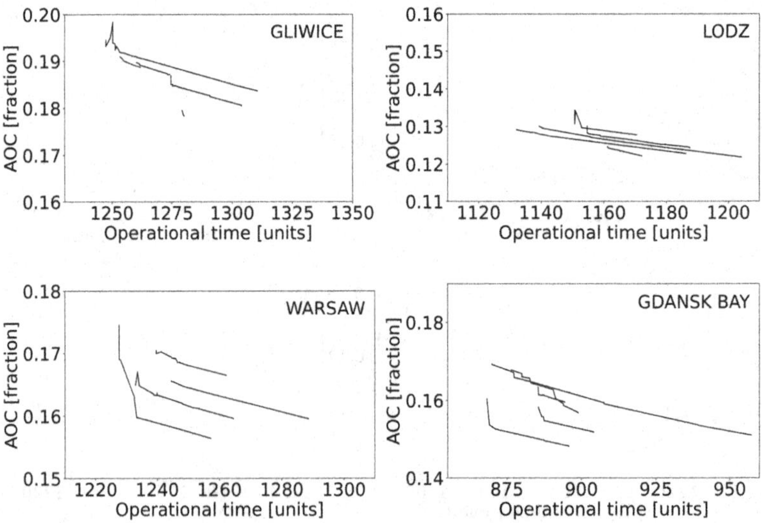

Fig. 7. The five most promising sets of solutions found by SPEA2 using random initial populations for the four areas

Experiments described in [13] show that for the Lodz and Warsaw areas, the perturbation operator LT:GT produced the best LS results. In our current experiments, for the Lodz area, both the best solutions were produced from the initial population generated by LT:GD. For the Warsaw area, the best makespan solution was in a set obtained from the initial population generated by LT:GD,

Table 1. Configurations of SPEA2 parameters: p_{cross}, p_{mut} and δ_{rew} for the experiments, which results are presented in Fig. 3 (the Gliwice area), Fig. 4 (the Lodz area) and Fig. 7

init	Gliwice area			Lodz area		
pop	p_{cross}	p_{mut}	δ_{rew}	p_{cross}	p_{mut}	δ_{rew}
LS[LD:GD]	0.50	0.75	0.050	0.25	0.50	0.050
	0.75	1.00	0.010	0.25	0.75	0.001
	1.00	0.50	0.005	0.50	0.50	0.001
	1.00	0.50	0.010	0.50	0.50	0.010
	1.00	0.50	0.050	1.00	0.50	0.010
LS[LD:GT]	0.75	0.75	0.010	0.25	0.50	0.005
	1.00	0.50	0.005	0.25	1.00	0.005
	1.00	0.75	0.005	0.75	0.25	0.001
	1.00	0.75	0.010	0.75	0.50	0.005
	1.00	1.00	0.005	0.75	1.00	0.005
LS[LT:GD]	0.25	0.75	0.005	0.50	0.75	0.005
	0.25	0.75	0.010	0.50	0.75	0.010
	0.50	0.75	0.010	0.75	0.50	0.005
	1.00	0.75	0.010	1.00	0.50	0.001
	1.00	1.00	0.005	1.00	0.50	0.010
LS[LT:GT]	0.50	0.75	0.005	0.25	0.50	0.005
	0.50	0.75	0.010	0.25	1.00	0.001
	0.75	0.75	0.010	0.50	0.75	0.005
	0.75	1.00	0.005	0.75	0.75	0.005
	1.00	0.75	0.050	1.00	0.50	0.001
randomly	0.75	0.75	0.050	0.50	1.00	0.001
	0.75	1.00	0.005	0.50	1.00	0.005
	1.00	0.50	0.005	0.75	0.75	0.001
	1.00	0.75	0.010	0.75	0.75	0.005
	1.00	1.00	0.005	1.00	0.75	0.005

while the best AOC solution was contained in a set obtained from the initial population generated by LD:GD. According to [13], for the Gdansk Bay area, operator LT:GD produced the best LS results. In our current experiments, the best makespan solution was in a set obtained from the initial population generated by LT:GD, while the best AOC solution was contained in a set obtained from the initial population generated by LD:GT. LD:GT produced the best LS results for the Gliwice area. In the current research, we got the best makespan solution using a random initial population. The best AOC solution was in a set

Table 2. Configurations of SPEA2 parameters: p_{cross}, p_{mut} and δ_{rew} for the experiments, which results are presented in Fig. 5 (the Warsaw (SW) area), Fig. 6 (the Gdansk Bay area) and Fig. 7

init	Warsaw (SW) area			Gdansk Bay area		
pop	p_{cross}	p_{mut}	δ_{rew}	p_{cross}	p_{mut}	δ_{rew}
LS[LD:GD]	0.75	0.50	0.001	0.25	0.50	0.005
	0.75	0.50	0.005	0.25	0.75	0.001
	0.75	0.75	0.050	0.50	0.75	0.001
	1.00	0.75	0.001	0.50	1.00	0.001
	1.00	0.75	0.010	0.75	0.75	0.001
LS[LD:GT]	0.50	0.75	0.010	0.25	0.50	0.001
	0.75	0.50	0.005	0.25	1.00	0.005
	0.75	1.00	0.005	0.50	0.75	0.001
	1.00	0.25	0.005	0.75	0.50	0.001
	1.00	0.50	0.001	0.75	0.75	0.001
LS[LT:GD]	0.25	0.75	0.010	0.25	0.75	0.001
	0.50	0.50	0.050	0.50	0.50	0.001
	0.75	0.75	0.010	0.50	1.00	0.001
	0.75	1.00	0.050	0.75	0.50	0.001
	1.00	0.75	0.010	0.75	0.75	0.001
LS[LT:GT]	0.25	1.00	0.050	0.25	0.50	0.001
	0.50	0.75	0.005	0.50	0.50	0.001
	0.50	0.75	0.010	0.50	0.75	0.005
	0.50	0.75	0.050	0.50	1.00	0.001
	0.75	0.75	0.005	0.75	0.50	0.001
randomly	0.50	0.75	0.010	0.25	1.00	0.001
	0.75	1.00	0.010	0.50	0.75	0.001
	1.00	0.50	0.010	1.00	0.50	0.001
	1.00	0.75	0.005	1.00	0.50	0.005
	1.00	1.00	0.010	1.00	0.75	0.001

obtained from the initial population generated by LD:GD. Thus, all five ways of generating an initial population for SPEA2 were useful for some data sets.

Tables 1 and 2 show configurations of SPEA2 parameters for experiments which results are presented in Figs. 3, 4, 5, 6, and 7. All used values of p_{cross}, p_{mut}, and δ_{rew} appear in these tables. Moreover, for each way of generating the initial solution and for each area, the five most promising results were obtained using five different triples of values (p_{cross}, p_{mut}, δ_{rew}). Among 100 values of p_{cross}, 1.0 appeared 29 times, 0.75 28 times, 0.5 26 times, and 0.25 only 17 times. Among 100 values of p_{mut}, the most common was 0.75 (45 times), while 0.25

was the least common (only twice), 0.5 appeared 32 times, and 1.0–21 times. Values of δ_{rew} 0.001, 0.005, 0.01, and 0.05 appeared 32, 34, 24, and 10 times respectively.

We can conclude from the aforementioned observations that almost all values of SPEA2 parameters can provide promising results for some data sets. Only the probability of mutation 0.25 should be avoided in future experiments.

Figure 8 shows the improvement of solutions returned by tested algorithms. While Local Search offered quite a lot of improvement, further optimization with SPEA2 managed to shorten these solutions by up to 20%.

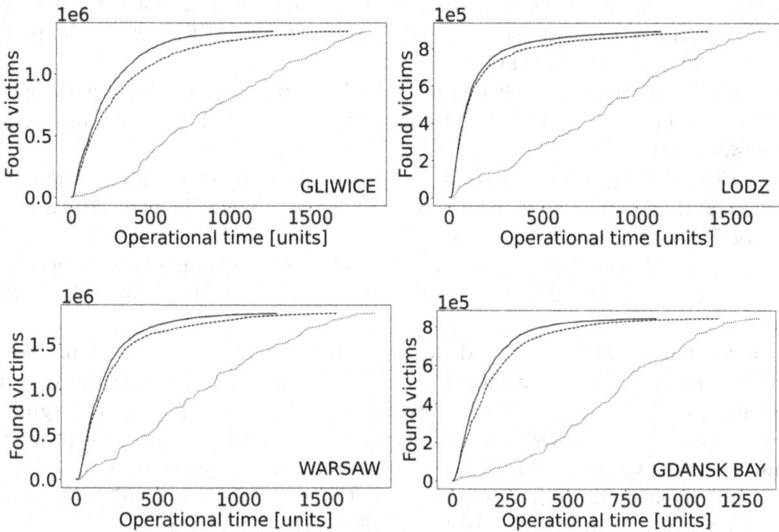

Fig. 8. Cumulative number of victims found over time for a randomly generated solution (dotted), a solution optimized by Local Search (dashed), and an example solution returned by SPEA2 (solid)

7 Conclusions

We applied the SPEA2 multiobjective optimization algorithm to the problem of complete coverage and path planning for emergency response by UAVs in disaster areas. We aimed to maximize the number of victims localized in the first minutes of the UAVs' operation and minimize the entire operation makespan.

For experiments, we used four test cases from TCG-CCPP based on the data from the Geostatistics Portal by Statistics Poland. Initial solutions for SPEA2 were created randomly or using LS with one of the four pairs of perturbation operators we described in our earlier paper. Our experiments generated promising sets of problem solutions.

We repeated the experiments with several parameters' values that determined the probability of crossover and mutation and the reward for a successful mutation. The results showed that almost all values used for these parameters could provide the most promising results for some data sets. The only restriction is that the mutation probability should be at least 0.5. The SPEA2 algorithm was able to significantly improve the entire operation makespan.

References

1. Basilico, N., Carpin, S.: Deploying teams of heterogeneous UAVs in cooperative two-level surveillance missions. In: 2015 IEEE/RSJ International Conference on Intelligent Robots and Systems (IROS), pp. 610–615. IEEE (2015). https://doi.org/10.1109/iros.2015.7353435
2. Cabreira, T., Brisolara, L., Ferreira, P.R., Jr.: Survey on coverage path planning with unmanned aerial vehicles. Drones **3**(1), 4 (2019). https://doi.org/10.3390/drones3010004
3. Emmerich, M.T.M., Deutz, A.H.: A tutorial on multiobjective optimization: fundamentals and evolutionary methods. Nat. Comput. **17**(3), 585–609 (2018). https://doi.org/10.1007/s11047-018-9685-y
4. Galceran, E., Carreras, M.: A survey on coverage path planning for robotics. Robot. Auton. Syst. **61**(12), 1258–1276 (2013). https://doi.org/10.1016/j.robot.2013.09.004
5. Goldberg, D.E., Lingle, Jr., R.: Alleles, loci, and the traveling salesman problem. In: Grefenstette, J.J. (ed.) The First International Conference on Genetic Algorithms and their Applications, pp. 154–159. L. Erlbaum Associates Inc., Pittsburgh (1985)
6. Grzeszczak, J., Trojanowski, K., Mikitiuk, A.: Test case generator for problems of complete coverage and path planning for emergency response by UAVs. In: Artificial Intelligence and Soft Computing. LNCS, vol. 14125, pp. 497–509. Springer, Cham (2023). https://doi.org/10.1007/978-3-031-42505-9_42
7. Kapanoglu, M., Alikalfa, M., Ozkan, M., Yazici, A., Parlaktuna, O.: A pattern-based genetic algorithm for multi-robot coverage path planning minimizing completion time. J. Intell. Manuf. **23**(4), 1035–1045 (2010). https://doi.org/10.1007/s10845-010-0404-5
8. Khan, A., Noreen, I., Habib, Z.: On complete coverage path planning algorithms for non-holonomic mobile robots: survey and challenges. J. Inf. Sci. Eng. **33**(1), 101–121 (2017). https://jise.iis.sinica.edu.tw/JISESearch/pages/View/PaperView.jsf?keyId=154_1997
9. Li, L., Shi, D., Jin, S., Kang, Y., Xue, C., et al.: Complete coverage problem of multiple robots with different velocities. Int. J. Adv. Robot. Syst. **19**(2), 17298 (2022). https://doi.org/10.1177/17298806221091685
10. Lin, H.Y., Huang, Y.C.: Collaborative complete coverage and path planning for multi-robot exploration. Sensors **21**(11), 3709 (2021). https://doi.org/10.3390/s21113709
11. Nasirian, B., Mehrandezh, M., Janabi-Sharifi, F.: Efficient coverage path planning for mobile disinfecting robots using graph-based representation of environment. Front. Robot. AI **8**, 1–19 (2021). https://doi.org/10.3389/frobt.2021.624333
12. Tan, C.S., Mohd-Mokhtar, R., Arshad, M.R.: A comprehensive review of coverage path planning in robotics using classical and heuristic algorithms. IEEE Access **9**, 119310–119342 (2021). https://doi.org/10.1109/access.2021.3108177

13. Trojanowski, K., Mikitiuk, A., Grzeszczak, J., Guinand, F.: Complete coverage and path planning for emergency response by UAVs in disaster areas. In: Computational Collective Intelligence. LNCS, vol. 14162, pp. 647–659. Springer, Cham (2023). https://doi.org/10.1007/978-3-031-41456-5_49
14. Zitzler, E., Laumanns, M., Thiele, L.: SPEA2: Improving the strength pareto evolutionary algorithm. TIK Report 103, ETH Zurich, Computer Engineering and Networks Laboratory (2001). https://doi.org/10.3929/ETHZ-A-004284029

Single-Scattering and Multi-scattering in Real-Time Volumetric Rendering of Clouds

Mikołaj Bajkowski and Dominik Szajerman(✉)

Institute of Information Technology, Lodz University of Technology, Łódź, Poland
dominik.szajerman@p.lodz.pl

Abstract. The aim of this work was to design an algorithm for rendering volumetric clouds in real time using a voxel representation. The results were verified using reference renders created with the *Blender* program using the *Principled Volume* shader. The important properties of the algorithm that were tried to be achieved are the ability to display clouds with different characteristics (thin and dense clouds) and the speed of operation enabling interactivity. We proposed a method consisting of two parameterizable image display algorithms with various performance and properties. The starting point was the single-scattering algorithm, which was extended with precalculation, and a simplified form of multi-scattering. Individual methods were compared with reference images. Methods performing similar tasks, depending on the purpose, generate single image frames at a rate ranging from several dozen hours to a few seconds. Using the described mechanisms, the proposed method allowed to achieve times between 1 and 200 milliseconds, depending on the method variant and quality settings.

Keywords: Single-scattering · Multi-scattering · Ray marching · Cloud rendering

1 Introduction

Being an inherent element of the Earth's landscape, clouds are an important element of the field of special effects, three-dimensional animation, and computer games that try to transfer at least a small part of the real world to virtual space. Unfortunately, the architecture of graphics accelerators is optimized for displaying graphics based on triangle meshes, which cannot properly reflect the complex structure of such a phenomenon.

For this reason, rendering realistic volumetric structures remains one of the most difficult challenges in high-performance computer graphics. This difficulty is caused by the great complexity of light phenomena that influence the appearance of such a structure. One of the frequently cited features of clouds is their high albedo. This causes the light to be dispersed multiple times in the cloud, which multiplies the number of physical interactions that must be simulated in order to generate a realistic image.

L. Franco et al. (Eds.): ICCS 2024, LNCS 14836, pp. 166–180, 2024.
https://doi.org/10.1007/978-3-031-63775-9_12

2 Related Work

Methods for realistic cloud rendering have been known for a long time [7]. One of them is synthesis using ray tracing, among others with support for Monte-Carlo [3] methods. In many scientific works, it serves as a reference technique to which the developed or optimized method must be as close as possible. Unfortunately, it is characterized by very high computational complexity. This precluded the method from being used in interactive applications for many years. Only recently, with the introduction of graphics cards supporting this technique in hardware, it has become possible to hybridly use ray tracing to enhance the effects obtained with traditional polygon rendering techniques in real time.

In the case of interactive solutions, where the observation point is only on the ground, it may still be a reasonable solution to implement a dome with a static or dynamically generated two-dimensional texture of clouds distributed in the sky [11]. This is a simple solution, and the appropriate lighting effects are synthesized by pre-processing the clouds to extract the data needed for appropriate shading operations. This method pays off with efficient rendering. This works particularly well when the lack of depth will not be perceived negatively, e.g. when the interactive experience effectively focuses the observer's attention on other objects in the virtual space.

If clouds are an object that takes up most of the rendered image or is a significant part of *gameplay*, then a different approach is needed. In games such as *War Thunder* or the *Flight Simulator* series, the player takes their feet off the ground, having the opportunity to observe clouds from different angles, lighting conditions, and the possibility of being inside such a phenomenon. Creators have at their disposal hybrid techniques based to some extent on *billboarding* [6]. A more advanced and increasingly popular solution is volumetric representation. The clouds in the above-mentioned titles look similar to those in Egor Yusov work [13]. A particle system was used here. The similarity of the spheres that make up clouds allowed, among other things, to optimize the display by precalculating the light scattering characteristics within such a volumetric domain. The problem in this case, however, is the relatively small variety of textures and shapes of clouds that can be obtained in this way.

There are techniques that compete with ray tracing. These are methods for rendering volumetric structures using voxels, but most of the work focuses on offline rendering solutions, often using methods simulating multi-scattering. D. Koerner et al. [10] used the *flux-limited diffusion* method, the effects of which were very similar to reference images created using the ray tracing method, and rendering reduced to below a second.

Another example is the work of Simon Kallweit et al. [8], which introduces a realistic synthesis of cloud images using a multilayer perceptron predicting the radiance distribution. There, it was also possible to achieve effects indistinguishable from the reference image generated by ray tracing, and the full-quality rendering time was reduced from several dozen hours to just a few seconds. Additionally, the results are compared with the *flux-limited diffusion* method, which in this case performs unfavorably when rendering volumetric structures with high albedo, such as clouds.

There are no works describing single- and multi-scattering methods using voxel-based raymarching of atmospheric clouds. This work aims to investigate this scenario and explore the possibility of extending the basic single-scattering algorithm to achieve render times that allow the technique to be used in interactive applications. At the same time, it will try to synthesize lighting effects characteristic of techniques with higher computational complexity.

3 Method

Later in the work, it is described how the following solutions were implemented in the rendering pipeline:

- Single-Scattering brute-force,
- Single-Scattering with cached photon map,
- Simplified Multi-Scattering with cached photon map based on Single-Scattering.

3.1 Dataset

The OpenVDB [5] format was used as the cloud density data source. It is a fixed-depth hierarchical structure that allows the use of voxels of a specific resolution and optimizes memory usage by excluding areas where the value can be considered equal to the background. In the presented case, these are all areas where the cloud density is zero. Standard OpenVDB structure settings were used: the root has a maximum of $(2^5)^3$ branches, each branch has a maximum of $(2^4)^3$ leaves and each leaf has a maximum of $(2^3)^3$ voxels. The data in the appropriate format was obtained from the 3D Art [2] website.

3.2 Single-Scattering

Single-scattering (SS) is understood as a situation in which light on its way from the source to the observer's eye is subject to a single scattering event. In the context of volumetric cloud rendering, this problem is solved by creating a ray originating from the point of view that penetrates the cloud mass, undergoing a single scattering event towards the light source, repeatedly along an axis that intersects the point of view.

In the SS implementation used, two parts can be basically distinguished: the primary ray and the shadow ray. The path of the primary ray begins at the point of observation and its direction is determined to pass through a given pixel on the screen. The step length is predetermined and should be a compromise between the algorithm's efficiency and the desired accuracy of mapping the cloud structure.

At each step of the primary ray, information about the cloud density is collected. The density of the sample determines the background occlusion factor and the amount of light reflected towards the camera. The shadow ray is taken into account at each step of the primary ray for which the cloud sample has a non-zero density (Fig. 1). It begins in the last position of the primary ray and is directed towards the light source. On the section from the beginning of this ray to the boundary of the cloud, the integral of the cloud mass is calculated. As in the case of the primary ray, a constant step length is defined here. The integration result is important to determine the amount of light absorbed by the cloud.

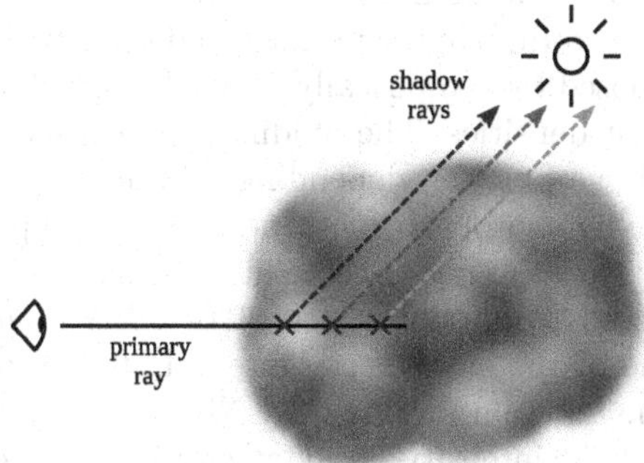

Fig. 1. Visualization of the rays used in the single-scattering algorithm

The Monte-Carlo [9] method was used to stabilize the integration results in the case of multiple sampling. It is important that the stochastic process in this case guarantees a uniform distribution of values in the considered interval over subsequent samples. This is because it is assumed that a light scattering event can occur with equal probability at any point along the main ray. In this case it was implemented by randomizing the length of the first iteration of the step for each ray.

The noise source is a previously generated blue noise texture whose offset is randomized for each call to the rendering pipeline. Blue noise has a high amount of high-frequency noise, which favors the integration of results and the use of noise removal techniques [4].

The background occlusion factor (final alpha of a given fragment) is calculated by the formula (1):

$$\alpha_x = \prod_{i=1}^{x} e^{-d_x l_g} \tag{1}$$

where x is the next step number, d_x is the cloud density at a given point in space resulting from the step number, and l_g is the length of the primary ray step.

The color of the pixel is calculated by the formula (2):

$$c_x = \sum_{i=1}^{x} \alpha_x c_{ss} i_{xss} (1 - e^{-d_x l_g}) \tag{2}$$

where c_{ss} is the base color of the cloud, and i_{xss} (3) is the intensity of the incoming light calculated from the shadow ray (4).

$$i_{xss} = L(d_v, \frac{10}{t_{xss}}) + l_b(t_{xss}) \tag{3}$$

$$t_{xss} = \sum_{j=1}^{\inf} d_j \cdot l_p \tag{4}$$

The function $L(d_v, d_p)$ is a modified approximate Lorenz-Mie phase function that depends on the dot product of the primary and shadow ray direction vectors. Nishita et al. [12] proposed two empirically derived approximations for nebulous atmospheres of different densities. The modification consists in modulating the shape and strength of the function depending on the integral value in order to imitate changes in the light propagation characteristics depending on the mass of the cloud integrated by the ray. The solution is sufficient for the presented application, as it is used to obtain the effect of *silverlining*[1] which occurs in areas with a low value of the shadow ray integral. The formula (5) contains a generalized form of the function proposed by Nishit et al. [12], modified so that there is no significant light intensity decay when the coefficient $d_p \approx 0$, and the primary and shadow ray vectors have opposite directions ($d_v \approx -1$).

$$L(d_v, d_p) = \frac{1}{4\pi} d_p \left(\frac{1 + d_v}{2}\right)^{d_p} \tag{5}$$

An additional isotropic factor (6) showing a lower degree of absorption imitates light scattering, assuming the existence of irradiance coming from the surrounding cloud mass. The coefficients modifying the actual weights of the input parameters were also selected empirically.

$$l_b(t_{xss}) = e^{-t_{xss}} + se^{\frac{t_{xss}}{10}} + \frac{1}{5} 2se^{-\frac{t_{xss}}{50}} \tag{6}$$

3.3 Single-Scattering Cached

In the case of a stationary light source, the value returned by the shadow ray depends only on the starting position of this ray. In addition to the density value, the structure can store the previously calculated integral value discretized in the space of a single voxel. This should largely reduce the computational complexity of the algorithm at the expense of limited resolution.

The theoretical increase in memory requirement according to the previous solution is twofold due to the need to store information about the sum of the integral in parallel with the cloud density data. In practice it is higher due to the

[1] A bright outline of the cloud observed "against the sun", characteristic of dense clouds. A light ray, passing through a small mass of cloud, is largely scattered in the original direction, which causes a large amount of light to reach the observer's eye, being perceived as intense [1].

unfavorable *padding* of the leaf cell array of the hierarchical structure. The final highest resolution cloud data size was 239 Megabytes compared to the original 62.

3.4 Multi-scattering

The multi-scattering (MS) algorithm was created on the basis of the Single-Scattering Cached algorithm and the photon map used in it. In order to achieve performance that will allow real-time interaction, a number of simplifications have been made to make this possible.

The implemented scenario is based on the following assumptions:

- The scene has one directional light source (sun).
- A single volumetric cloud is rendered.
- The properties of the cloud located only at a short distance from the point under consideration are important

In relation to the SS algorithm, in addition to a single light scatter, a point cloud is introduced at a short distance from the branching point, which simulates the irradiance resulting from the scattering of light coming from the immediate vicinity of the sampled point. In addition to the value of the integral taken from a given cloud point, the integral between the initial sampling point and the cloud point is approximated by taking the density value at both points (7). Indeed, unlike a single scattering in the basic form of the algorithm, two such events occur, which are additionally repeated for different variants of the point arrangement at which the second scattering occurs.

$$t_{xms} = \sum_{j=1}^{n} |\boldsymbol{x} - \boldsymbol{x}_j| \frac{d_x + d_{xj}}{2} + t_{xjss} \tag{7}$$

The set of points that is part of the sampling cloud has the following properties:

- The scattering direction vector is uniformly random.
- In the case of a cloud consisting of N points, the distance of subsequent points from the local center of the cloud coordinate system is $r_x = \frac{Rx}{N}$. Where R is the maximum radius of the sphere of influence.

The MS factor undergoes a different transformation of the obtained intensity into the color of the cloud. In this case, the phase function used is isotropic. Additionally, the underlying SS algorithm is still in use, but its absorption coefficient is higher to effectively illuminate only directly illuminated parts of the cloud. This action increases the dynamic range deep in the cloud, where the MS algorithm plays a greater role.

4 Experiments and Results

In order to compare the results, images generated for different camera angles and parameters affecting the performance and quality of the render were compared separately for each method. Additionally, images of differences between selected scenarios have been processed to show in which parts of the images they are the largest. The difference in speed and accuracy of the single-scattering algorithm depending on the presence of the precalculation mechanism is shown. Based on a precalculated integral, single-scattering and multi-scattering are compared to reference method. Reference images are obtained from Blender using *Principled Volume* shader.

The generated images have a resolution of 1024 × 1024 pixels. Renders of the same shots are compared (identical camera position, cloud position, light direction). Frame render time was also documented. For each scenario, there are settings that change from the default ones and affect performance and image quality (Table 1).

Table 1. Rendering settings. Columns "hq" show hight quality settings and columns "hp" show high performance settings

Parameter	SSBF		SSC		MS	
	hq	hp	hq	hp	hq	hp
Primary ray lenght	0.20	0.60	0.05	0.60	0.05	0.60
Shadow ray lenght	0.11	0.60	0.60	0.60	0.60	0.60
Integral multiplier	22.00				5.00	
Intensity multiplier	3.573				6.414	
Lorenz-Mie function multiplier	1.00				0.00	
Influence factor exponent	n/a				2.00	
Radius of the sphere of influence	n/a				0.97	
Points	n/a				11	8
Rand on	n/a				1	0

This part presents the results of the implemented algorithms compared with reference images. Each scene shows an identical cloud and light source setting. Different scenarios arise due to changes in camera position and cloud density, which should introduce sufficient variability in the algorithm's input data.

The following terminology is used hereinafter:

- SS – Single Scattering – Single scattering algorithms in a general sense.
- SSBF – Single Scattering Brute Force – An algorithm implementing single scattering without precalculation support.
- SSC – Single Scattering Cached – Single scattering algorithm supported by a precalculated photon map.

- MS/MSC – Multi Scattering Cached – Multiple scattering algorithm supported by the same photon map.
- Reference – Render using *Principled Volume* shader.

Full image quality (shown in all subsequent renders) is achieved after generating 100 frames of a still image. All time measurements show the time it takes to generate a single frame.

4.1 Single Scattering Brute Force

This section contains visual (Fig. 2) and performance (Fig. 3) results for the SSBF method with high quality and high performance settings settings. Cloud rendering with a lower density multiplier has a longer single frame render time. There is a visible reduction in the detail of the generated image and an increase in noise for high performance settings.

Fig. 2. SSBF, high quality (3 columns on the left) vs. high performance (3 columns on the right), various points of view (one per column), various density multipliers d

4.2 Single Scattering Cached

Table 1 shows the settings for this algorithm. Unlike the SSBF algorithm, no shadow ray is used because its results have been precalculated and discretized within a single voxel.

Figure 4 shows visual results for the method. For high quality settings the images obtained are very similar to those generated by the SSBF method.

The high performance setting is characterized by an increased primary ray length. As in the case of SSBF, there is a visible reduction in the level of detail in

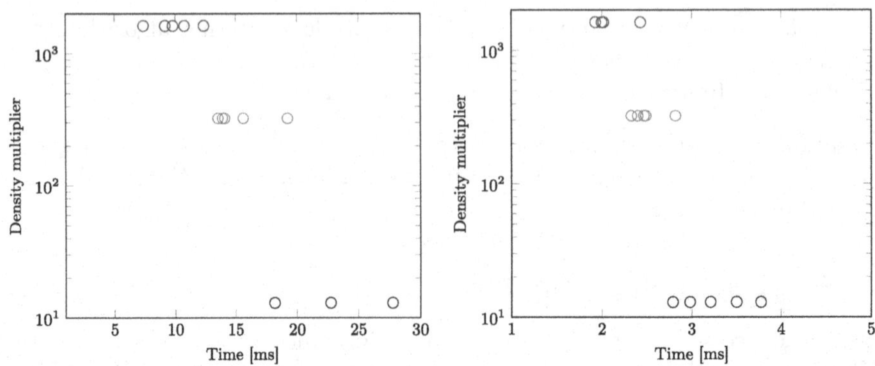

Fig. 3. SSBF, high quality (left) and high performance (right). Frame generation time vs. cloud density multiplier d

similar areas. Moreover, in the case of dense areas visible from the observation point in the direction of incidence of light, a significant darkening of directly illuminated areas is visible. The frame generation time is inversely proportional to the cloud density multiplier (Fig. 5).

Fig. 4. SSC, high quality (3 columns on the left) vs. high performance (3 columns on the right), various points of view (one per column), various density multipliers d

4.3 Multi Scattering Cached

Table 1 contains the settings for the algorithm. The length of the primary ray and the number of sampled points have an influence here. In each frame of the

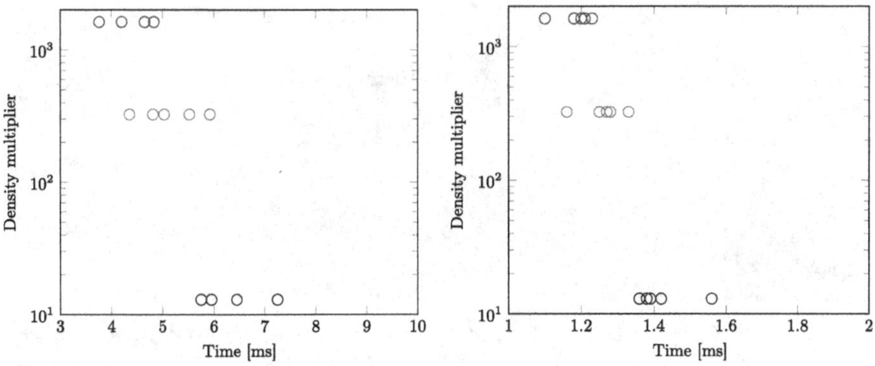

Fig. 5. SSC, high quality (left) and high performance (right). Frame generation time vs. cloud density multiplier d

animation, the position of the sampling points is randomized. For high quality settings, unlike SS renders, details and increased brightness of cloud fragments that are not directly lit are visible (Fig. 6).

For high performance settings, the length of the primary ray has been increased and the number of sampled points has been reduced. Randomization of points per frame is disabled to speed up image stabilization.

High-performance render is characterized by a reduction in the detail of the cloud mass representation, similar to the rest of the algorithms. The details created by MS operation do not differ significantly. However, it can be suggested that a larger number of points and their randomization in each frame (the results are stabilized by the TAA algorithm) are not necessary in the case of high performance settings in order to obtain a detailed representation of the cloud.

The frame generation time compared to SS algorithms has been significantly increased (Fig. 7). It is inversely proportional to the density multiplier of the generated volumetric structure.

4.4 Comparison with the Reference Method

In Fig. 8 (top) there is a big difference in contrast due to the different color management system. In addition, in the case of the SS algorithm, the cloud is perceptually denser. In dark parts of the cloud, the SS algorithm highlights less details than the reference image. The texture of the low-density cloud (Fig. 8 bottom) is very similar for both algorithms. The backlight view (Fig. 9 top) of a dense cloud highlights the differences in the *silverlining* effect of both methods, which is more intense in the case of reference. The same shot for a thin cloud (Fig. 9 bottom) shows a similar effect, with the difference that it is visible throughout the entire volume of the cloud.

The use of the MS algorithm (Fig. 10 top) shows a similar nature of details in the dark parts of the cloud to those present in the reference image, but their intensity is lower. In the case of thin clouds (Fig. 10 bottom), the effect is similar

Fig. 6. MS, high quality (3 columns on the left) vs. high performance (3 columns on the right), various points of view (one per column), various density multipliers d

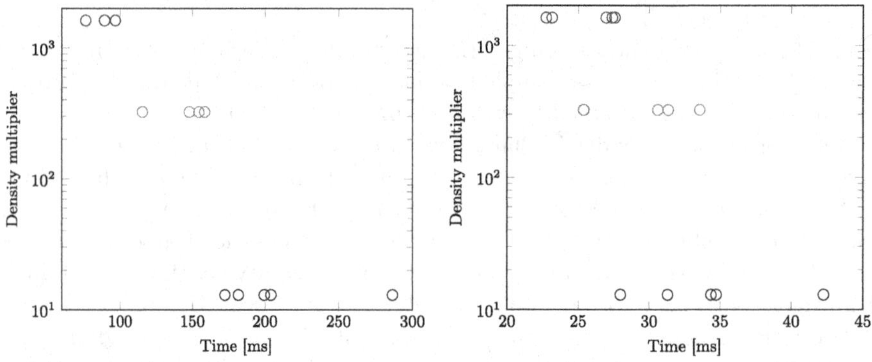

Fig. 7. MS, high quality (left) and high performance (right). Frame generation time vs. cloud density multiplier d

to the SS algorithm (Fig. 8 bottom). The application of the MS algorithm (Fig. 11 top) shows a similar difference to that in Fig. 10 (top). The effect of *silverlining* does not change because it comes from the underlying SS algorithm. In the case of thin clouds (Fig. 11 bottom) the situation is similar to Fig. 9 (bottom).

4.5 Summary

The algorithm based solely on traditional single scattering effectively reflects the appearance of low-density clouds and fragments of dense clouds that are directly illuminated by a light source. Areas hidden in their own shadow are characterized by a lack of contrast and a low level of detail. Using our multi scattering approximation, details are obtained in these parts of the structure. It

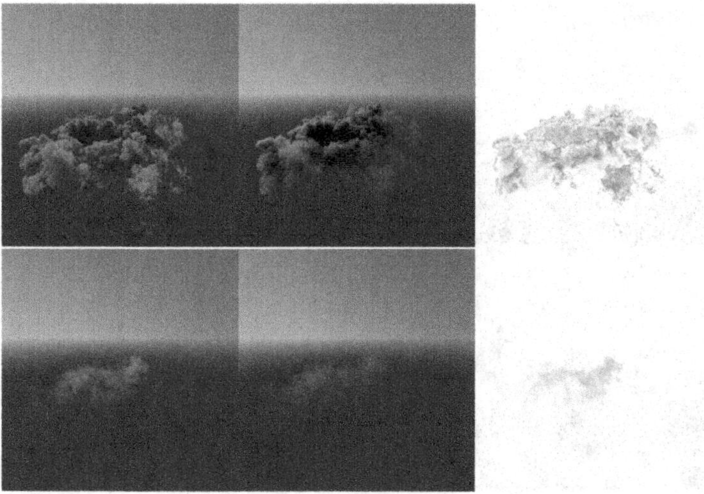

Fig. 8. Comparison of SS with the reference, and difference, $d = 1620$ (top) $d = 12.96$ (bottom), the first point of view

Fig. 9. Comparison of SS with the reference, and difference, $d = 1620$ (top) $d = 12.96$ (bottom), the second point of view

also produces a visual effect that is less different from the reference images than the initial SS algorithm.

The lengthening of the primary ray reduces the effectiveness of reflecting the cloud density, and the longer shadow ray in the case of the SSBF algorithm reduces the detail of determining the brightness of a given cloud fragment.

Fig. 10. Comparison of MS with the reference, and difference, $d = 1620$ (top) $d = 12.96$ (bottom), the first point of view

Fig. 11. Comparison of MS with the reference, and difference, $d = 1620$ (top) $d = 12.96$ (bottom), the second point of view

For each implemented algorithm, a performance degradation is observed in proportion to the reduction of the cloud density multiplier (Table 2). In thin clouds, the difference between settings with different performance is also reduced. What is important from the point of view of the MS algorithm is the number of points and randomization in each frame for the final effect. There are visible differences, although they are small.

Table 2. Average frame render times (in milliseconds) for various density multipliers

density	SSBF		SSC		MS	
	hq	hp	hq	hp	hq	hp
1620	9.90	2.08	4.24	1.18	89.66	25.56
324	15.27	2.50	5.13	1.26	144.90	29.26
12.96	22.93	3.25	6.23	1.42	208.89	34.12

5 Discussion

Slow performance was observed for each algorithm when rendering thinner clouds. This results from the use of optimization, which samples the denser structure with the primary ray only to the point where it is profitable during the operation *alpha* is accumulated. Exceeding the limit value (here it is 95%) stops sampling because the influence of further samples on the final color is negligible. Thinner clouds are less sensitive to step length in terms of final render quality, so increasing the step length should allow to recover the performance lost due to the above-mentioned reason.

The use of multi-scattering is profitable only in the case of dense clouds. Below a certain density, the capabilities of the single-scattering algorithm do not differ much from those of reference renderers.

The generated images were created as a result of a different *tone mapping* process, therefore the reference images are more contrasting than those generated using the presented methods. Nevertheless, when comparing the performance of the MS and SS algorithms, it can be concluded that the characteristics of the details are more similar to the reference images in the case of the MS implementation. MS algorithm, while generating an image slower than any SS method, it is still significantly faster than ray tracing used in *Principled Volume* shader.

The use of precalculation results in a significant increase in performance at the expense of increased demand for GPU memory. To avoid this problem, a different data arrangement would be needed in the SSBO structure to avoid unfavorable *padding*.

6 Conclusions

The assumption of constant lighting conditions allowed the use of precalculation of SS calculations, while the analysis of the characteristics of phenomena occurring in high-albedo clouds and other algorithms based on MS allowed the creation of a simplified method that can be used in interactive applications.

The resulting solution renders the volumetric structure in a way closer to the reference method than the initial algorithm (SS). However, it was not possible to achieve effects identical to the images generated in *Blender*. The main problem is the use of a different color transformation at the post-processing stage in both programs, which is the reason for differences in contrast and color tone.

Differences in the nature of clouds with different densities were also noticed. In the case of thin clouds, the SS algorithm is sufficient to generate images similar to the reference methods. Thin clouds also have lower step length requirements for *ray marching*. Dense clouds, however, benefit from optimization based on their properties. On the other hand, they require the use of an MS-based algorithm to generate more realistic results.

Rendering times allows the use of the proposed methods for real-time rendering, especially for the Single Scattering Cached algorithm where the times are single milliseconds per frame. For Multi Scattering Cached, the high performance setting reaches times of approximately 30 milliseconds, which also allows for attempts to be used in real-time rendering. The high quality setting in MSC gives times above 100 milliseconds. Future works could consider improving these times.

References

1. Silver lining and cloud iridescence. http://ww2010.atmos.uiuc.edu/(Gh)/guides/mtr/opt/wtr/ir.rxml. Accessed 3 Feb 2024
2. 3D Art: Free VDB clouds by VFX assets. https://www.3dart.it/en/free-vdbclouds/. Accessed 19 Dec 2022
3. Deng, H., Wang, B., Wang, R., Holzschuch, N.: A practical path guiding method for participating media. Comput. Vis. Media **6**(1), 37–51 (2020). https://doi.org/10.1007/s41095-020-0160-1
4. Eric Heitz, L.B.: Distributing monte carloerrors as a blue noise in screen space by permutingpixel seeds between frames (2019). https://hal.archives-ouvertes.fr/hal-02158423/le/blueNoiseTemporal2019slides.pdf. Accessed 2 Jan 2023
5. Academy Software Foundation: Openvdb. https://www.openvdb.org/. Accessed 19 Dec 2022
6. Harris, M.J., Lastra, A.: Real-time cloud rendering. Comput. Graph. Forum **20**(3), 76–85 (2001). https://doi.org/10.1111/1467-8659.00500
7. Kajiya, J.T., Von Herzen, B.P.: Ray tracing volume densities. ACM SIGGRAPH Comput. Graph. **18**(3), 165–174 (1984). https://doi.org/10.1145/964965.808594
8. Kallweit, S., Müller, T., Mcwilliams, B., Gross, M., Novák, J.: Deep scattering: rendering atmospheric clouds with radiance-predicting neural networks. ACM Trans. Graph. **36**(6), 1–11 (2017). https://doi.org/10.1145/3130800.3130880
9. Keller, A.: Quasi-Monte Carlo methods in computer graphics: Theglobal illumination problem. https://www.semanticscholar.org/paper/Quasi-Monte-Carlo/579bca8c938f1e0f25474a02a493f5b6ea8886. Accessed 23 Jan 2024
10. Koerner, D., Portsmouth, J., Sadlo, F., Ertl, T., Eberhardt, B.: Flux-limited diffusion for multiple scattering in participating media. Comput. Graph. Forum **33**(6), 178–189 (2014). https://doi.org/10.1111/cgf.12342
11. Mukhina, K., Bezgodov, A.: The method for real-time cloud rendering. Procedia Comput. Sci. **66**, 697–704 (2015). https://doi.org/10.1016/j.procs.2015.11.079
12. Nishita, T., Miyawaki, Y., Nakamae, E.: A shading model for atmospheric scattering considering luminous intensity distribution of light sources. In: Proceedings of 14th SIGGRAPH 1987. ACM (1987). https://doi.org/10.1145/37401.37437
13. Yusov, E.: High-performance rendering of realistic cumulus clouds using pre-computed lighting (2014). https://doi.org/10.2312/HPG.20141101

Modeling the Dynamics of a Multi-planetary System with Anisotropic Mass Variation

Alexander Prokopenya[1(✉)], Mukhtar Minglibayev[2,3], and Aiken Kosherbayeva[2,3]

[1] Warsaw University of Life Sciences–SGGW, Nowoursynowska 159, 02-776 Warsaw, Poland
alexanderprokopenya@sggw.edu.pl

[2] Fesenkov Astrophysical Institute, Observatoriya 23, 050020 Almaty, Kazakhstan

[3] Al-Farabi Kazakh National University, Al-Farabi avenue 71, 050040 Almaty, Kazakhstan

Abstract. A classical non-stationary $(n+1)$-body planetary problem with n bodies of variable mass moving around the central star on quasi-elliptic orbits is considered. In addition to the mutual gravitational attraction, the bodies may be acted on by reactive forces arising due to anisotropic variation of their masses. The problem is analyzed in the framework of Newtonian's formalism and the differential equations of motion are derived in terms of the osculating elements of aperiodic motion on quasi-conic sections. These equations can be solved numerically and their solution will describe the motion of the bodies in detail. However, due to the orbital motion of the bodies the perturbing forces include many terms describing short-period oscillations. Therefore, to obtain the solution with high precision one needs to choose very small step size or to use an adoptive step size method and this increase a time of calculation substantially. As we are interested in the long-term behaviour of the system it will be necessary to perform additional calculations in order to extract a secular part of the solution. To simplify the calculations we expand the perturbing forces into power series in terms of eccentricities and inclinations which are assumed to be small and average these equations over the mean longitudes of the bodies. Finally, we obtain the differential equations describing the evolution of orbital parameters over a long period of time. As an application, we have solved the evolution equations numerically in the case of $n = 3$ and demonstrated an influence of the mass variation on the motion of the bodies. All the relevant symbolic and numeric calculations are performed with the aid of the computer algebra system Wolfram Mathematica.

Keywords: Multi-planetary system · variable mass · equations of motion · reactive forces · long-term evolution · Wolfram Mathematica

This research is partly funded by the Committee of Science of the Ministry of Science and Higher Education of the Republic of Kazakhstan (Grant No. AP14869472).

L. Franco et al. (Eds.): ICCS 2024, LNCS 14836, pp. 181–196, 2024.
https://doi.org/10.1007/978-3-031-63775-9_13

1 Introduction

The classical many-body problem is a famous model of celestial mechanics that is applied for studying an orbital motion in the planetary systems (see [1,2]). Recall that it describes the dynamical behaviour of the bodies P_0, P_1,... P_n of masses m_0, m_1,... m_n, respectively, attracting each other according to Newton's law of universal gravitation. Such a model provides good approximation for the motion of planets $P_j, (j = 1, 2, \ldots, n)$ around a parent star P_0 if the bodies are spherically symmetric and their masses are constant. Note that applying Newton's second law, one can easily write out the equations of motion of the $(n+1)$-body system but their general solution cannot be found in the case of three or more interacting bodies.

The mass of the parent star in the planetary system is usually much greater than the masses of planets and so in the first approximation the planets move around the star along Keplerian orbits determined by the corresponding exact solution of the two-body problem. Mutual attraction of the planets disturbs their motion and enforces their orbital parameters to change. However, application of the perturbation theory that has been developed quite well enables to investigate these effects accurately (see [3,4]). This approach turned out to be very successful for understanding a satellite motion in the Sun-planet and binary star systems when all parameters of the system remain constant and the stationary perturbation theory is used for its analysis (see, for example, [5,6]).

Real celestial bodies are not always stationary and their characteristics such as mass, size, shape, and internal structure, may vary with time (see, for example, [7–9]). The bodies masses influence essentially on their interaction and motion and so it is natural to study the dynamics of the many-body system with variable masses. Investigation of the simplest such system composed of two bodies has shown that the mass variability affects essentially its dynamic evolution (see [10–14]). Later these investigations were generalized to the system of three bodies of variable masses although works in this field are not numerous (see [15,16]).

Note that the problem of two bodies of variable masses is not integrable, in general. Therefore, the perturbation theory based on the exact solution of the two-body problem cannot be applied in the case of variable masses. However, one can modify the equations of motion in the problem of two bodies of variable mass in such a way that their general solution can be written in symbolic form for arbitrary law of mass variation of the bodies (see [17]). This solution describes aperiodic motion of a body on a quasi-conic section and may be considered as unperturbed motion. Such approach was exploited in a series of works [18–22], where the problem of three bodies of variable masses was investigated in the framework of the Hamiltonian formalism. Recently the three-body problem was investigated in the framework of Newton's formalism what enables to obtain directly differential equations for the orbital elements (see [23]).

The present work is an extension of [23] and is devoted to the study of dynamical evolution of multi-planetary system of $(n+1)$ bodies when n planets P_1, P_2,..., P_n move around a central star P_0 on quasi-elliptic orbits which are assumed to not intersect. The problem is studied in the framework of the perturbation theory where an aperiodic motion on quasi-conic sections is considered as the unperturbed motion. Mutual attraction of the bodies P_1, $P_2, \ldots$, P_n and reactive forces arising in the case of anisotropic mass variation enforce the orbital elements to change. Differential equations determining the perturbed motion of the bodies are obtained in terms of the osculating elements of aperiodic motion on quasi-conic sections in the framework of Newton's formalism. In the case of small eccentricities and inclinations of the orbits the perturbing forces may be expanded in series in these parameters up to any desired order but here we consider only the first order terms what is sufficient to obtain the results corresponding to the accuracy of the observations. Averaging the equations of the perturbed motion over mean longitudes of the bodies P_1, $P_2, \ldots$, P_n in the absence of mean-motion resonances, we obtain the differential equations describing the evolution of orbital elements over long periods of time. These equations are solved numerically for different laws of the masses change in the case of $n = 3$. All relevant symbolic and numerical calculations are performed here with the aid of the computer algebra system Wolfram Mathematica [24].

The paper is organized as follows. In Sect. 2 we describe the model under consideration and obtain the equations of motion in the osculating elements of aperiodic motion on quasi-conic sections. Then in Sect. 3 derive the evolutionary equations which are solved numerically in Sect. 4 in the case of $n = 3$. At last, we summarize the results in Conclusion.

2 Equations of Motion

In a relative coordinate system with the origin at the center of parent star P_0 of mass $m_0(t)$ the equations of motion of the planets P_1, $P_2, \ldots$, P_n of masses $m_1(t), m_2(t), \ldots, m_n(t)$, respectively, may be written in the form (see [17,23])

$$\frac{d^2\boldsymbol{r}_j}{dt^2} + G(m_0 + m_j)\frac{\boldsymbol{r}_j}{r_j^3} - \frac{\ddot{\gamma}_j}{\gamma_j}\boldsymbol{r}_j = \boldsymbol{F}_j, \; j = 1, 2, \ldots, n. \tag{1}$$

Here G is the gravitational constant, $\boldsymbol{r}_j$ is the radius-vector of the planet P_j and the twice differentiable functions $\gamma_j(t)$ are defined by

$$\gamma_j(t) = \frac{m_{00} + m_{j0}}{m_0(t) + m_j(t)}, \; j = 1, 2, \ldots, n, \tag{2}$$

where $m_{00} = m_0(t_0)$, $m_{j0} = m_j(t_0)$ are the masses of the bodies P_0, P_j, $(j = 1, 2, \ldots, n)$, respectively, at the initial instant of time. The forces $\boldsymbol{F}_j$ in the right-hand side of (1) are given by

$$\boldsymbol{F}_j = G \sum_{k=1(k\neq j)}^{n} m_k \left(\frac{\boldsymbol{r}_k - \boldsymbol{r}_j}{r_{jk}^3} - \frac{\boldsymbol{r}_k}{r_k^3} \right) - \frac{\ddot{\gamma}_j}{\gamma_j}\boldsymbol{r}_j + \boldsymbol{Q}_j, \tag{3}$$

where

$$r_{jk} = \sqrt{(x_k - x_j)^2 + (y_k - y_j)^2 + (z_k - z_j)^2}, \; r_j = \sqrt{x_j^2 + y_j^2 + z_j^2}, \tag{4}$$

and the reactive forces $\boldsymbol{Q}_j$ are determined by the expressions (see [25])

$$\boldsymbol{Q}_j = \frac{\dot{m}_j}{m_j}\boldsymbol{V}_j - \frac{\dot{m}_0}{m_0}\boldsymbol{V}_0, \; j = 1, 2, \ldots, n. \tag{5}$$

The dot above a symbol in (3) – (5) denotes the total time derivative of the corresponding function, and $\boldsymbol{V}_j$, $(j = 0, 1, 2, \ldots, n)$ are the relative velocities of the particles leaving the body P_j or falling on it.

2.1 Unperturbed Motion

Note that in the case of constant masses when $\gamma_j(t) = 1, (j = 1, 2, \ldots, n)$ equations (1) reduce to the well-known equations determining relative motion of the bodies in the classical $(n+1)$-body problem. These equations are not integrable and are usually studied by methods of perturbation theory using an exact solution of the two-body problem as the first approximation (see, for example, [3]).

To apply similar approach to the case of variable masses we add the terms $\ddot{\gamma}_j/\gamma_j \boldsymbol{r}_j$ in the left-hand side of equations (1) and in expressions (3) for the forces $\boldsymbol{F}_j$ in the right-hand side of (1). This does not change the equations of relative motion (1) but enables to get integrable differential equations from (1) at $\boldsymbol{F}_j = 0$ for arbitrary laws of mass variation of the bodies.

Indeed, at $\boldsymbol{F}_j = 0, (j = 1, 2, \ldots, n)$ equations (1) become independent of each other and each of them has an exact solution that describes aperiodic motion of the body $P_j, (j = 1, 2, \ldots, n)$ on a quasi-conic section (see [17]); it can be written as

$$\begin{aligned}
x_j = \gamma_j a_j \Big(&(\cos E_j - e_j)(\cos\omega_j \cos\Omega_j - \sin\omega_j \sin\Omega_j \cos i_j) - \\
&- \sqrt{1 - e_j^2} \sin E_j (\sin\omega_j \cos\Omega_j + \cos\omega_j \sin\Omega_j \cos i_j) \Big), \\
y_j = \gamma_j a_j \Big(&(\cos E_j - e_j)(\cos\omega_j \sin\Omega_j + \sin\omega_j \cos\Omega_j \cos i_j) - \\
&- \sqrt{1 - e_j^2} \sin E_j (\sin\omega_j \sin\Omega_j - \cos\omega_j \cos\Omega_j \cos i_j) \Big), \\
z_j = \gamma_j a_j \Big(&(\cos E_j - e_j) \sin\omega_j + \sqrt{1 - e_j^2} \sin E_j \cos\omega_j \Big) \sin i_j.
\end{aligned} \tag{6}$$

The constants a_j, e_j, i_j, Ω_j and ω_j in (6) are analogues of the well-known Keplerian orbital elements and are determined from the initial conditions of motion (see [17]). An analogue of the eccentric anomaly E_j is determined by the well-known Kepler equation

$$E_j - e_j \sin E_j = M_j, \tag{7}$$

where analog of the mean anomaly M_j is given by

$$M_j = \frac{\sqrt{\kappa_j}}{a_j^{3/2}} (\Phi_j(t) - \Phi_j(\tau_j)), \;\; \Phi_j(t) = \int_0^t \frac{dt}{\gamma_j^2(t)}, \tag{8}$$

where $\kappa_j = G(m_{00} + m_{j0}), (j = 1, 2, \ldots, n)$. By τ_j in (8) we denote an analog of the time when the body P_j passes through the pericenter.

Note that solutions (6) differ from the corresponding solutions to the two-body problem with constant masses only by the presence of a time-dependent scaling coefficient $\gamma_j(t)$. Besides, the mean anomaly M_j is not a linear function of time but it is an increasing function of time (see (8)). If the laws of masses variation $m_j(t)$ are known the functions $\gamma_j(t)$ define the mean anomalies $M_j(t)$ and equation (7) enables to find the eccentric anomalies $E_j(t)$ as functions of time. Therefore, solutions (6) define the unperturbed motion of the planets P_j in terms of the time and $6n$ constants of integration $a_j, e_j, i_j, \Omega_j, \omega_j$, and $\tau_j, j = 1, 2 \ldots, n$, which may be considered as analogues of the Keplerian orbital elements (see [17]).

2.2 Perturbed Motion

Mutual attraction and reactive forces (5) arising in the case of anisotropic mass variation of the bodies P_j affect their motion and the orbital elements must necessarily vary with the time. To obtain the differential equations, determining the dependence of the orbital parameters on time, one can use the method of the variation of arbitrary constants that is well-known in the theory of differential equations. Assuming the orbital parameters are functions of time and substituting solutions (6) into (1), we obtain $3n$ differential equations for $6n$ unknown functions $a_j(t)$, $e_j(t)$, $i_j(t)$, $\Omega_j(t)$, $\omega_j(t)$, $M_j(t), (j = 1, 2, \ldots, n)$. Additional $3n$ equations are usually obtained from the condition that the coordinates x_j, y_j, z_j and the corresponding velocity components at time t are determined by functions (6) and their derivatives with respect to time under the condition that the orbital elements are constant. As a result, the perturbed coordinates and velocity components of the bodies P_j yield the instantaneous orbital elements $a_j, e_j, i_j, \Omega_j, \omega_j$, and M_j given by formulas (6)–(8). Such instantaneous elements are known as the osculating elements (see, for example, [1,2]).

By performing the corresponding symbolic calculations (see details in [23]), we obtain the following system of differential equations for finding the dependence of the orbital elements on time:

$$\frac{da_j}{dt} = \frac{2a_j^{3/2}\gamma_j(t)}{\sqrt{\kappa_j}(1 - e_j \cos E_j)} \left(e_j \sin E_j \ F_{rj} + \sqrt{1 - e_j^2} F_{\tau j} \right), \tag{9}$$

$$\frac{de_j}{dt} = \frac{\sqrt{a_j(1 - e_j^2)\gamma_j(t)}}{\sqrt{\kappa_j}(1 - e_j \cos E_j)} \left(\sqrt{1 - e_j^2} \sin E_j \ F_{rj} + \right.$$
$$\left. + \ (2 \cos E_j - e_j - e_j \cos^2 E_j) F_{\tau j} \right), \tag{10}$$

$$\frac{di_j}{dt} = \frac{\sqrt{a_j}\gamma_j(t)}{\sqrt{\kappa_j(1 - e_j^2)}} F_{nj} \left((\cos E_j - e_j) \cos \omega_j - \sqrt{1 - e_j^2} \sin \omega_j \sin E_j \right), \tag{11}$$

$$\frac{d\Omega_j}{dt} = \frac{\sqrt{a_j}\gamma_j(t)}{\sqrt{\kappa_j(1-e_j^2)}} \frac{F_{nj}}{\sin i_j}\left((\cos E_j - e_j)\sin\omega_j + \sqrt{1-e_j^2}\cos\omega_j\sin E_j\right), \quad (12)$$

$$\frac{d\omega_j}{dt} = -\frac{\sqrt{a_j}\gamma_j(t)\cot i_j}{\sqrt{\kappa_j(1-e_j^2)}} F_{nj}\left((\cos E_j - e_j)\sin\omega_j + \sqrt{1-e_j^2}\cos\omega_j\sin E_j\right) -$$

$$-\frac{\sqrt{a_j}\gamma_j(t)}{e_j\sqrt{\kappa_j}(1-e_j\cos E_j)}\left((\cos E_j - e_j)\sqrt{1-e_j^2}F_{rj} - \right.$$

$$\left. -(2-e_j^2-e_j\cos E_j)\sin E_j F_{\tau j}\right), \quad (13)$$

$$\frac{dM_j}{dt} = \frac{\sqrt{a_j}\gamma_j(t)}{e_j\sqrt{\kappa_j}(1-e_j\cos E_j)}\left(\sqrt{1-e_j^2}(-2+e_j^2+e_j\cos E_j)\sin E_j F_{\tau j} + \right.$$

$$\left. + \left((1+3e_j^2)\cos E_j - e_j(3+e_j^2\cos(2E_j))\right) F_{rj}\right) + \frac{\sqrt{\kappa_j}}{a_j^{3/2}\gamma_j^2(t)}. \quad (14)$$

The forces F_{rj}, $F_{\tau j}$, and F_{nj} in the right-hand sides of (9)–(14) are the radial, transversal and normal components of the forces $\boldsymbol{F}_j$, respectively, determined by expressions (3), (5). The reactive forces $\boldsymbol{Q}_j$ (see (5)) are usually determined in the orbital systems of coordinates of the bodies P_j, so the forces $\boldsymbol{F}_j$ are also written in these systems of coordinates. The direction cosines of the unit vectors $\boldsymbol{e}_{rj} = (e_{xj}, e_{yj}, e_{zj})$, $\boldsymbol{e}_{\tau j} = (\tau_{xj}, \tau_{yj}, \tau_{zj})$, and $\boldsymbol{e}_{nj} = (n_{xj}, n_{yj}, n_{zj})$ along the radial, transversal, and normal directions, respectively, can be easily written on the basis of solutions (6):

$$e_{xj} = \frac{x_j}{\gamma_j a_j},\ e_{yj} = \frac{y_j}{\gamma_j a_j},\ e_{zj} = \frac{z_j}{\gamma_j a_j}, \quad (15)$$

$$n_{xj} = \sin\Omega_j \sin i_j,\ n_{yj} = -\cos\Omega_j \sin i_j,\ n_{zj} = \cos i_j, \quad (16)$$

$$\tau_{xj} = n_{yj}e_{zj} - n_{zj}e_{yj},\ \tau_{yj} = n_{zj}e_{xj} - n_{xj}e_{zj},\ \tau_{zj} = n_{xj}e_{yj} - n_{yj}e_{xj}. \quad (17)$$

Denoting the components of the relative velocities of particles leaving the body $P_j, (j = 1, \ldots, n)$ or falling on them along the radial, transversal, and normal directions in the orbital system of coordinates related to the body P_j by V_{rj}, $V_{\tau j}$, V_{nj} and using (3), (5), we obtain

$$F_{rj} = \boldsymbol{F}_j \cdot \boldsymbol{e}_{rj} = G \sum_{k=1(k\neq j)}^{n} m_k \left(\left(\frac{r_k}{r_{jk}^3} - \frac{1}{r_k^2}\right)(\boldsymbol{e}_{rk}\cdot\boldsymbol{e}_{rj}) - \frac{r_j}{r_{jk}^3}\right) - \frac{\ddot{\gamma}_j}{\gamma_j}r_j + Q_{rj},$$

$$F_{\tau j} = \boldsymbol{F}_j \cdot \boldsymbol{e}_{\tau j} = G \sum_{k=1(k\neq j)}^{n} m_k \left(\frac{r_k}{r_{jk}^3} - \frac{1}{r_k^2}\right)(\boldsymbol{e}_{rk}\cdot\boldsymbol{e}_{\tau j}) + Q_{\tau j},$$

$$F_{nj} = \boldsymbol{F}_j \cdot \boldsymbol{e}_{nj} = G \sum_{k=1(k\neq j)}^{n} m_k \left(\frac{r_k}{r_{jk}^3} - \frac{1}{r_k^2}\right)(\boldsymbol{e}_{rk}\cdot\boldsymbol{e}_{nj}) + Q_{nj}, \quad (18)$$

where the corresponding components of the reactive forces $\boldsymbol{Q}_j$ are given by

$$Q_{rj} = \frac{\dot{m}_j}{m_j} V_{rj} - \frac{\dot{m}_0}{m_0} \left(V_{r0} \left(\boldsymbol{e}_{r1} \cdot \boldsymbol{e}_{rj} \right) + V_{\tau 0} \left(\boldsymbol{e}_{\tau 1} \cdot \boldsymbol{e}_{rj} \right) + V_{n0} \left(\boldsymbol{e}_{n1} \cdot \boldsymbol{e}_{rj} \right) \right),$$

$$Q_{\tau j} = \frac{\dot{m}_j}{m_j} V_{\tau j} - \frac{\dot{m}_0}{m_0} \left(V_{r0} \left(\boldsymbol{e}_{r1} \cdot \boldsymbol{e}_{\tau j} \right) + V_{\tau 0} \left(\boldsymbol{e}_{\tau 1} \cdot \boldsymbol{e}_{\tau j} \right) + V_{n0} \left(\boldsymbol{e}_{n1} \cdot \boldsymbol{e}_{\tau j} \right) \right), \quad (19)$$

$$Q_{nj} = \frac{\dot{m}_j}{m_j} V_{nj} - \frac{\dot{m}_0}{m_0} \left(V_{r0} \left(\boldsymbol{e}_{r1} \cdot \boldsymbol{e}_{nj} \right) + V_{\tau 0} \left(\boldsymbol{e}_{\tau 1} \cdot \boldsymbol{e}_{nj} \right) + V_{n0} \left(\boldsymbol{e}_{n1} \cdot \boldsymbol{e}_{nj} \right) \right). \quad (20)$$

The relative velocities $\boldsymbol{V}_0$ in (19) of the particles leaving the body P_0 or falling on it are assumed to be given in the orbital system of coordinates related to the body P_1. If the relative velocities $\boldsymbol{V}_0$ and $\boldsymbol{V}_j$ and laws of variation of body masses are known, equations (9) – (14) completely determine the perturbed motion of the bodies $P_j, j = 1, 2, \ldots, n$.

3 Evolutionary Equations

Differential equations (9) – (14) describe the perturbed motion of the planets in terms of the osculating orbital elements but they are not integrable and their exact solution cannot be found. However, in many problems of celestial mechanics, eccentricities and inclinations of body orbits are small (see [1,4]). Here we consider this practically important case of small eccentricities $e_j << 1$ and inclinations $i_j << 1, (j = 1, 2, \ldots, n)$ and expand the right-hand sides of equations (9) – (14) in power series in these parameters. Note that applying the computer algebra system *Mathematica* (see [24]), one can calculate such expansions with any required accuracy but the corresponding expressions become very cumbersome in higher order terms. Here we restrict ourselves to computations up to the first order and obtain the following differential equations for the secular perturbations of the orbital elements of the body P_1:

$$\frac{da_1}{dt} = \frac{2a_1^{3/2}\gamma_1}{\sqrt{\kappa_1}} \left(\frac{\dot{m}_1}{m_1} V_{\tau 1} - \frac{\dot{m}_0}{m_0} V_{\tau 0} \right),$$

$$\frac{de_1}{dt} = -\frac{3\sqrt{a_1}}{2\sqrt{\kappa_1}} e_1 \gamma_1 \left(\frac{\dot{m}_1}{m_1} V_{\tau 1} - \frac{\dot{m}_0}{m_0} V_{\tau 0} \right) + \sum_{s=2}^{n} \frac{G m_s e_s}{\sqrt{a_1 \kappa_1}} \Pi_{12}^{1\,s} \sin(\omega_1 - \omega_s + \Omega_1 - \Omega_s),$$

$$\frac{di_1}{dt} = -\frac{3\sqrt{a_1}}{2\sqrt{\kappa_1}} e_1 \gamma_1 \left(\frac{\dot{m}_1}{m_1} V_{n1} - \frac{\dot{m}_0}{m_0} V_{n0} \right) \cos\omega_1 + \sum_{s=2}^{n} \frac{G m_s i_s}{4\sqrt{a_1 \kappa_1}} B_1(\alpha_{1\,s}) \sin(\Omega_1 - \Omega_s),$$

$$\frac{d\Omega_1}{dt} = -\frac{3\sqrt{a_1}}{2\sqrt{\kappa_1}} e_1 \gamma_1 \left(\frac{\dot{m}_1}{m_1} V_{n1} - \frac{\dot{m}_0}{m_0} V_{n0} \right) \frac{\sin\omega_1}{i_1} -$$

$$-\sum_{s=2}^{n} \frac{G m_s}{4\sqrt{a_1 \kappa_1}} B_1(\alpha_{1\,s}) \left(1 - \frac{i_s}{i_1} \cos(\Omega_1 - \Omega_s) \right),$$

$$\frac{d\omega_1}{dt} = \frac{3\sqrt{a_1}}{2\sqrt{\kappa_1}} e_1\gamma_1 \left(\frac{\dot{m}_1}{m_1}V_{n1} - \frac{\dot{m}_0}{m_0}V_{n0}\right) \frac{\sin\omega_1}{i_1} + \frac{\sqrt{a_1}}{\sqrt{\kappa_1}}\gamma_1 \left(\frac{\dot{m}_1}{m_1}V_{r1} - \frac{\dot{m}_0}{m_0}V_{r0}\right) -$$

$$-\frac{3a_1^{3/2}}{2\sqrt{\kappa_1}}\gamma_1\ddot{\gamma}_1 + \sum_{s=2}^{n} \frac{Gm_s}{\sqrt{a_1\kappa_1}} \left(\Pi_{11}^{1\,s} - \frac{1}{4}B_1(\alpha_{1\,s})\left(1 + \frac{i_s}{i_1}\cos(\Omega_1 - \Omega_s)\right)\right) +$$

$$+\sum_{s=2}^{n} \frac{Gm_s}{\sqrt{a_1\kappa_1}} \frac{e_s}{e_1} \Pi_{12}^{1\,s} \cos(\omega_1 - \omega_s + \Omega_1 - \Omega_s), \; \alpha_{1\,s} = \frac{a_1\gamma_1}{a_s\gamma_s} < 1. \quad (21)$$

Remind that reactive forces (19) acting on the star P_0 are determined in the orbital coordinate system of the body P_1. Due to this equations (21) differ a little bit of the differential equations for the secular perturbations of the orbital elements of the bodies $P_2, \ldots, P_n$ which are given by

$$\frac{da_j}{dt} = \frac{2a_j^{3/2}\gamma_j}{\sqrt{\kappa_j}} \frac{\dot{m}_j}{m_j} V_{\tau j},$$

$$\frac{de_j}{dt} = -\frac{3\sqrt{a_j}}{2\sqrt{\kappa_j}} e_j\gamma_j \frac{\dot{m}_j}{m_j} V_{\tau j} + \frac{3\sqrt{a_j}}{2\sqrt{\kappa_j}} e_j\gamma_j \frac{\dot{m}_0}{m_0} \left(e_1 V_{\tau 0}\cos(\omega_1 - \omega_j + \Omega_1 - \Omega_j) +\right.$$

$$\left. + e_1 V_{r0}\sin(\omega_1 - \omega_j + \Omega_1 - \Omega_j) + i_1 V_{n0}\cos(\omega_j - \Omega_1 + \Omega_j) - i_j V_{n0}\cos\omega_j\right) -$$

$$-\sum_{s=1}^{j-1} \frac{Gm_s e_s}{\sqrt{a_j\kappa_j}} \Pi_{12}^{sj} \sin(\omega_s - \omega_j + \Omega_s - \Omega_j) +$$

$$+\sum_{s=j+1}^{n} \frac{Gm_s e_s}{\sqrt{a_j\kappa_j}} \Pi_{12}^{js} \sin(\omega_j - \omega_s + \Omega_j - \Omega_s),$$

$$\frac{di_j}{dt} = -\frac{3\sqrt{a_j}}{2\sqrt{\kappa_j}} e_j\gamma_j \left(\frac{\dot{m}_j}{m_j}V_{nj} - \frac{\dot{m}_0}{m_0}V_{n0}\right)\cos\omega_j -$$

$$-\sum_{s=1}^{j-1} \frac{Gm_s i_s}{4\sqrt{a_j\kappa_j}} B_1(\alpha_{sj})\sin(\Omega_s - \Omega_j) + \sum_{s=j+1}^{n} \frac{Gm_s i_s}{4\sqrt{a_j\kappa_j}} B_1(\alpha_{js})\sin(\Omega_j - \Omega_s),$$

$$\frac{d\Omega_j}{dt} = -\frac{3\sqrt{a_j}}{2\sqrt{\kappa_j}} e_j\gamma_j \left(\frac{\dot{m}_j}{m_j}V_{nj} - \frac{\dot{m}_0}{m_0}V_{n0}\right)\frac{\sin\omega_j}{i_j} -$$

$$-\sum_{s=1}^{j-1} \frac{Gm_s}{4\sqrt{a_j\kappa_j}} B_1(\alpha_{sj})\left(1 - \frac{i_s}{i_j}\cos(\Omega_s - \Omega_j)\right) -$$

$$-\sum_{s=j+1}^{n} \frac{Gm_s}{4\sqrt{a_j\kappa_j}} B_1(\alpha_{js})\left(1 - \frac{i_s}{i_j}\cos(\Omega_j - \Omega_s)\right),$$

$$\frac{d\omega_j}{dt} = -\frac{3a_j^{3/2}}{2\sqrt{\kappa_j}}\gamma_j\ddot{\gamma}_j + \frac{3\sqrt{a_j}}{2\sqrt{\kappa_j}} e_j\gamma_j \left(\frac{\dot{m}_j}{m_j}V_{nj} - \frac{\dot{m}_0}{m_0}V_{n0}\right)\frac{\sin\omega_j}{i_j} +$$

$$+\frac{\sqrt{a_j}}{\sqrt{\kappa_j}}\gamma_j\frac{\dot{m}_j}{m_j}V_{rj}+\frac{3\sqrt{a_j}}{2e_j\sqrt{\kappa_j}}\gamma_j\frac{\dot{m}_0}{m_0}\left(V_{n0}(i_j\sin\omega_j-i_1\sin(\omega_j-\Omega_1+\Omega_j))-\right.$$

$$\left.-V_{r0}e_1\cos(\omega_1-\omega_j+\Omega_1-\Omega_j)+V_{\tau 0}e_1\sin(\omega_1-\omega_j+\Omega_1-\Omega_j)\right)+$$

$$+\sum_{s=1}^{j-1}\frac{Gm_s}{\sqrt{a_j\kappa_j}}\left(\Pi_{22}^{sj}-\frac{1}{4}B_1(\alpha_{sj})\left(1+\frac{i_s}{i_j}\cos(\Omega_s-\Omega_j)\right)+\right.$$

$$\left.+\frac{e_s}{e_j}\Pi_{12}^{sj}\cos(\omega_s-\omega_j+\Omega_s-\Omega_j)\right)+$$

$$+\sum_{s=j+1}^{n}\frac{Gm_s}{\sqrt{a_j\kappa_j}}\left(\Pi_{11}^{js}-\frac{1}{4}B_1(\alpha_{js})\left(1+\frac{i_s}{i_j}\cos(\Omega_j-\Omega_s)\right)+\right.$$

$$\left.+\frac{e_s}{e_j}\Pi_{12}^{js}\cos(\omega_j-\omega_s+\Omega_j-\Omega_s)\right),\quad j=2,3,\ldots,n. \tag{22}$$

Here

$$\Pi_{12}^{ik}=\frac{1}{8}\left(9B_0(\alpha_{ik})+B_2(\alpha_{ik})\right)-\frac{1+\alpha_{ik}^2}{8\alpha_{ik}}\left(9C_0(\alpha_{ik})-3C_2(\alpha_{ik})\right)+$$

$$+\frac{3}{16}\left(7C_1(\alpha_{ik})+C_3(\alpha_{ik})\right),$$

$$\Pi_{11}^{ik}=-\frac{3}{4}\alpha_{ik}\left(B_0(\alpha_{ik})+2C_1(\alpha_{ik})\right)+\frac{6\alpha_{ik}^2+15}{8}C_0(\alpha_{ik})-\frac{9}{8}C_2(\alpha_{ik}),$$

$$\Pi_{22}^{ik}=-\frac{3}{4\alpha_{ik}}\left(B_0(\alpha_{ik})+2C_1(\alpha_{ik})\right)+\frac{6+15\alpha_{ik}^2}{8\alpha_{ik}^2}C_0(\alpha_{ik})-\frac{9}{8}C_2(\alpha_{ik}), \tag{23}$$

and $B_0(\alpha_{ik}), B_1(\alpha_{ik}), B_2(\alpha_{ik}), C_0(\alpha_{ik}), C_1(\alpha_{ik}), C_2(\alpha_{ik}), C_3(\alpha_{ik})$ are the Laplace coefficients (see [4,23]). As orbital parameters of the bodies are assumed to satisfy the conditions $a_1\gamma_1 < a_2\gamma_2 < \ldots < a_n\gamma_n$ the arguments of the Laplace coefficients in (21)– (23) are smaller than 1:

$$\alpha_{ik}=\frac{a_i\gamma_i}{a_k\gamma_k}<1,\quad 1\le i<k\le n. \tag{24}$$

Equations (21), (22) determine the secular perturbations of the orbital elements of the planets $P_1,\ldots,P_n$. Although we take into account only linear terms in the power expansions of the right-hand sides of equations (9) – (14) in terms of eccentricities e_j and inclinations i_j, the Eqs. (21), (22) are very complicated and we cannot find their solution in symbolic form. However, we can choose some realistic laws of the masses variations and find their numerical solution. In this way we can investigate an influence of the masses variation on the dynamics of the $(n+1)$-body planetary system.

4 Simulation

To test the model, let us consider the case of three planets P_1, P_2, P_3 orbiting the parent star P_0. To solve equations (21), (22) numerically, it is expedient to use the dimensionless variables. For example, we use initial values of the semi-major axis $a_{10} = a_1(t_0)$ and the mass m_{00} of body P_0 as units of distance and mass, respectively, and define dimensionless distance a_j^*, mass m_j^* and time t^* by

$$a_j^* = \frac{a_j}{a_{10}}, \; m_0^* = \frac{m_0}{m_{00}}, \; m_j^* = \frac{m_j}{m_{00}}, \; t^* = t\frac{\sqrt{\kappa_1}}{a_{10}^{3/2}}, \; j = 1, 2, 3. \tag{25}$$

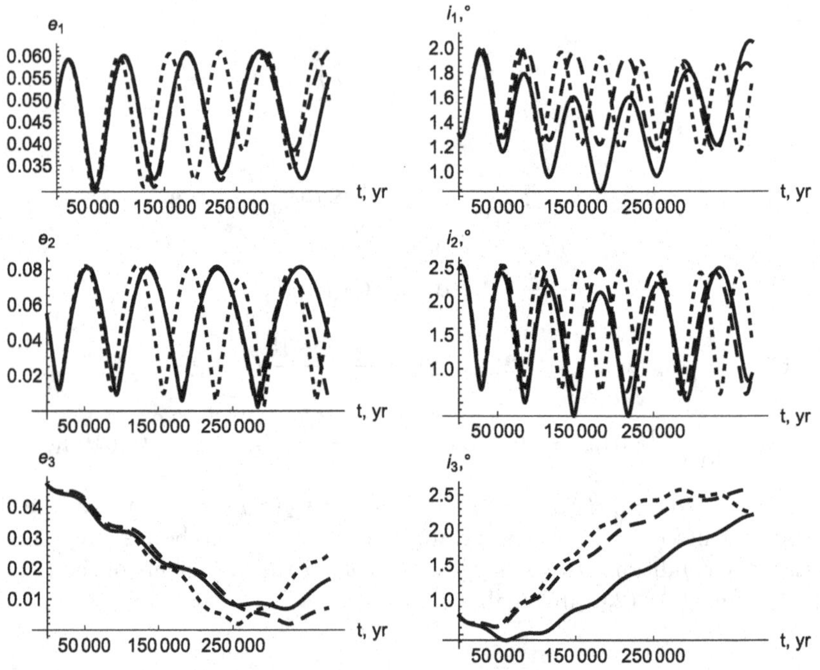

Fig. 1. Eccentricities e_j and inclinations i_j of the bodies P_1, P_2, P_3 (short dashing – constant masses, long dashing – isotropic mass changes, solid curves – anisotropic mass changes, $V_{n0} = 1/2, V_{\tau 1} = -1$)

The mass of the parent star P_0 is assumed to decrease according to the Eddington-Jeans law

$$m_0^*(t^*) = \left((m_{00}^*)^{1-n_j} - \beta_0(1-n_0)(t^* - t_0^*)\right)^{\frac{1}{1-n_0}}, \tag{26}$$

where $n_0 = 2$, $\beta_0 = 1/300000$, while the mass of the planet P_1 increases with time at constant dimensionless rate $\dot{m}_1 = 2,277 \cdot 10^{-12}$. Masses of the planets P_2, P_3 are assumed to be constant.

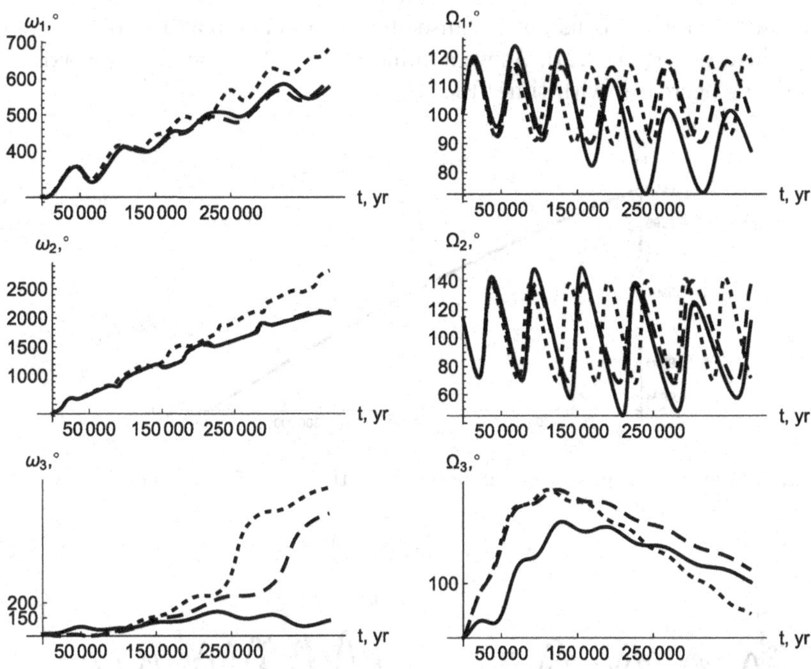

Fig. 2. Parameters ω_j and Ω_j of the bodies P_1, P_2, P_3 (short dash – constant masses, long dash – isotropic mass changes, solid curves – anisotropic mass changes, $V_{n0} = 1/2$, $V_{\tau 1} = -1$)

As a test system, we consider the Sun, Jupiter, Saturn, and Uranus as bodies P_0, P_1, P_2, and P_3, respectively, and choose the following initial values for orbital elements (see [4]):

$$m_{00} = 1,9891 \times 10^{30} kg,\ m_{10} = 1,8982 \times 10^{27} kg,\ m_{20} = 5,6852 \times 10^{26} kg,$$
$$m_{30} = 8,6843 \times 10^{25} kg,\ a_{10} = 5,2034 AU,\ a_{20} = 9,5371 AU,\ a_{30} = 19,191 AU,$$
$$e_{10} = 0,0484,\ e_{20} = 0,0541,\ e_{30} = 0,0472,\ i_{10} = 1,304^\circ,\ i_{20} = 2,485^\circ,$$
$$i_{30} = 0,772^\circ,\ \Omega_{10} = 100,56^\circ,\ \Omega_{20} = 113,72^\circ,\ \Omega_{30} = 74,23^\circ,$$
$$\omega_{10} = 273,98^\circ,\ \omega_{20} = 338,71^\circ,\ \omega_{30} = 96,73^\circ.$$

In the case of constant masses of the bodies Eqs. (21), (22) describe the secular perturbations of the orbital elements in the framework of the classical four-body problem (see [2,4]). Taking into account the isotropic mass variation of the body P_0 according to the Eddington-Jeans law and linear isotropic increase of mass of the body P_1 when reactive forces do not arise results in only some quantitative changes of solutions to (21), (22) (see Figs. 1 and 2). However, the anisotropic mass variation with only two nonzero dimensionless velocities $V_{n0} = 1/2, V_{\tau 1} = -1$ modifies substantially behaviour of the inclinations i_j and the longitudes of the ascending node Ω_j of all three planets. The semi-major axes a_2 and a_3 remain constant because only transversal reactive forces $Q_{\tau j}$ can change them (see (21), (22)). As we assume $V_{\tau 1} = -1$ the semi-major axis a_1 of the body P_1 decreases with time (Fig. 3).

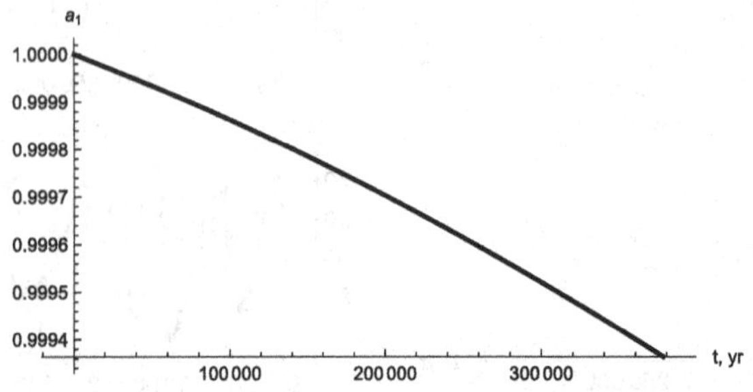

Fig. 3. Semi-major axis a_1 of the body P_1 in the case of $V_{n0} = 1/2, V_{\tau 1} = -1$)

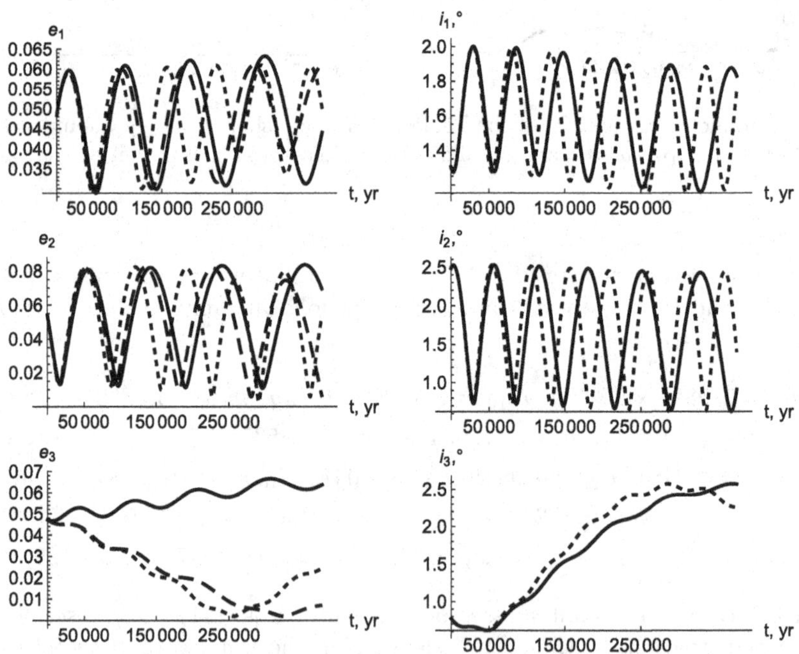

Fig. 4. Eccentricities e_j and inclinations i_j of the bodies P_1, P_2, P_3 (short dash – constant masses, long dash – isotropic mass changes, solid curves – anisotropic mass changes, $V_{r0} = 1$)

If only one component of the relative velocity V_{r0} of the particles leaving the most massive body P_0 along the radial direction is greater than zero ($V_{r0} =$

1) dependance of the eccentricities e_j and arguments of pericenter ω_j on time changes (see Figs. 4 and 5). Again orbital elements of the body P_3 are the most sensitive to appearance of the radial reactive force because its mass is the smallest one. These results demonstrate that even very small changes of the masses of celestial bodies can influence essentially on their long-term evolution.

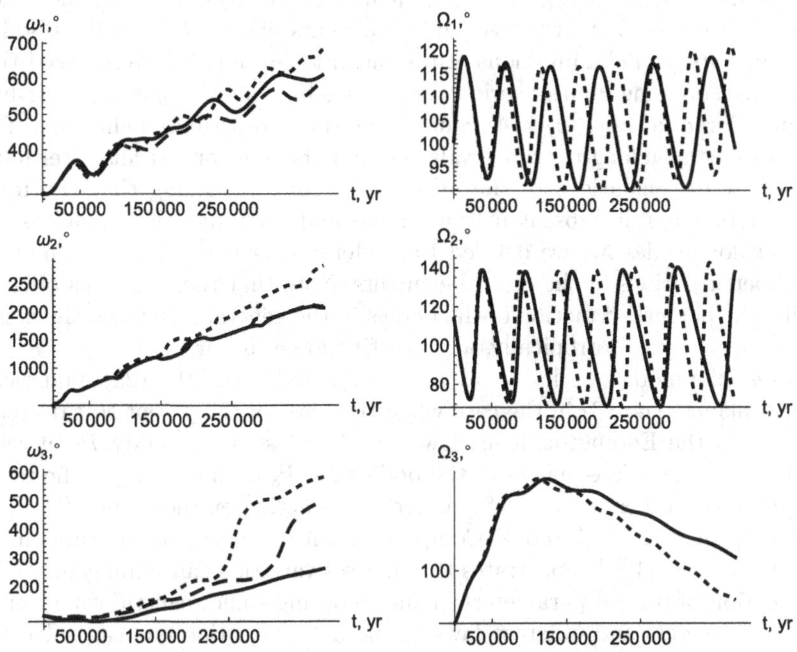

Fig. 5. Parameters ω_j and Ω_j of the bodies P_1, P_2, P_3 (short dash – constant masses, long dash – isotropic mass changes, solid curves – anisotropic mass changes, $V_{r0} = 1$)

5 Conclusion

In this paper, we investigate a multi-planetary problem of many bodies of variable masses that attract each other according to Newton's law of universal gravitation. We assume that the bodies may be acted on by the reactive forces arising due to anisotropic variation of the bodies masses. Using Meshcherskii equation, we have defined the reactive forces explicitly and derived the differential equations of motion of the bodies in the relative system of coordinates with the most massive body P_0 located at the origin in the framework of Newton's formalism. Equations of motion (1) are presented in the form which enables to find an exact solution (6) to the two-body problem of variable masses in the case of $\boldsymbol{F}_j = 0$. Using the exact solution (6) and applying the method of variation of constants, we derived differential equations of the perturbed motion in terms of the osculating elements of the aperiodic motion on quasi-conical section. It should be emphasized that the obtained Eqs. (9)–(14) are valid for any laws of the mass variation of the bodies and completely determine the perturbed motion of the bodies P_j, $(j = 1, 2, \ldots, n)$.

In the case of small eccentricities and inclinations of orbits, we have expanded the right-hand sides of Eqs. (9)–(14) in power series in terms of the orbital elements up to the first order. As the coefficients of e_j and i_j in the obtained expressions are periodic functions of the mean longitudes λ_j, we replaced them by the corresponding Fourier series. Finally, we have shown that the right-hand sides of differential Eqs. (9)–(14) contain the terms describing behaviour of the orbital elements on long-time intervals and quite cumbersome terms determining the short-term oscillations of the orbital elements. Assuming that the mean-motion resonances are absent in the system and averaging the equations over the mean longitudes λ_j, we derived the differential equations determining the secular perturbations of the orbital elements. Note that the equations obtained describe the perturbed motion of the bodies in the general case when the masses of all bodies vary anisotropically, and reactive forces occur.

To test the model, we have solved the averaged Eqs. (21), (22) numerically for three planets ($n = 3$) in the case when the mass of the parent star decreases according to the Eddington-Jeans law and the mass of the body P_1 increases linearly with time while masses of the bodies P_2, P_3 do not change. The results obtained in two different cases of reactive forces acting on the bodies P_0, P_1 are shown on Figs. 1, 2, 3, 4 and 5. Comparison with the case of constant masses (see, for example, [4]) demonstrates that masses variation can significantly affect the evolution of orbital parameters. Thus, choosing some realistic values of the system parameters and different laws of the mass variation one can solve Eqs. (21), (22) numerically and investigate an influence of the masses variation on the dynamics of multi-planetary systems. In our future work, we plan to study exoplanetary systems of variable masses when the number of celestial bodies exceeds four.

References

1. Brouwer, D., Clemence, G.M.: Methods of Celestial Mechanics, p. 602. Academic Press, New York (1961)
2. Roy, A.E.: Orbital Motion, p. 504. 4th edn. Institute of Physics Publishing, Bristol and Philadelphia (2005). ISBN: 0750310154
3. Cilletti, A.: Stability and Chaos in Celestial Mechanics, p. 264. Springer Praxis Books, Berlin, Heidelberg (2010). https://doi.org/10.1007/978-3-540-85146-2
4. Murray, C.D., Dermott, S.F. : Solar System Dynamics, p. 604. Cambridge University Press (1999). ISBN: 9780521575973
5. Lidov, M.L., Vashkov'yak, M.A.: On quasi-satellite orbits in a restricted elliptic three-body problem. Astron. Lett. **20**(5), 676–690 (1994)
6. Ford, E.B., Kozinsky, B., Rasio, F.A.: Secular evolution of hierarchical triple star systems. Astron. J. **535**, 385–401 (2000)

7. Schulz, N.S.: The Formation and Early Evolution of Stars. Springer-Verlag, Berlin, Heidelberg (2012). https://doi.org/10.1007/978-3-642-23926-7
8. Omarov, T.B. (Ed.): Non-Stationary Dynamical Problems in Astronomy, p. 260. Nova Science Publishing Inc., New-York (2002). ISBN:1-59033-331-4
9. Eggleton, P.: Evolutionary Processes in Binary and Multiple Stars, p. 332. Cambridge University Press, Cambridge (2006). ISBN: 9780511536205
10. Omarov, T.B.: Two-body motion with corpuscular radiation. Astron. Zh. **40**(5), 921–928 (1963)
11. Hadjidemetriou, J.D.: Two-body problem with variable mass: a new approach. Icarus **2**, 440–451 (1963). https://doi.org/10.1016/0019-1035(63)90072-1
12. Bekov, A.A., Omarov, T.B.: The theory of orbits in non-stationary stellar systems. Astron. Astrophys. Trans. **22**(2), 145–153 (2003). https://doi.org/10.1080/1055679031000084803
13. Rahoma, W.A., Abd El-Salam, F.A., Ahmed, M.K.: Analytical treatment of the two-body problem with slowly varying mass. J. Astrophys. Astron. **30**(3–4), 187–205 (2009)
14. Veras, D., Hadjidemetriou, J.D., Tout, C.A.: An exoplanet's response to anisotropic stellar mass-loss during birth and death. Mon. Not. R. Astron. Soc. **435**(3), 2416–2430 (2013). https://doi.org/10.1093/mnras/stt1451
15. Michaely, E., Perets, H.B.: Secular dynamics in hierarchical three-body systems with mass loss and mass transfer. Astrophysical J. **794**(2), 122–133 (2014). https://doi.org/10.1088/0004-637X/794/2/122
16. Veras, D.: Post-main-sequence planetary system evolution. Royal Soc. open sci. **3**, 150571 (2016). https://doi.org/10.1098/rsos.150571
17. Minglibayev, M.Z.: Dynamics of gravitating bodies with variable masses and sizes [Dinamika gravitiruyushchikh tel s peremennymi massami i razmerami], p. 224. LAP LAMBERT Academic Publishing (2012). ISBN: 978-3-659-29945-2
18. Prokopenya, A.N., Minglibayev, M.Z., Beketauov, B.A.: Secular perturbations of quasi-elliptic orbits in the restricted three-body problem with variable masses. Int. J. Non-Linear Mech. **73**, 58–63 (2015). https://doi.org/10.1016/j.ijnonlinmec.2014.11.007
19. Minglibayev, M.Z., Prokopenya, A.N., Mayemerova, G.M., Imanova, Z.U.: Three-body problem with variable masses that change anisotropically at different rates. Math. Comput. Sci. **11**, 383–391 (2017). https://doi.org/10.1007/s11786-017-0306-4
20. Prokopenya, A.N., Minglibayev, M.Z., Mayemerova, G.M., Imanova, Z.U.: Investigation of the restricted problem of three bodies of variable masses using computer algebra. Program. Comput. Softw. **43**(5), 289–293 (2017). https://doi.org/10.1134/S0361768817050061
21. Minglibayev, M., Prokopenya, A., Shomshekova, S.: Applications of computer algebra in the study of the two-planet problem of three bodies with variable masses. Program. Comput. Softw. **45**(2), 73–80 (2019). https://doi.org/10.1134/S0361768819020087
22. Minglibayev, M., Prokopenya, A., Shomshekova, S.: Computing perturbations in the two-planetary three-body problem with masses varying non-isotropically at different rates. Math. Comput. Sci. **14**(2), 241–251 (2020). https://doi.org/10.1007/s11786-019-00437-0
23. Imanova, Z.U., Minglibayev, M.Z., Prokopenya, A.N.: Modelling the evolution of the two-planetary three-body system of variable masses. Math. Model. Anal. **28**(4), 636–652 (2023). https://doi.org/10.3846/mma.2023.18453

24. Wolfram, S.: An elementary introduction to the Wolfram Language, 2nd edn., p. 340. Wolfram Media, New York (2016)
25. Meshcherskii, I.V.: Works on mechanics of bodies with variable masses. GITTL, Moscow (1952)

Best of Both Worlds: Solving the Cyclic Bandwidth Problem by Combining Pre-existing Knowledge and Constraint Programming Techniques

Guillaume Fertin[1], Eric Monfroy[2], and Claudia Vasconcellos-Gaete[2](✉)

[1] Nantes Université, École Centrale Nantes, CNRS, LS2N, UMR 6004, 44000 Nantes, France
guillaume.fertin@univ-nantes.fr

[2] LERIA, Université d'Angers, Angers, France
{eric.monfroy,claudia.vasconcellos}@univ-angers.fr

Abstract. Given an optimization problem, combining knowledge from both (i) structural or algorithmic known results and (ii) new solving techniques, helps gain insight and knowledge on the aforementioned problem by tightening the gap between lower and upper bounds on the sought optimal value. Additionally, this gain may be further improved by iterating (i) and (ii) until a fixed point is reached.

In this paper, we illustrate the above through the classical Cyclic Bandwidth problem, an optimization problem which takes as input an undirected graph $G = (V, E)$ with $|V| = n$, and asks for a labeling φ of V in which every vertex v takes a unique value $\varphi(v) \in [1; n]$, in such a way that $B_c(G, \varphi) = \max\{\min_{uv \in E(G)}\{|\varphi(u) - \varphi(v)|, n - |\varphi(u) - \varphi(v)|\}\}$, called the *cyclic bandwidth of* G, is minimized.

Using the classic benchmark from the Harwell-Boeing sparse matrix collection introduced in [16], we show how to combine (i) previous results from the Cyclic Bandwidth literature, and (ii) new solving techniques, which we first present, and then implement, starting from the best results obtained in step (i). We show that this process allows us to determine the optimal cyclic bandwidth value for half of the instances of our benchmark, and improves the best known bounds for a large number of the remaining instances.

Keywords: Graph Labeling · Cyclic Bandwidth · Lower Bounds · Upper Bounds · Constraint Programming · Solver

1 Introduction

The classical Cyclic Bandwidth problem is an optimization problem that takes as input an undirected graph $G = (V, E)$ with $|V| = n$ and asks for a labeling φ of V in which every vertex v takes a unique value $\varphi(v) \in [1; n]$, in such a way that $B_c(G, \varphi) = \max\{\min_{uv \in E(G)}\{|\varphi(u) - \varphi(v)|, n - |\varphi(u) - \varphi(v)|\}\}$ (i.e., the *cyclic bandwidth of* G) is minimized.

L. Franco et al. (Eds.): ICCS 2024, LNCS 14836, pp. 197–211, 2024.
https://doi.org/10.1007/978-3-031-63775-9_14

The Cyclic Bandwidth problem was first presented in [11] within the framework of creating a ring interconnection network for a group of computers. This problem can be regarded as a modification of the widely recognized Bandwidth Minimization problem, originally proposed by Harper [6] in 1964. For a comprehensive historical overview, readers can refer to the survey by Chinn et al. [2]. The Bandwidth Minimization problem also asks for a labeling φ, i.e., a bijection from V to $[1;n]$, using a computed value of $B(G,\varphi) = \max_{uv\in E}\{|\varphi(u) - \varphi(v)|\}$. The Bandwidth Minimization problem asks for a labeling φ^* such that $B(G,\varphi^*)$ is minimized. It can be seen that both problems are related in the sense for any graph G, we have $B_c(G) \leq B(G)$.

The Cyclic Bandwidth has been extensively studied. Its complexity has been established as NP-hard, even in the scenario of trees with a maximum degree of 3 [13]. Furthermore, the specific value of $B_c(G)$ has been ascertained for graphs within distinct categories, including paths, cycles, Cartesian products of paths (or cycles, or a combination of both), full k-ary trees, complete graphs, complete bipartite graphs, and unit interval graphs [3,8,12,13].

More recent papers are concerned with designing efficient heuristics for Cyclic Bandwidth (see, e.g., [19,20,22]) or Bandwidth Minimization (see, e.g., [15,17,18]). For both problems, execution time, upper bounds, or lower bounds are considered, examined, and tested on a subset of the classical Harwell-Boeing sparse matrix collection[1].

Other investigations studied the correlation between $B_c(G)$ and $B(G)$, with a specific focus on identifying conditions that guarantee the equality $B(G) = B_c(G)$ [4,9,14]. Another set of results focuses on establishing bounds for $B_c(G)$, particularly lower bounds, in the context of general graphs [4,25]. Initially, it is evident that for any graph G, $B_c(G) \geq \frac{\Delta(G)}{2}$, where $\Delta(G)$ represents the maximum degree of G. Furthermore, across all graphs G, the relationship $\frac{B(G)}{2} \leq B_c(G) \leq B(G)$ holds, with the leftmost bound stemming from [14]. Various other lower bounds have been derived in the literature, many of which are grounded in density ("propagation") considerations, or a relevant cycle basis of the examined graph (refer to, e.g., [4,25]).

In this paper, we propose the Recycling algorithm, orchestrating some existing results, both experimental and theoretical, based on bandwidth and cyclic bandwidth. For the 113 Harwell-Boeing graph instances proposed in [16], the Recycling algorithm is either able to tighten bounds, prove optimization, and find the optimum of the bandwidth or cyclic bandwidth.

Constraint programming (CP) [23] is a problem-solving paradigm for solving combinatorial problems using techniques issued from artificial intelligence, computer science, and operations research. In CP, rather than describing how to solve the problem, users formulate the problem by defining decision variables interconnected by constraints. As said by E. Freuder, "Constraint Programming represents one of the closest approaches computer science has yet made to the Holy Grail of programming: the user states the problem, the computer solves it."

[1] see, e.g., https://math.nist.gov/MatrixMarket/collections/hb.html and https://sparse.tamu.edu/HB

The constraints encompass various types of variables (such as Boolean, bounded integers, etc.) and constraint types (including linear, non-linear arithmetic, symbolics, etc.). In CP, a model represents a problem, and an instance is given by a model and some data. In our case, the problem consists of generating a graph labeling minimizing the cyclic bandwidth of graphs, and an instance is the instantiation of the model with a given graph. The direct modeling in CP of the cyclic bandwidth problem is very close to its mathematical formulation. However, this model is not well-suited for CP, and not efficiently solved.

Here, we propose a more original model, taking advantage of the strengths of CP. This model is based on constraints in extension [1], i.e., table constraints [10] considering the candidate labelings of two vertices linked by an edge to respect a given cyclic bandwidth value. This model is thus a satisfiability model for computing a labeling value less or equal to a given k. This k can easily be minimized by an efficient dichotomy algorithm considering a property of the problem: if there is a cyclic labeling of value k, there is also one of size $k+1$, and if there is no labeling of value k, none of value $k-1$ exists. This model is solved more efficiently than the direct model and may improve the results of the Recycling algorithm, which in turn offers the possibility to better solve the CP model. A fixed-point application of the sequence Recycling algorithm, CP model solving, is thus beneficial.

Experimentally, on the 113 instances selected in [16], our method proves highly powerful, as it optimally solves half of the 113 instances. More precisely, the fixed point of our Recycling algorithm and CP model solving proves the optimality of 63 instances, determines the optimal value of 56 instances, and improves the bounds of one instance.

This paper is organized as follows: Sect. 2 presents the CYCLIC BANDWIDTH problem along with some existing results for it, or for its related problem, BANDWIDTH MINIMIZATION. Section 3 presents the Recycling algorithm and how it orchestrates the existing results from literature. Section 4 presents an optimization function for the Recycling Algorithm, using constraint programming. Finally, Sect. 5 presents the results obtained using the Recycling algorithm and its optimization.

2 The Cyclic Bandwidth Problem and Existing Results

2.1 The Problem

The CYCLIC BANDWIDTH problem is a graph labeling problem that can be formulated as follows. Let $G(V, E)$ be a finite undirected graph (called the *guest graph*) of order n, and $C_n(V', E')$ be a cycle graph (called the *host graph*) with $|V'| = n$. An embedding of G in C_n is an injection $\phi : V \to V'$. The *cyclic distance* d_c between two vertices $u, v \in V$ linked by an edge of E is defined by:

$$d_c(u, v) = min\{|\phi(u) - \phi(v)|, n - |\phi(u) - \phi(v)|\} \quad (1)$$

The cyclic bandwidth of an embedding $\phi : V \rightarrow V'$ is the maximum distance between two vertices:

$$B_c(G, \phi) = max_{u,v \in E}\{d_c(u, v)\} \tag{2}$$

The cyclic bandwidth problem consists in finding an embedding ϕ^* among the set $\mathcal{E}$ of embeddings from G to C_n such that $B_c(G, \phi^*)$ is minimum:

$$B_C^*(G) = min_{\phi \in \mathcal{E}}\{B_c(G, \phi)\} \tag{3}$$

2.2 Metaheuristics Results

Cyclic Bandwidth and Bandwidth Minimization are both NP-complete problems. Solving any of these two problems requires exploring very large search spaces, hence it is not surprising that most of the methods developed to solve them are metaheuristics. In this section, we present some relevant algorithms and results that lately will become inputs for our Recycling algorithm. The first three metaheuristics solve Bandwidth Minimization (also known as Matrix Bandwidth Minimization, or MBMP), while the last one is a recent algorithm developed for Cyclic Bandwidth.

Martí's enhanced Branch-and-Bound [15]. This is one of the few exact methods proposed to solve MBMP. It takes advantage of a solution provided by a GRASP method to reduce the size of the tree to explore, focusing only in the branch for $b_t = b_{up} - 1$, with b_{up} being the solution provided by GRASP. It is also the first exact method to provide upper and lower bounds for some large instances ($|V| > 500$).

Mladenovic's VNS. In [17] the authors propose a local search algorithm known as Variable Neighborhood Search (VNS) that combines reduced neighborhoods, fast local search procedures, and specific neighborhood structures (from [21]). Additionally, they use the *number of critical vertices* as a secondary objective function, to deal with several neighborhoods having the same bandwidth. Their results outperform several other heuristics in terms of solution quality and computing time required, and they improve the best-known solutions for 42 (out of the 113) Harwell-Boeing instances.

Pop's genetic algorithm [18]. Using a list of interchange of rows or columns as individuals and a problem-specific genetic operator called pruning, this algorithm was successfully tested in the 113 Harwell-Boeing instances. Their results improve the best-known bandwidth values for almost a third of the graphs, and outperforming methods like the already described VNS [17] and Martí's branch-and-bound [15].

Ren's New Iterated Local Search (NILS). This is a metaheuristic proposed in [20] to solve the Cyclic Bandwidth problem. Starting from a random initial solution, it iterates over a local optimization (a dedicated Tabu search, DTS) and adds two perturbation strategies, to escape local optimum traps and explore unvisited areas. When the DTS stagnates, a Direct perturbation is triggered to modify the current solution, using a randomized shift-insert operator. If this perturbation also stagnates, then a Strong perturbation is triggered to apply a destruction-reconstruction heuristic, moving some uncritical vertices closer to the critical ones.

The NILS algorithm was applied to 85 standard graphs (paths, cycles, caterpillars, etc.) and 28 Harwell-Boeing instances; from these 28 graphs, NILS improved 4 existing results (compared against [19,22]) and matched all other best results obtained.

2.3 Theoretical Structural Results

We summarize here three structural properties about $B_c^*(G)$ (and possibly $B^*(G)$) that apply to any graph G, and that rely on polynomial time computations [4]. They allow to: (i) obtain lower bounds for $B_c^*(G)$, (ii) determine conditions under which $B_c^*(G) = B^*(G)$ and, in certain conditions, (iii) given a labeling for CYCLIC BANDWIDTH, provide a labeling for BANDWIDTH MINIMIZATION, of same value. All these properties will be used in our Recycling algorithm (see Sect. 3).

Extended density. Theorem 1 in [4] gives a lower bound for $B_c^*(G)$ based on the neighborhood "up to distance i" for any vertex $u \in V(G)$ and any (relevant) value of i. It can be seen as a generalization of the obvious $\frac{\Delta(G)}{2}$ lower bound. Theorem 3 in [4] is in the same spirit, but considers the neighborhood (up to distance i) of any any *pair* of vertices connected by an edge in G.

Cycle basis considerations. Lower bounds on $B_c^*(G)$ can also be obtained by computing the length ℓ of the longest cycle in a cycle basis of G (the notion of cycle basis being a classical graph-theoretical notion, see e.g. [5]). This is the purpose of Theorem 9 in [4], which actually contains two results: a lower bound for $B_c^*(G)$ based on ℓ, and a condition under which $B_c^*(G) = B^*(G)$.

Relabeling. Another interesting result from [4] is its Algorithm 1, which provides a labeling ϕ' for the BANDWIDTH MINIMIZATION problem, given a labeling ϕ for the CYCLIC BANDWIDTH problem, in such a way that $B(G, \phi') \leq B_c(G, \phi)$. Note that this result is only guaranteed under some conditions described in Lemma 8 in [4]. This result may be useful as it may decrease upper bounds on $B^*(G)$, which in turn, may lead us to conclude that $B^*(G) = B_c^*(G)$ (see Algorithms 2 and 4 in Sect. 3 for more details).

3 Recycling Algorithm: Orchestration of Existing Results

The Recycling algorithm we propose here is based on results of previous works, and part of its structure is directed by some theorems of [4] (see Sect. 2.3). For sake of readability, from now on (and if clear from the context), we will denote by B_c^* (resp. B^*) the optimal value $B_c^*(G)$ (resp. $B^*(G)$). The inputs for the Recycling algorithm are described in Table 1, whereas the Recycling algorithm is described in Algorithms 1 to 4.

Table 1. Inputs for the Recycling algorithm

	input	description
Graph	G	A graph
	n	Order of G
	ℓ	Length of the longest cycle in a cycle basis of G
Bandwidth	lb_{B_Marti}	Bandwidth lower bound from Martí et al. [15]
	ub_{B_Marti}	Bandwidth upper bound from Martí et al. [15]
	ub_{B_Pop}	Bandwidth upper bound from Pop et al. [18]
	$ub_{B\_Mladenovic}$	Bandwidth upper bound from Mladenovic et al. [17]
Cyclic Bandwidth	ub_{C_NILS}	Cyclic Bandwidth upper bound from Ren et al. [20]
	$lb\_density$	Cyclic Bandwidth lower bound from the extended density notion [4]

4 Cyclic Bandwidth as an Optimization Constrained Problem

Constraint programming [23] (CP) is a paradigm for solving combinatorial problems using a wide range of methods issued from artificial intelligence, computer science and operations research. In CP, users focus on the "what", not on the "how": this means that users declaratively state the problem, not how to solve it. Hence, a problem is described as a constraint satisfaction problem (CSP) or constrained optimization problem (COP): a CSP is defined by some decision variables, each one with its domain (its candidate values) and constraints (relations) linking these variables; a COP is given by a CSP and an objective function to be optimized.

To model the Cyclic Bandwidth problem, we consider finite domain decision variables, i.e., bounded integer variables, and arithmetic constraints. A label corresponds to a finite domain variable ranging from 1 to n, i.e., $\mathcal{N} = \{1, \ldots, n\}$ is a set of n labels, and for each $v \in V$, the variable ϕ_v represents the label of v:

$$\forall v \in V, \phi_v \in \mathcal{N} \tag{4}$$

Algorithm 1. The Recycling Algorithm

```
                         ▷ Collect all knowledge about lower and upper bounds for B* and B*_c
lb_C ← max{lb_density, min{lb_B_Marti, ⌈n/ℓ⌉}}
ub_C ← ub_C_NILS
lb_B ← lb_B_Marti
ub_B ← min{ub_B_Marti, ub_B_Pop, ub_B_Mladenovic}
                         ▷ Compare the lower (resp. upper) bounds and adjust
if lb_B < lb_C then
    lb_B ← lb_C
end if
if ub_B < ub_C then
    ub_C ← ub_B
end if
                         ▷ If ub_C < lb_B then B*_c ≠ B*, otherwise check whether equality holds
if ub_C ≥ lb_B then
    eq_or_unknown()
end if
```

Algorithm 2. eq_or_unknown()

```
if ub_B ≤ ⌈n/ℓ⌉ then
    update_bounds()      ▷ In that case, B*_c = B* and bounds should be updated
else
    try_relabeling()     ▷ B*_c =? B* Relabeling is then tested.
end if
```

4.1 Arithmetic (or Direct) Model

This model is a direct translation of the mathematical definition of the problem. The constraints of the arithmetic model are:

- All the labels must be unique:

$$\texttt{AllDifferent}\{\phi_v | v \in V\} \tag{5}$$

  with `AllDifferent` [7] being the standard global constraint[1] which states that all variables in this constraint must be pairwise different.
- Cyclic bandwidth of the current labeling ϕ:

$$B_c(G, \phi) = \max_{(u,v)\in E} \{d_c(\phi_u, \phi_v)\} \tag{6}$$

  where $B_c(G, \phi)$ is a finite domain variable ranging from 1 to $n - 1$. Note that d_c has been defined in Sect. 2.1.

[1] A global constraint provides some abstractions to improve expressiveness, but also are treated more efficiently by the solver using some dedicated algorithms.

Algorithm 3. `update_bounds()`

▷ Note: this algorithm is invoked only if $B^* = B_c^*$

▷ Note: if ub_C improved ub_B, relabeling gives a labeling of value ub_B for bandwidth

```
ubB, ubC ← min{ubB, ubC}
lbB, lbC ← max{lbB, lbC}

if ubB = lbB then                 ▷ We have reached the optimum for both problems
    B*, Bc* ← lbC
end if
```

Algorithm 4. `try_relabeling()`

```
if ubB > ubC then                              ▷ Relabeling useful only in this case
    ub'B ← apply_relabeling()                  ▷ Algorithm 1 from [4] – see Section 2.3
    if ub'B < ubB then                         ▷ Relabeling has improved bandwidth upper bound
        ubB ← ub'B
        eq_or_unknown()                        ▷ We can test again whether B* = Bc*
    end if
end if
```

– Optimization to find the labeling ϕ^* that minimizes the cyclic bandwidth:

$$B_C^*(G) = minimize\ B_C(G, \phi) \tag{7}$$

where $B_C(G)$ is a finite domain variable ranging from 1 to $n - 1$.

The arithmetic model $\mathcal{M}_A^\star$ is thus:

$$\mathcal{M}_A^\star = (4) \wedge (5) \wedge (6) \wedge (7)$$

However, in terms of efficiency, a constraint solver uses dynamically and intensively the constraints to prune/reduce the search space, and it cannot take full advantage of this formulation since most of the problem is in the objective function.

4.2 Finite Domain Extensional Constraint Model for Satisfiability

We now consider finite domain extensional constraints (see, e.g., [1]), also known as table constraints [10]: a constraint is defined by enumerating the allowed (resp. forbidden) tuples of constants satisfying (resp. violating) it. Then, we enforce a tuple of variables to be an element of this table (using the *in* keyword). There are several types of table constraints and we use the classic one.

We have first to change the problem into a satisfiability problem: given an integer k, find a labeling ϕ such that $B_C(G, \phi) \leq k$. Now, let us consider $\mathcal{L}(k)$, the set of possible pairs of labels for pairs of vertices linked by an edge:

$$\mathcal{L}(k) = \{(\ell, \ell') \mid \ell, \ell' \in \mathcal{N}^2, \ell \neq \ell', \min\{|\ell - \ell'|, n - |\ell - \ell'|\} \leq k\}$$

The finite domain variables we need are the same as for the $\mathcal{M}_A$ model. The constraints are the following:

- All labels must be different. For this, we use the `AllDifferent` constraints as above (see (5)).
- Labels must respect the cyclic bandwidth of cost k w.r.t. distance d_c, i.e.

$$\forall (u, v) \in E,\ (\phi_u, \phi_v)\ in\ \mathcal{L}(k) \tag{8}$$

Thus, the finite domain extensional constraint satisfiability model is:

$$\mathcal{M}_E = (4) \wedge (5) \wedge (8)$$

4.3 Possible Improvements

To break symmetries, i.e., to remove some symmetric solutions that could be derived from remaining solutions, we can add the constraint:

$$\phi_{u_\Delta} = 1 \tag{9}$$

where u_Δ is the vertex of highest degree. This constraint removes $n-1$ cyclic permutation solutions that can be recovered later on (if necessary) by rotation of the remaining solution. This constraint thus reduces the search space. Note that we could assign 1 to the label of any vertex. However, intuitively, we feel that fixing the vertex of highest degree is more efficient.

As the labeling is cyclic, this means that we can turn one way (counterclockwise for example) or the other (clockwise). This symmetry can be broken to enforce one direction by ordering any two labels:

$$\phi_u \geq \phi_v \tag{10}$$

Note that u and v can be any label, but fixing $u = u_\Delta$ and v the vertex having the second highest degree seems a good intuition. Constraints (9) and (10) can be added to models $\mathcal{M}_E$ and $\mathcal{M}_A^\star$.

Some redundant constraints can also be added. For example, consider all the cycles of size 3 in G. Then, we can build a table of 3-uples representing the "legal" labeling of 3 vertices for a cyclic bandwidth of cost k. Although too expensive in the general case, this kind of redundant constraint can be added beneficially only around the vertex of the highest degree for example.

4.4 From Satisfiability to Optimization Models

It is obvious that if there is a cyclic labeling of cost k, there is also one of cost $k+1$ and, using the contraposition, if there is no labeling of cost k, then none of cost $k-1$ exists. We propose to use the satisfiability model $\mathcal{M}_E$ inside a dichotomy algorithm (see Algorithm 5) benefiting from the above property.

This `optimization` function (as described in Algorithm 5) can be called with the lower bound $lb = \lceil \Delta/2 \rceil$ (where Δ is the maximum degree of G) and the upper bound $ub = \lceil n/2 \rceil$. When knowing better bounds, such as the ones returned by our Recycling Algorithm, this function will be more efficient and

Algorithm 5. `optimization()`

```
k_best ← ub
while lb < ub do
    k ← (ub + lb) div 2
    L ← {(ℓ, ℓ') ∈ N² | ℓ ≠ ℓ', min{|ℓ − ℓ'|, n − |ℓ − ℓ'| ≤ k}}
    if solve(M_E, L) is SAT then
        ub ← k
        k_best ← ub
    else
        lb ← k + 1
    end if
end while
return k
```

will succeed more often in a reasonable time. The `solve` function creates the required model and solves it with an appropriate solver, i.e., a finite domain (FD) solver for $\mathcal{M}_E$.

Iterating Recycling and Optimization algorithms. Both the Recycling and Optimization algorithms aim at reducing the search interval, whose extremities are respectively the lower and upper bounds for $B_c^*(G)$, for any instance G. Obviously, if this interval is reduced to one value, then we have determined the optimal value $B_c^*(G)$. If not, then it is possible to iteratively apply Algorithms 1 and 5, aiming at further reducing the search interval, until a fixed point is reached.

5 Results

The benchmark we use here to evaluate performances of both our Recycling algorithm (Algorithm 1) and our Optimization algorithm (Algorithm 5) on the Cyclic Bandwidth problem, is the classic 113 Harwell-Boeing benchmark [16], extracted from the Harwell-Boeing sparse matrix collection.

The results are shown Table 2 (33 medium-sized instances of our benchmark), along with Tables 3 and 4 (remaining 80 instances). In these three tables, we describe, in the three first columns, the name of each instance, its number of vertices and edges. The four subsequent columns, collectively called *inputs*, respectively provide the best lower and upper bounds for the Bandwidth Minimization and the Cyclic Bandwidth problems. These four values correspond to the ones obtained by applying lines 1–4 of Algorithm 1. The two following columns, collectively called *Recycling algorithm*, respectively indicate whether $B_c^* = B^*$, and the search interval (lower bound/upper bound) for B_c^* obtained after applying Algorithm 1. Finally, the rightmost column *value* indicates either the optimal value B_c^* computed by Algorithm 5, or a lower/upper bound interval for B_c^*. In this column, '–' indicates that the Optimization algorithm timed out.

Results obtained by our Recycling algorithm. Algorithm 1 (lines 1–4), by simply gathering the best knowledge from literature and (when applicable) by relying on the fact that $B_c^* = B^*$, can provide optimal values B_c^*, or drastically

Table 2. Results for 33 medium Harwell-Boeing instances. The results with an * in the Recycling algorithm column indicates a discrepancy between the Recycling algorithm and the optimization function

Instance	vertices	edges	Inputs				Recycling algorithm		Opt. algorithm
			lb_B	ub_B	lb_C	ub_C	$B_c^* = B^*$?	value	value
pores_1	30	103	7	7	7	7	*yes*	7	7
ibm32	32	90	11	11	8	9	*no*	[8, 9]	9
bcspwr01	39	46	5	5	4	4	*no*	4	4
bcsstk01	48	176	16	16	12	12	*no*	12	12
bcspwr02	49	59	7	7	7	7	*yes*	7	7
curtis54	54	124	10	10	8	8	*no*	8	8
will57	57	127	6	6	6	6	*yes*	6	6
impcol_b	59	281	19	19	14	17	*no*	[14, 17]	17
steam3	80	424	7	7	7	7	*yes*	7	7
ash85	85	219	9	9	9	9	*yes*	9	9
nos4	100	247	10	10	10	10	*yes*	10	10
gent113	104	549	25	25	20	23	*no*	[20, 23]	23
bcsstk22	110	254	9	10	6	6	*no*	6	6
gre__115	115	267	20	22	20	23	*yes**	[20, 23]	23
dwt__234	117	162	11	11	11	11	*yes*	11	11
bcspwr03	118	179	9	9	9	10	*yes**	[9, 10]	10
lns__131	123	275	18	20	18	20	*yes*	[18, 20]	20
arc130	130	715	63	63	62	63	*unknown*	[62, 63]	63
bcsstk04	132	1758	36	37	33	37	*unknown*	[33, 37]	37
west0132	132	404	23	28	23	31	*unknown*	[23, 31]	31
impcol_c	137	352	23	27	23	24	*unknown*	[23, 24]	24
can__144	144	576	13	13	7	7	*no*	7	7
lund_a	147	1151	19	23	19	23	*yes*	[19, 23]	23
lund_b	147	1147	19	23	19	23	*yes*	[19, 23]	23
bcsstk05	153	1135	19	20	19	20	*yes*	[19, 20]	20
west0156	156	371	33	33	23	32	*no*	[23, 32]	–
nos1	158	312	3	3	3	3	*yes*	3	3
can__161	161	608	18	18	18	18	*yes*	18	18
west0167	167	489	31	34	28	34	*unknown*	[28, 34]	34
mcca	168	1662	32	32	32	37	*yes*	32	–
fs_183_1	183	701	52	58	52	58	*unknown*	[52, 58]	58
gre__185	185	650	17	19	17	21	*yes**	[17, 21]	21
will199	199	660	55	65	34	50	*no*	[34, 50]	–

reduce the search interval. Indeed, we immediately conclude optimality for 28 of the 113 studied instances. Among the 85 remaining instances, the search interval is of length 2 (resp. 3) for 12 (resp. 9) instances.

Table 3. Results for 80 large Harwell-Boeing instances (*part 1*)

Instance	vertices	edges	Inputs				Recycling algorithm		Opt. algorithm
			lb_B	ub_B	lb_C	ub_C	$B^*_c = B^*$?	value	value
impcol_a	206	557	24	32	23	32	*unknown*	[23, 32]	–
dwt__209	209	767	20	23	20	23	*yes*	[20, 23]	23
gre_216a	216	660	17	21	17	21	*yes*	[17, 21]	21
dwt__221	221	704	11	13	11	13	*yes*	[11, 13]	13
impcol_e	225	1187	34	42	34	42	*unknown*	[34, 42]	–
saylr1	238	445	12	14	12	14	*yes*	[12, 14]	14
steam1	240	1761	32	44	32	44	*yes*	[32, 44]	–
dwt__245	245	608	21	21	21	21	*yes*	21	21
nnc261	261	794	22	24	22	24	*yes*	[22, 24]	24
bcspwr04	274	669	23	24	23	24	*yes*	[23, 24]	–
ash292	292	958	16	19	16	19	*yes*	[16, 19]	–
can__292	292	1124	34	36	34	38	*yes*	[34, 36]	–
dwt__310	310	1069	11	12	11	12	*yes*	[11, 12]	12
gre__343	343	1092	23	28	23	28	*yes*	[23, 28]	–
dwt__361	361	1296	14	14	14	14	*yes*	14	14
plat362	362	2712	28	34	28	34	*yes*	[28, 34]	–
plskz362	362	880	14	18	14	18	*yes*	[14, 18]	–
str____0	363	2446	87	116	58	91	*unknown*	[58, 91]	–
str__200	363	3049	90	124	65	99	*unknown*	[65, 99]	–
str__600	363	3244	95	132	71	103	*unknown*	[71, 103]	–
west0381	381	2150	117	151	86	113	*no*	[86, 113]	–
dwt__419	419	1572	22	25	22	25	*yes*	[22, 25]	–
bcsstk06	420	3720	37	45	37	45	*yes*	[37, 45]	–
bcsstm07	420	3416	37	42	37	45	*yes*	[37, 42]	–
impcol_d	425	1267	36	40	24	35	*no*	[24, 35]	–
hor__131	434	2138	46	55	46	55	*yes*	[46, 55]	–
bcspwr05	443	590	25	27	25	27	*yes*	[25, 27]	26
can__445	445	1682	45	52	45	46	*unknown*	[45, 46]	–
pores_3	456	1769	13	13	13	13	*yes*	13	13
bcsstk20	467	1295	8	13	8	13	*unknown*	[8, 13]	–
nos5	468	2352	52	63	52	63	*yes*	[52, 63]	–
west0479	479	1889	81	118	80	105	*unknown*	[80, 105]	–
mbeacxc	487	41686	246	260	243	243	*no*	243	–
mbeaflw	487	41686	246	261	243	243	*no*	243	–
mbeause	492	36209	249	254	245	246	*no*	[245, 246]	–
494_bus	494	586	25	29	25	28	*unknown*	[25, 28]	–
west0497	497	1715	69	85	69	81	*unknown*	[69, 81]	–
dwt__503	503	2762	29	40	29	41	*yes*	[29, 40]	–
lns__511	503	1425	33	44	33	44	*yes*	[33, 44]	–
gre__512	512	1680	30	36	30	36	*yes*	[30, 36]	–

Table 4. Results for 80 large Harwell-Boeing instances (*part 2*)

Instance	vertices	edges	Inputs				Recycling algorithm		Opt. algorithm
			lb_B	ub_B	lb_C	ub_C	$B_c^* = B^*$?	value	value
fs_541_1	541	2466	270	270	270	270	*unknown*	270	–
sherman4	546	1341	21	27	21	27	*yes*	[21, 27]	–
dwt__592	592	2256	22	28	22	29	*yes*	[22, 28]	–
steam2	600	6580	54	63	54	63	*yes*	[54, 63]	–
nos2	638	1272	3	3	3	3	*yes*	3	3
west0655	655	2841	109	160	94	149	*unknown*	[94, 149]	–
662_bus	662	906	36	39	36	38	*unknown*	[36, 38]	–
shl____0	663	1682	211	226	211	212	*unknown*	[211, 212]	–
shl__200	663	1720	220	231	220	220	*unknown*	220	–
shl__400	663	1709	213	230	213	215	*unknown*	[213, 215]	–
nnc666	666	2148	33	40	33	41	*yes*	[33, 40]	–
nos6	675	1290	15	16	15	16	*yes*	[15, 16]	16
fs_680_1	680	1464	17	17	17	17	*yes*	17	17
saylr3	681	1373	35	47	35	46	*yes*	[35, 46]	–
sherman1	681	1373	35	47	35	46	*yes*	[35, 46]	–
685_bus	685	1282	30	32	30	32	*yes*	[30, 32]	–
can__715	715	2975	54	72	52	60	*unknown*	[52, 60]	–
nos7	729	1944	43	65	43	65	*yes*	[43, 65]	–
mcfe	731	15086	112	126	112	126	*yes*	[112, 126]	[125, 126]
fs_760_1	760	3518	36	38	36	38	*yes*	[36, 38]	–
bcsstk19	817	3018	13	14	13	14	*yes*	[13, 14]	–
bp_____0	822	3260	174	236	174	207	*unknown*	[174, 207]	–
bp___200	822	3788	186	258	186	218	*unknown*	[186, 218]	–
bp___400	822	4015	188	268	188	220	*unknown*	[188, 220]	220
bp___600	822	4157	190	272	189	229	*unknown*	[189, 229]	229
bp___800	822	4518	197	278	190	239	*unknown*	[190, 239]	239
bp__1000	822	4635	197	287	191	241	*unknown*	[191, 241]	241
bp__1200	822	4698	197	291	193	241	*unknown*	[193, 241]	241
bp__1400	822	4760	199	290	193	242	*unknown*	[193, 242]	242
bp__1600	822	4809	199	293	192	241	*unknown*	[192, 241]	241
can__838	838	4586	75	86	57	58	*no*	[57, 58]	58
dwt__878	878	3285	23	25	23	24	*yes*	[23, 24]	24
orsirr_2	886	2542	62	84	62	85	*yes*	[62, 84]	–
gr_30_30	900	3422	31	31	31	32	*yes*	31	31
dwt__918	918	3233	27	32	27	32	*yes*	[27, 32]	–
jagmesh1	936	2664	24	27	20	20	*no*	20	20
nos3	960	7442	43	43	43	43	*yes*	43	43
jpwh_991	983	2678	82	90	82	88	*yes*	[82, 88]	–
west0989	989	3500	123	210	123	217	*unknown*	[123, 217]	–
dwt__992	992	7876	35	35	35	35	*yes*	35	35

Results obtained by our Recycling algorithm along with the Optimization function. Tightening the search interval through Algorithm 1 is obviously advantageous for our `optimization()` function (Algorithm 5), as it is based on dichotomy. As a matter of fact, the results we obtain through "Recycling + optimization", prove very efficient. Indeed, among the 113 initial instances, Algorithm 5 determines the optimal B_c^* value for half of them (56 cases), and the search interval is reduced in one case. Note also that, every time $B_c^* = B^*$, determining B_c^* or tightening its search interval also improves knowledge on B^*. All experiments are run on a computer equipped with an Intel Xeon ES 2630, 2.66 GHz processor, and coded in Python using the PyCSP^3 v2.2 library [24]. Each instance has up to 120 h to complete the `optimization()` function, which can lead to several calls to the CSP solver (same instance, different k values).

6 Conclusions and Future Work

This paper shows how the combination of pre-existing knowledge and some solving techniques can help to improve results for some hard combinatorial problems, such as the Cyclic Bandwidth problem. In our case, we leverage the close relation between the Bandwidth Minimization and the Cyclic Bandwidth problems to propose the Recycling algorithm. Our algorithm takes advantage of existing knowledge and uses it in a way that no solver is required to tighten bounds, prove optimization, or find the optimum at almost zero computational cost. For instances that require long computational runtimes, the bound tightening is a direct gain in the quest to find (or validate) optimums. In a second step, we propose an optimization function, based on constraint programming, to test the remaining values in the lower and upper bounds interval obtained in the first step. The experimental results validated the proposed approach by obtaining optimums for 56 (out of the 113) Harwell-Boeing instances and tightening bounds for another instance. We also found three instances presenting discrepancies between the Recycling algorithm and the optimization function; upon reviewing the data and running a CP-based bandwidth minimization function, it seems there may be errors in the reported values from Pop's article. Our next steps will involve: 1) a revision of bandwidth minimization values and, 2) the study of graph properties or different modeling approaches in CP to overcome the specificities of certain instances for which the solver could not find a solution.

References

1. Apt, K.R., Monfroy, É.: Constraint programming viewed as rule-based programming. Theory Pract. Log. Program. **1**(6), 713–750 (2001)
2. Chinn, P.Z., Chvátalová, J., Dewdney, A.K., Gibbs, N.E.: The bandwidth problem for graphs and matrices - a survey. J. Graph Theor. **6**(3), 223–254 (1982)
3. Chung, F.R.: Labelings of graphs. Sel. Top. Graph Theor. **3**, 151–168 (1988)
4. Déprés, H., Fertin, G., Monfroy, E.: Improved lower bounds for the cyclic bandwidth problem. In: Paszynski, M., Kranzlmüller, D., Krzhizhanovskaya, V.V., Dongarra, J.J., Sloot, P.M.A. (eds.) ICCS 2021. LNCS, vol. 12742, pp. 555–569. Springer, Cham (2021). https://doi.org/10.1007/978-3-030-77961-0_45

5. Harary, F., Manvel, B.: On the number of cycles in a graph. Matematický časopis **21**(1), 55–63 (1971)
6. Harper, L.H.: Optimal assignments of numbers to vertices. J. Soc. Ind. Appl. Math. **12**(1), 131–135 (1964)
7. van Hoeve, W.J.: The all different constraint: a survey. CoRR **cs.PL/0105015** (2001). https://arxiv.org/abs/cs/0105015
8. Hromkovič, J., Müller, V., Sýkora, O., Vrťo, I.: On embedding interconnection networks into rings of processors. In: Etiemble, D., Syre, J.-C. (eds.) PARLE 1992. LNCS, vol. 605, pp. 51–62. Springer, Heidelberg (1992). https://doi.org/10.1007/3-540-55599-4_80
9. Lam, P.C.B., Shiu, W.C., Chan, W.H.: Characterization of graphs with equal bandwidth and cyclic bandwidth. Discret. Math. **242**(1–3), 283–289 (2002)
10. Lecoutre, C.: Optimization of simple tabular reduction for table constraints. In: Stuckey, P.J. (ed.) CP 2008. LNCS, vol. 5202, pp. 128–143. Springer, Heidelberg (2008). https://doi.org/10.1007/978-3-540-85958-1_9
11. Leung, J.Y., Vornberger, O., Witthoff, J.D.: On some variants of the bandwidth minimization problem. SIAM J. Comput. **13**(3), 650–667 (1984)
12. Lin, Y.: A level structure approach on the bandwidth problem for special graphs. Ann. N. Y. Acad. Sci. **576**(1), 344–357 (1989)
13. Lin, Y.: The cyclic bandwidth problem. Syst. Sci. Math. Sci. **7** (1994)
14. Lin, Y.: Minimum bandwidth problem for embedding graphs in cycles. Networks **29**(3), 135–140 (1997)
15. Martí, R., Campos, V., Piñana, E.: A branch and bound algorithm for the matrix bandwidth minimization. Eur. J. Oper. Res. **186**, 513–528 (2008)
16. Martí, R., Laguna, M., Glover, F., Campos, V.: Reducing the bandwidth of a sparse matrix with tabu search. Eur. J. Oper. Res. **135**, 450–459 (2001)
17. Mladenovic, N., Urosevic, D., Pérez-Brito, D., García-González, C.: Variable neighbourhood search for bandwidth reduction. Eur. J. Oper. Res. **200**(1), 14–27 (2010)
18. Pop, P., Matei, O., Comes, C.A.: Reducing the bandwidth of a sparse matrix with a genetic algorithm. Optimization **63**(12), 1851–1876 (2014)
19. Ren, J., Hao, J., Rodriguez-Tello, E.: An iterated three-phase search approach for solving the cyclic bandwidth problem. IEEE Access **7**, 98436–98452 (2019)
20. Ren, J., Hao, J.K., Rodriguez-Tello, E., Li, L., He, K.: A new iterated local search algorithm for the cyclic bandwidth problem. Knowl.-Based Syst. **203**, 106136 (2020)
21. Rodriguez-Tello, E., Hao, J.K., Torres-Jimenez, J.: An improved simulated annealing algorithm for bandwidth minimization. Eur. J. Oper. Res. **185**(3), 1319–1335 (2008)
22. Rodriguez-Tello, E., Romero-Monsivais, H., Ramírez-Torres, G., Lardeux, F.: Tabu search for the cyclic bandwidth problem. Comput. Oper. Res. **57**, 17–32 (2015)
23. Rossi, F., van Beek, P., Walsh, T. (eds.): Handbook of Constraint Programming, Foundations of Artificial Intelligence, vol. 2. Elsevier (2006)
24. XCSP3 Team: PyCSP^3 v2.2 (2023). https://www.pycsp.org/
25. Zhou, S.: Bounding the bandwidths for graphs. Theoret. Comput. Sci. **249**(2), 357–368 (2000)

A Novel Bandwidth Occupancy Forecasting Method for Optical Networks

Paweł Krysztofik[1], Bartłomiej Grzelak[2], Piotr Śliwka[3], Sławomir Sujecki[2,4], and Stanisław Kozdrowski[1(✉)]

[1] Department of Computer Science, Warsaw University of Technology, Nowowiejska 15/19, 00-665 Warsaw, Poland
stanislaw.kozdrowski@pw.edu.pl
[2] Faculty of Electronics, Military University of Technology, Warsaw, Poland
[3] Department of Computer Science, Cardinal Wyszynski University, Warsaw, Poland
[4] Telecommunications and Teleinformatics Department, Wroclaw University of Science and Technology, Warsaw, Poland

Abstract. In this contribution, we developed a software tool for collecting information on the data traffic via control plane of an operating optical network. From this data, demand matrix elements were calculated and used to numerically estimate the edge occupancy in the optical network studied. For this purpose, a detailed network model was formulated with cost function and constraints. The formulated network model leads to an optimization problem, which was efficiently solved by meta-heuristic algorithms. Finally, statistical methods were used to model forecasting, in terms of the probability of the edge occupancy, under a Markov process approximation. Additionally, on the basis of the numerical results obtained, the scalability of the applied heuristic and statistical methods was analyzed.

Keywords: Optical network · Modeling · Markov chain · Optimization · Network congestion · Heuristic algorithms

1 Introduction

Developing an optical network infrastructure is a complex process involving the participation of many stakeholders. Due to the ever-increasing demands for high bandwidth in optical networks, an important need is a rational use of bandwidth and its optimization [1,7] in Space Division Multiplexing SDM [6,8], Passive Optical Networks (PON) [14] and in 5G and beyond technologies [2].

The primary stakeholder is the Network Operator (NO), i.e., the company which owns and runs an optical network. NO is interested in minimizing the operational expenditure (opex) and capital expenditure (capex) while maintaining a high level of commercially offered services. NO interacts with optical network Equipment Providers (EP) when planning Optical Network (ON) maintenance activities and expansion. Before making any decisions on network expansion, the

L. Franco et al. (Eds.): ICCS 2024, LNCS 14836, pp. 212–226, 2024.
https://doi.org/10.1007/978-3-031-63775-9_15

NO needs to have a detailed view of the current use of the available resources and estimates of the future traffic. For this purpose tools are needed that directly monitor the traffic between network nodes, calculate the percentage use of the resources, and estimate the future trends. In this contribution, therefore, we developed a software suite that collects traffic information in real-time directly from the ON via the control plane and creates a database. Then, from the information collected in the database, we derive the elements of the demands matrix and apply optimization methods to minimize capital expenditures and estimate the usage of resources. Unfortunately, when used with standard optimization techniques, the developed methodology results in long computation times [13,16] including 5G ready networks [12]. Therefore, in this contribution, we propose using specially tailored metaheuristic methods to reduce the calculation time.

In the last step, we perform an analysis and prediction of the capacity resources of the analyzed optical network using statistical methods based on a Markov process modeling with a discrete state space and discrete time ([3,5,17,19]). In particular, the developed software predicts the magnitude of change in the edge occupancy (an increase or decrease) expressed as a percentage or number of occupied DWDM frequency slices in the available band for a given edge in subsequent periods.

The developed software suite consisting of a network traffic monitoring tool, network edge occupancy modeling tool, and the probability of the network edge occupancy forecasting tool presented in this paper is potentially very useful to telecommunication network operators, as it allows for optimal use of the allocated resources and aids the process of network expansion planning. This is because the results obtained allow assessing the need for additional investment into the optical network infrastructure. Hence, the developed model predicts the network edges that are most likely to be subjected to traffic congestion. In practice, operators of large networks pay particular attention to network utilization levels, which are as follows: a safe utilization level is from 0 to 50% (green light), then the range between 50 and 70% is a warning level at which edge expansion should consider (yellow light), while above 70% utilization is at an alarming level (red light), ordering an immediate network upgrade. These assumptions are taken into account in our tests. This allows the network operator to plan in advance the network expansion and allocate appropriate financial resources for the necessary capital expenditure [15].

Taking the above into account, the main contribution of this work is the optimization of network resources using metaheuristics and then, using the results of the optimization, performing an edge-occupancy prediction by applying Markov chain approach (Markov chain states represent the network utilization level). Detailed contributions of this work are the following:

1. network model formulation with cost function and constraints,
2. development of efficient meta-heuristic algorithms for solving the optimisation problem,

3. application of the modeling for prediction the probability of the edge occupancy states using Markov process, and
4. presentation of numerical results evaluating the scalability of the effectiveness of the heuristics and statistical results.

Considering the paper structure, after the introduction, we give the problem formulation. In the 3 Section, we present the proposed metaheuristic algorithms. The 4 Section gives a description of the software tool that collects the data from ON. In the results Section for an optical network, we show the superiority of the proposed metaheuristic approach. Finally, Sect. 6 provides conclusions and open issues and future work.

2 Problem Formulation

An optical network can be modeled by an undirected graph with vertices representing the individual cities and edges as the optical fibers connecting them. The main task of the network is to enable data transmission between all pairs of cities with an expected minimum throughput. The problem is solving the task of deploying enough devices in network nodes to meet the traffic demand while minimizing their cost. This task can be described using mathematical equations derived in [9]. To formulate the problem, the following sets were defined: N - the set of nodes, E - the set of edges, T - the set of transponders, S - the set of frequency slices, $P_{(n,n')}$ - the set of all paths between nodes $n, n' \in N$; $p \subseteq E$, S_t - the set of all frequency slices that can be used as starting frequencies for transponders $t \in T$; $S_t \subseteq S$.

A binary decision variable $x_{tnn'ps}$ equals 1 if a transponder t between nodes n and n' on a path p starting from a frequency slice s is installed and 0 otherwise. For the sake of clarity, it is noted that transponders are installed only at the start and end nodes of a path p. The objective of the optimization problem is to minimize the total cost of transponder installation $\xi(t)$ and the cost of bandwidth usage:

$$\min\Big(\sum_{t\in T}\xi(t)\sum_{n,n'\in N}\sum_{p\in P_{(n,n')}}\sum_{s\in S_t}x_{tnn'ps}\Big) \tag{1}$$

The optimization is performed subject to constraints. 1. Demands realization between each pair of nodes, 2. Adequate power levels for each optical channel, realized by transponders, and 3. Appropriate and unique allocation of frequency slices for each channel. Their exact mathematical formulation can be found in the literature [10].

Homogeneous Markov chains or Markov Set Chains (if the homogeneity condition was not met) were used to predict the level of slice occupancy.

3 Optimisation Algorithms

To solve this problem presented in Sect. 2, two metaheuristics based on evolutionary algorithms $\mu + \lambda$ and bee colony (BC) are proposed.

3.1 $\mu + \lambda$ Algorithm

The $\mu + \lambda$ algorithm is a basic evolutionary algorithm used to find the best or good enough solution relative to the objective function. It is based on an initial population of μ individuals created randomly. In each iteration of the algorithm, a λ of new offspring solutions are created based on the parent population. Each offspring is created by intersecting the parameters of two randomly selected parents. For each of them, there is a small probability that it will be subjected to a mutation operator that slightly changes its parameters. From the parent population and the offspring population, μ solutions are selected to make up the new parent population. Individuals with a better value of the objective function have a higher probability of being selected for the new population. The algorithm performs successive iterations until a stop condition occurs, usually finding a good enough solution or being permanently stuck in the local optimum. The pseudocode of the $\mu + \lambda$ algorithm is shown in the Algorithm 1.

Algorithm 1. $\mu + \lambda$ Algorithm

procedure MuPlusLambda(μ, λ) ▷ μ - population size, λ - number new solutions per iteration
    $P \leftarrow$ RandomInitialization(μ) ▷ P - population
    Evaluate(P)
    best $\leftarrow$ ReturnBest(P) ▷ *best* - solution with the best goal function value
    while stop condition is not met **do**
        P' $\leftarrow$ GenerateOffsprings(P, λ) ▷ P' - new solutions
        Evaluate(P')
        P $\leftarrow$ SelectNewPopulation($P \cup P'$, μ)
        best $\leftarrow$ ReturnBest($P \cup best$)
    end while
    return *best*
end procedure

During crossover, the parent with the higher value of the objective function can be favored by increasing the probability p_c that it is its parameters that will be passed on to the created individual. When selecting solutions for a new population, selection pressure is influenced by how strongly we favor the selection of better individuals - it is lowest when the selection is completely random and highest when we select only the best individuals. With higher selection pressure, the algorithm will find better solutions in less time, but it may also get stuck in the local maximum sooner.

3.2 Bee Colony Algorithm

Another heuristic algorithm used for the study is the bee colony (BC) algorithm. Compared to the μ+λ algorithm, it searches a larger region of the state space and also employs a mechanism to protect against getting stuck in a local maximum.

The BC algorithm is based on an initially randomly generated population of N individuals. During each iteration, new individuals are generated in the following:

1. m of the best individuals are selected from the population. Each of these can be drawn as a parent for a new offspring. A new solution is created based on a single parent - it has the same parameters, but some of them are subjected to the mutation operator. By randomly selecting parents, k_1 new solutions are created;
2. from a population of m individuals, e those with the best value of the objective function are selected. For each of these, k_2 offsprings solutions are created;
3. $N - m$ of new individuals are randomly generated.

Next, N solutions are selected from the initial population, and all the individuals created to form the new initial population. Since the algorithm creates new, completely random individuals in each iteration, the algorithm always has a chance of finding a better solution than the one it encountered in the local maximum. Selection pressure will be determined by the selection methods and the parameters m, e, and k_2. Decreasing the parameters m and e and increasing the parameter k_2 will increase the selection pressure. The operation of the BC algorithm is shown in the Algorithm 2 pseudocode below.

Algorithm 2. BC Algorithm

```
procedure BeeColonyAlg(N, m, e, k) ▷ N - pop. size, m - best bees nb, e - elite bees nb., k1
- new bees created from best bees, k2 - nb of new bees per one elite bee
    P ← RandomInitialization(μ)                                    ▷ P - population
    Evaluate(P)
    best ← ReturnBest(P)               ▷ best - solution with the best goal function value
    while stop condition is not met do
        bestBees ← SelectBest(P, m)
        eliteBees ← SelectBest(bestBees, e)
        bestSol ← RandomNeighbors(bestBees, k1)
        eliteSol ← EliteNeighbors(bestBees, k2)
        randomSol ← RandomNeighbors(P, N − m)
        P ← SelectNewPopulation(P ∪ bestSol ∪ eliteSol ∪ randomSol, N)
        best ← ReturnBest(P ∪ best)
    end while
    return best
end procedure
```

4 Data Collection and Statistical Analysis

As part of presenting the possibilities of utilizing available resources for specific customer connections, data on Performance Monitoring (PM15) was collected every 15 min. The study was based on a segment of the Dense Wavelength Division Multiplexing (DWDM) network, belonging to one of the Polish telecommunications operators. The analyzed network segment consists of 7 nodes and 10 edges (E_{xy}), as illustrated by the network diagram in Fig. 1.

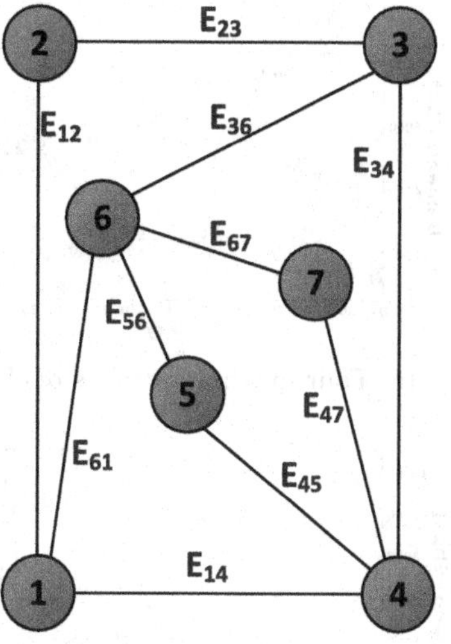

Fig. 1. Diagram of the DWDM network used for the tests.

The network used in the study relies on NOKIA devices on the 1830 PSS (Photonic Service Switch) platform, and is managed by the Network Management System (NMS). This system enables the management of equipment at the stage of service creation, monitoring the network's state, and collecting network data in real-time. With NMS, it is possible to gather data related to Performance Monitoring (PM). To collect PM15 data for the analyzed nodes, it was necessary to enable the PM15 data collection function for all services at these nodes through NMS. Once activated, PM15 data are recorded in .csv files every two hours and stored on the NMS server for 24 h. After this period, the PM15 data files are overwritten by data from the next day. This mechanism prevents the server's disk memory from overflowing, so it is important to timely download data files to prevent them from being overwritten and to maintain a continuous record of past network performance.

Four levels of data filtering were adopted, based on PM15 measurements. The first method involves determining four maximum values from a given day for the time intervals: 00:00–06:00 (Time of day 1), 06:00–12:00 (Time of day 2), 12:00–18:00 (Time of day 3) and 18:00–00:00 (Time of day 4). The second method presents a single maximum value from the entire measured day. The third method identifies the maximum value from an entire week. The fourth and final method indicates the maximum value from a given month. Data filtering was conducted using a proprietary program written in Python. Thanks to the application of appropriate criteria, this program is a flexible tool for data analysis.

Specifically, we selected from 7 nodes network a 4 nodes subnetwork consisting of nodes: 1,2,3,4 only and considered traffic exchanged between these nodes only. The data was collected for 365 days 24-h periods from 18 January 2023 to 17 January 2024 ($\mathbb{T} = \{1, 2, \ldots, 365\}$). The input data as a function of time for the one service implemented between nodes 1 and 2 are shown in Fig. 2 - for sampling rate 4 maximum values of the day (Fig. 2a), sampling rate of one maximum value of the day (Fig. 2b), sampling rate of one maximum value of the week(Fig. 2c) and sampling rate of one maximum value of the month (Fig. 2d).

Concerning the application of Markov chain formalism first for each communication channel (i.e. the corresponding time series) between any two distinct network nodes stationarity testing was performed using the Phillips-Perron test with hypothesis H_0: the time series has a unit root ["no stationary"]. The test was unequivocally rejected. For H_0 the P-P test yields for service 1: D-F $Z_\alpha = -1410.2$ with p-value < 0.01 and for service 2: D-F $Z_\alpha = -1420.7$ with p-value < 0.01). Consequently, it can be stated that the time series under

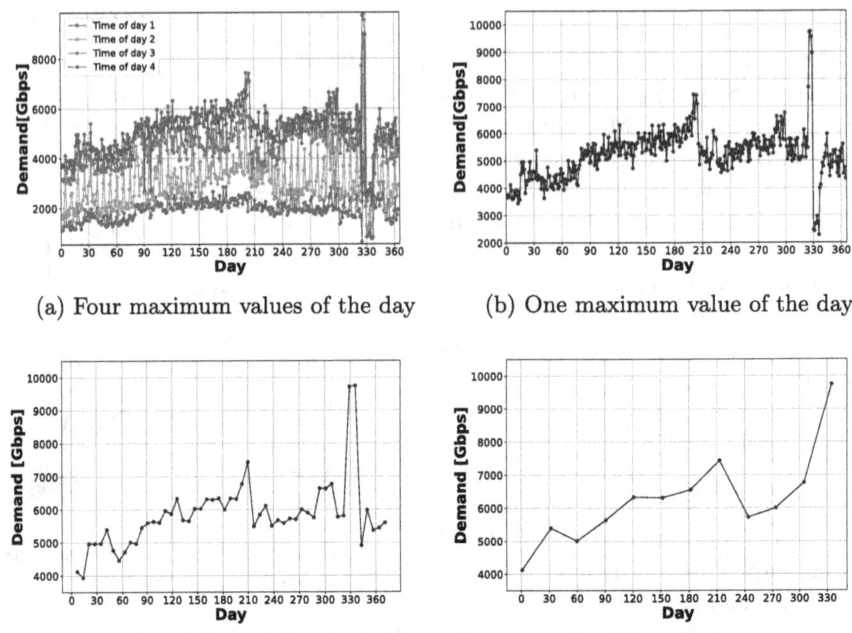

(a) Four maximum values of the day (b) One maximum value of the day.

(c) One maximum value of the week. (d) One maximum value of the month.

Fig. 2. Analyzed input.

consideration is stationary. This fact makes it easier to model the series, as it suffices to determine a probability distribution of the time series values, which does not change with time. With this in mind for the analysis of the time series a Markov chain was proposed, i.e., a Markov process with a discrete state space $\mathbb{S}$ and a discrete-time $\mathbb{T}$, based on a vector of random variables $X_{t_1}, X_{t_2}, \ldots, X_{t_n}$ defined on a common state space $\mathbb{S} = \{s_1, s_2, \ldots, s_k\}$ and satisfying the Markov condition: $P(X_{t_n} = s_j | X_{t_{n-1}} = s_i, X_{t_{n-2}} = s_{i_{n-2}}, \ldots, X_{t_0} = s_{i_0}) = P(X_{t_n} = s_j | X_{t_{n-1}} = s_i)$, which means the transition of the chain from state s_i to state s_j (designation: $p_{ij}(t) = P(X_{t_n} = s_j | X_{t_{n-1}} = s_i)$).

For the purpose of this study, the following state space was defined: $\mathbb{S} = \{s_1, s_2, s_3\}$, where s_1, denotes "high" level of bandwidth occupancy i.e. above 70% of occupied slices, s_2, is "medium" bandwidth occupancy and corresponds to between 51% and 70% of all slices used, while s_3, is "low" bandwidth occupancy with up to 50% of occupied slices in a given edge. The transition probability matrix $P(t)$ was defined as:

$$P(t) = \begin{bmatrix} p_{11}(t) & p_{12}(t) & p_{13}(t) \\ p_{21}(t) & p_{22}(t) & p_{23}(t) \\ p_{31}(t) & p_{32}(t) & p_{33}(t) \end{bmatrix}, \tag{2}$$

under the assumptions: $\sum_{j \in S} p_{ij}(t) = 1, i \in \mathbb{S}$ in the period $(t-1, t]$. Finding a minimum of (1) yields optimal bandwidth occupancy for each time instance, which was then translated into the Markov chain state space $\mathbb{S}$ and treated as microdata n_{ij}, from which the $\hat{p}_{ij}$ were determined [11,20]:

$$\hat{p}_{ij} = \frac{\sum_{t \in T} n_{ij}(t)}{\sum_{t \in T} \sum_{j \in \mathbb{S}} n_{ij}(t)}, \quad i, j \in \mathbb{S}. \tag{3}$$

If $\hat{p}_{ij}$ does not depend on time $t \in \mathbb{T}$ (formula (3)), it is assumed that the studied phenomenon is described by a homogeneous Markov chain.

Based on available empirical data, this research considered the following periods: Jan 2023 - Jan 2024, Jan 2023 - Jul 2023 and Aug 2023 - Jan 2024. However, due to the increased network traffic in the second half of 2023 and the space limitations, the rest of the article focuses mainly on the period Aug 2023 - Jan 2024.

5 Results and Discussion

First we apply the proposed heuristic methods to the considered problem of optimizing the frequency band occupation (band slices used) at the edges of an optical network subject to known demands matrix (cf. an example shown in Table 1). The simulation has to be repeated for each time instance considered. So, if we consider 365 samples taken daily over the entire year then the optimisation procedure has to be repeated 365 times. Thus any acceleration of the optimisation calculations is of paramount importance for the analysis performed. Once optimisation simulations are completed we analyse the calculated results using a methodology based on Markov chains.

Table 1. Example of a demand matrix from a specific day.

	2	3	4
1	5300	0	0
2	—	0	4450
3	—	—	0

The results of optimisation performed using heuristic methods have been compared with a deterministic Mixed Integer Programming (MIP) reference method based on integer programming and available through the CPLEX package [4]. The traffic demands (demand matrix elements expressed in Gbps) were calculated using statistical methods as described in Sect. 4. The calculations were carried out using a linear solver engine of CPLEX 12.8.0.0 on a 2.1 GHz Xeon E7-4830 v.3 processor with 256 GB RAM running under the Linux Debian operating system. Table 2a summarizes the sets used by the optimization procedures, while Table 2b lists modeling parameters for performing computations.

Table 2c shows the values of the objective function for the methods analyzed. The results obtained with the proposed metaheuristics do not differ from the reference MIP method. Table 2c also contains the computation time needed to complete calculations using the proposed metaheuristics $(\mu + \lambda)$ and BC and for the deterministic MIP method. An important finding is that the computation time of the proposed methods is much less than that of the deterministic MIP method. The results confirm that the calculation times for the proposed metaheuristics methods are at least two orders of magnitude lower than for the MIP reference method. Such improvement in computational efficiency allows effective

Table 2. Parameters description and fitness function results.

Set	Set settings
$\mathcal{N}$	4
$\mathcal{E}$	4
$\mathcal{S}$	384 slots
$\mathcal{T}$	3 transponders
$\mathcal{S}_t$	$\mathcal{S}_1 = \{1 \ldots 384\}$
	$\mathcal{S}_2 = \{1 \ldots 382\}$
	$\mathcal{S}_3 = \{1 \ldots 380\}$

(a) Set settings.

Constant	Settings
bitrate [*Gbps*]	$v(1) = 10$
	$v(2) = 100$
	$v(3) = 400$
$\xi(t)$	$\xi(1) = 1$
	$\xi(2) = 3$
	$\xi(3) = 12$

(b) Constant settings.

Analyzed Case	Fitness function MIP	BC	$(\mu + \lambda)$
day average	205	205	205
week average	215	215	215
Analyzed Case	**Comp. time [sec.] MIP**	**BC**	$(\mu + \lambda)$
day average	99.0	0.9	**0.8**
week average	99.8	0.9	**0.8**

(c) Fitness and time.

implementation of the proposed design methodology too much more complex networks than in the case of MIP. Thus, these results encourage further research on designing realistic networks for large telecom operators using this methods.

Figure 3 shows bandwidth occupation for edge E21 of the analyzed network as a function of time each day (whole period), calculated by heuristic methods while Table 3 shows the calculated average of the occupation for each edge expressed as a percentage of the total bandwidth. It can be seen that, for the analyzed services, the average result does not exceed the alarming level of bandwidth occupancy (70%). However, results from Fig. 3 show that the alarming level is reached in many time instances of the second half of the year, cf.in particular Fig. 3a and 3d. The warning level (50%) is reached already in the first half of the year. Such results in practice would suggest a telecom operator to think about expanding the network.

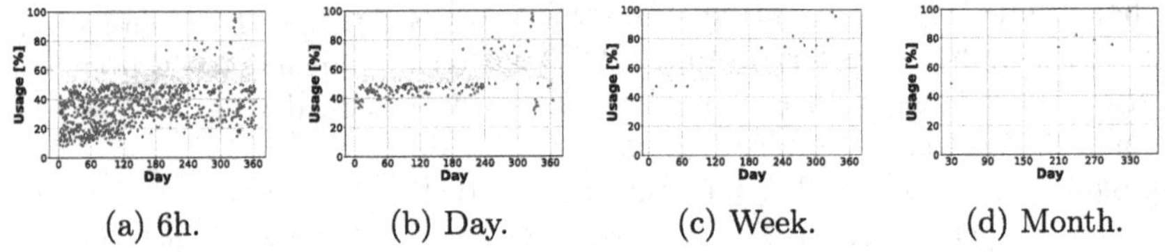

(a) 6h. (b) Day. (c) Week. (d) Month.

Fig. 3. Edge occupancy (E12) in the analyzed period for different sampling rates.

Table 3. Results of the slice occupation

Edges	MIP slice av.	BC slice av.	$(\mu + \lambda)$ slice av.
	%	%	%
E12	50,40	50,76	50,66
E23	50,00	50,32	50,61
E34	50,00	50,32	50,61
E14	49,20	49,72	50,46

In the next step we applied Markov chain methods to the results obtained by the heuristic optimisation algorithms. The application of the homogeneous and non-homogeneous Markov chain to the model described in the Sect. 4 for each edge $(E_{12}, E_{23}, E_{34}, E_{14})$ of the graph made it possible to determine the transition probability matrix P (or $P(t)$ given in the (2), depending on the BC algorithm used (P^{BC}) or $\mu + \lambda$ $(P^{\mu+\lambda})$ taking into account the Markov chain running time. In this case, the stochastic process was observed from August 2023 to January 2024, and 6-h, 1-day, 1-week and 1-month network traffic measurements were used to construct the P matrix. Due to the limited number of pages, the rest of the article includes results only for the E_{12} edge.

Thus, based on (4)–(5):

$$\widehat{P^{BC}_{6h,E_{12}}} = \begin{bmatrix} 0.23 & 0.09 & 0.68 \\ 0.04 & 0.38 & 0.58 \\ 0.03 & 0.32 & 0.66 \end{bmatrix}, \quad \widehat{P^{BC}_{Da}} = \begin{bmatrix} 0.35 & 0.55 & 0.1 \\ 0.1 & 0.73 & 0.17 \\ 0.03 & 0.56 & 0.42 \end{bmatrix}, \quad \widehat{P^{BC}_{We}} = \begin{bmatrix} 0.44 & 0.44 & 0.11 \\ 0.36 & 0.57 & 0.07 \\ 0 & 0.5 & 0.5 \end{bmatrix} \tag{4}$$

and

$$\widehat{P^{\mu+\lambda}_{6h,E_{12}}} = \begin{bmatrix} 0.29 & 0.05 & 0.67 \\ 0.05 & 0.36 & 0.59 \\ 0.02 & 0.32 & 0.66 \end{bmatrix}, \quad \widehat{P^{\mu+\lambda}_{Da}} = \begin{bmatrix} 0.28 & 0.61 & 0.11 \\ 0.11 & 0.72 & 0.17 \\ 0.02 & 0.45 & 0.52 \end{bmatrix}, \quad \widehat{P^{\mu+\lambda}_{We}} = \begin{bmatrix} 0.33 & 0.67 & 0 \\ 0.28 & 0.72 & 0 \\ 0 & 1 & 0 \end{bmatrix} \tag{5}$$

one can determine the probability of transitioning $p_{ij}(i, j \in \mathbb{S})$ between states $\mathbb{S} = \{high, medium, low\}$, as well as staying in one of the states by knowing the initial distribution of d_0 relative to the occupancy of % slices at time t_0.

For example, if we assume that the matrix $\widehat{P^{BC}_{6h,E_{12}}}$ was generated from state transitions $\mathbb{S}$ for edge E_{12} in the period August 2023 - January 2024, then - assuming the homogeneity of the Markov chain - in 3 periods P will take the form:

$$(\widehat{P^{BC}_{6h,E_{12}}})^3 = \begin{bmatrix} 0.048 & 0.321 & 0.644 \\ 0.042 & 0.332 & 0.638 \\ 0.042 & 0.336 & 0.645 \end{bmatrix}. \tag{6}$$

The matrix P obtained in (6) is close to the ergodic matrix E (stationary state) and means that it will remain in the *high* state in period t, provided that in the previous periods t-1 the process was also in state *high*.

The ergodic distribution $e = (e_{high}, e_{med}, e_{low})$ for edge E_{12}, with the proposed algorithms for the optimization methods $(BC, \mu + \lambda)$ and the accuracy is presented in Table 4.

Since the Markov chain used for the empirical data described in Sect. 4 is irreducible, non-periodic and has an ergodic distribution e, the mean recurrence time r_j to state j can be determined according to the formula: $r_j = \frac{1}{e_j}$. Thus calculated mean recurrence times r_j are presented in Table 5.

Table 4. Ergodic distribution **e**

Edges	BC			$\mu + \lambda$		
	e_{high}	e_{med}	e_{low}	e_{high}	e_{med}	e_{low}
$e_{6hAuJa,E_{12}}$	0.04	0.33	0.63	0.04	0.33	0.64
$e_{DaAuJa,E_{12}}$	0.12	0.67	0.21	0.11	0.64	0.25
$e_{WeAuJa,E_{12}}$	0.33	0.52	0.15	0.29	0.71	0

Table 5. The mean recurrence time r_j

Edges	BC			$\mu + \lambda$		
	r_{high}	r_{medium}	r_{low}	r_{high}	r_{medium}	r_{low}
$r_{6hAuJa,E_{12}}$	25.43	3.03	1.59	26.61	3.08	1.57
$r_{DaAuJa,E_{12}}$	8.55	1.49	4.75	9.58	1.56	3.93
$r_{WeAuJa,E_{12}}$	3.0	1.92	6.75	3.4	1.42	Inf

The mean first passage time (expected first return time from state j to i) of Markov chain $M = [m_{ij}]$ where the elements of the matrix M are taken from a recursive formula: $m_{ij} = p_{ij} + \sum_{k \neq j} p_{ik}(m_{kj} + 1)$ and $i, j \in S = \{high, medium, low\}$ (Table 6).

Table 6. The Mean First Passage Time $m_{i,j}$

Edges	BC		$\mu + \lambda$	
	$medium \rightarrow high$	$low \rightarrow high$	$medium \rightarrow high$	$low \rightarrow high$
$m_{ij_{6h,E12}}$	31.1	31.7	34.7	35.9
$m_{ij_{Da,E12}}$	11.4	12.6	11.7	13.2
$m_{ij_{We,E12}}$	–	–	–	–

As mentioned, the original data came from Jan 2023 to Jan 2024. Due to the increased network traffic in the second half of 2023, this period was included in the above analyses. By increasing the frequency of data observation (e.g. every 15 min), you can notice that there are periods of increased traffic (e.g. in the evening) compared to other times of the day. In such situations, the condition of the uniformity of the Markov chain may not be met. Then we propose Markov Set Chains (the algorithm for determining the upper e^{HI} and lower e^{LO} bounds for the elements of the ergodic distribution e of the transition probability matrices $P_1, P_2, \ldots, P_k$ is included in, among others, the papers: [3,18]). The values of the constraints e^{HI} and e^{LO} for the ergodic distribution e of the relevant transition probability matrices for the studied period are included in Table 7.

Taking into account the number of transponders used $N = \{n_{40}, n_{100}, n_{400}\}$, the cost of 40T, 100T, 400T transponders: $pr \in Price = \{1, 3, 12\}$ USD respectively, the ergodic distribution e for the states $S = \{high, medium, low\}$ with respect to the edges $k \in \{E_{12}, E_{23}, E_{34}, E_{14}\}$, it is possible to determine the cost C related to the charge on the paths E_{12}, E_{23}, E_{34} and paths E_{14}: $C_{E_{12},E_{23},E_{34}} = \sum_{j \in N} \sum_{i \in S} \sum_{k \in \{E_{12},E_{23},E_{34}\}} n_j * e_{ik} * pr_j$ and $C_{E_{14}} = \sum_{j \in N} \sum_{i \in S} n_j * e_{iE_{14}} * pr_j$ where e_{ik} is the i-th coordinate of the ergodic distribution e for the edge k. Assuming that the edge occupancy in both paths should be equal, a measure $M_D = \frac{C_{E_{12},E_{23},E_{34}}}{C_{E_{14}}}$ can be proposed, where $D = \{6h, Da, We, Mo\}$ denotes the frequency of network monitoring (M_D closer to 1 indicates the choice of method). In the case of empirical data, the indicator values are: $M_{6h} = \frac{32.886}{10.948} = 3.004$, $M_{Da} = \frac{33.733}{11.246} = 2.999$, $M_{We} = \frac{34.027}{11.341} = 3$, and $M_{Mo} = \frac{34.322}{11.355} = 3.023$ regardless of the algorithm used: $BC, \mu + \lambda$ $(M_{D,BC} = M_{D,\mu+\lambda})$. Taking into account $min\{M_{6h}, M_{Da}, M_{We}, M_{Mo}\}$, the lowest value is M_{Da}, so the recommended frequency of network traffic monitoring is daily.

Table 7. Markov Set Chains - Ergodic distribution $e_{HI} \geq e \geq e_{LO}$

Edges	BC			$\mu + \lambda$		
	e_{high}	e_{med}	e_{low}	e_{high}	e_{med}	e_{low}
$e^{LO}_{6h,E_{12}}$	0.01	0.2	0.53	0.01	0.2	0.52
$e^{HI}_{6h,E_{12}}$	0.1	0.44	0.75	0.12	0.45	0.75
$e^{LO}_{Da,E_{12}}$	0	0.12	0	0	0.38	0.01
$e^{HI}_{Da,E_{12}}$	0.38	0.99	0.82	0.3	0.96	0.51
$e^{LO}_{We,E_{12}}$	0	0	0	0	0	0
$e^{HI}_{We,E_{12}}$	1	1	1	1	1	1

Based on the results obtained, the following observations can be made:

1. regardless of the algorithm used, P matrices achieve ergodic distribution relatively quickly (usually after 5–6 periods with an accuracy of 3 decimal places), which is mainly due to the small number of S states,
2. in the case of data from the period Jan'23-Jan'24 and Aug'23-Jan'24, the probability of exceeding 70% is slightly higher using the BC algorithm than $\mu + \lambda$ for data 6h, Da and We without depending on the analyzed edge,
3. in the case of data from Jan'23 - Jan'24, the probability of reaching the *high* state is usually twice lower than in the period Aug'23–Jan'24. This increased network traffic in the second period, probably dictated by the autumn parliamentary elections, is an indication for the operator to be prepared for a potential increase in network traffic in the event of the next elections (e.g., securing the appropriate number and power of transponders),
4. obtained in the Table 5 the mean recurrence time r_j confirms the previous conclusion: the time to return to the *high* state after leaving it is shorter in

the case of data from the period Aug'23 - Jan'24 and usually slightly shorter in the case of the BC algorithm,

5. the first passage time to the *high* state from the *medium, low* state is usually slightly shorter in the case of the BC algorithm,
6. the previously mentioned significant increase in network traffic in the second half of 2023 may suggest using a non-homogeneous Markov chain. Transition probability matrices $P_{Jan23-Jul23}$ and $P_{Aug23-Jan24}$ verified by the χ^2 compliance test with the $\chi^2_{P_{Jan23-Jul23},P_{Aug23-Jan24}}$ test of $H_0 : \hat{P}_{Jan23-Jul23} = \hat{P}_{Aug23-Jan24}$ differ significantly (e.g. for $Da, \mu+\lambda, E_{12} : \chi^2_{emp} = 127.55, df = 8, p-value < 0.001 \Rightarrow H_0$ rejected). In such a situation, we propose to model the network traffic using Markov Set Chains (HI-LO, [3]), i.e. we determine the intervals in which the proper ergodic distribution is realized: $e_{LO} \leq e \leq e_{HI}$ (results in Table 7),
7. based on the results in the Table 7, it can be seen that:
   - with a decrease in the frequency of data downloading, the span of the interval covering the coordinates of the $e = (e_{high}, e_{med}, e_{low})$ increases, which means less precision in determining the estimates r_i and m_{ij},
   - HI-LO intervals around e from the period from Aug'23 to Jan'24 are usually narrower than those determined based on the period from Jan'23 to Jan'24 (similarly in the case of r_{high} - return time to *high* and $m_{i,high}$ - the time of reaching *high* for the first time from state i), therefore it is advisable to use different Markov chains for different periods to model network traffic.
8. Based on the M_* measure, we propose using daily data.

Once we developed the Markov chain formalism, we can use the determined transition probability matrices P to forecast the bandwidth occupancy for 365 days ahead (Fig. 4), based on the ergodic distribution for the state space $\mathbb{S}$. Consequently, the predicted expected bandwidth usage does not change in time for each state, which limits the model's accuracy. However, the model is still fully capable of predicting the probability of the state of the particular edge occupancy in the future days, which is the key point illustrated by the results shown in Fig. 4.

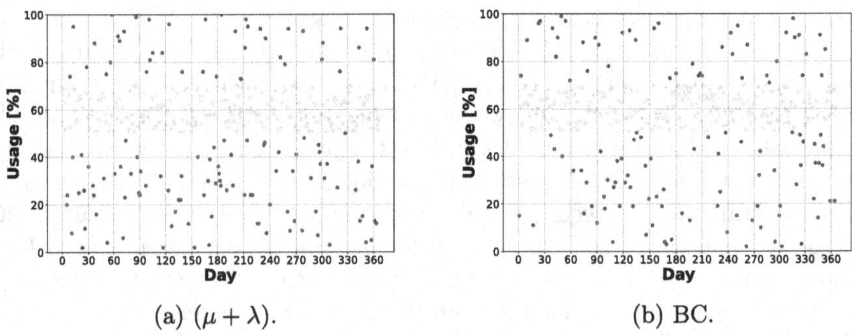

(a) $(\mu + \lambda)$. (b) BC.

Fig. 4. Forecasted bandwidth occupancy at the edge E_{12} for a year ahead, and two proposed methods

6 Conclusions

This contribution presents a methodology for an analysis of bandwidth occupancy in an optical network. The presented approach has been applied to real data collected via a management system from an operating network. For the calculation of bandwidth occupancy within a considered time period optimization metaheuristic methods are used to accelerate the numerical calculations. The results obtained confirm that the calculation times for the proposed metaheuristics methods are at least two orders of magnitude lower than for the MIP reference method. This improvement in computational efficiency allows an effective implementation of the proposed analysis methodology. Finally, the forecasting needed for optical network expansion planning has been performed using the Markov chain approach. The derived Markov transition probability matrices are used to predict traffic in network edges for future days using the data calculated by the metaheuristic methods for the preceding time period. Using the calculated Markov chain transition probability matrices, we have calculated a bandwidth occupancy forecast for a year ahead and thus demonstrated the practical relevance of the presented approach.

References

1. Chatterjee, B.C., Fadini, W., Oki, E.: A spectrum allocation scheme based on first-last-exact fit policy for elastic optical networks. J. Netw. Comput. Appl. **68**, 164–172 (2016)
2. Ejaz, W., Sharma, S.K., Saadat, S., Naeem, M., Anpalagan, A., Chughtai, N.: A comprehensive survey on resource allocation for Cran in 5G and beyond networks. J. Netw. Comput. Appl. **160**, 102638 (2020)
3. Hartfiel, D.: Markov Set-Chains. Springer, New York (1998). https://doi.org/10.1007/BFb0094586
4. ILOG: CPLEX 11.0 User's Manual. ILOG (2007)
5. Kemeny, J.G., Snell, J.L.: Finite Markov Chains. Springer, New York (1976)
6. Khodashenas, P.S., et al.: Comparison of spectral and spatial super-channel allocation schemes for SDM networks. J. Lightwave Technol. **34**(11), 2710–2716 (2016). https://opg.optica.org/jlt/abstract.cfm?URI=jlt-34-11-2710
7. Klinkowski, M., Walkowiak, K.: An efficient optimization framework for solving RSSA problems in spectrally and spatially flexible optical networks. IEEE/ACM Trans. Netw. **27**(4), 1474-1486 (2019). https://doi.org/10.1109/TNET.2019.2922761
8. Klinkowski, M., Lechowicz, P., Walkowiak, K.: Survey of resource allocation schemes and algorithms in spectrally-spatially flexible optical networking. Opt. Switch. Netw. **27**, 58–78 (2018)
9. Kozdrowski, S., Żotkiewicz, M., Sujecki, S.: Ultra-wideband WDM optical network optimization. Photonics **7**(1) (2020). https://doi.org/10.3390/photonics7010016
10. Kozdrowski, S., Sliwka, P., Sujecki, S.: Modeling traffic forecasts with probability in DWDM optical networks. In: Paszynski, M., Kranzlmüller, D., Krzhizhanovskaya, V.V., Dongarra, J.J., Sloot, P.M.A. (eds.) ICCS 2021. LNCS, vol. 12745, pp. 365–378. Springer, Cham (2021). https://doi.org/10.1007/978-3-030-77970-2_28

11. Lee, T., Judge, G., Zellner, A.: Estimating the Parameters of the Markov Probability Model from Aggregate Time Series Data. North-Holland Pub. Co., Amsterdam (1970)
12. Khorsandi, B.M., Raffaelli, C.: BBU location algorithms for survivable 5G c-ran over WDM. Comput. Netw. **144** (2018). https://doi.org/10.1016/j.comnet.2018.07.026
13. Khorsandi, B.M., Tonini, F., Raffaelli, C.: Design methodologies and algorithms for survivable c-ran, 106–111 (2018). https://doi.org/10.23919/ONDM.2018.8396115
14. Nakayama, Y., et al.: Efficient DWBA algorithm for TWDM-PON with mobile fronthaul in 5G networks. In: GLOBECOM 2017 - 2017 IEEE Global Communications Conference, pp. 1–6. IEEE Press (2017). https://doi.org/10.1109/GLOCOM.2017.8254768
15. Patri, S.K., Autenrieth, A., Elbers, J.P., Machuca, C.M.: Planning optical networks for unexpected traffic growth. In: 2020 European Conference on Optical Communications (ECOC), pp. 1–4 (2020). https://doi.org/10.1109/ECOC48923.2020.9333215
16. Poturała, A., Konieczka, M., Śliwka, P., Sujecki, S., Kozdrowski, S.: Numerical and statistical probability distribution transformation for modeling traffic in optical networks. In: Groen, D., de Mulatier, C., Paszynski, M., Krzhizhanovskaya, V.V., Dongarra, J.J., Sloot, P.M.A. (eds.) ICCS 2022. LNCS, vol. 13352, pp. 383–397. Springer, Cham (2022). https://doi.org/10.1007/978-3-031-08757-8_32
17. Salman, S., Alaswad, S.: Alleviating road network congestion: traffic pattern optimization using Markov chain traffic assignment. Comput. Oper. Res. **99**, 191–205 (2018). https://www.sciencedirect.com/science/article/pii/S0305054818301692
18. Sliwka, P.: Markov (set) chains application to predict mortality rates using extended Milevsky-Promislov generalized mortality models. J. Appl. Stat. **49**(15), 3868–3888 (2022). https://doi.org/10.1080/02664763.2021.1967891
19. Sliwka, P.: Markov (set) chains application to predict mortality rates using extended Milevsky-Promislov generalized mortality models. J. Appl. Stat. (2021). https://doi.org/10.1080/02664763.2021.1967891
20. Sliwka, P., Swistowska, A.: Economic Forecasting Methods with the R Package. UKSW, Warszawa (2019)

Automatic Gradient Estimation for Calibrating Crowd Models with Discrete Decision Making

Philipp Andelfinger(✉) and Justin N. Kreikemeyer

Institute for Visual and Analytic Computing, University of Rostock, 18059 Rostock, Germany
{philipp.andelfinger,justin.kreikemeyer}@uni-rostock.de

Abstract. Recently proposed gradient estimators enable gradient descent over stochastic programs with discrete jumps in the response surface, which are not covered by automatic differentiation (AD) alone. Although these estimators' capability to guide a swift local search has been shown for certain problems, their applicability to models relevant to real-world applications remains largely unexplored. As the gradients governing the choice in candidate solutions are calculated from sampled simulation trajectories, the optimization procedure bears similarities to metaheuristics such as particle swarm optimization, which puts the focus on the different methods' calibration progress per function evaluation. Here, we consider the calibration of force-based crowd evacuation models based on the popular Social Force model augmented by discrete decision making. After studying the ability of an AD-based estimator for branching programs to capture the simulation's rugged response surface, calibration problems are tackled using gradient descent and two metaheuristics. As our main insights, we find 1) that the estimation's fidelity benefits from disregarding large jumps inherent to the Social Force model, and 2) that the common problem of inferring a parameter's posterior distribution given some data obviates the need for AD across the Social Force calculations, allowing gradient descent to excel.

Keywords: calibration · parameter synthesis · automatic differentiation · gradient descent · agent-based simulation · social force

1 Introduction

Agent-based crowd models are widely used in urban planning [13,23], to study disease spread [15] or to optimize strategies for emergency evacuations [6,18,34]. In contrast to coarse-grained models that consider groups of people in aggregate, agent-based models operate on the microscopic level, giving each agent its own state, perception of its environment, and decision making. To generate meaningful simulation results, the calibration of the model parameters to empirical data, e.g., gathered from video footage, is a crucial prerequisite.

Often, metaheuristics such as genetic algorithms or particle swarm optimization are applied [30,33], which permit a straightforward parallelization to

L. Franco et al. (Eds.): ICCS 2024, LNCS 14836, pp. 227–241, 2024.
https://doi.org/10.1007/978-3-031-63775-9_16

tackle the computational cost of evaluating microscopic simulations across high-dimensional parameter spaces. Surrogate models generated by sampling the simulation response can support a faster calibration but may require enormous numbers of samples to capture an original model's dynamics [25].

However, all these methods typically operate on black-box observations of the simulation output. If it is possible to also determine its partial derivatives wrt. the parameters, local search strategies based on gradient descent can steer the calibration toward a local optimum. Calibration using Bayesian inference [4, 11], which also offers uncertainty information instead of sheer point estimates, could particularly benefit from the resulting increase in sampling efficiency [9]. Unfortunately, in the presence of models of discrete decision making, agent-based simulations form stochastic functions with discrete jumps. For these functions, simple averaging over derivatives determined using the established methods and tools for automatic differentiation (AD) [14,21] yields biased estimates [12].

Recently, there has been renewed interest in gradient estimation over stochastic functions with discrete jumps. Rooted in infinitesimal perturbation analysis, this line of research has produced AD-based estimators tailored to specific domains [1,7,29], for programs involving random sampling from discrete probability distributions [2], and for general imperative programs [19]. The recent generic estimators combine pathwise derivatives with the contributions of jumps, the former being determined using traditional AD, and the latter based on a priori knowledge of the distributions or density estimations. As an AD-free alternative, modern finite differences estimators compute gradient estimates from series of function evaluations on stochastically perturbed inputs [24]. These recent estimators' reliance on sampling raises the question whether the function evaluations permitted by a time budget are better spent on obtaining gradient estimates, or to directly evaluate a set of candidate solutions as part of a metaheuristic.

Here, we explore the suitability of gradient descent for calibrating crowd evacuation models based on Treiber's popular Social Force model [16] augmented by discrete decisions. We assess an AD-based and a stochastic finite differences-based estimator [19,24] compared to a genetic algorithm and particle swarm optimization. Our main contributions are threefold:

- We present an alternative derivation of our gradient estimator DiscoGrad Gradient Oracle [19] starting from the concept of stratified derivatives.
- We study the fidelity of sampling-based gradient estimates over an evacuation scenario with continuous or discrete objective.
- The calibration progress is evaluated for three problems, one being a distribution fitting problem over a 20-dimensional parameter space, showing that this problem class permits fast convergence via gradient descent.

The remainder of the paper is structured as follows: In Sect. 2, we briefly introduce methods for (automatic) differentiation across discrete jumps. In Sect. 3, we introduce the existing AD-based gradient estimator DGO. Section 4 describes the considered simulation model and scenarios. In Sect. 5, we present our experiment results and discuss their implications. Section 6 provides an interpretation of our results and concludes the paper.

2 Background and Related Work

Simulation models of crowd dynamics typically combine low-level models of pedestrian movement with 1) models of discrete decisions, e.g., for path planning, and 2) stochastic components accounting for uncertainty and variability in initial conditions and pedestrian behaviors [6,18]. In effect, the models thus take the form of stochastic functions $\mathcal{P}\colon \mathbb{R}^n \to \mathbb{R}$ involving discrete jumps, which poses challenges to traditional gradient estimation methods. In the following, we briefly introduce the existing work on gradient estimation across discontinuities and concrete estimators for this purpose.

To begin, we briefly recapitulate the widely employed concept of *automatic differentiation* (AD) [14,21]. This method views the execution of $\mathcal{P}$ as a composition of operators $\mathcal{P}_1 \circ \mathcal{P}_2 \circ \cdots$. By repeatedly applying the chain rule, the partial derivatives wrt. the inputs can be determined from the intermediate derivatives and values at the operators $\mathcal{P}_i$. Implementations can be grouped into *reverse* and *forward* modes. Whereas the former implements AD as a second (reverse) pass over previously stored intermediate values retrieving one row of the Jacobian per pass, the latter propagates derivatives through the forward execution retrieving one column per pass. The partial derivatives obtained this way are *pathwise* in the sense that they only capture the operation sequence of a single program execution, disregarding alternative branches.

When estimating gradients of stochastic programs, we are typically interested in the partial derivatives of the expected value of $\mathcal{P}$ wrt. a parameter vector θ:

$$\frac{\partial}{\partial \theta}\mathbb{E}\left[\mathcal{P}(\theta)\right] \tag{1}$$

As is common in the literature [12,28], we introduce an additional parameter ω. This allows the explicit consideration of the stochasticity of $\mathcal{P}$, so that $\mathcal{P}(\omega;\theta)$ refers to a specific realization of the stochastic function, e.g., as determined by a pseudo-random generator's seed. Without loss of generality, we sometimes consider only $n = 1$, while $n > 1$ follows directly from separately calculating the partial derivative for each dimension. One important case occurs if $\mathcal{P}$ exhibits discontinuities (jumps), whose positions depend on θ and/or ω. In the last decades, several methods have been developed to deal with this situation, an overview of which is given in [10].

Among the earliest is the *infinitesimal perturbation analysis* (IPA) estimator [17]. Relying on an interchange of the differentiation and expectation operators in Eq. 1 it can be computed by averaging over pathwise derivatives. However, this estimator is biased for discontinuous $\mathcal{P}$, as then the requirements for the interchange of operators are not satisfied. To still account for jumps, *smoothed perturbation analysis* (SPA) [12] employs a method inspired by Conditional Monte Carlo. Based on the law of total expectation, $\mathbb{E}[\mathcal{P}]$ is calculated as $\mathbb{E}[\mathbb{E}[\mathcal{P}|z]]$ for some tailored characterization z of $\mathcal{P}$'s execution. If z is chosen correctly, this allows the use of pathwise derivatives as in IPA.

Recently, inspired by the success of AD on deterministic programs, automatic methods to calculate Eq. 1 gained new interest. The *StochasticAD* estimator [2] builds on SPA and AD to allow the automatic differentiation of programs sampling from discrete parametric distributions. Other methods include: the *straight-through estimator* [3]; *interpolation* using continuous approximations of discontinuous operators $\mathcal{P}_i$ in $\mathcal{P}$ [1,8]; and *abstract interpretation* [5], which propagates distributions symbolically through $\mathcal{P}$ to smooth over discontinuities.

Another approach to the calculation of Eq. 1 are black-box estimators like REINFORCE [32] and *randomized finite-differences* schemes [27]. A notable candidate from the latter category is proposed in [24] building on [26, Chapter 3.4], which we adopt here as follows under the name Polyak Gradient Oracle (PGO):

$$\nabla\mathcal{P}(\theta) \approx \textstyle\sum_{s=1}^{S}(\mathcal{P}(\theta+\sigma\mathbf{u},\omega_s)-\mathcal{P}(\theta,\omega))\sigma^{-1}\mathbf{u}/S, \tag{2}$$

where $\mathbf{u}$ is a vector of i.i.d. standard normal variates and σ a "smoothing factor". By introducing random perturbations on θ, this estimator can provide a full gradient estimate from one sample. Note that introducing such perturbations is possible (or even required) with many estimators, allowing their application to deterministic programs with discontinuities.

3 DiscoGrad Gradient Oracle

In this publication, we evaluate practical applications of the recently proposed DiscoGrad Gradient Oracle (DGO) [19]. The following provides an alternative derivation starting from the abstract concept of the "stratified derivative" from [28]. The latter is constructed around the concept of a *critical event* A, which occurs if $|\mathcal{P}(\omega;\theta+\epsilon/2)-\mathcal{P}(\omega;\theta-\epsilon/2)| > B|\epsilon|$ for some bound $B>0$, i.e., when a jump is observed in an ϵ-neighborhood around θ. Then, it holds that

$$\frac{\partial\mathbb{E}\left[\mathcal{P}(\theta)\right]}{\partial\theta} = \mathbb{E}\left[\frac{\partial}{\partial\theta}\mathcal{P}(\theta)\right] + \mathbb{E}\left[\Delta_{\mathcal{P}}\right]p'_\theta. \tag{3}$$

Here, $\Delta_{\mathcal{P}}$ denotes the distribution of the jump's magnitudes conditioned on A, and p'_θ is the *critical rate*, defined as $\lim_{\epsilon\downarrow 0}\frac{1}{\epsilon}\mathbb{P}(A)$. The DGO estimates the above for the special case of imperative programs with conditional branches.

Let us consider a program $\mathcal{P}$ with scalar input and output, including a single branch of the form "if $\mathrm{G}(\theta) < \mathrm{d}$", with d a constant and the value of $\mathcal{P}$ depending on the path taken. Defining $C(\theta) := G(\theta) - d$, the branching condition can be rewritten as $C(\theta) < 0$. Thus, a realization of C's sign indicates the chosen branch.

The term $\mathbb{E}\left[\frac{\partial}{\partial\theta}\mathcal{P}(\theta)\right]$ in Eq. 3 can be trivially estimated by sampling pathwise derivatives using AD as in IPA. Considering p'_θ, we first note that $\mathbb{P}(|\mathcal{P}(\omega;\theta+\epsilon/2)-\mathcal{P}(\omega;\theta-\epsilon/2)| > B|\epsilon|) = \mathbb{P}(C(\omega;\theta+\epsilon/2)\cdot C(\omega;\theta-\epsilon/2) < 0)$. This is the probability of a sign change in an ϵ-neighborhood around θ. We assume all jumps to originate from branches of the above form, which still allows many other discontinuous functions such as the minimum or absolute value to be expressed.

For $\epsilon\to 0$, we obtain $p'_\theta = f_{C(\theta)}(0)\mathbb{E}\left[\frac{\partial}{\partial\theta}C(\theta)\right]$, where $f_{C(\theta)}$ is the probability density function of $C(\theta)$. DGO estimates p'_θ by gathering realizations $C(\omega_s;\theta)$ and calculating a density estimation, which is evaluated at the origin. $\mathbb{E}\left[\frac{\partial}{\partial\theta}C(\theta)\right]$ as well as $\mathbb{E}\left[\Delta_{\mathcal{P}}\right]$ are estimated using the realizations of $C(\theta)$ closest to 0.

Estimating gradients across programs with several branches requires additional considerations. In the presence of sequential branches, a branch condition's distribution can depend on whether previous branches have been taken. Hence, the density estimation must distinguish the control flow path along which a branch is reached. We uniquely identify each branch b by the path along which it is encountered, corresponding to the sequence of condition signs at all previous branches. Now, we can express DGO for programs with $\mathfrak{B} \in \mathbb{N}$ branches:

$$\frac{\partial \mathbb{E}\left[\mathcal{P}(\theta)\right]}{\partial \theta} = \lim_{S\to\infty} \frac{1}{S} \sum_{s=1}^{S} \frac{\partial \mathcal{P}(\omega_s;\theta)}{\partial \theta} + \sum_{b=1}^{\mathfrak{B}} \left(\mathcal{P}(\omega_b^+;\theta) - \mathcal{P}(\omega_b^-;\theta)\right) \lambda_b \hat{f}_{C_b(\theta)}(0) \frac{\partial C_b^\epsilon}{\partial \theta}$$

where ω_b^+ and ω_b^- select the samples s corresponding to the positive and negative realizations of $C_b(\theta)$ closest to the branching point, λ_b is the proportion of samples that encountered the branch, $\hat{f}_{C_b(\theta)}$ is an estimate of $C_b(\theta)$'s probability density function, and $\frac{\partial}{\partial\theta}C_b^\epsilon$ is the partial derivative of the condition near the branching point.

We note that, in contrast to PGO (cf. Sect. 2), whose calculation of directional derivatives relies on perturbations of the parameters, DGO can operate on an original program without introducing external stochasticity. However, by reducing the ruggedness of the objective function, smoothing via external perturbations can contribute to faster convergence of gradient descent.

An implementation of DGO exists as part of the DiscoGrad tool [19], which permits the differentiation across a subset of C++ programs with conditional branches. The implementation is available publicly[1].

4 Crowd Model and Scenarios

Our experiments build on the typical constituents of evacuation studies: a force-based model of crowd mobility in a two-dimensional continuous space and per-pedestrian discrete decision making. The mobility is modeled using Treiber's popular Social Force model [16], in which a person's effective acceleration vector is calculated as a sum of three forces. The *internal* force reflects a person i's intention to move in a straight line towards its goal location in direction e_i^0 with desired velocity v_i^0, $v_i(t)$ being the current velocity. The adaptation time is scaled according to a characteristic time τ_i. *Interaction* forces f_{ij} between person i and each other person j in his or her vicinity exert a repellent effect, reflecting avoidance maneuvers and maintenance of personal distance. Finally, *obstacle* forces f_{iW} repel the person from any nearby wall W, leading to the overall force equation for a person i with mass m_i and scaling coefficients w_1, w_2, w_3:

$$m_i \frac{dv_i}{dt} = w_1 m_i \frac{v_i^0 e_i^0(t) - v_i(t)}{\tau_i} + w_2 \sum_{j\neq i} f_{ij} + w_3 \sum_{W} f_{iW}$$

[1] https://github.com/DiscoGrad/DiscoGrad.

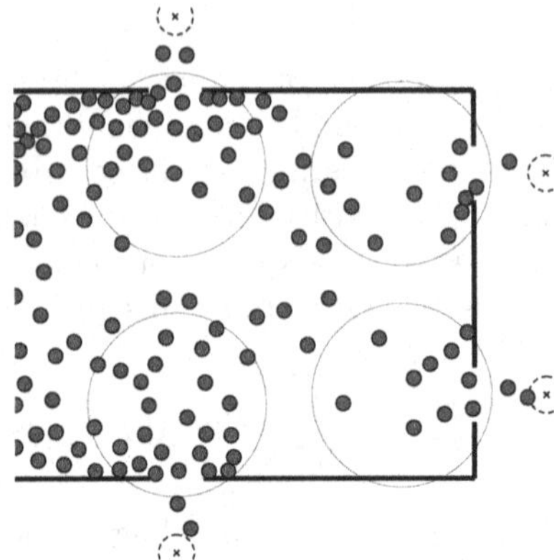

Fig. 1. Scenario for calibrating exit selection coefficients. The crowd enters from the left-hand side, aiming to evacuate by reaching the circular waypoints (dashed circles), each agent periodically reconsidering the targeted exit by weighing its distance against the number of agents in its vicinity (gray circles) (Color figure online).

We consider two scenarios, the first representing a bottleneck in an evacuation situation, similar to [11]. The simulation space is a 30×30 m square separated in the center by a single wall with a door 4m in width. As in [11], we calibrate the weight coefficients that determine the strengths of the forces experienced by pedestrians aiming to pass through the door. The output of the simulation to be calibrated is either the pedestrians' average horizontal position or the number of pedestrians evacuated after 20 s of simulation time.

The second scenario (cf. Fig. 1) combines low-level mobility via the Social Force model with discrete exit selections as in existing work such as [31]. A crowd comprised of 1 000 pedestrians gradually enters the scenario from the left-hand side, aiming to exit the building via any of the four available doors, each 3m wide. Each pedestrian selects its target door by weighing its distance against the congestion level at the door as measured by the number of agents nearby. To be able to react to changing circumstances, each agent reconsiders the previous decision every 15 s. The weight coefficient underlying the decision is drawn from a probability distribution supplied as model parameters in the form of a histogram ranging in 20 steps from 0.1, where the decision is dominated by the congestion level, to 1.0, where the decision is made solely based on distance. The simulation output to be minimized is the Wasserstein distance of the histogram of observed evacuation times to a reference histogram after a warm-up time of 100 s, spanning 20 steps from 10 s to 75 s.

The scenarios differ fundamentally in their implications for gradient estimation. While the first scenario involves explicit conditional branches only in the counting of evacuations, discrete jumps are created by the Social Force model itself, making this a challenging scenario for DGO. In contrast, the calibration of the second scenario leads to gradients entirely defined by conditional branches, which prevents their estimation via AD alone but is well-suited for DGO.

5 Experiments

The goal of our experiments is to determine whether gradient descent using sampling-based gradient estimators can outperform genetic algorithms and particle swarm optimization in the calibration of evacuation models. We approach this objective by first studying the degree to which the estimators are able to capture Social Force's dynamics. We then turn to the calibration problems and carry out hyperparameter sweeps in order to shed light on the relative performance of the different optimization methods.

The simulation models were implemented in C++ within DiscoGrad [19], closely following PEDSIM[2] for the Social Force model and its parametrization. As the genetic algorithm implementation, we used pyeasyga[3], and for the particle swarm optimization we employ the pyswarms library [22]. All simulations use Leapfrog integration with a time step of 0.1 s. The calibration experiments were carried out on two identical machines, each equipped with an AMD EPYC 9754 processor with 256 threads and 768 GB RAM, running Ubuntu 22.04.4 LTS, each machine executing at most 256 calibration runs in parallel.

5.1 Automatically Differentiating the Social Force Model

While the Social Force model itself does not call for explicit conditional branches, slight perturbations in the parameters can cause large discrete jumps in acceleration. For instance, the interaction force increases exponentially as the distance between two agents decreases, and its direction depends discretely on the distance and angle difference between two agents. If the distance between two agents is small, minor changes in parameters can cause a change in direction and thus extreme changes in forces. While AD alone correctly determines the gradient at a given point of the parameter space, it cannot capture such jumps. DGO can treat these jumps as explicit conditional branches. However, its estimation relies on sampled derivatives of intermediate branch conditions (cf. Sect. 3), which may suffer from high variance when pairs of agents come in close mutual proximity.

To assess the differences in the AD-based and black-box estimators' ability to estimate gradients of crowd simulations, we carried out a parameter sweep across the weight coefficients w_0, w_1, w_2 that govern the intensity of the internal, interaction, and obstacles forces for the single-exit evacuation scenario. For competitiveness with optimization procedures other than gradient descent, we are particularly interested in the gradients' fidelity with small numbers of samples. Thus, we study the estimation error when varying the number of samples in comparison to reference gradients calculated from 100 000 function evaluations via PGO, which delivers unbiased estimates of the smoothed gradient.

Figure 2 shows the partial derivatives wrt. each of the coefficients being varied separately in 300 steps while keeping the others fixed at values of 0.6, 5.5, and 5.5, respectively. The grey curves show the simulation output, which is the squared

[2] https://github.com/chgloor/pedsim.

[3] https://github.com/remiomosowon/pyeasyga.

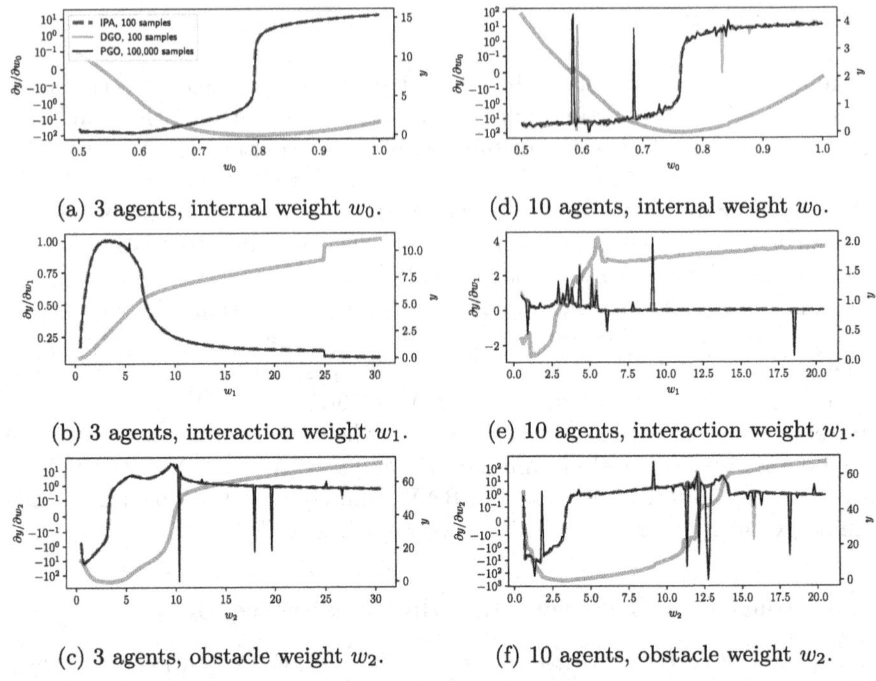

(a) 3 agents, internal weight w_0.

(b) 3 agents, interaction weight w_1.

(c) 3 agents, obstacle weight w_2.

(d) 10 agents, internal weight w_0.

(e) 10 agents, interaction weight w_1.

(f) 10 agents, obstacle weight w_2.

Fig. 2. Derivative estimates with respect to the weight of the three types of forces in scenarios populated with 3 agents (a–c) and 10 agents (d–f) with the fit in the agents' final coordinates' as output and $\sigma = 0.001$. PGO with a large number of samples is used as the reference. DGO captures most of the derivatives' spikes, whereas the IPA estimate only reflects the general curvature.

error in the average final agent positions compared to the pre-defined reference. For comparability to PGO, the input is perturbed by Gaussian noise with $\sigma = 0.001$. For the partial derivative estimates, we evaluate averaging across plain AD gradients (IPA) and DGO against the reference produced by PGO. With 3 agents, both AD-based estimators closely track the derivatives' curvature. In (b), we can see that the jump in the simulation output at about 24.5 is not visible in the derivative at the chosen resolution along the w_1 axis. Importantly, we observe that the IPA curve does not follow the sharp downward spikes in $\partial y/\partial w_2$, while they are accurately captured by DGO. The results with 10 agents follow a similar trend, but due the increased number of force calculations, the simulation output becomes substantially more rugged, resulting in noisier estimates using DGO.

Figure 3 assesses the same scenario focusing on w_0 after changing the output to the squared error in the number of evacuations compared to a reference value. This entirely discrete objective is smoothed only by the parameter's perturbations, whose standard deviation we set to 0.01 and 0.1, observing the expected increase in smoothness with the larger value. In this problem, the simulation output is gathered by counting the number of agents that have passed the exit,

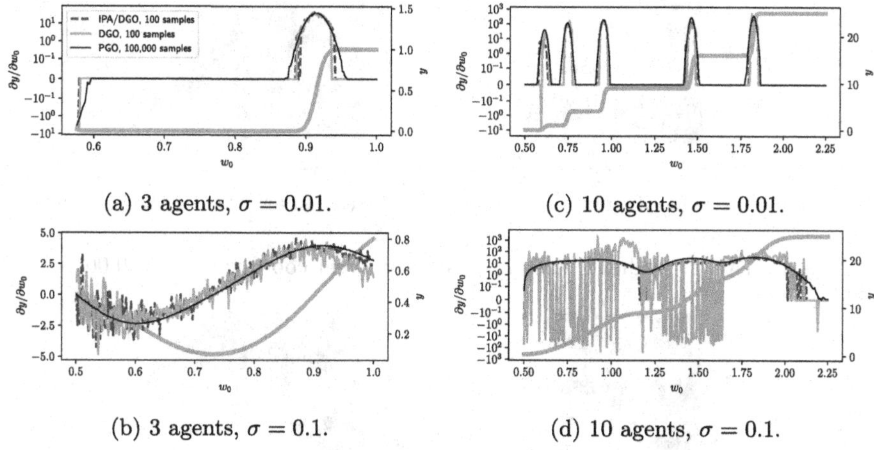

(a) 3 agents, $\sigma = 0.01$.

(c) 10 agents, $\sigma = 0.01$.

(b) 3 agents, $\sigma = 0.1$.

(d) 10 agents, $\sigma = 0.1$.

Fig. 3. Derivative estimates with respect to the weight of the internal force, the simulation output being the fit in the number of evacuations. In the larger scenario, DGO's estimates suffer from substantial noise as jumps in the mobility derivatives translate to biased and high-variance derivative estimates of the simulation output. When ignoring jumps in the Social Force model (IPA/DGO), estimates observe some bias but capture the trends well.

which in the model's source code translates to a series of conditional branches on the agent positions. Considering DGO, the estimation of the branches being taken and their effects on the overall derivatives are now subject to any noise in the positions' partial derivatives, leading to extremely noisy and often inaccurate derivative estimates particularly with 10 agents.

In the previous scenario, we have seen that the IPA estimates reflect the main curvature of the derivatives. However, since IPA alone yields zero-valued derivatives with this discrete objective, it cannot be applied here. Instead, we combine IPA and DGO by disregarding any jumps in Social Force while still accounting for the effects of branches using DGO. In this combination, while some noise and slight deviations from the reference are observed, the tendencies of the reference are reflected much more accurately with 100 function evaluations.

To quantify the gradient estimates' fidelity, we consider the mean absolute error compared to PGO with 100 000 function evaluations for w_0. Figure 4 supports the previous observations: As the source of IPA's deviation from the reference is its inability to capture jumps, its estimates do not improve with additional samples, while the estimates with 10 samples observe a similar error compared to the other estimators. In contrast, DGO improves somewhat with more evaluations but consistently outperforms PGO only in the smaller scenario. PGO reliably approaches the reference when increasing the sample count. However, we note that at low sample counts, the AD-based estimators are highly competitive.

The results for the same scenario with the discrete objective are shown in Fig. 5. Here, similar error levels are observed for all estimators with 3 agents,

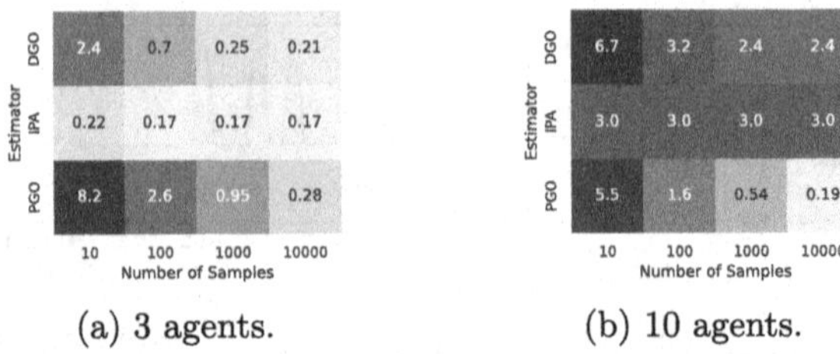

(a) 3 agents. (b) 10 agents.

Fig. 4. MAE of gradients wrt. w_0 fitting the agent coordinates, $\sigma = 0.001$.

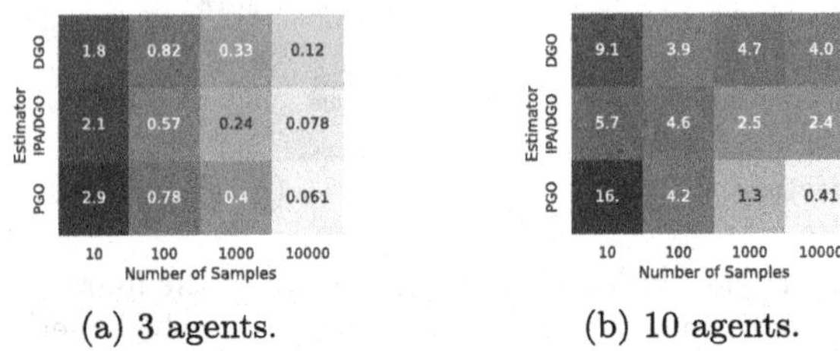

(a) 3 agents. (b) 10 agents.

Fig. 5. MAE of gradients wrt. w_0 fitting the evacuation count, $\sigma = 0.001$.

while with 10 agents, PGO is superior at 1 000 function evaluations and beyond. Again, the AD-based estimators are competitive up to 100 evaluations.

5.2 Calibrating Force Coefficients

We now compare the practical capabilities of the different gradient estimators, PSO and GA. To achieve a reasonably fair comparison among the calibration progress, we carried out a sweep across a range of sensible hyperparameters, with 10 microreplications of the program and 20 macroreplications of each hyperparameter configuration (cf. Table 1).

For each estimator, PSO, and GA, we select the configuration that produced the best average solution at the end of the time budget of 5 min wall time and report the mean over macroreplications of the crisp simulation output, i.e., without any perturbations, which we evaluated in a post-processing step over 1 000 microreplications. In addition to the progress over wall time, we also show the progress over function evaluations, each of which reflects one simulation

Table 1. Hyperparameter sampling ranges. Legend: σ sample size, lr learning rate, (c_1, c_2, w) social, cognitive, and inertia parameters, LHC latin hyper cube.

DGO, PGO, and IPA		PSO		GA	
samples	1, 10, 50	particles	10, 50	population	10, 50
σ	0, 0.01, 0.1, 0.5, 1.0	c_1, c_2, w	10 LHC samples	elitism	yes, no
lr	0.01, 0.1, 0.5, 1.0	neighbors	3, 6, all	mutation	x=$\mathcal{U}(0, 10)$, x+=$\mathcal{N}(0, 0.1)$

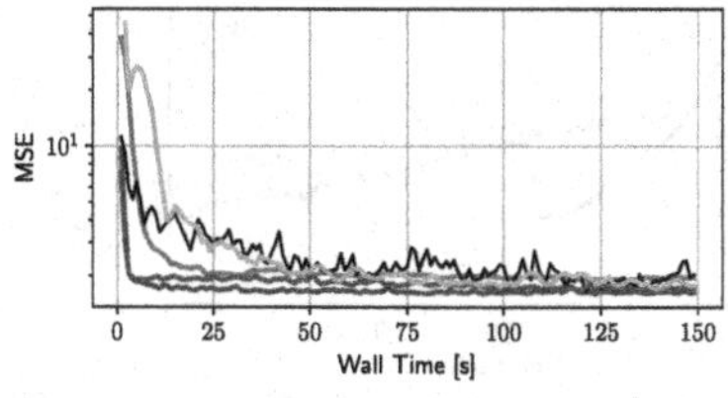

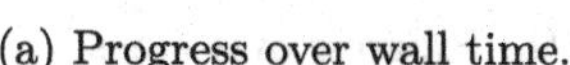
(a) Progress over wall time.

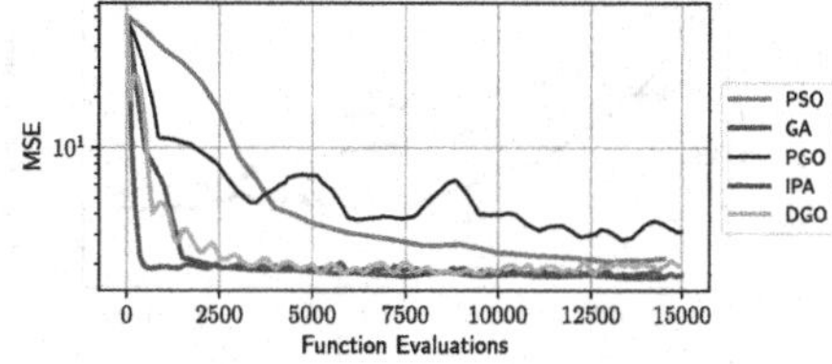

(b) Progress over function evaluations.

Fig. 6. Calibration progress over steps, wall time and function evaluations during calibration of the final agent coordinates via the Social Force weights.

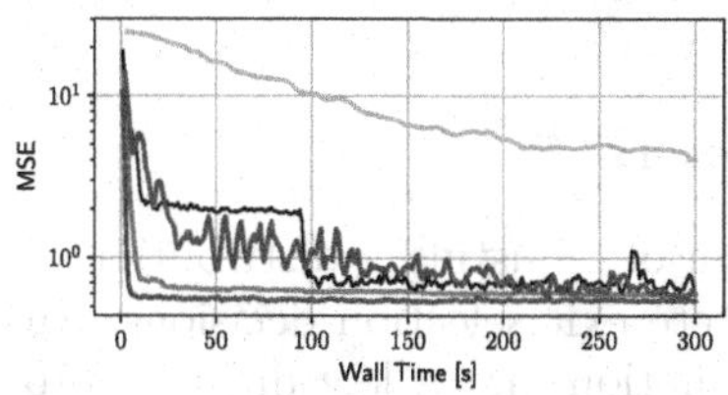

(a) Progress over wall time.

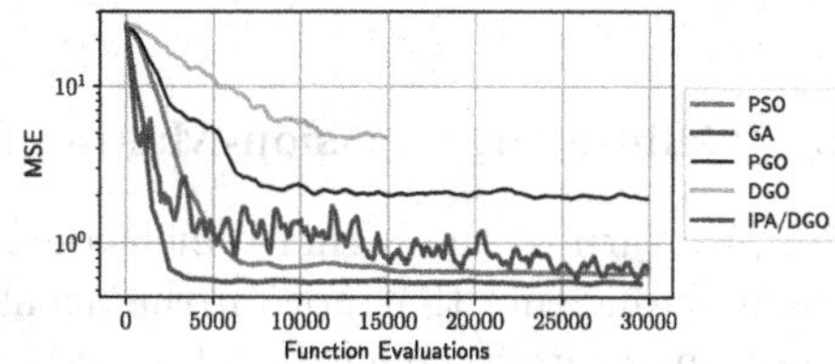

(b) Progress over function evaluations.

Fig. 7. Calibration progress over steps and wall time during calibration of the number of evacuations via the Social Force weights.

trajectory executed by a sample (gradient estimators), particle (PSO), or population member (GA).

Figure 6 shows the solution quality over function evaluations and wall time for the first problem. As expected, we observe that all approaches converge quickly for this problem. The fastest progress over time is achieved by GA, which converged after only a few seconds. Considering the gradient estimators, DGO and the finite differences-based PGO exhibit comparatively high variance and are the slowest to converge. DGO's progress per function evaluation is similar to GA, but is slowed in wall time by the gradient estimation overhead. Remarkably, IPA, which fared worst in terms of gradient fidelity (cf. Fig. 4), yields the fastest convergence, which suggests that capturing the general curvature of the objective function suffices to quickly identify a local minimum.

In the second calibration problem shown in Fig. 7, GA outperforms the other methods both in terms of function evaluations and time. Of the gradient estimators, PGO makes the fastest progress in wall time, with a plateau between about 10 s to 100 s stemming from a lack of initial progress in one of the 20 macroreplications. Again, the AD-based gradient estimation benefits from disregarding jumps in the force calculations, which allows IPA/DGO to overtake PGO over function evaluations, albeit encumbered by the AD overhead.

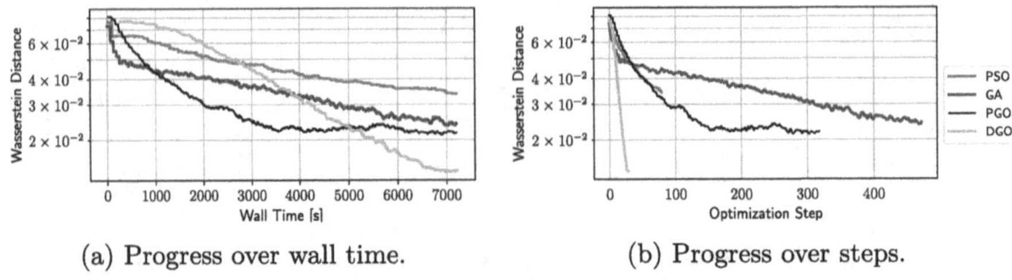

(a) Progress over wall time.

(b) Progress over steps.

Fig. 8. Calibration progress over optimization steps, function evaluations, and wall time for the exit selection scenario. Gradient descent using the sampling-based estimators PGO and DGO achieved the best fit to the reference.

5.3 Calibrating Decision-Making Parameters

Next, we turn to the higher-dimensional problem of adjusting a distribution of weight coefficients that govern the agents' discrete exit selection decisions. An important property regarding the gradient estimation and a key difference to the previous problems lies in drawing the individual agents' coefficients from the input distribution as part of a simulation run. Each coefficient is determined by inverse transform sampling on the distribution specified in discretized form by the parameters. After drawing $u \sim U(0,1)$, we iterate over the histogram's normalized per-bin probabilities b_i and select the coefficient according to the lowest bin index i with $u \geq \sum_{i=1}^{20} b_i$. Drawing each pedestrian's coefficient thus involves a sequence of conditional branches on the cumulative sum of normalized bin weights. The branch conditions are functions of the input parameters and are thus taken into account as part of DGO's gradient estimation. However, the remainder of each simulation trajectory, including all force calculations, exit selection decisions, and the calculation of the distance to the reference output distribution, is a direct consequence of the branches taken while drawing coefficients. Since the pathwise derivatives wrt. parameters thus extend only to the initial coefficient calculation, DGO's gradient estimates are based solely on the critical events generated by these initial branches. Hence, AD-based derivatives across the force calculations do not contribute to DGO's estimates, which eliminates the main source of noise observed in the previous problems.

Figure 8 shows the calibration progress for this problem based on a subset of the hyperparameters from Table 1, and setting the number of microreplications to 1 or 10. As one function evaluation corresponds to about 1 s of wall time, we omit the progress over function evaluations and instead show the progress over optimization steps, each of which can cover several evaluations depending on the method's hyperparameters. Here, gradient descent using PGO and DGO identified the best solutions within the time budget. We note the steepness of DGO's progress over steps in the best-performing configuration of 10 simulation replications and 25 samples per step, in contrast to PGO's 1 replication and 25 samples. DGO overtakes PGO at around 5 000 steps, whereas PGO stagnates.

PSO and GA show similar progress, and neither reaches convergence by the end of the time budget, with GA achieving a similar solution quality as PGO.

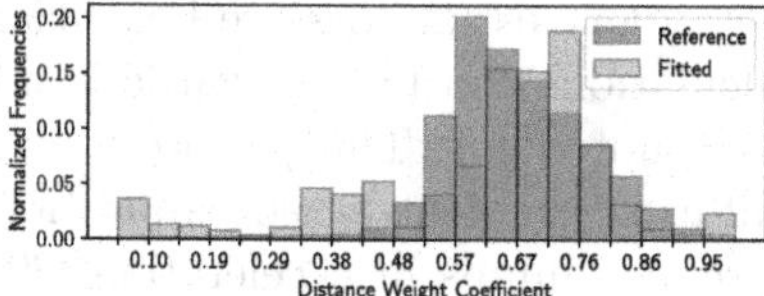

(a) Input Distribution over Exit Selection Coefficients

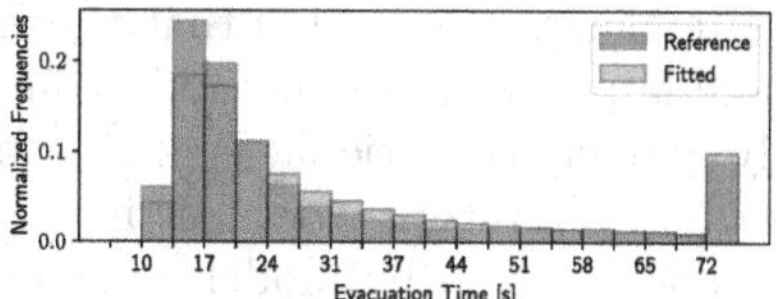

(b) Output Distribution over Evacuation Times

Fig. 9. Best calibration results generated by DGO. While the output matches the reference distribution well, the identified input distribution is similar in mean to the reference but differs in shape.

Finally, we consider the solution identified by DGO in its best-performing hyperparameter combination and macroreplication. Figure 9 shows the calibrated input histogram over exit selection coefficients, and the output histogram over evacuation times. As expected, a good fit is achieved with respect to the distribution of evacuation times. In contrast, the input distribution still somewhat deviates from the reference in shape, although the tendency is captured. This result shows that for this problem, high-quality solutions can be achieved via differently shaped input distributions. Hence, the calibration for real-world purposes would likely benefit from additional criteria to increase identifiability.

6 Conclusions

Our study of the crowd model calibration via gradient descent using sampling-based estimators shows both the challenges and the promise of the approach.

Firstly, even in the absence of explicit conditional branching, crowd simulations based on Social Force can observe large jumps in the simulation output. While our AD-based gradient estimator is capable of accounting for such jumps, scenarios of non-trivial size can generate sufficiently rugged output so that capturing the jumps based on only few samples becomes challenging. The general function curvature seems to be largely independent of the jumps, which suggests using either simple pathwise derivatives for the Social Force portion of a simulation, or resorting to black-box gradient estimators.

Our largest calibration problem involved finding a posterior distribution over per-agent decision parameters, from which we draw by inverse transform sampling using a series of conditional branches. A key insight is that in this formulation of simulation-based inference, our AD-based method reduces to estimating these branches' effects. This can positively affect both performance, by reducing the AD overhead to a minimum, and fidelity, by reducing the dependence on potentially noisy intermediate pathwise gradients. We consider this an encouraging result demonstrating a gainful integration of AD into inference workflows [9].

In a calibration problem across 20 input dimensions, gradient descent using the sampling-based estimators outperformed the gradient-free methods. The general tendency observed is that the AD-based gradient estimators are beneficial where large jumps in the underlying force calculations can be disregarded and only explicit conditional branches must be accounted for. Thus, a promising direction for future work lies in further facilitating gradient-based calibration and optimization by model refinements that reduce jumps in acceleration [20] while maintaining a realistic representation of real-world crowd behavior.

Acknowledgments. Funded by the Deutsche Forschungsgemeinschaft (DFG, German Research Foundation), grant no. 497901036 (PA) and 320435134 (JK).

Disclosure of Interests. The authors have no competing interests to declare that are relevant to the content of this article.

References

1. Andelfinger, P.: Towards differentiable agent-based simulation. ACM Trans. Model. Comput. Simul. **32**(4), 1–26 (2023)
2. Arya, G., Schauer, M., Schäfer, F., Rackauckas, C.: Automatic differentiation of programs with discrete randomness. In: Koyejo, S., Mohamed, S., Agarwal, A., Belgrave, D., Cho, K., Oh, A. (eds.) Advances in Neural Information Processing Systems, vol. 35, pp. 10435–10447. Curran Associates, Inc. (2022)
3. Bengio, Y., Léonard, N., Courville, A.: Estimating or propagating gradients through stochastic neurons for conditional computation. arXiv preprint arXiv:1308.3432 (2013)
4. Bode, N.: Parameter calibration in crowd simulation models using approximate bayesian computation. arXiv preprint arXiv:2001.10330 (2020)
5. Chaudhuri, S., Solar-Lezama, A.: Smooth interpretation. ACM SIGPLAN Not. **45**(6), 279–291 (2010)
6. Chen, X., Zhan, F.B.: Agent-based modelling and simulation of urban evacuation: relative effectiveness of simultaneous and staged evacuation strategies. J. Oper. Res. Soc. **59**(1), 25–33 (2008)
7. Chopra, A., et al.: DeepABM: scalable and efficient agent-based simulations via geometric learning frameworks - a case study for COVID-19 spread and interventions. In: Winter Simulation Conference, pp. 1–12. IEEE (2021)
8. Christodoulou, S., Naumann, U.: Differentiable programming: efficient smoothing of control-flow-induced discontinuities. arXiv preprint arXiv:2305.06692 (2023)
9. Cranmer, K., Brehmer, J., Louppe, G.: The frontier of simulation-based inference. Proc. Natl. Acad. Sci. **117**(48), 30055–30062 (2020)
10. Fu, M.C.: Chapter 19: gradient estimation. In: Henderson, S.G., Nelson, B.L. (eds.) Simulation, Handbooks in Operations Research and Management Science, vol. 13, pp. 575–616. Elsevier (2006)
11. Gödel, M., Bode, N., Köster, G., Bungartz, H.J.: Bayesian inference methods to calibrate crowd dynamics models for safety applications. Saf. Sci. **147**, 105586 (2022)

12. Gong, W.B., Ho, Y.C.: Smoothed (conditional) perturbation analysis of discrete event dynamical systems. IEEE Trans. Autom. Control **32**(10), 858–866 (1987)
13. González-Méndez, M., Olaya, C., Fasolino, I., Grimaldi, M., Obregón, N.: Agent-based modeling for urban development planning based on human needs. Conceptual basis and model formulation. Land Use Policy **101**, 105110 (2021)
14. Griewank, A., Walther, A.: Evaluating derivatives: principles and techniques of algorithmic differentiation. SIAM (2008)
15. Hackl, J., Dubernet, T.: Epidemic spreading in urban areas using agent-based transportation models. Future Internet **11**(4), 92 (2019)
16. Helbing, D., Molnar, P.: Social force model for pedestrian dynamics. Phys. Rev. E **51**(5), 4282 (1995)
17. Ho, Y.C., Cassandras, C.: A new approach to the analysis of discrete event dynamic systems. Automatica **19**(2), 149–167 (1983)
18. Kasereka, S., Kasoro, N., Kyamakya, K., Goufo, E.F.D., Chokki, A.P., Yengo, M.V.: Agent-based modelling and simulation for evacuation of people from a building in case of fire. Procedia Comput. Sci. **130**, 10–17 (2018)
19. Kreikemeyer, J.N., Andelfinger, P.: Smoothing methods for automatic differentiation across conditional branches. IEEE Access **11**, 143190–143211 (2023)
20. Kreiss, S.: Deep social force. arXiv preprint arXiv:2109.12081 (2021)
21. Margossian, C.C.: A review of automatic differentiation and its efficient implementation. Wiley Interdisc. Rev. Data Min. Knowl. Discov. **9**(4), e1305 (2019)
22. Miranda, L.J.: PySwarms: a research toolkit for particle swarm optimization in Python. J. Open Source Softw. **3**(21), 433 (2018)
23. Motieyan, H., Mesgari, M.S.: An agent-based modeling approach for sustainable urban planning from land use and public transit perspectives. Cities **81**, 91–100 (2018)
24. Nesterov, Y., Spokoiny, V.: Random gradient-free minimization of convex functions. Found. Comput. Math. **17**, 527–566 (2017)
25. Pietzsch, B., et al.: Metamodels for evaluating, calibrating and applying agent-based models: a review. J. Acad. Soc. Sci. Stud. **23**(2) (2020)
26. Polyak, B.: Introduction to Optimization. Optimization Software, New York (1987)
27. Scheinberg, K.: Finite difference gradient approximation: to randomize or not? INFORMS J. Comput. **34**(5), 2384–2388 (2022)
28. Seyer, R.: Differentiable Monte Carlo Samplers with piecewise deterministic markov processes. Master's thesis, Chalmers University of Technology (2023)
29. Son, S., Qiao, Y.L., Sewall, J., Lin, M.C.: Differentiable hybrid traffic simulation. ACM Trans. Graph. (TOG) **41**(6), 1–10 (2022)
30. Voloshin, D., Rybokonenko, D., Karbovskii, V.: Optimization-based calibration for micro-level agent-based simulation of pedestrian behavior in public spaces. Procedia Comput. Sci. **66**, 372–381 (2015)
31. Wang, X., Mohcine, C., Chen, J., Li, R., Ma, J.: Modeling boundedly rational route choice in crowd evacuation processes. Saf. Sci. **147**, 105590 (2022)
32. Williams, R.J.: Simple statistical gradient-following algorithms for connectionist reinforcement learning. Mach. Learn. **8**, 229–256 (1992)
33. Wolinski, D., Guy, S.J., Olivier, A.H., Lin, M., Manocha, D., Pettré, J.: Parameter estimation and comparative evaluation of crowd simulations. Comput. Graph. Forum **33**(2), 303–312 (2014)
34. Yin, W., Murray-Tuite, P., Ukkusuri, S.V., Gladwin, H.: An agent-based modeling system for travel demand simulation for hurricane evacuation. Transp. Res. Part C: Emerg. Technol. **42**, 44–59 (2014)

Parameter Tuning of the Firefly Algorithm by Standard Monte Carlo and Quasi-Monte Carlo Methods

Geethu Joy[1,2(✉)], Christian Huyck[1], and Xin-She Yang[1]

[1] School of Science and Technology, Middlesex University, London NW4 4BT, UK
gj219@live.mdx.ac.uk

[2] Computer Engineering and Informatics, Middlesex University Dubai, Dubai Knowledge Park, P. O. Box 500697, Dubai, United Arab Emirates

Abstract. Almost all optimization algorithms have algorithm-dependent parameters, and the setting of such parameter values can significantly influence the behavior of the algorithm under consideration. Thus, proper parameter tuning should be carried out to ensure that the algorithm used for optimization performs well and is sufficiently robust for solving different types of optimization problems. In this study, the Firefly Algorithm (FA) is used to evaluate the influence of its parameter values on its efficiency. Parameter values are randomly initialized using both the standard Monte Carlo method and the Quasi Monte-Carlo method. The values are then used for tuning the FA. Two benchmark functions and a spring design problem are used to test the robustness of the tuned FA. From the preliminary findings, it can be deduced that both the Monte Carlo method and Quasi-Monte Carlo method produce similar results in terms of optimal fitness values. Numerical experiments using the two different methods on both benchmark functions and the spring design problem showed no major variations in the final fitness values, irrespective of the different sample values selected during the simulations. This insensitivity indicates the robustness of the FA.

Keywords: Algorithm · Firefly algorithm · Parameter tuning · Monte Carlo method · Optimization

1 Introduction

Many problems in engineering design and industry can be formulated as optimization problems with a main design objective, subject to multiple nonlinear constraints. Best design options correspond to the optimal solutions to such design optimization problems. To find such optimal solutions requires the use of sophisticated optimization algorithms and techniques [1,18]. A recent trend is to use nature-inspired algorithms to solve engineering design optimization problems because nature-inspired algorithms tend to be effective, flexible and easy to implement.

L. Franco et al. (Eds.): ICCS 2024, LNCS 14836, pp. 242–253, 2024.
https://doi.org/10.1007/978-3-031-63775-9_17

Almost all optimization algorithms and techniques, including nature-inspired algorithms, have algorithm-dependent parameters, and these parameters need to be properly tuned. Tuning algorithm-specific parameters play a crucial role in determining the effectiveness of an algorithm, and the way these parameters are configured can significantly influence the performance of the algorithm under consideration [8,15]. Consequently, fine-tuning algorithmic parameters is a vital aspect of implementing and applying algorithms for solving problems in real-world scenarios [13,21,22].

Ideally, an efficient method should be used for tuning or setting the parameters for a given algorithm so that the algorithm can obtain better results. However, such methods are not yet available, and thus tuning is largely empirical or experience-based. Therefore, tuning parameters for a given algorithm can still be a challenging task, especially for tuning nature-inspired optimization algorithms [2]. In addition, parameter tuning can be problem-specific, and even with advanced tuning tools, a finely-tuned algorithm for one problem may not generalize well to other problems, leading to the need for re-tuning for each new problem or problem type and thus making it a time-consuming task in optimization.

In this study, two different methods, namely, the standard Monte Carlo (MC) and Quasi-Monte Carlo (QMC) methods, will be used to tune the parameters of the Firefly Algorithm (FA). The performance of the tuned FA will then be evaluated by comparing the fitness values obtained by FA using both MC and QMC. These numerical experiments may give some insights into the tuning efficiency of MC and QMC by investigating both the fitness values of problems with two different tuning methods and over different optimization problems.

2 Literature Review of Parameter Tuning

The literature of parameter tuning for evolutionary algorithms and metaheuristics is expanding, especially in the context of tuning parameters of nature-inspired metaheuristic algorithms. Here, a brief review is carried out on different approaches to parameter tuning, including online and offline approaches [18].

From the perspective of parameter tuning, for a given algorithm, a tuning tool (or a tuner) should be used to tune the algorithm first, and then use the tuned algorithm to solve a set of problems. Thus, there are three key components here: an algorithm, a tuner, and a problem set. Since these three components are involved simultaneously in tuning, it is possible that the parameter values tuned may depend on both the algorithm under consideration and the problems to be solved. Therefore, parameter settings can be algorithm-specific as well as problem-specific.

For a given algorithm, its parameters can be tuned first before it is used for solving optimization problems. This approach is usually called offline tuning. Other studies also indicate that it may be advantageous to vary parameters during iterations, and this approach is often referred to as online tuning [8,13]. Parameter tuning can be carried out either in sequence or in parallel and these different methods can be loosely divided into ten different categories:

- Manual or brute force method
- Tuning by systematic scanning
- Empirical tuning as parametric studies
- Monte Carlo based method
- Tuning by design of experiments (DOE)
- Machine learning based methods
- Adaptive and automation methods
- Self-tuning method [9]
- Heuristic tuning with parameter control
- Other tuning methods

Other parameter tuning methods include sequential optimization approaches, multi-objective optimization approaches, self-parameterization, fuzzy methods as well as dynamic parameter adaptation approaches and hyper-parameter optimization [10,11]. Although extensive studies have been dedicated to exploring parameter tuning methods, both offline and online, a lack of comprehensive understanding persists regarding these methods. Sometimes, these methods may not perform as well as expected, and the reasons behind these unexpected outcomes remain elusive, underscoring the need for deeper insights. Some key issues in parameter tuning include [15]

1. *Non-universality.* It is not clear whether tuned parameters are inherently problem-specific and algorithm-specific, which may limit their generalization to different problem sets.
2. *High computational efforts.* Parameter tuning tends to be a computationally intensive task. This poses a significant barrier to effective parameter tuning, necessitating the development of methods to minimize computational efforts.
3. *Lack of theoretical insights.* Despite a diverse spectrum of tuning methods available in the current literature, most rely on heuristic approaches without theory-based guidelines, lacking a clear understanding of their mechanisms and optimal conditions for applications.

There are still some open problems related to parameter tuning. For example, it is not clear how to tune parameters in the most effective way and how mathematical theories can be applied to parameter tuning. In addition, the practical implications of well-tuned parameters on algorithm convergence remain another open problem, thus highlighting the need for further research in this area.

3 Tuning Parameters by MC and QMC

Before the details of tuning parameters using the MC and QMC methods are discussed, the main idea of FA and its parameters are outlined.

3.1 Firefly Algorithm

The Firefly Algorithm (FA) is a nature-inspired algorithm that was developed by Xin-She Yang in 2008, based on the flashing characteristics and flying patterns of tropical fireflies [19]. FA has been applied to a diverse range of applications, including multi-modal optimization, multi-objective optimization [17,19], clustering [20], software testing [3], vehicle routing problems [21], multi-robot swarming [22] and others.

For a given optimization problem, its solution vector $\mathbf{x}$ is encoded as the locations of fireflies. Thus, the locations of two fireflies i and j correspond to two solution vectors $\mathbf{x}_i$ and $\mathbf{x}_j$, respectively. The main updating equation of firefly locations or solution vectors is

$$\mathbf{x}_i^{t+1} = \mathbf{x}_i^t + \beta e^{-\gamma r_{ij}^2}(\mathbf{x}_j - \mathbf{x}_i) + \alpha \epsilon_i^t, \tag{1}$$

where the random number vector ϵ_i^t is drawn from a Gaussian normal distribution. In addition, the distance r between two solutions is given by the Euclidean distance or L_2-norm

$$r_{ij} = ||x_i^t - x_j^t||. \tag{2}$$

The parameters to be tuned are the attractiveness parameter β, the scaling parameter γ and the randomization strength parameter α. In most FA implementations, parameter α is further rewritten as

$$\alpha = \alpha_0 \theta^t, \tag{3}$$

where α_0 is its initial value, which can be set to $\alpha_0 = 1$. Here, t is the pseudo-time or iteration counter, and $0 < \theta < 1$ is the parameter to be tuned, instead of α.

3.2 Monte Carlo Method

One approach to offline tuning is to use MC-based methods. In this study, all the parameters in the FA are initialized randomly using MC and pseudo-random numbers that are uniformly distributed. Pseudo-random numbers are random numbers generated using generators, which are used in computer programs. They are not truly random numbers and are generated in a deterministic way with some sophisticated permutations.

In essence, the MC method is a statistical sampling method with statistical foundations and its errors tend to decrease as $O(1/\sqrt{N})$ where N is the number of samples [12]. Though this inverse-square convergence may be slow, it can work well in practice [15], in comparison with manual or brute force methods.

In the current simulation for parameter tuning, the parameters of the FA are randomly initialized by drawing random samples from uniform distributions in a specific range of parameter values. Then, the discrete random samples are used as the parameter setting of the FA. With such settings, the FA is executed to solve the given optimization problems, such as the benchmark functions and the spring design problem [1,14].

3.3 Quasi-Monte Carlo Method

To obtain better estimates, the standard MC method requires a large number of samples. Theoretical analysis and studies from various applications suggest that a quasi-Monte Carlo method can potentially speed up the convergence because its errors decrease as $O(1/N)$ under certain conditions. Such QMC methods use low-discrepancy sequences or quasi-random numbers, and such sequences require some careful generation and random scrambling of the initial sequences [4,12, 14]. Therefore, this study also uses QMC to tune parameters in the FA and comparison with the standard MC will be carried out.

For the generation of quasi-random numbers, there are efficient algorithms such as van der Corput sequence, Sobol sequence and Halton sequence. Most of these sequences will generate quasi-random numbers in the interval between 0 and 1. In the current simulation, the Sobol sequence with affine scramble and digital shift will be used [4,5,16], which is a standard implementation in Matlab.

4 Experiment Setup and Benchmarks

To investigate the possible effect of two different tuning methods on the performance of the FA, two benchmark functions and a design problem are used in this study. The two benchmark functions are the sphere function and Rosenbrock's banana function. The former is a convex, separable function, whereas the latter is a non-convex, non-separable function. The design problem is a non-convex, nonlinear spring design problem, subject to four constraints.

4.1 Experimental Setup for FA Parameters

In the standard FA, there are three parameters θ, β and γ to be tuned. These parameters of the FA can typically take the following values:

- Population size: $n = 20$ to 40 (up to 100 if necessary).
- $\beta = 0.1$ to 1, $\gamma = 0.01$ to 10, though typically, $\beta = 1$ and $\gamma = 0.1$.
- $\alpha_0 = 1$, $\theta = 0.9$ to 0.99 (typically, $\theta = 0.97$). where $\alpha = \alpha_0\theta^t$.
- Number of iterations: $t_{\max} = 100$ to 1000.

For simplicity in this study for both MC and QMC, the ranges of these parameter values will be further narrowed down, as shown in Table 1.

4.2 Benchmark Functions

Optimization algorithms are typically assessed using a diverse set of standard benchmark functions to validate their efficiency and reliability. Researchers evaluate these algorithms by comparing their performance across a wide range of more than two hundred benchmark functions. The choice of benchmarks lacks standardized criteria, but it is essential to use a diverse range of benchmark problems, including different modes, separability, dimensionality, linearity and

Table 1. Experimental setting for MC and QMC Simulations

Initialization Values	Monte Carlo	Quasi-Monte Carlo
Population size	20	20
Number of MC/QMC runs	10	10
Number of iterations	1000	1000
Parameter ranges of		
θ	$[0.9, 1.0]$	$[0.9, 1.0]$
β	$[0, 1]$	$[0, 1]$
γ	$[0.5, 2.5]$	$[0.5, 2.5]$

nonlinearity. Numerous benchmark collections, including CEC suites, and those referenced in articles, for example, Jamil and Yang [6], are available online.

While test functions are usually unconstrained, real-world benchmark problems originate from various applications with complex constraints and large datasets. This present work will assess the FA's parameter settings using three test benchmarks.

1. The Rosenbrock function is a nonlinear benchmark [7], which is not convex in the D-dimensional space. It is written as
$$f(\mathbf{x}) = (1 - x_1)^2 + \sum_{i=1}^{D-1} \left[100\left(x_{i+1} - x_i^2\right)^2\right], \quad \mathbf{x} \in \mathbb{R}^D, \tag{4}$$
where
$$-30 \le x_i \le 30, \quad i = 1, 2, ..., D. \tag{5}$$
Its global minimum is located at $\mathbf{x}^* = (1, \ldots, 1)$ with $f_{\min}(\mathbf{x}^*) = 0$.
2. The sphere function is a convex benchmark in the form
$$f(\mathbf{x}) = \sum_{i=1}^{D} x_i^2, \quad \mathbf{x} \in \mathbb{R}^D, \tag{6}$$
where
$$-10 \le x_i \le 10, \quad i = 1, 2, ..., D. \tag{7}$$
Its global minimum is located at $\mathbf{x}^* = (0, \ldots, 0)$ with $f(\mathbf{x}^*) = 0$.
3. The spring design is an engineering design benchmark with three decision variables and four constraints [1].
$$\text{Minimize} \quad f(\mathbf{x}) = (2 + x_3)x_1^2 x_2, \tag{8}$$

subject to

$$g_1(x) = 1 - \frac{x_2^3 x_3}{71785 x_1^4} \leq 0,$$

$$g_2(x) = \frac{4x_2^2 - x_1 x_2}{12566(x_2 x_1^3 - x_1^4)} + \frac{1}{5108 x_1^2} - 1 \leq 0,$$

$$g_3(x) = 1 - \frac{140.45 x_1}{x_2^2 x_3} \leq 0,$$

$$g_4(x) = \frac{x_1 + x_2}{1.5} \leq 0.$$

The simple bounds for design variables are

$$0.05 \leq x_1 \leq 2.0, \quad 0.25 \leq x_2 \leq 1.3, \quad 2.0 \leq x_3 \leq 15.0. \tag{9}$$

The best optimal solution found so far in the literature is

$$\mathbf{x}_* = [0.051690,\ 0.356750,\ 11.287126], \quad f_{\min}(\mathbf{x}_*) = 0.012665. \tag{10}$$

It is worth pointing out that the constraints in this optimization problem are handled by using the standard penalty method.

5 Results and Hypothesis Testing

To test the possible effects of different tuning methods on the performance of the FA, a set of 10 runs have been carried out using both MC and QMC methods over three different optimization problems. All problems and runs use the same maximum number of 1000 iterations. Based on the numerical experiments, two hypotheses are proposed and the paired Student's t-tests will be used for comparison.

The two hypotheses to be tested are as follows

Hypothesis H1: Parameter Tuning methods (MC or QMC) have no significant effect on the fitness values obtained, for a given optimization problem.

Hypothesis H2: For a given algorithm, its performance on different problems is not affected by the parameter tuning method used.

5.1 Testing the First Hypothesis

For the MC simulation, the parameters of the FA (θ, β and γ) are taken from uniform distributions in the ranges given in Table 1. Similarly, for the QMC simulation, the parameters are taken from a scrambled Sobol sequence and then mapped into the proper ranges of the parameters. For every objective function, the optimal fitness value obtained along with the corresponding optimal solution,

Table 2. Rosenbrock function

Fitness Values	Monte Carlo	Quasi-Monte Carlo
Run 1	4.0075e−01	2.0416e−01
Run 2	3.9088e−01	2.7798e−01
Run 3	7.2409e−02	7.0090e−02
Run 4	5.1595e−02	3.9830e−01
Run 5	3.3661e−01	3.2203e−02
Run 6	8.3948e−02	4.5692e−01
Run 7	1.3567e−01	7.1474e−02
Run 8	5.4517e−02	3.7071e−01
Run 9	1.1233e−01	1.7295e−01
Run 10	2.1764e−01	9.7198e−01
Mean	0.1856	0.3027
Std Deviation	0.1408	0.2776
Test h-value	0	
p-value	0.2498	

and the parameter values from MC and QMC are recorded for post-processing and hypothesis testing.

The t-tests are then used to test the hypotheses. According to the standard t-test criteria, the h and p values obtained from the paired t-tests will determine whether the null hypothesis should be rejected or not. The values for 10 runs obtained from the MC and QMC simulations for each objective function are listed in Table 2, Table 3 and Table 4.

1. *Rosenbrock Function.* From the results for the Rosenbrock function summarized in Table 2, a paired t-test has been carried out. As the p-value is much larger than the threshold value 0.05, the first hypothesis cannot be rejected. That is to say, there is no strong evidence to say that the fitness values obtained are affected by different tuning methods.
2. *Sphere Function.* For the sphere function, the paired t-test has been carried out. As clearly seen in Table 3, the high p-value means that there are no significant differences in the results obtained by two tuning methods.
3. *Spring design problem.* For a realistic design problem with highly nonlinear constraints, a similar hypothesis test, based on the results in Table 4, indicates again no significant differences in the results between MC and QMC.

5.2 Testing the Second Hypothesis

The tests of the first hypothesis for all three problems give a consistent conclusion that there are no significant differences in the results obtained by the FA whatever the tuning methods were used. To see if this is consistent with the

Table 3. Sphere function

Fitness Values	Monte Carlo	Quasi-Monte Carlo
Run 1	3.0614e−003	0.0000e+000
Run 2	0.0000e+000	2.7361e−242
Run 3	1.4977e−001	1.7983e−003
Run 4	3.1305e−002	0.0000e+000
Run 5	7.4281e−316	0.0000e+000
Run 6	7.6838e−003	3.0621e−139
Run 7	1.3688e−003	2.3227e−003
Run 8	0.0000e+000	0.0000e+000
Run 9	9.1990e−110	0.0000e+000
Run 10	0.0000e+000	1.0116e−254
Mean	1.9318e−02	4.1211e−04
Std Deviation	4.6842e−02	8.7754e−04
Test h-value	0	
p-value	0.2181	

Table 4. Spring design

Fitness Values	Monte Carlo	Quasi-Monte Carlo
Run 1	0.016590	0.030303
Run 2	0.035127	0.226819
Run 3	0.033739	0.0171432
Run 4	0.026841	0.047312
Run 5	0.119560	0.039970
Run 6	0.017505	0.021963
Run 7	0.027859	0.053268
Run 8	0.021918	0.148673
Run 9	0.021212	0.0193273
Run 10	0.021681	0.071922
Mean	0.0342	0.0567
Std Deviation	0.0306	0.0621
Test h-value	0	
p-value	0.3179	

group means, the mean fitness values from three separate problems are tested using the same t-test.

The t-test results are summarized in Table 5, which again shows that the null hypothesis (H2) holds. That is to say, there is no statistically significant support for one tuning method being better than the other.

Table 5. Mean fitness values for three different problems

Problem	Monte Carlo	Quasi-Monte Carlo
Rosenbrock	0.1856	0.3027
Sphere	0.0193	0.0004
Spring Design	0.0342	0.0567
Test h-value	0	
p-value	0.7259	

This conclusion is a bit surprising from the perspective that the QMC method usually produces better results than the standard MC method for multiple dimensional numerical integrals. This study seems to show that both MC and QMC methods produce similar results for the parameter tuning purpose.

5.3 F-Test for Variances

The hypothesis tests so far show that no significant differences were found in the mean objective values obtained by MC and QMC. However, the same level of mean values does not necessarily give the same level of variances. Thus, it is useful to carry out the test of variances. For this purpose, we use the two-sample F-test to see if the variances for the spring design problem obtained by MC and QMC are equal.

The F-test using the same data shown in Table 4 gives $h = 1$ and p-value $p = 0.0262$, which is smaller than the critical value 0.05. This means that there is a sufficient difference in variances to reject the null hypothesis. Therefore, it can be concluded that there are no significant differences in mean values obtained by the FA using MC and QMC, but there are some statistically noticeable differences in their corresponding variances of the objective values.

However, it is worth pointing out that the statistical tests that have been carried out here are mainly to test the differences in means using paired t-tests. In addition, the sample size of 10 is relatively small, thus it may be possible that further more extensive tests may reveal that more comprehensive results may not be completely consistent with this preliminary conclusion.

6 Conclusion and Future Work

From the simulation results of the three different optimization benchmarks, there is not enough evidence to reject the null hypotheses. For hypothesis 1, surprisingly, there was no significant difference in the fitness values obtained via MC and QMC.

For Hypothesis 2, the fitness values obtained by MC and QMC simulations also fail to reject the null hypothesis. For all three benchmark functions, similar orders of fitness values were obtained for both tuning methods. The QMC method does not produce significantly better results, when compared to the standard MC method.

The preliminary study consists of only a small number of optimization problems, it may be the case that other benchmark problems and other algorithms may not show such robustness. Therefore, a further study is required to determine whether the parameter settings of the FA using these two parameter tuning methods exhibit the same property. Furthermore, some detailed statistical analysis and theoretical analysis will be needed to gain insights into the effect of parameter tuning and its potential link to the convergence behavior observed in these numerical experiments. These will form part of the authors' further research topics.

References

1. Cagnina, L.C., Esquivel, S.C., Coello, C.A.: Solving engineering optimization problems with the simple constrained particle swarm optimizer. Informatica **32**(3), 319–326 (2008)
2. de Lacerda, M.G.P., de Lima Neto, F.B., Ludermir, T.B., Kuchen, H.: Out-of-the-box parameter control for evolutionary and swarm-based algorithms with distributed reinforcement learning. Swarm Intel. **17**, 173–217 (2023)
3. Srivastava, P.R., Malikarjun, B., Yang, X.-S.: Optimal test sequence generation using firefly algorithm. Swarm Evol. Comput. **8**, 44–53 (2013)
4. He, Z., Wang, X.: Convergence analysis of quasi-Monte Carlo sampling for quantile and expected shortfall. Math. Comput. **90**(327), 303–319 (2021)
5. Hong, H.S., Hickernell, F.J.: Algorithm 823: implementing scrambled digital sequences. ACM Trans. Math. Softw. **29**(2), 95–109 (2003)
6. Jamil, M., Yang, X.-S.: A literature survey of benchmark functions for global optimization problems. Int. J. Math. Model. Numer. Optim. **4**(2), 150–194 (2013)
7. Rosenbrock, H.H.: An automatic method for finding the greatest or least value of a function. Comput. J. 175–184 (1960)
8. Eiben, A.E., Hinterding, R., Michalewicz, Z.: Parameter control in evolutionary algorithms. IEEE Trans. Evol. Comput. **3**(2), 124–141 (1999)
9. Yang, X.-S., Deb, S., Loomes, M., Karamanoglu, M.: A framework for self-tuning optimization algorithm. Neural Comput. Appl. **23**, 2051–2057 (2013)
10. Tatsis, V.A., Parsopoulos, K.E.: Dynamic parameter adaptation in metaheuristics using gradient approximation and line search. Appl. Soft Comput. **74**, 368–384 (2019)
11. Yoo, Y.: Hyperpameter optimization of deep neural network using univariate dynamic encoding algorithm for searches. Knowl.-Based Syst. **178**(1), 74–83 (2019)
12. Fishman, G.S.: Monte Carlo: Concepts, Algorithms and Applications. Springer, New York (1996). https://doi.org/10.1007/978-1-4757-2553-7
13. Yang, X.-S., He, X.-S.: Mathematical Foundations of Nature-Inspired Algorithms. Springer, Cham (2019). https://doi.org/10.1007/978-3-030-16936-7
14. Yang, X.-S.: Introduction to Computational Mathematics, 2nd edn. World Scientific Publishing Company, Singapore (2014)
15. Joy, G., Huyck, C., Yang, X.-S.: Review of parameter tuning methods for nature-inspired algorithms. In: Yang, X.-S. (ed.) Benchmarks and Hybrid Algorithms in Optimization and Applications. Springer Tracts in Nature-Inspired Computing, pp. 33–47. Springer, Singapore (2023). https://doi.org/10.1007/978-981-99-3970-1_3

16. Sobol, I.M.: A Primer for the Monte Carlo Method. CRC Press, Boca Raton (2017)
17. Yang, X.-S., Slowik, A.: Firefly algorithm (chapter 13). In: Swarm Intelligence Algorithms. Swarm Intelligence Algorithms: Modifications and Applications. CRC Press, Boca Raton (2020)
18. Yang, X.-S.: Nature-Inspired Optimization Algorithms, 2nd edn. Academic Press, London (2020)
19. Yang, X.-S.: Firefly algorithms for multimodal optimization. In: Watanabe, O., Zeugmann, T. (eds.) SAGA 2009. LNCS, vol. 5792, pp. 169–178. Springer, Heidelberg (2009). https://doi.org/10.1007/978-3-642-04944-6_14
20. Senthilnath, J., Omkar, S.N., Mani, V.: Clustering using firefly algorithm: performance study. Swarm Evol. Comput. **1**(3), 164–171 (2011)
21. Osaba, E., Yang, X.-S., Diaz, F., Onieva, E., Masegosa, A., Perallo, A.: A discrete firefly algorithm to solve a rich vehicle routing problem modelling a newspaper distribution system with recycling policy. Soft. Comput. **21**(18), 5295–5308 (2017)
22. Palmieri, N., Yang, X.-S., Rango, F.D., Santmaria, A.F.: Self-adaptive decision-making mechanisms to balance the execution of multiple tasks for a multi-robots team. Neurocomputing **306**(1), 17–36 (2018)

Generative AI and Large Language Models (LLMs) in Advancing Computational Medicine

Quantifying Similarity: Text-Mining Approaches to Evaluate ChatGPT and Google Bard Content in Relation to BioMedical Literature

Jakub Klimczak[1,2](✉) and Ahmed Abdeen Hamed[2](✉)

[1] Faculty of Computer Science, AGH, Kawiory 31, 30-055 Cracow, Poland
jklimczak@student.agh.edu.pl
[2] Clinical Data Science - Network Medicine and AI,
Sano Centre for Computational Medicine, Nawojki 11, 30-072 Cracow, Poland
a.hamed@sanoscience.org

Abstract. The emergence of generative AI tools, empowered by Large Language Models (LLMs), has shown power in generating content. The assessment of the usefulness of such content has become an interesting research question. Using prompt engineering, we assess the similarity of such contents to real literature produced by scientists. In this exploratory analysis, we prompt-engineer ChatGPT and Google Bard to generate clinical content to be compared with medical literature, and we assess the similarities of the generated contents by comparing them with biomedical literature. Our approach is to use text-mining methods to compare documents and bigrams and to use network analysis to check the centrality. The experiments demonstrated that ChatGPT outperformed Google Bard in different similarity and term network centrality methods, but both tools achieved good results compared to the baseline.

Keywords: Generative AI · LLM · Content Assessment · Google Bard · ChatGPT · Text Mining · Provoking Questions

1 Introduction and Related Work

In 2022, our world witnessed an epic event an OpenAI launching pre-trained, and transformer-based Large Language Model. The tool is known to be conversational, generative, pre-trained, transformer-based, and hence its name [2]. The creation of ChatGPT started a new phenomenon in the IT world and implies the appearance of a lot of new models, with various architectures, like Google Bard with PaLM [1]. ChatGPT is known for its capability to receive prompts in natural languages. It also provides the human language responses [11]. LLMs can process extensive amounts of text for tasks such as translation [12], question answering, and content generation [4]. LLMs are used in health and biomedicine. Thirunavukarasu et al. describe usages of LLMs in medicine [16], such tools are very popular as chatbots in the biomedical domain, but with mixed results. A different case is dental medicine [5]. Eggmann et al. describe LLMs as a tool

L. Franco et al. (Eds.): ICCS 2024, LNCS 14836, pp. 257–265, 2024.
https://doi.org/10.1007/978-3-031-63775-9_18

for finding and extracting information from giant amounts of medical data and structuring medical Electronic Health Records (EHR). One promising direction is the Clinical Decision Support System. Scientists experiment with using LLM in such systems. Singhal et al. introduced MultiMedQA [14], the benchmark of medical questions to evaluate LLMs, as a tool that could be used in CDSS.

Before the launch of LLMs, and the generation of massive data, Real-World Data (RWD) played a recognizable role in CDSS and diagnostic applications [3]. The Food and Drug Administration defined RWD as "*data regarding the usage, or the potential benefits or risks, of a drug derived from a variety of sources other than traditional clinical trials*" [17]. PubMed articles can also be RWD. We think that LLMs can be a source of RWD and that checking their capabilities to generate such data is important. It is reasonable to infer that the natural language capabilities provided by generative AI could be used to build the next CDSS. As a result, researchers have explored their potential integration with such systems [10]. LLMs tools are associated with a lack of credibility [7]. Scientists point out a need for the detection of potential harms [8] of the generated data. The users of those tools confuse such responses for truth without questioning the harm. There are a lot of new AI projects, focused on decision-making. Scientists want to publish their system as fast as possible. Shortliffe claims, that despite the decision making we should focus on the evaluation [13]. Here we deal with the problem of initial verification. We are performing the exploratory analysis to evaluate the similarity to real data. In the following sections, we describe in detail the process for generating content that can be compared with articles. Comparing real and generated data is a popular topic. Researchers are looking for methods to detect and check the similarity of generated data [6].

2 Data

In this work, we use different datasets related to the prostate cancer topic:

(1) 10,000 **biomedical abstracts**, extracted from The PubMed web portal using the search keyword: "prostate cancer treatment" which is our baseline of comparison. PubMed contains science documents, but also clinical trials and reviews, and all of these types of documents have summary abstracts. We obtained the first 10000 abstracts of the most frequently cited.

(2) two generated datasets of 100 **abstracts**, received by prompt-engineering ChatGPT and Google Bard. We used LLMs to produce documents similar to PubMed abstracts (a random ID, a title, an abstract). LLMs use everything on the internet, so these documents are in abstract format but have sources on the whole internet. We have limited the amount of generated articles because, at the time of the data collection, we did not have access to the APIs.

2.1 Large Language Models

The data generation process was performed in May and June of 2023. During this time there were available versions of large language models: **(a) ChatGPT-3.5**

turbo - this is an upgraded version of the 3.5 version, with faster processing and response time. At the moment there is a newer version available: ChatGPT-4.0. **(b) Google Bard** - experimental version of Google chatbot, the precursor of the modern model: Google Gemini. The information about specifications and specific subversion of the model is not accessible by the website, and it is impossible to receive this information directly. The API provides such information and allows one to choose one from a list of specific sub-models.

2.2 Prompt Engineering

This work aims to perform prompt engineering in ChatGPT and Google Bard to generate content related to prostate cancer treatment. Algorithm 1 shows the steps of the prompt engineering process to generate what we call real-world data.

Algorithm 1 Prompt-engineering for generating abstract-like documents

Require: The number n of simulated articles.
Require: The number w of words in each article.
[Content:] Generate a list of n real-world data reports with titles and abstracts.
[Specs:] For each abstract that contains three fields – GPT-ID, Title, and Abstract – make it to m words.
[Specs:] Make the GPT-ID random, containing at most five letters and numbers.
[Format:] A valid JSON format returned as an array of valid JSON records.
[Topic:] Investigating prostate cancer treatment.

3 Methods

The abstracts are analyzed using text mining and network analysis. We have performed two text mining methods (1) document similarity using the Cosine and Jaccard similarity [15], (2) bigram frequency comparison with Term Frequency-Inverse Term Frequency (TF-IDF) [9]. The network analysis methods are derived from the bigrams forming networks that can be analyzed to compare modularity and term centrality [18].

3.1 Text Mining Similarity Analysis

Following the step of generating the reports, we further compare the content of both tools using traditional text mining. This includes: (1) comparing the documents against other documents, (2) extracting and comparing bigrams of words, and (3) constructing networks of bigrams with identifying novel links. We conducted a comparison using random samples of documents against the PubMed corpus. for the **Document Similarity Analysis** – we count the similarity between real medical abstracts and the reports that were generated using

LLMs. We are looking for the most matching pairs of generated-real articles. For this task, we used two metrics: (1) the Cosine, and the Jaccard. On the other hand, in **Bigram Analysis** we use bigrams. Bigram is a sequence of two next elements in the text, usually composed of letters, syllables, or words. For example, the following words are bigrams: ('prostate cancer', 'cancer cells'). This measures the frequency similarity of bigrams extracted from literature and documents generated by generative AI tools. Bigrams can be used for the creation of graphs, that offer a model that can be used to explore topology and structural property. Here, we use the TF-IDF method with the different datasets of bigrams to count the importance of bigrams within the documents. In each case we are comparing also PubMed to PubMed articles, to use it as a baseline to interpret results. These Cosine and Jaccard methods are syntactic, they compare the structure of the text, and the number of words. But they can be used in other methods, to compare the semantic similarity. TF-IDF uses Cosine, but it is semantic similarity. This statistical-based method counts the importance of words in the text.

3.2 Networks Analysis

Bigrams can construct interesting networks that can be analyzed for their topology properties. The common words act as a linking node to connect more than one bigram. These approaches enable us to dissect the structural relationships in different terms in the generated and real data. Bigram networks are a popular approach for text analysis [7]. Such networks can be used in various tasks, including text classification, sentiment analysis, pattern recognition in text, and topic modeling. We select the top 50 most frequent bigrams from the entire corpus of documents from PubMed, ChatGPT, and Google Bard and build the bigram networks as rigorous models of comparison. The type of analysis we present is degree and closeness centrality as common measures that demonstrate the differences. The degree presents how many connections a specific unigram (word) has in the graph [19]. The closeness presents how close a given word is to all others in the same graph [19]. By comparing centrality metrics across different data sources, we can get valuable knowledge about similarities and differences in the texts. Centrality comparison allows to comparison of semantic relations between words in the text.

4 Results

We performed similarity experiments to measure the similarity of LLMs data and a sample of PubMed documents with PubMed article abstracts. We performed experiments on different dataset sizes (10, 25, 50, 75, and 100).

4.1 Document Similarity Analysis

We use two different methods to compare document similarity: (1) Cosine similarity, and (2) Jaccard similarity. Table 1 shows the result of the scores.

Table 1. Combined Analysis Results

Size	Source	Document Similarity		Bigram Analysis	Centrality	
		Cosine	Jaccard	TF-IDF	Degree	Closeness
10	Bard	0.3435	0.1954	0.3999	0.0349	0.1618
	GPT	0.3802	0.2286	0.4182	0.0256	0.0471
	Pubmed	0.7683	0.1711	0.3524	0.0264	0.0590
25	Bard	0.3389	0.1914	0.40473	0.0392	0.1640
	GPT	0.3604	0.2139	0.44100	0.0273	0.0633
	Pubmed	0.8801	0.1914	0.3512	0.0219	0.0608
50	Bard	0.3336	0.1873	0.37391	0.0483	0.2223
	GPT	0.3612	0.2157	0.46699	0.0250	0.0456
	Pubmed	0.8804	0.1978	0.3636	0.0219	0.0461
75	Bard	0.3205	0.1775	0.3814	0.0505	0.2415
	GPT	0.3595	0.2147	0.4556	0.0250	0.0484
	Pubmed	0.8681	0.1956	0.3441	0.0256	0.0744
100	Bard	0.3202	0.1775	0.36766	0.0425	0.1821
	GPT	0.3531	0.2093	0.42217	0.0286	0.0829
	Pubmed	0.8421	0.1833	0.3217	0.0264	0.0722

Results in Table 1 show that the average cosine similarity scores for this method are very high, between 32–38% for both solutions. ChatGPT shows more similarity than Bard. ChatGPT's similarity is around 35–38%, and Bard's average similarity is around 32–34%. In the plot 1) we can see the trend, that for every sample, the average similarity score is higher for ChatGPT.

The Jaccard analysis is performed with a word-bag representation of text, that counts the number of common words. The similarity is not so high, but the generated texts are shorter. The average scores (Table 1 are hovering around 18–23%. This result indicates a good connection between the generated and real data. ChatGPT shows more similarity in average scores, than Bard. ChatGPT's

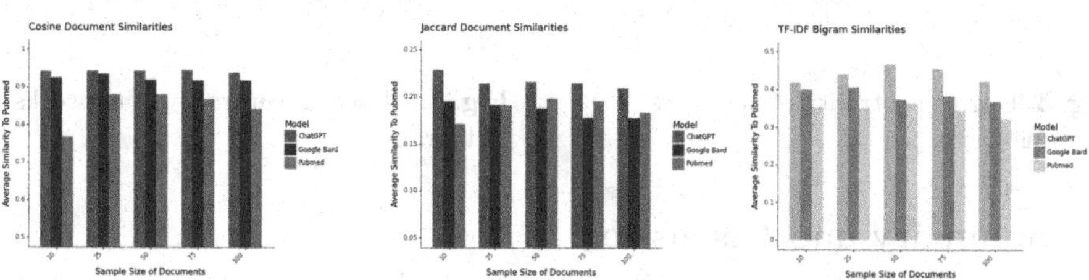

Fig. 1. Cosine similarity of documents

Fig. 2. Jaccard similarity of documents

Fig. 3. TF-IDF similarity of bigrams

similarity is around 21–23%, and Bard's average similarity is 17–19%. Plot 2) demonstrates the advantage of ChatGPT. The similarity between a random sample of PubMed articles with the whole dataset is at a similar level, between ChatGPT and Google Bard.

4.2 Bigram Similarity – TF-IDF Bigram Frequency Analysis

Both ChatGPT and Bard, show a meaningful level of average similarity (Fig. 3), hovering around the 37–47% mark. This shows a visible connection between the RWD generated by these models and existing medical research from sources like PubMed. The Fig. 3 plot shows that the difference is bigger with a bigger sample of documents. With this method, values of similarity between the PubMed sample and the whole PubMed corpus are a little lower in comparison to ChatGPT and Google Bard. That also speaks for the good quality of generated data.

4.3 Bigram Networks Analysis

The **degree centrality** (plot Fig. 4), shows a structural similarity between the PubMed and ChatGPT. Bard exhibits much higher values. This suggests that ChatGPT's values are closer to PubMed's. ChatGPT and PubMed centrality have prevalent degree centrality, between 0.02 and 0.03, and Bard is between 0.035 and 0.05. The **closeness centrality** (plot Fig. 5), shows a correlation between PubMed and ChatGPT. Bard has higher values, and ChatGPT's values are close to PubMed, showing that the networks are more similar. PubMed closeness is between 0.05 and 0.08, and for ChatGPT is between 0.05 and 0.89, for Google Bard it is between 0.16 and 0.24. The numbers support ChatGPT's better fit.

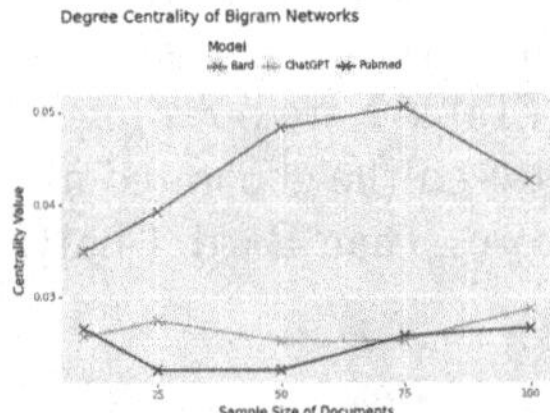

Fig. 4. Degree centrality of networks of bigrams

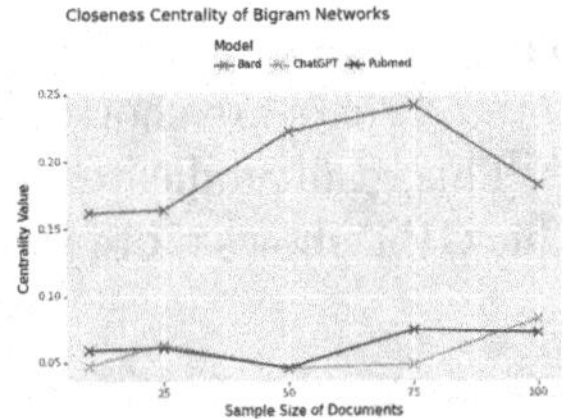

Fig. 5. Closeness centrality of networks of bigrams

5 Summary and Discussion

We presented a text-mining and network analysis approach to count the similarity between generated and real biomedical data. With random samples of documents, we observed that ChatGPT scored a closer similarity than Bard. This analysis in three different measures favors ChatGPT over Google Bard.

The network analysis offers us another field to compare the similarities. The results of the experiments show that ChatGPT graphs exhibit closer similarities in structure and centrality. Table 2 shows top bigrams. We observed that ChatGPT offered 7, and Google Bard offered 3 bigrams that overlapped. While both LLMs have "prostate cancer" as the first bigram, they vary in the rest of the common bigrams. The "cancer patients" which was 7 in PubMed is 4 in ChatGPT. The "quality life", which was 11 in PubMed is 5 in the ChatGPT dataset. This could indicate that ChatGPT was trained on data related to patient wellness, while PubMed data is more about the clinical aspects of the diseases.

Table 2. Bigram ranks in Pubmed and generated datasets

Rank	PubMed Bigrams	GPT	Bard	Rank	PubMed Bigrams	GPT	Bard
1	prostate cancer	1	1	7	cancer patient	4	47
2	radiation therapy	8	–	8	specific antigen	–	–
3	radical prostatectomy	28	–	9	external beam	–	–
4	localized prostate	32	–	10	free survival	43	–
5	prostate specific	–	–	11	quality life	5	44
6	androgen deprivation	–	–	12	patient treated	–	–

6 Conclusions and Future Directions

Gathering new datasets related to various diseases is one of the directions to further study. These results are within the scope of "prostate cancer", but there is a need to check other domains (e.g., diabetes, depression, cardiovascular) and newer models. For ChatGPT, we used version 3.5 (not 4.0), and for Google Bard, we used the version before Google Gemini, which now empowers Google Bard. Different benchmarks show that there is a huge difference between models and their successors. We are aware that such LLMs are black boxes and we do not know the details of them. Due to that future work will include testing such models in answering clinical questions, with MultiMedQA [14]. This popular benchmark is often used to test the quality and progress of new language models.

Acknowledgements. This publication is partially supported by the European Union's Horizon 2020 research and innovation programme under grant agreement Sano No. 857533 and carried out within the International Research Agendas programme of the Foundation for Polish Science, co-financed by the European Union under the European Regional Development Fund. Additionally is partially created as part of the Ministry of Science and Higher Education's initiative to support the activities of Excellence Centers established in Poland under the Horizon 2020 program based on the agreement No MEiN/2023/DIR/3796.

References

1. Google bard. https://bard.google.com/. Accessed 03 Aug 2023
2. Openai chatgpt. https://chat.openai.com/. Accessed 03 Aug 2023
3. Baumfeld Andre, E., et al.: The current landscape and emerging applications for real-world data in diagnostics and clinical decision support and its impact on regulatory decision making. Clinical Pharmacol. Therapeut. **112**(6), 1172–1182 (2022)
4. Chung, J., Kamar, E., Amershi, S.: Increasing diversity while maintaining accuracy: text data generation with large language models and human interventions, pp. 575–593. Association for Computational Linguistics (ACL) (2023). https://doi.org/10.18653/v1/2023.acl-long.34
5. Eggmann, F., Weiger, R., Zitzmann, N.U., Blatz, M.B.: Implications of large language models such as chatGPT for dental medicine (2023). https://doi.org/10.1111/jerd.13046
6. Gao, C.A., et al.: Comparing scientific abstracts generated by chatGPT to real abstracts with detectors and blinded human reviewers. NPJ Digit. Med. **6**(1), 75 (2023)
7. Hamed, A.A., Wu, X.: Improving detection of chatGPT-generated fake science using real publication text: introducing xfakebibs a supervised-learning network algorithm (2023)
8. Hamed, A.A., Zachara-Szymanska, M., Wu, X.: Safeguarding authenticity for mitigating the harms of generative AI: issues, research agenda, and policies for detection, fact-checking, and ethical AI. IScience (2024)
9. Kim, S.W., Gil, J.M.: Research paper classification systems based ON TF-IDF and LDA schemes. Human-Centric Comput. Inf. Sci. **9** (12 2019). https://doi.org/10.1186/s13673-019-0192-7
10. Liao, Z., Wang, J., Shi, Z., Lu, L., Tabata, H.: Revolutionary potential of chatGPT in constructing intelligent clinical decision support systems (2023). https://doi.org/10.1007/s10439-023-03288-w
11. Moro, A., Greco, M., Cappa, S.F.: Large languages, impossible languages and human brains. Cortex **167**, 82–85 (2023). https://doi.org/10.1016/j.cortex.2023.07.003
12. Mu, Y., et al.: Augmenting large language model translators via translation memories, pp. 10287–10299. Association for Computational Linguistics (ACL) (2023). https://doi.org/10.18653/v1/2023.findings-acl.653
13. Shortliffe, E.H.: Role of evaluation throughout the life cycle of biomedical and health AI applications. BMJ Health Care Inform. **30**(1), e100925 (2023). https://doi.org/10.1136/bmjhci-2023-100925
14. Singhal, K., et al.: Large language models encode clinical knowledge. Nature **620**, 172–180 (2023). https://doi.org/10.1038/s41586-023-06291-2
15. Thada, V., Jaglan, V.: Comparison of Jaccard, dice, cosine similarity coefficient to find best fitness value for web retrieved documents using genetic algorithm. Int. J. Innov. Eng. Technol. **2**, 202–205 (2013). http://www.dknmu.org/uploads/file/6842.pdf
16. Thirunavukarasu, A.J., Ting, D.S.J., Elangovan, K., Gutierrez, L., Tan, T.F., Ting, D.S.W.: Large language models in medicine (2023). https://doi.org/10.1038/s41591-023-02448-8
17. U.S. Food and Drug Administration: Framework for FDA's real-world evidence program (Year of Publication). https://www.fda.gov/media/120060/download. Accessed 27 Oct 2023

18. Wang, G., Shen, Y., Luan, E.: Measure of centrality based on modularity matrix. Progr. Nat. Sci. **18** (2008). https://doi.org/10.1016/j.pnsc.2008.03.015
19. Zhang, J., Luo, Y.: Degree Centrality, Betweenness Centrality, and Closeness Centrality in Social Network. Atlantis Press (2017). https://doi.org/10.2991/msam-17.2017.68

ClinLinker: Medical Entity Linking of Clinical Concept Mentions in Spanish

Fernando Gallego[1,2(✉)], Guillermo López-García[1,2], Luis Gasco-Sánchez[3], Martin Krallinger[3], and Francisco J. Veredas[1,2]

[1] Departamento de Lenguajes y Ciencias de la Computación, Universidad de Málaga, Málaga, Spain
fgallegodonoso@uma.es

[2] Research Institute of Multilingual Language Technologies, Universidad de Málaga, Málaga, Spain

[3] Barcelona Supercomputing Center, Barcelona, Spain

Abstract. Advances in natural language processing techniques, such as named entity recognition and normalization to widely used standardized terminologies like UMLS or SNOMED-CT, along with the digitalization of electronic health records, have significantly advanced clinical text analysis. This study presents ClinLinker, a novel approach employing a two-phase pipeline for medical entity linking that leverages the potential of in-domain adapted language models for biomedical text mining: initial candidate retrieval using a SapBERT-based bi-encoder and subsequent re-ranking with a cross-encoder, trained by following a contrastive-learning strategy to be tailored to medical concepts in Spanish. This methodology, focused initially on content in Spanish, substantially outperforming multilingual language models designed for the same purpose. This is true even for complex scenarios involving heterogeneous medical terminologies and being trained on a subset of the original data. Our results, evaluated using top-k accuracy at 25 and other top-k metrics, demonstrate our approach's performance on two distinct clinical entity linking Gold Standard corpora, DisTEMIST (diseases) and MedProcNER (clinical procedures), outperforming previous benchmarks by 40 points in DisTEMIST and 43 points in MedProcNER, both normalized to SNOMED-CT codes. These findings highlight our approach's ability to address language-specific nuances and set a new benchmark in entity linking, offering a potent tool for enhancing the utility of digital medical records. The resulting system is of practical value, both for large scale automatic generation of structured data derived from clinical records, as well as for exhaustive extraction and harmonization of predefined clinical variables of interest.

Keywords: Encoder-only large language model · Contrastive learning · Biomedical text mining · Medical entity linking · SNOMED-CT

L. Franco et al. (Eds.): ICCS 2024, LNCS 14836, pp. 266–280, 2024.
https://doi.org/10.1007/978-3-031-63775-9_19

1 Introduction

In the medical domain, significant advancements in natural language processing (NLP) have proven essential for the effective analysis of medical texts. The integration of NLP in healthcare has opened new avenues for patient care and research, allowing for more efficient and accurate analysis of large volumes of patient data. This progress is not just technological but also represents a paradigm shift in how medical data is processed and understood. Techniques such as named entity recognition (NER) and entity linking (EL), also known as entity normalization in the clinical NLP field, play a crucial role in the interpretation and utilization of electronic health records (EHR). NER, for instance, enables the identification of critical medical terms or clinical variables within unstructured data, while EL facilitate the mapping, normalization or harmonization of these terms to standardized medical controlled vocabularies. This process is vital for ensuring consistency and accuracy in data interpretation, as well as enabling data standardization, harmonization and semantic interoperability.

These technological advances align with the ongoing digital transformation in healthcare, highlighting the growing importance of EHRs in clinical practice. EHRs are more than digital versions of patient charts; they are comprehensive, interactive records that offer a holistic view of a patient's medical history. Their adoption has been pivotal in improving patient outcomes and healthcare efficiency. EHRs are invaluable to medical informatics due to their ability to consolidate diverse patient data, transforming unstructured data into a structured, analyzable knowledge base (KB). The richness of data in EHRs, including everything from clinical notes to laboratory results, presents both an opportunity and a challenge for NLP applications. The efficient extraction and analysis of this data have significant implications for clinical decision-making and patient care. However, integrating EHRs with NLP presents several challenges that must be addressed. One of these is the need for these systems to be adaptable and scalable to various healthcare settings and requirements.

The primary challenge for medical entity linking (MEL) involves handling heterogeneous mentions, where a controlled vocabulary concept is mentioned in practice through a diversity of written expressions or phrases. This heterogeneity can lead to misinterpretations and inconsistencies in data analysis, posing a risk to patient safety and care quality. Another important challenge arises when the mentions in the text do not have an exact match or have only a partial correspondence with the terms present in the KBs. This issue is particularly prevalent in free-text clinical notes, where the context and nuances of language can greatly affect meaning. It is also pivotal to account for variations, identifying multiple valid mentions for a single medical entity. These variations might include synonyms, acronyms, and different linguistic expressions, further complicating the text mining process. An overarching issue is data scarcity, particularly the lack of adequately annotated public medical records, being most of the current resources limited exclusively to content in English, thus making it even more urgent to generate resources and tools also for other languages. Moreover, the limited availability of datasets in less commonly spoken languages exacerbates the challenge of developing truly global NLP solutions. Figure 1 presents a MEL example from the DisTEMIST shared task [15], illustrating disease entities linked to concepts from the SNOMED-CT terminology.

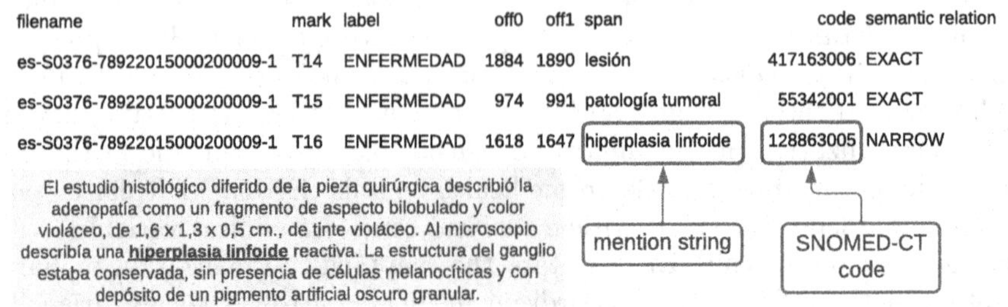

Fig. 1. DisTEMIST-linking subtrack: requires automatically finding disease mentions in published clinical cases and assigning, to each mention, a SNOMED-CT term.

This backdrop motivates many recent developments in MEL, focusing on embedding similarity rather than conventional classification problems. Embedding-based approaches, leveraging deep learning (DL) techniques, have shown promise in capturing the semantic relationships between medical terms, enhancing the model's ability to understand and process complex medical language. Recent research has introduced multilingual language models designed as generic solutions for such challenges. These multilingual models are not only innovative from a technical point of view, but are also essential for bridging the language gap in medical informatics. Examples of these models include SapBERT [12], which starts from BERT-based language models—such as PubMedBERT [7], BioBERT [9], ClinicalBERT [1] or XLM-RoBERTa [3], pre-trained with generic or domain-specific corpora, either monolingual or multilingual—, and continues with the pre-training of these models following a contrastive-learning strategy that allows self-aligning the spatial representation of biomedical entities, thus reaching the state-of-the-art (SOTA) on various MEL benchmarks. For this purpose, the authors used the entire Unified Medical Language System (UMLS) database[1], a meta-thesaurus of +4M medical concepts, to propose a new scalable learning framework for MEL task-focused models. SapBERT's ability to comprehend and interpret medical terminology across different languages represents a significant step forward in the field.

Other studies aim to tailor these models to specific languages or standards using dedicated KBs. Such customization is essential for ensuring that the models are accurately aligned with local medical practices and terminologies. Recently, a number of evaluation campaigns have emerged that seek to encourage the development of approaches to specific corpora, such as DisTEMIST (see Fig. 1)

[1] Unified Medical Language System®(UMLS®): https://www.nlm.nih.gov/research/umls/knowledge_sources/metathesaurus/index.html.

and MedProcNER, focused on the detection and normalization of disease and procedure mentions in medical texts, respectively. These targeted approaches, while effective in specific contexts, highlight the need for more versatile and comprehensive solutions that can adapt to a broader range of medical texts and terminologies.

The UMLS Metathesaurus offers vocabularies in multiple languages, with Spanish having the second highest number of resources after English, providing extensive exploitation opportunities. This diversity in languages is key to developing NLP systems that are truly inclusive and globally applicable. Consequently, in this study we propose some MEL models, based on the SapBERT contrastive-learning strategy for self-alignment of medical concepts, which have trained solely on Spanish mentions from the entire UMLS (+1M medical concepts), including SNOMED-CT[2], ICD-10, and other terminological resources present in the metathesaurus, aiming to outperform generic language models on Spanish corpora and advance the SOTA. Our approach, focusing on a specific language edition, paves the way for more tailored and effective biomedical text mining applications in diverse linguistic environments.

2 Related Work

In the past decade, MEL has seen significant growth, primarily driven by the development of Transformer-based algorithms and increased computational capacity enabling their deployment. Additionally, various approaches have been proposed for obtaining valid candidates to map given entity mentions to standards like UMLS or SNOMED-CT. Among these methods, those ones that use KBs to enrich biomedical corpora win domain-specific information have shown promising results for MEL [8]. This strategy aims to surpass the limitations of conventional language models by incorporating structured data from biomedical KBs. The main advantage of these approaches is their ability to comprehend and process the complexities and nuances of medical texts. Techniques include adapting pre-trained language models and using contrastive learning objectives for mention encoding, improving not only machine learning (ML) convergence but also adaptability and scalability. However, a major limitation of these strategies is accessing high-quality KBs, which are often inaccessible, biased, or lacking in content to provide models with essential knowledge. In our work, this information is typically provided by a corpus-specific gazetteer and, in some cases, combined with other corpora from generic domains.

Another trend gaining prominence in recent years is combining techniques like prompt learning and generative artificial intelligence (AI). The impressive performance of these generative large language models (LLMs) has led to their adoption in biomedicine. The work of Yuan and his collaborators [17] details the development of one of the first generative models for the biomedical domain, adapting the traditional BART model proposed by Lewis et al. [10] with PubMed abstracts, showing significant performance improvement in multiple language

[2] SNOMED Clinical Terms®(SNOMED-CT®): http://www.snomed.org/snomed-ct.

generation tasks. An advancement in this area is using prompt learning to fine-tune these models' responses. Thus, Ding et al. [4] employ this method to enhance EL, especially in scenarios with few or zero samples (few- or zero-shot scenarios). For its part, the work of Yuan et al. [18] demonstrates the appropriateness of combining these two approaches, proposing training a generative model with a KB and creating synthetic samples with synonyms and definitions. The limitation of these approaches is, on the one hand, the need for high quality prompts and, on the other hand, the insufficient performance of these models in zero-shot inference setups. To date, there is not enough information on prompt optimization techniques to determine if this line of work is most suitable.

The predominant trend in which ClinLinker is focused is a two-phase approach: in the first stage a model generates the space with concept candidates for each entity mention; in the second stage another model selects the most relevant concepts/codes among the candidates by re-ranking them in order of their relevance. Some of-the-shell approaches use candidate generation based on a bi-encoder model and the subsequent re-ranking with a cross-encoder model. The work of Yu and collaborators [16] emphasizes the importance of using cross-lingual models for improving long-text information retrieval and highlights the significance of employing specific pretraining and language adaptation for information retrieval tasks. A key advantage of our proposed approach over these methods is that, with an international version of the UMLS standard available, cross-lingual adaptation is unnecessary as the information is already contextually adapted. Additionally, techniques like cross-lingual typically rely on mapping more generic words rather than biomedical domain-specific ones. Furthermore, the training conducted on the cross-encoder proposed in this paper, based on hard triplets generated by the bi-encoder, enables the model to differentiate the closest optimal concept among more complex candidates, improving results both generally, with the entire corpus, and with previously unseen codes.

3 Methods

The following sections describe the corpora used for the training and evaluation of the MEL models, as well as the methodology followed for the MEL approach developed in this article.

3.1 Description of the Corpora

For the DisTEMIST-linking subtrack [15] the organizers supplied the participants with a training dataset consisting of 750 annotated clinical cases. They also supplied a test dataset consisting of 250 unannotated clinical cases, together with a larger collection of 2,800 "background" clinical cases, to avoid manual corrections. For an example of an annotated document, see Fig. 1. The MedProcNER [11] corpus is a collection of 1,000 clinical cases in Spanish from different medical specialties such as cardiology, oncology, otorhinolaryngology, dentistry, pediatrics, primary care, allergology, radiology, psychiatry, ophthalmology, and urology annotated with clinical procedure mentions. Every mention both in the MedProcNER and DisTEMIST corpus has been normalized using SNOMED-CT terminology.

3.2 ClinLinker: Bi-encoder+Cross-encoder Pipeline for MEL

Our approach, which we call ClinLinker, consists of a two-stage pipeline for MEL of medical texts in Spanish: a first stage of candidate retrieval, using a SapBERT bi-encoder, and subsequent stage of re-ranking, employing a SapBERT cross-encoder (see Fig. 2). This approach addresses an EL problem by leveraging the similarity between the input mention and each candidate given by a bi-encoder previously trained. Our main contribution in this study is training the bi-encoder by using only medical concepts/codes (CUI[3]) in Spanish from the UMLS corpus (+1M medical concepts), with two different approaches: 1) excluding (Spanish-SapBERT) and 2) including obsolete concepts/codes from the latest 2023 version (Spanish-SapBERT-oc). These comprehensive approaches ensure that our model is well-versed in both current and historical medical terminologies, enhancing its applicability in a wide range of clinical scenarios. To train a bi-encoder this way, once the entire Spanish UMLS corpus is pre-processed, for each CUI the full standard name (FSN) is used as an entity mention and every description or synonym of the CUI are used as positive candidates. This methodology aligns with the original scenario proposed in the SapBERT-model paper [12], a pretraining scheme that self-aligns the representation space of biomedical entities. Like SapBERT, we utilize *multi-similarity-loss* (from the *Pytorch Metrics Learning* library[4]) as the loss function, so that for each *(mention, positive candidate)* pair (M_i, C_i), the negative candidates concepts/codes are sourced from the same training batch using a hard-triplet mining strategy. The incorporation of negative candidates during the training of the bi-encoder—thus following what is called a contrastive-learning approach—sourced by using the model's own predictions, is critical as it teaches the model to discern between candidates that are close to but not exactly the correct match, a key aspect in fine-tuning its accuracy for MEL. The model chosen as the basis for the bi-encoder in this study is *roberta-base-biomedical-clinical-es* [2], a Transformer language model based on RoBERTa [14] and chosen in this work for having been pre-trained on medical texts in Spanish, thereby offering high efficiency for processing our specific corpora. The resulting bi-encoders are coined as Spanish-SapBERT models.

The bi-encoder models were trained for a single epoch, with a batch size of 256, a learning rate of 2e−5, a maximum input length of 256 tokens, and a hard triple mining margin of 0.2. These parameters were selected to balance the trade-off between training time and model performance. The large batch size was chosen to provide more data for the extraction of negative candidates, enhancing the learning process by presenting a wider variety of contrasts and contexts. On the other hand, for performance comparison with the Spanish-SapBERT models,

[3] In UMLS, CUI is the Concept Unique Identifier.

[4] https://kevinmusgrave.github.io/pytorch-metric-learning/.

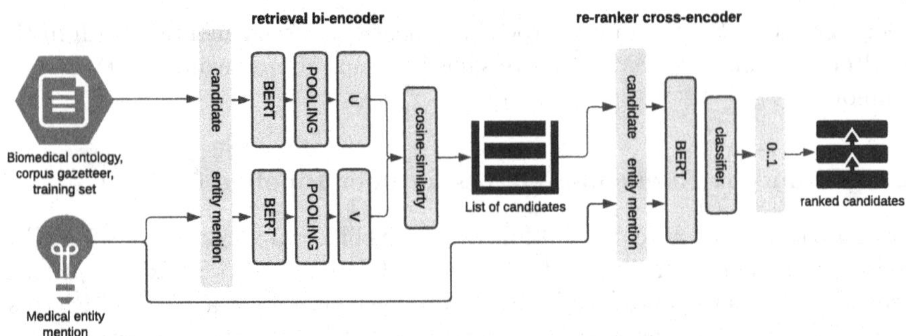

Fig. 2. ClinLinker's two-stage pipeline for MEL: a first stage of candidate retrieval, using a bi-encoder, and subsequent stage of re-ranking, employing a cross-encoder.

we also followed an alternative strategy to train domain-adapted RoBERTa models for each corpora by using input triplets composed as follows: positive terms extracted from the descriptions provided in the gazetteers—supplied together with the DisTEMIST and MedProcNER corpora by the challenge organizers—and a random subset as negative terms (five negative terms per each positive term were chosen). For this corpus-specific bi-encoders, we use triplet-margin-loss implemented in the *Sentence Transformer* library. Finally, to compare the performance of the proposed Spanish-SapBERT bi-encoders with a baseline language model, we have also worked with a domain-adapted language RoBERTa model not trained for MEL tasks, but for other general-purpose tasks.

An overview of the bi-encoder models analyzed in this study is the following:

- **Spanish-SapBERT** variants: these models are based on the *Roberta-base-biomedical-clinical-es* [2] architecture and were trained with Spanish UMLS triplets (+1M medical concepts). The *standard variant Spanish-SapBERT* excludes obsolete terms, focusing on current biomedical and clinical contexts in Spanish, while the *obsolete-codes variant Spanish-SapBERT-oc* includes obsolete terms to ensure comprehensive coverage of both contemporary and historical medical terminology. This model is publicly available in HuggingFace[5]
- **Corpus-specific** bi-encoders: the *DisTEMIST-biencoder* and *MedProcNER-biencoder*, both grounded in the *Roberta-base-biomedical-clinical-es* [2] model, were trained with corpus-specific triplets. The former focuses on disease entity recognition using DisTEMIST data [15], whereas the latter focuses on medical procedure identification with data from MedProcNER, showcasing their specialized applications in clinical text analysis.
- **SapBERT-XLMR** multilingual models: they are the *SapBERT-XLMR-base* and *SapBERT-XLMR-large* models [12], leveraging the XLM-Roberta architectures [3]. Both models underwent training with the entire UMLS dataset

[5] https://huggingface.co/BSC-NLP4BIA/SapBERT-from-roberta-base-biomedical-clinical-es.

(+4M medical concepts) and feature cross-lingual adaptation to the languages contained within UMLS, enhancing their global applicability.
- **Roberta-base-biomedical-clinical-es**: this model (abbreviated as *Roberta-base-bio-cli* in the tables an figures below), derived from XLM-Roberta-base [3], was refined with a comprehensive biomedical corpus in Spanish [2], setting a robust foundation for adaptation to various clinical tasks.

The SapBERT-XLMR bi-encoders were trained on a node with 8 NVIDIA A100 GPUs, while the Spanish-SapBERT variants, the corpus-specific bi-encoders and XLM-Roberta-base models were trained on a single GPU, thus highlighting the improvement in performance and computational requirements for these less complex models.

Additionally, we used the FAISS library [5] for retrieving the K-nearest neighbors candidates from the embeddings provided by the bi-encoders, enhancing the base results obtained in the candidate-retrieval stage. FAISS is critical for efficiently managing the vast amount of data involved in this process, ensuring large scale-processing. FAISS improves efficiency in both time and space, essential for handling large sets of embeddings and enabling rapid retrieval of nearest neighbors, decisive for our candidate re-ranking process (see Eq. 1).

$$\forall M_i \in \{M_1, M_2, \ldots, M_N\}, \quad C_i = \text{FAISS}(\text{bi-encoder}(M_i)), \quad |C_i| = k \tag{1}$$

For the candidate re-ranking stage of ClinLinker, we built upon the work done by these same authors for the SympTEMIST-linking subtask [6], where initial explorations with the Sentence Transformer's cross-encoder were conducted. This phase is crucial as it refines the initial selection made by the bi-encoder, ensuring that the most relevant candidates are prioritized. After several experiments, we found that the optimal way to train the cross-encoder was using the candidates generated by the bi-encoder, leveraging the similarity between the correct mention and each candidate (see Fig. 2). This similarity comparison is vital for the cross-encoder to learn the subtle differences between similar candidates, thereby improving its ability to select the most appropriate candidates for a given entity mention. This method is known as training with hard triplets.

For cross-encoder training, we generated candidates from the training set of DisTEMIST and MedProcNER corpora, without using the gazetteer provided by the organizers of those shared tasks, creating the mentioned triplets—*(anchor, positive, negative)*—, always initializing the cross-encoder's parameters with each corresponding bi-encoder's weights. This approach ensures that the cross-encoder is exposed to a wide range of scenarios, including challenging cases where the distinction between candidates is not immediately obvious. At the outset, the effectiveness of fine-tuning domain-adapted models such as *roberta-base-biomedical-clinical-es* [2]—which, unlike SapBERT, do not start from language models pre-trained for the self-alignment of entities—has been also evaluated in this study; however, we noted that this approach did not surpass the performance observed in the bi-encoder+cross-encoder pipelines proposed herein.

The complete ClinLinker's inference process (see Fig. 2 and Eqs. 2 and 3) for each test set mention involves generating candidates with the Spanish-SapBERT bi-encoder, this time using the gazetteer supplied within the given corpora, and obtaining scores for each (M_i, C_i) pair. By scoring these pairs, we quantitatively assess the relevance of each candidate (C_i) to the mention (M_i). This allowed us to group them by mentions and sort them according to the obtained scores.

$$\text{Score}_{ij} = \text{CrossEncoder}(M_i, C_{ij}) \quad \forall C_{ij} \in C_i \tag{2}$$

$$C'_i = \text{rerank}(C_i, \text{Score}_{ij}) \quad \forall M_i \in \{M_1, M_2, \ldots, M_N\} \tag{3}$$

The final candidate rankings provided by ClinLinker are a mixture of advanced NLP techniques and domain-specific knowledge, ensuring high accuracy in EL across diverse medical texts. All the results obtained with ClinLinker are analyzed in the following section, in which we distinguish between the results obtained for the two corpora, DisTEMIST and MedProcNER. We also separate the analysis of the results obtained on the *gold standard* datasets—which contains all the codes from the provided test sets—from those obtained on the *unseen codes* datasets—encompassing only those codes in the test sets not present in the training sets supplied by the shared-tasks organizers.

4 Results and Discussion

To assess the advancements made in this study, two distinct corpora were employed: DisTEMIST [15], a dataset for automatic detection of disease mentions in clinical cases, and MedProcNER, a dataset for the automated identification of findings and procedures in clinical scenarios. These datasets represent a diverse collection of medical texts, providing a comprehensive testing ground for our models.

The chosen metric for comparative evaluation was top-k accuracy at 25. This metric is particularly relevant for practical applications where a balance between accuracy and a broad range of candidate suggestions is necessary. However, calculations were also performed for the top-k accuracy at 1, 5, 50, and 100. This range of metrics provides a detailed view of the model's performance across different levels of specificity. The selection of these metrics is rooted in the current limitation of models in accurately determining the linkage to a code with a single candidate. Yet, for higher candidate counts, such as @20 (k = 20) or @25 (k = 25), the predictions exhibit robust performance, demonstrating the model's effectiveness in providing a wide selection of relevant candidates. This robustness is essential in practical scenarios, where identifying an exact match as a unique code among candidates is often challenging due to the presence of cross-synonyms among different codes within the terminology, a situation that is particularly prevalent in highly granular vocabularies such as SNOMED-CT.

The results achieved are in line with the anticipated ideal behavior of the trained models, indicating the effectiveness of our training approach and the

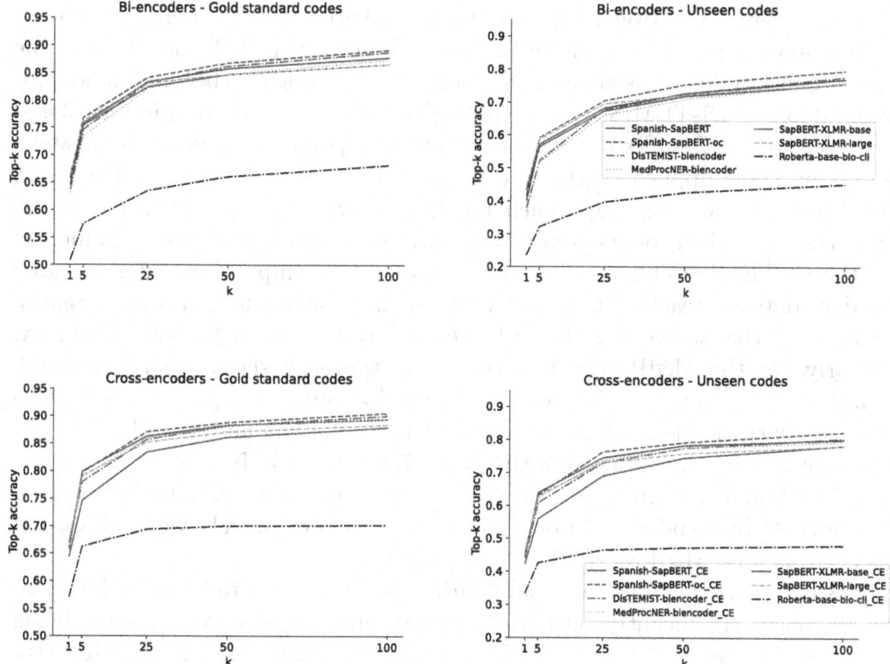

Fig. 3. Performance comparison of the bi-encoder and bi-encoder+cross-encoder ("_CE") models on the DisTEMIST corpus: efficacy of the models across various retrieval thresholds (top-k accuracy) for both the validated gold-standard annotations and the unseen-codes subsets. (Note that the figures are in different scales to show the differences between the performance of the different models in both subsets.)

suitability of ClinLinker for practical applications. In both MedProcNER and DisTEMIST datasets, the Spanish-SapBERT bi-encoders, trained exclusively with concept descriptions in Spanish from UMLS, surpass the generic multilingual SapBERT-XLMR base and large models[6], which are currently the benchmarks from the associated shared tasks. Furthermore, the cross-encoder strategy successfully reorders the candidates, providing a refined list where higher-ranked candidates are more likely to be the correct match, thus enhancing the practical utility of the system in clinical settings.

Figures 3 and 4 show the performance of trained bi-encoder models on the DisTEMIST and MedProcNER corpora. We compare our Spanish-SapBERT bi-encoders used in our ClinLinker pipeline against the multilingual SapBERT-XLMR, highlighting performance improvements, especially with Spanish-SapBERT-oc using outdated terms in the pretraining step. This comparison underscores the superior efficacy of our language-specific bi-encoders over the generic multilingual alternatives.

[6] *cambridgeltl/SapBERT-UMLS-2020AB-all-lang-from-XLMR-large* and *cambridgeltl/SapBERT-UMLS-2020AB-all-lang-from-XLMR* [13].

The difference between the ClinLinker's triplet-based bi-encoder+cross-encoder models (Spanish-SapBERT_CE and Spanish-SapBERT-oc_CE) and the current benchmarks—based on the Sapbert-XLMR models—, becomes more pronounced for the DisTEMIST and MedProcNER datasets, as shown in Figs. 3 and 4, where the "_CE" suffix denotes the complete pipeline, i.e. the cross-encoder's re-ranking stage applied to the output candidates of each specific bi-encoder. CinLinker, our two-stage approach for MEL in Spanish, consistently outperforms the SOTA biencoders across both datasets in top-k accuracy at 25 metric. For the DisTEMIST-linking task, we achieve a 4.8% improvement for the gold standard dataset, reaching 0.871, and a 9.6% increase for the zero-shot scenario, reaching a 0.763, surpassing all top-k accuracy values except for top-1 accuracy. Similarly, for the MedProcNER corpus, our approach shows a similar trend, slightly outperforming models used in SOTA. We improve top-k accuracy at 25 by 5.1% for the gold-standard dataset and 14.8% for the unseen-codes dataset, achieving noteworthy values of 0.894 and 0.752, respectively.

Tables 1 and 2 summarize the performance results for ClinLinker and all the analyzed bi-encoder and bi-encoder+cross-encoder models for MEL on the DisTEMIST and MedProcNER corpora, respectively.

Results clearly demonstrate that ClinLinker, specifically tailored to the Spanish language, significantly outperforms the multilingual onset, despite being trained on a subset of the data used to train the multilingual XLM-Roberta

Table 1. Comparative results (top-k accuracy) of selected bi-encoder (last 7 rows) and bi-encoder+cross-encoder ("_CE", first 7 rows) models on the DisTEMIST *gold standard* and *unseen codes* dataset. (†: ClinLinker's bi-encoder alone; ‡: ClinLinker's bi-encoder+cross-encoder. **Bold**: best result for each top-k accuracy; underlined: second best; *italic*: current benchmark).

MODEL'S NAME	gold standard					unseen codes				
	@1	@5	@25	@50	@100	@1	@5	@25	@50	@100
Spanish-SapBERT_CE‡	.661	**.799**	.863	.883	.894	.450	**.639**	.745	.782	.800
Spanish-SapBERT-oc_CE‡	.656	.797	**.871**	**.888**	**.905**	.455	.632	**.763**	**.791**	**.823**
DisTEMIST-biencoder_CE	.670	.780	.857	.882	.900	.440	.609	.730	.774	.804
MedProcNER-biencoder_CE	.658	.789	.860	.882	.894	.452	.624	.737	.774	.794
SapBERT-XLMR-base_CE	.644	.746	.834	.861	.879	.424	.560	.690	.744	.782
SapBERT-XLMR-large_CE	**.674**	.791	.852	.871	.884	**.463**	.629	.730	.758	.781
Roberta-base-bio-cli_CE	.570	.662	.694	.700	.702	.334	.426	.465	.472	.479
Spanish-SapBERT†	.647	.757	.833	.858	.878	.403	.572	.683	.727	.771
Spanish-SapBERT-oc†	.658	.768	.841	.868	.892	.417	.592	.705	.754	.796
DisTEMIST-biencoder	.637	.746	.832	.862	.888	.380	.523	.678	.727	.777
MedProcNER-biencoder	.631	.734	.822	.848	.874	.368	.518	.661	.710	.761
SapBERT-XLMR-base	.665	.754	.824	.847	.866	.434	.565	.674	.719	.757
SapBERT-XLMR-large	*.663*	*.762*	*.831*	*.847*	*.866*	*.445*	*.587*	*.696*	*.723*	*.757*
Roberta-base-bio-cli	.509	.575	.635	.661	.682	.236	.322	.397	.426	.452

Table 2. Comparative results (top-k accuracy) of selected bi-encoder (last 7 rows) and bi-encoder+cross-encoder ("_CE", first 7 rows) models on the MedProcNER *gold standard* and *unseen codes* dataset. (†: ClinLinker's bi-encoder alone; ‡: ClinLinker's bi-encoder+cross-encoder. **Bold**: best result for each top-k accuracy; underlined: second best; *italic*: current benchmark).

MODEL'S NAME	gold standard					unseen codes				
	@1	@5	@25	@50	@100	@1	@5	@25	@50	@100
Spanish-SapBERT_CE‡	.675	.816	.886	.904	.918	.361	.578	.729	.771	.794
Spanish-SapBERT-oc_CE‡	**.695**	**.828**	**.894**	**.908**	**.920**	**.373**	**.606**	**.752**	**.778**	**.795**
DisTEMIST-biencoder_CE	.678	.799	.873	.890	.901	.370	.540	.709	.733	.748
MedProcNER-biencoder_CE	.659	.805	.880	.902	.914	.343	.548	.696	.743	.767
SapBERT-XLMR-base_CE	.669	.784	.866	.890	.905	.302	.492	.687	.731	.760
SapBERT-XLMR-large_CE	.671	.815	.886	.904	.912	.339	.578	.742	.775	.794
Roberta-base-bio-cli_CE	.623	.695	.730	.734	.740	.234	.328	.382	.390	.396
Spanish-SapBERT†	.681	.788	.865	.888	.906	.314	.501	.669	.714	.756
Spanish-SapBERT-oc†	.685	.799	.870	.888	.908	.318	.523	.693	.726	.768
DisTEMIST-biencoder	.647	.755	.831	.859	.882	.245	.428	.586	.647	.703
MedProcNER-biencoder	.665	.774	.853	.879	.902	.263	.447	.615	.675	.727
SapBERT-XLMR-base	.683	.786	.855	.878	.895	.310	.484	.640	.685	.718
SapBERT-XLMR-large	*.689*	*.788*	*.851*	*.876*	*.897*	*.323*	*.522*	*.655*	*.708*	*.755*
Roberta-base-bio-cli	.543	.613	.671	.696	.715	.133	.197	.281	.314	.351

based models. This superiority underscores the critical importance of linguistic adaptation in model training, particularly when dealing with complex clinical data. Furthermore, the effectiveness of our model is not confined to its linguistic specificity; it is sufficiently versatile to surpass the SOTA in two different corpora annotated with SNOMED-CT codes. This finding emphasizes the broad applicability and efficacy of ClinLinker, highlighting its potential utility across a diverse range of clinical and linguistic contexts.

These advancements are not just incremental but represent a considerable leap forward in the field of NLP applied to clinical text exploitation. The improvement in unseen codes is particularly noteworthy as it indicates the robustness and generalizability of our model, fundamental for real-world clinical applications where unpredictability is the norm. This success can be attributed to the focused approach of our model, which, by prioritizing linguistic specificity, demonstrates the profound impact of language-tailored solutions in the domain of clinical informatics. Additionally, the improvement in gold standard performance suggests that our model is highly effective in identifying correct codes, even among a vast array of possibilities, which is paramount for practical clinical usage.

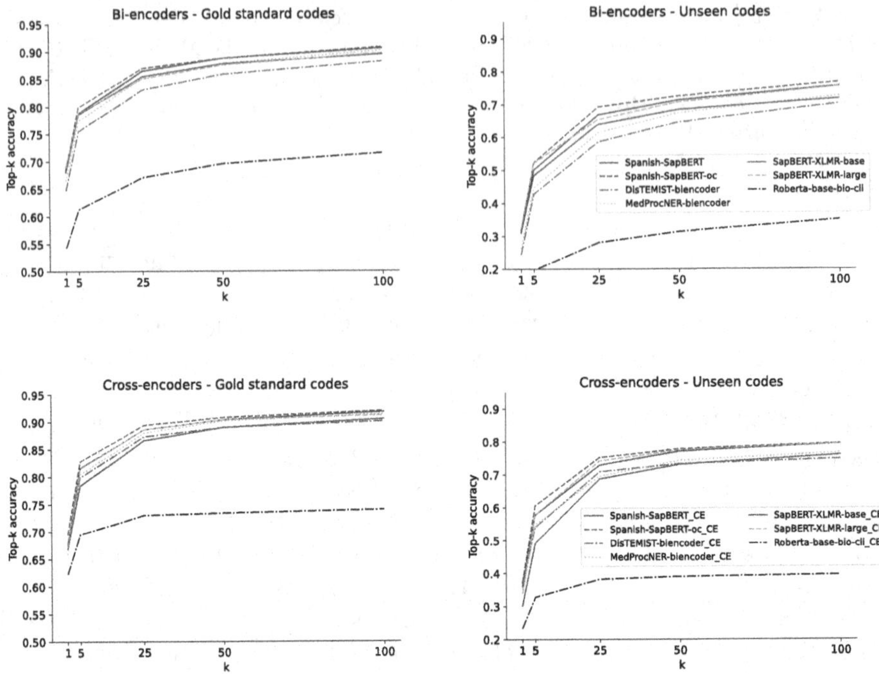

Fig. 4. Performance comparison of the bi-encoder and bi-encoder+cross-encoder ("_CE") models on the MedProcNER corpus: efficacy of the models across various retrieval thresholds (top-k accuracy) for both the validated gold-standard annotations and the unseen-codes subsets. (Note that the figures are in different scales to show the differences between the performance of the different models in both subsets.)

5 Conclusions

Our study marks a significant leap in NLP applied to medical text analysis. Utilizing a two-phase approach with a bi-encoder for candidate retrieval, followed by a cross-encoder for candidate re-ranking, with models trained with contrastive-learning strategies and specifically tailored for the Spanish language, ClinLinker has demonstrated superior performance compared to existing multilingual models for MEL tasks. Moreover, our models trained with the entire set of terms in Spanish from the UMLS ontology show superior performance rates than those ones trained with the specific corpora, proving to facilitate their reuse for new corpora tagged with UMLS subsets such as SNOMED-CT. Also, the results presented in this paper underscore the high impact of our models in clinical settings, offering high accuracy for both top-k accuracy at 5 and top-k accuracy at 25 metrics. This level of precision is more than acceptable for semi-automatic annotation processes, underscoring the practical applicability of our approach in real-world clinical environments. These findings not only highlight the importance of linguistic adaptation in model training but also underscore the potential of our approach in enhancing the utility of digital medical records across various clinical contexts. The robustness and generalizability of our MEL models, particularly in handling unfamiliar codes, pave the way for future research focused on language-specific NLP solutions in clinical informatics. The presented approach showed promising results and could be adapted also for concept normalization of other types of clinical relevant entities such as drugs, chemical compounds and proteins or for clinical variable extraction and data structuring applied to more tailored use cases for instance in cardiology or other medical specialities.

For future research, we aim to work not only with mentions but also with the context of these mentions, endowing MEL language models with information usable for both generating and ordering candidate medical concepts to be matched to the target mentions. Additionally, we plan also to further explore combinations with more traditional approaches such as classification—due to the computational cost of training the models used in this paper—, aiming to make it scalable to other languages, and also the use of knowledge-graph (KG) enhanced language models, by leveraging KGs obtained from UMLS or SNOMED-CT to enrich MEL models with in-domain semantic information.

Acknowledgments. The authors acknowledge the support from the Spanish Ministerio de Ciencia e Innovación (MICINN) under projects PID2020-116898RB-I00, PID2020-119266RA-I00 and BARITONE (TED2021-129974B-C22). This work is also supported by the European Union's Horizon Europe Co-ordination & Support Action under Grant Agreement No 101080430 (AI4HF), Grant Agreement No 101058779 (BIOMATDB) as well as Grant Agreement No 101057849 (DataTool4Heartproject).

Disclosure of Interests. The authors have no competing interests to declare that are relevant to the content of this article.

References

1. Alsentzer, E., et al.: Publicly available clinical BERT embeddings, April 2019
2. Carrino, C.P., et al.: Biomedical and clinical language models for Spanish: on the benefits of domain-specific pretraining in a mid-resource scenario (2021)
3. Conneau, A., et al.: Unsupervised cross-lingual representation learning at scale (2019)
4. Ding, N., et al.: Prompt-learning for fine-grained entity typing. In: Goldberg, Y., Kozareva, Z., Zhang, Y. (eds.) Findings of the Association for Computational Linguistics: EMNLP 2022, pp. 6888–6901. Association for Computational Linguistics, Abu Dhabi, United Arab Emirates (2022). https://doi.org/10.18653/v1/2022.findings-emnlp.512, https://aclanthology.org/2022.findings-emnlp.512
5. Douze, M., et al.: The Faiss library (2024)
6. Gallego, F., Veredas, F.J.: ICB-UMA at BioCreative VIII @ AMIA 2023 task 2 SYMPTEMIST (symptom TExt mining shared task). In: Islamaj, R., et al. (eds.) Proceedings of the BioCreative VIII Challenge and Workshop: Curation and Evaluation in the era of Generative Models (2023)
7. Gu, Y., et al.: Domain-Specific language model pretraining for biomedical natural language processing. ACM Trans. Comput. Healthc. **3**(1), 1–23 (2021)

8. Lai, T.M., Zhai, C., Ji, H.: KEBLM: knowledge-enhanced biomedical language models. J. Biomed. Inform. **143**, 104392 (2023)
9. Lee, J., et al.: BioBERT: a pre-trained biomedical language representation model for biomedical text mining. Bioinformatics **36**(4), 1234–1240 (2020)
10. Lewis, M., et al.: BART: denoising sequence-to-sequence pre-training for natural language generation, translation, and comprehension. arXiv preprint arXiv:1910.13461 (2019)
11. Lima-López, S., et al.: Overview of MedProcNER task on medical procedure detection and entity linking at bioasq 2023. In: Conference and Labs of the Evaluation Forum (2023). https://api.semanticscholar.org/CorpusID:264441740
12. Liu, F., Shareghi, E., Meng, Z., Basaldella, M., Collier, N.: Self-alignment pretraining for biomedical entity representations. In: Proceedings of the 2021 Conference of the North American Chapter of the Association for Computational Linguistics: Human Language Technologies, pp. 4228–4238. Association for Computational Linguistics, Online (2021)
13. Liu, F., Vulić, I., Korhonen, A., Collier, N.: Learning domain-specialised representations for cross-lingual biomedical entity linking. In: Proceedings of ACL-IJCNLP 2021 (2021)
14. Liu, Y., et al.: RoBERTa: a robustly optimized BERT pretraining approach (2019)
15. Miranda-Escalada, A., et al.: Overview of DisTEMIST at BioASQ: automatic detection and normalization of diseases from clinical texts: results, methods, evaluation and multilingual resources. In: Working Notes of Conference and Labs of the Evaluation (CLEF) Forum. CEUR Workshop Proceedings (2022). https://ceur-ws.org/Vol-3180/paper-11.pdf
16. Yu, P., Fei, H., Li, P.: Cross-lingual language model pretraining for retrieval. In: Proceedings of the Web Conference 2021, pp. 1029–1039 (2021)
17. Yuan, H., Yuan, Z., Gan, R., Zhang, J., Xie, Y., Yu, S.: BioBART: pretraining and evaluation of a biomedical generative language model. arXiv preprint arXiv:2204.03905 (2022)
18. Yuan, H., Yuan, Z., Yu, S.: Generative biomedical entity linking via knowledge base-guided pre-training and synonyms-aware fine-tuning. arXiv preprint arXiv:2204.05164 (2022)

Stylometric Analysis of Large Language Model-Generated Commentaries in the Context of Medical Neuroscience

Jan K. Argasiński[1,2(✉)], Iwona Grabska-Gradzińska[1], Karol Przystalski[1], Jeremi K. Ochab[1,3], and Tomasz Walkowiak[4]

[1] Jagiellonian University, Kraków, Poland
jan.argasinski@uj.edu.pl
[2] Sano - Centre for Computational Medicine, Kraków, Poland
[3] M. Kac Center for Complex Systems Research, Jagiellonian University, Kraków, Poland
[4] Faculty of Information and Communication Technology, Wroclaw University of Science and Technology, Wroclaw, Poland

Abstract. This study investigates the application of Large Language Models (LLMs) in generating commentaries on neuroscientific papers, with a focus on their stylometric differences from human-written texts. Utilizing three papers from reputable journals in the field of medical neuroscience, each accompanied by published expert commentaries, we compare these with commentaries generated by state-of-the-art LLMs. Through quantitative stylometric analysis and qualitative assessments, we aim to be a part of the discussion around the viability of LLMs in augmenting scientific discourse within the domain of medical neuroscience.

Keywords: Large Language Models · Stylometry · Text generation

1 Introduction

In the context of the application of Large Language Models (LLMs) in medicine, the ability to distinguish between human-written and computer-generated text becomes critically important for several reasons [5].

First, it ensures the integrity and trustworthiness of medical research and its implementation [37]. Human experts often provide nuanced insights based on years of experience and tacit knowledge that LLMs, despite their advanced capabilities, might not fully replicate. Identifying the text's origin allows readers to weigh the insights accordingly, appreciating the depth of human expertise or the data-driven breadth of LLMs.

Second, distinguishing between these sources of text helps in maintaining a high standard of ethical transparency [52]. As the medical field relies heavily on evidence-based practice, the clear labelling of human versus AI contributions

L. Franco et al. (Eds.): ICCS 2024, LNCS 14836, pp. 281–295, 2024.
https://doi.org/10.1007/978-3-031-63775-9_20

upholds the ethical standards of research dissemination and consumption. It ensures that practitioners and researchers are fully informed about the nature of the information they are engaging with, facilitating informed decision-making processes.

Lastly, this distinction aids in the ongoing evaluation and improvement of LLMs themselves. Researchers can identify areas where LLMs excel or fall short by comparing human and AI-generated papers, guiding further development and training efforts. This not only enhances the utility of LLMs in medical applications but also ensures that these tools are used in a manner that complements, rather than supplants, human expertise [41].

In this paper, we compare artificially generated papers with human-written scientific literature. We do this by matching LLM-produced text with published commentaries on existing medical papers. Medical commentaries serve as a vital component of the scientific communication ecosystem, offering insights, critiques, and expanded discussion on published research findings. Unlike original research articles that require the presentation of new experimental data, statistical analyses, and results, commentaries primarily rely on the interpretation and discussion of existing studies. The distinction makes commentaries an ideal genre for exploring the potential of LLMs in medical literature generation, as the focus shifts from providing new, verifiable data to synthesising and discussing existing knowledge.

Opting for the generation of medical commentaries not only aligns with the strengths of LLMs but also offers a pragmatic pathway to evaluate their potential in medical literature. It allows for both qualitative and quantitative assessments of the generated texts, and enables the stylometric analyses – giving insights into their linguistic structure. Supplemented with a qualitative evaluation that is particularly suited to the nuanced and interpretative nature of commentaries – our study aims to explore the capabilities of LLMs in generating medical commentaries, comparing them with human-written texts to assess their potential and limitations. This approach not only highlights the current capabilities of LLMs but also sets the stage for future advancements and applications in the generation of medical texts.

2 Related Works

Comparison of various LLMs should consider the aim of the source text processing and analysis: not only text summarisation and data extraction [29,43,50] but also simplification [18,31], semantic similarity and reasoning, critical commentary and quality evaluation [7,22]. Depending on the objectives of the source text analysis, fine-tuned domain-specific language models remain the first choice, rather than general LLM models [7]. The comparison of different models in the context of the aim of their use is shown in [7,16]. The models used in the present paper are described in [9,20,44,53].

There are some problems which should be addressed in the context of LLMs comparison. One of them is the text structure of scientific papers. Scientific works, due to different structuring standards and the use of data not only in the form of continuous text but also tabular data and graphs can be treated as unstructured texts [11,30,51] when analysed using LLMs. Another problem is the length of the article considered as a single statement. Studies have been carried out comparing models with long source texts [55]. The problem of hallucination in natural language generation is adressed in [24,40].

Comparing the performance of LLMs to the effect of human labour can emphasize the factual correctness of the generated text [15,17] as well as the stylistic (in)distinctiveness of the generated response [4,46]. In the latter context, it becomes convenient to use linguistic and stylometric tools [35], which have long been used for problems of text attribution, verifying text's authorship and investigating characteristics of their style [42]. Moreover, research efforts are increasingly focused on the development of automated methods for detecting text generated by LLMs [8,32].

Producing fake scientific papers has a long tradition, notably with 1996 Sokal's hoax and a later computer-science paper generator SCIgen [45], and so has the detection of gibberish or computer-generated papers [26,48]. Since then, whole benchmarks for the task of detecting automatically generated academic papers have appeared [27,34], which however may undergo a fast deterioration due to the exponential growth of LLM capabilities and the fact that authors may utilise LLMs as auxiliary tools in their writing [1]. More recent works focus specifically on detecting LLM-generated texts [19,34], and the methods include stylometric analysis [54].

3 Methodology

3.1 Stylometry

Stylometry, as an academic field, constitutes an area of study within computational linguistics and digital humanities, focusing on the quantitative analysis of textual features to deduce metadata such as authorship, chronology, and stylistic evolution. This discipline leverages statistical and machine learning methodologies to scrutinize the stylistic fingerprints left by authors in their texts, thereby facilitating a deeper understanding of literary corpora beyond the capabilities of traditional approaches.

The potential utility of stylometry extends into the realm of analyzing LLMs, offering a lens through which to examine the generative capabilities and inherent biases of models.

Methods and Setup. For the unsupervised quantitative analysis of text style, two baseline methods from the R package 'Stylo' [13] were used: the principal component analysis (PCA) of the covariance matrix of feature frequencies and bootstrap consensus trees (BCT). The correlation matrix was less useful since the real commentaries (number one and two) stand out mainly due to their larger length. The features checked included the normalised occurrence frequencies of the most frequent words (MFWs; we chose the typical range between 100 and 1000 MFWs and, for BCT, they were iterated by 100) and their N-grams (again, following the typical choice from 1 to 3). The features were 'culled' in the range from 0% to 25% (which means the minimum percentage of texts in which a feature must occur; higher values were leaving too few features to compute a BCT). The consensus trees used the cosine delta distance, commonly considered in stylometry the reliable choice [14].

As a more modern alternative, the authors' own pipeline for interpretable stylometric analysis [36] was used. The pipeline used Spacy [33] 'en_core_web_lg' model for preprocessing steps (including tokenisation, named entity recognition, dependency parsing, and part-of-speech annotation), LighGBM [25] as the state-of-the-art DART boosted trees classifier, SHAP [28] for computing explanations, and Scikit-learn [38] for feature counting and cross-validation. The texts were chunked into 50-token samples. The extracted features included: 1–3-grams of lemmas and parts of speech, dependency-based bigrams of lemmas, NER entity types, and morphological annotations. Binary classification was performed between the real and fake commentary for each LLM. The baseline corresponds to a dummy classifier with the strategy of choosing the most frequent class. Stratified 10-fold cross-validation (i.e., in each fold the ratio training: test set size was 9:1) was repeated 10 times to collect more reliable statistics. Within the training sets, 10% data were used for model validation.

Both stylometric analyses excluded: the abstract, keywords, figures or tables and their captions, the lists of authors, affiliations, footnotes, and references, as well as sections like acknowledgements. Preprocessing included also removal of hyphenations at line breaks, leaving line breaks only between paragraphs, converting numerical citations to in-text citations (author, year), normalising quotation marks, etc.

3.2 Qualitative Criteria for Text Evaluation

In conducting the study, specific heuristics were selected for the qualitative comparison of texts: the annotators answered if the generated texts provided an:

1. accurate summarization and referencing of original research,
2. correct references to real academic papers,
3. proper abstraction of relevant knowledge from the cited papers,
4. coherent argumentation of presented arguments,
5. realistic numerical results, tables, or figures,
6. strict scientific knowledge – in terms of factual correctness,
7. strict scientific knowledge – in terms of being state-of-the-art,
8. fitting structure/argumentation as expected from a commentary,
9. pertinent tone/style as expected from a commentary,
10. qualitatively new insight with respect to the original paper.

The responses: "Yes", "No" or "Partly/Not applicable" are summarized in Table 2. "Partly" was used in several cases, including partial correctness of the text, but also only paraphrasing the information already given in the prompt or in the original paper (hence, assumed to be correct, e.g., in terms of factual correctness) without any new information produced.

Two authors of this paper, JKA and JKO, were the annotators. The papers matched the annotators' expertise. They had access to the original papers, the real commentaries, and the LLM-generated ones and could read them and respond to the criteria at their own discretion and in any order.

3.3 Large Language Models Used

The Large Language Models considered for the generation of commentary were: GPT-4-1106-preview (general purpose) [2], Google Gemini (general purpose) [47], and MED-PaLM2 (medical purpose) [44].

The GPT and Gemini were chosen as one the best LLMs for general purpose. We tried to use LLMs dedicated to medical cases, but only MED-PaLM2 was a ready to use LLMs. Meditron [9] and MedAlpaca [21] are also designed for medical purposes, but are not fine-tuned and for this reason both were excluded for this research.

Limitations of Considered LLMs. The final two models mentioned in the previous subsection, namely Meditron and MedAlpaca, are publicly available on HuggingFace[1] platform. Meditron requires that users agree to share their contact information. These models are based on the Llama 2 model and have been trained on medical data, making them potentially suitable for the analysis outlined in this study. However, a detailed analysis reveals two significant limitations that make them unsuitable for use in this research:

The initial issue concerns the fine-tuning process in constructing these LLMs. The publicly accessible Meditron has not undergone fine-tuning; it remains a raw pre-trained model. Although it has been exposed to articles and can generate content similar to them, it lacks comprehension when it comes to the specific task of commentary article generation. However, MedAlpaca has been fine-tuned, but only for question-type tasks, excelling in providing answers related to medical issues. However, it fails to grasp the task of article generation, leading to very poor results that do not resemble an academic article at all.

The second issue relates to the challenge of context length. The prompts used to test the LLM (see Sect. 4.2) consist of more than 9,000 tokens. However, Meditron-7B and MedAlpaca-7B are limited to 2048 tokens each, while Meditron-70B has a capacity of 4096 tokens. Consequently, both the query and the response must comply with these constraints (2048, 4096). While it is feasible to trim the query, any adjustments must consider the model's limitations and allocate space for the response (e.g., accommodating up to 1024 tokens).

In summary, above mentioned open-source models pre-trained on medical data are still far from having the capability to generate commentaries resulting in their exclusion from the study.

[1] https://huggingface.co/.

4 Data

The experimental data consists of three original research papers [23,39,49], together with their commentary articles [3,6,10] and the corresponding LLM-generated commentaries.

Only the commentaries were analysed, while the original research papers were only used in LLM prompts.

4.1 The Source Papers and Commentaries

The selection criteria for the source papers were: existing commentary article in the same journal, highly ranked journal within the research area of the paper, publication within the last 5 years.

4.2 The Generation of LLM's Commentary (Prompts)

The prompt used to generate the commentaries included the information on the type of academic paper to be written (a commentary), the journal in which it should be published (the same as original), the focus (criticism of methodology and the interpretation of the results; not to summarise the whole paper), the method (to cite the scientific papers following given arguments), and lastly – the text of the original research paper (without references).

Prompt structure The prompt has following structure:

1. Paragraph: *Given the following article, write a commentary article to be published in the same journal. Consider only the criticism of the methodology and the interpretation of the results. Do not summarise the whole text. Cite the scientific papers with your arguments. Use only real, published scientific work:*
2. Citation of the original paper including title and full journal name.
3. Phrase: *The original article is provided below:*
4. The text of the original research paper with abstract, highlights (when apply) etc. but without references.

5 Results

5.1 Quantitative Analysis

The unsupervised methods visualised in Fig. 1 show that (i) the texts mostly cluster according to which paper they were commenting, (ii) GPT-4's output consistently clusters with the real texts, while the other two models form separate clusters. The figure presents only one chosen set of parameters (the features were 1000 MFWs for PCA and 100-1000 MFWs for the BCT), but the results were stable over other MFW ranges, their N-grams, and culling.

Table 1 shows the results of the LGBM classifier with around 3000 features. Although the accuracy scores are close to the baseline due to a large class imbalance (the real commentaries were longer, hence, they provided a larger number

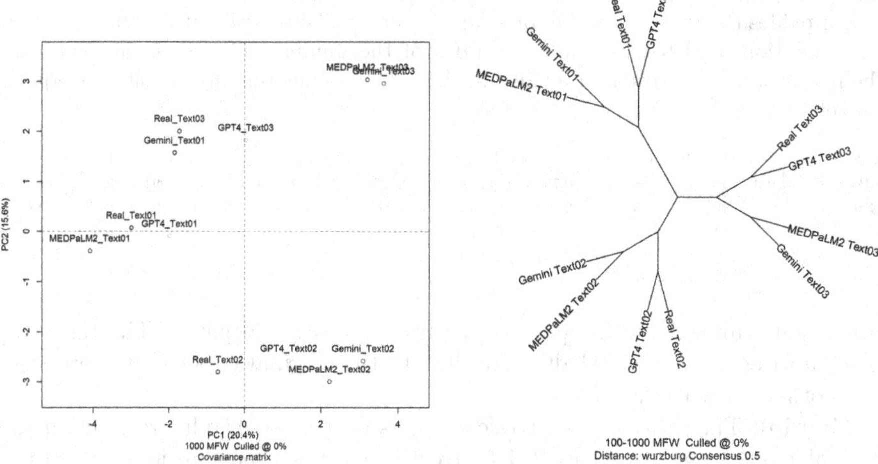

Fig. 1. Unsupervised visualisation: (left) Covariance PCA, (right) Bootstrap Consensus Tree.

of samples), the F1 scores show a reasonable performance, given that the samples are short and come in small numbers. Such scores allow us to use SHAP explanations to obtain an idea about the textual features that make the LLM-generated texts imperfect. As shown in Fig. 2(middle), the features responsible for detection can be as simple 'SPACE', which in fact corresponds to the number of paragraphs delimited by line breaks (large in Gemini and MED-PaLM2); another example is the usage of plurals, 'Number:Plur' (over the top in Gemini and MED-PaLM2, but underused in GPT-4 with respect to the real texts).

5.2 Qualitative Assessment

[Paper 1] GPT-4: The generated text correctly referred to specific paragraphs or in the original paper including some numerical results. The structure, register and tone were all acceptable for a commentary article, with the exclusion of the commentary title. The argumentation heavily relied on the discussion section of the original paper but it did not introduce any new ideas. The arguments were all reasonable (except, perhaps, a plea for consensus in the field rather than correction of the paper) but since they followed the limitations mentioned by the original paper's authors, they would not constitute a reason to publish the commentary. There were 10 references, out of which nine appeared in the original paper, five were on the reference list (three correct, one not cited in the text, one had incorrect journal/volume/page, one had an incorrect but existing author and incorrect year/page), four in-text citations were not on the list.

MED-PaLM2: The model misunderstood the task and produced responses to a hypothetical reviewer's comments. No new numbers or tables were generated. The generated text and all the scientific statements it contained referred

Table 1. LGBM classification results of 50-token samples. The left-hand side of the table provides the median number of samples (across all cross-validation runs) in training, validation, and test sets and the ratio of the numbers of real to fake samples. The performance metrics are provided against the baseline dummy classifier in square brackets.

LLM	Train	Val	Test	Imbalance	Accuracy [baseline]	F1 [baseline]	Recall
GPT-4	100	12	12	1.8	0.75+/−0.11 [0.646+/−0.025]	0.7+/−0.1 [0.3924+/−0.0094]	0.82+/−0.15 [0]
Gemini	88	10	11	2.8	0.844+/−0.083 [0.735+/−0.022]	0.79+/−0.12 [0.4234+/−0.0071]	0.70+/−0.28 [0]
MED-PaLM2	96	11	12	2.1	0.78+/−0.09 [0.673+/−0.018]	0.74+/−0.12 [0.4021+/−0.0063]	0.61+/−0.19 [0]

to and paraphrased specific paragraphs from the original paper. The reference list comprised only a PubMedCentral link to the original paper, but there were seven other in-text citations.

Gemini: The generated text referred to specific paragraphs or items in the original paper, e.g., to tables 2–4 (correctly quoting their captions) or to the numerical value of median age [correctly: mean]. It also contained a fictitious quote (together with the page, where it should be found in the article), although one close in meaning to several existing sentences. The model made an untrue statement that the authors of the original paper “do not provide any evidence ... specific examples or data” to support their claims. There were no citations, a reference list nor new numbers or tables. The argumentation consisted of repeated general statements that repeated limitations mentioned in the paper itself; it did not contain any other piece of specific scientific knowledge. The tone and convention were acceptable with the exclusion of the section titles.

[Paper 2] GPT-4: The produced text accurately cited particular sections and numerical findings from the initial paper. The format, style, and tone were suitable for an analytical piece. However, the discussion didn’t present any novel concepts, adhering closely to the constraints acknowledged by the authors of the original study. While the arguments made were logical, their reliance on the original paper’s limitations means they do not provide a sufficient basis for the publication of the commentary. Furthermore, although all references were accurately cited, they were directly drawn from the information provided in the prompt.

MED-PaLM2: The model undertook a secondary task of summarizing the text and identifying its strengths and limitations. The strengths and limitations highlighted were directly extracted from the prompt provided. There are no quotes, footnotes, or references to numerical data beyond the most basic information, such as the number of study participants or the number of experiments conducted. The structure of the output does not resemble that of an article but rather that of a summary. While the discourse maintains a scientific tone, it lacks any element of novelty. This approach reflects a concise synthesis and critique of the original content, focusing on presenting a distilled overview rather than expanding on the data or introducing new interpretations.

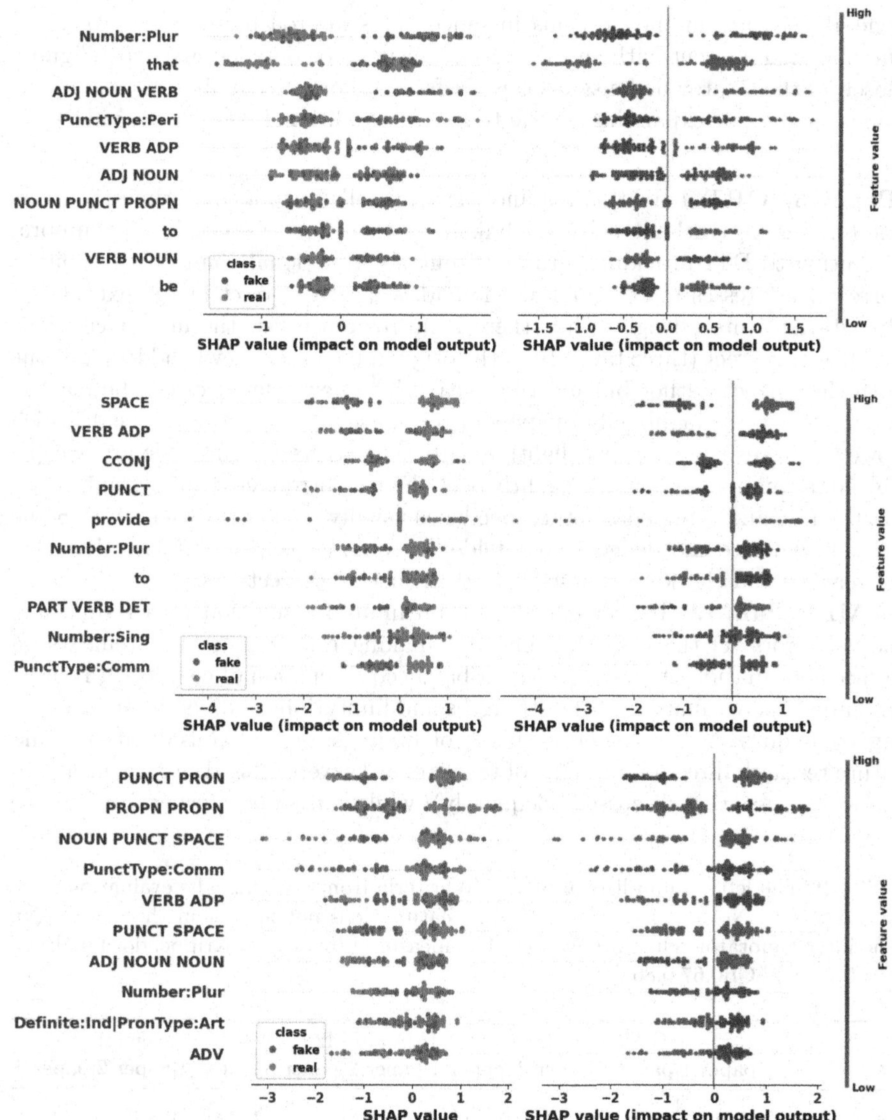

Fig. 2. SHAP values of the first 10 features most important for classifying real commentary vs (top) GPT-4, (middle) Gemini, and (bottom) MED-PaLM2. Each point is a 50-token text sample coloured (left) by its class membership or (right) by its feature intensity. Positive SHAPs point toward real texts, and negative toward fake ones.

Gemini: The model performed a task very similar to that of MED-PaLM2 as described above. It lacks original observations, and the entirety of its output is limited to a summary of the text prompt devoid of quantitative data and references. While the style remains appropriate, the structure is even more generalized than in the previous instance. This approach suggests a focus on summarizing content without adding new insights or detailed analysis, aligning closely with the provided instructions yet falling short of contributing to a deeper understanding or expanding on the topic with additional context or evidence.

[Paper 3] GPT-4: In this instance, the model's output is at times incoherent (e.g., going without a logical link from the issue of spatial and temporal resolution to DTI providing structural but not functional connectivity), but it correctly addresses the original article and is mostly correct in contextualising the citation. Interestingly, while there is no reference list, the in-text citations are mostly correct (three taken from the original paper, six newly added, and one with the correct author but incorrect year). The new references can be tracked to exist: they are highly cited papers of recognisable authors in the field, which however makes the citations slightly dated. The structure and style are acceptable for a commentary, with a slightly over-the-top introductory paragraph, not a fitting title, and comments not as specific as usually seen in commentaries. Some remarks indicate knowledge from outside the commented paper and involve careful reading and checking, to ascertain their factual correctness.

MED-PaLM2: The model focused on limited discussion rather than any factual errors (cf the real commentary), making rather general statements on "critical evaluation of study quality, a balanced discussion of conflicting results, a comprehensive analysis of the strengths and limitations of different neuroimaging techniques [...] the generalizability of findings, ethical considerations, and future research directions". Some of these remarks were missed (e.g., neuroimaging techniques were discussed adequately), while some were new with regard to

Table 2. The left column lists abbreviated criteria from Sec. 3.2. The evaluation scale was: Yes – ✓, No – ✗, Partly – ✳, N/A (feature does not appear in the text) – ⊗. The inter-annotator reliability was good as measured by ordinal Krippendorff's alpha, $\alpha = 0.77$, 95%CI[0.67,0.86].

	GPT-4			MED-PaLM2			Gemini		
	paper 1	paper 2	paper 3	paper 1	paper 2	paper 3	paper 1	paper 2	paper 3
1. summary	✓,✓	✓,✓	✓,✓	✓,✓	✓,✓	✳,✳	✓,✳	✓,✳	✳,✳
2. references	✳,✳	✓,✓	✓,✓	✳,✳	⊗,⊗	⊗,⊗	⊗,⊗	⊗, ⊗	⊗,⊗
3. citing	✓,✳	✳,✳	✳,✳	✳,✳	✓,⊗	⊗,⊗	⊗,⊗	✳,⊗	⊗,⊗
4. coherence	✓,✓	✓,✓	✓,✳	✓,✓	✓,✓	✳,✓	✳,✗	✓,✓	✳,✓
5. numbers	✳,⊗	⊗,⊗	⊗,⊗	⊗,⊗	⊗⊗	⊗,⊗	✳,⊗	⊗,⊗	⊗,⊗
6. factuality	✳,✳	✓,✓	✓,✓	✳,✳	✳,✳	✳,✳	✳,⊗	✳,✳	✳,✳
7. SOTA	✳,✳	✓,✓	✓,✳	✳,✳	✳,⊗	⊗,⊗	✳,⊗	✳,⊗	✗,⊗
8. structure	✓,✓	✓,✓	✓,✓	✗,✗	✳,✗	✗,✗	✓,✗	✗,✗	✳,✓
9. tone	✓,✓	✓,✓	✓,✓	✗,✗	✳,✳	✳,✳	✓,✳	✓,✳	✗,✳
10. novelty	✗,✗	✗,✳	✳,✳	✗,✗	✗,✳	✗,✳	✗,✗	✗,✳	✗,✳

the original paper (e.g., ethical considerations). The overall tone and structure reminded that of a reviewer's comments rather than a commentary article. There were no citations, numbers, etc. From underneath the generalities, there were only glimpses of specific pieces of scientific knowledge.

Gemini: The model's output is very similar to that of MED-PaLM2 above. Again, it mostly comprises general statements on the original paper's limited discussion. As before, some criticisms are poorly grounded (e.g., that individual modalities had been discussed in isolation, as the original text mentions a few comparisons between the neuroimaging techniques – although, admittedly, part of this information had been presented in a figure not included in the prompt).

6 Discussion

Qualitatively, current LLMs produce grammatically and semantically correct texts. The generated papers' content usually does not exceed the content of the original research paper, and as such can be merely considered a summary of the paper's limitations or discussion sections. The best model, however, can write a compelling commentary article that references relevant studies, and whose tone and structure agree with editorial standards for that particular article type, and thus hypothetically it might incur some editorial costs before rejection.

Classic stylometric techniques provide a reliable classification of bag-of-word (BoW) text samples for sizes > 2000 tokens in corpora of 100 novels [12]. In our easier setup (binary classification and text segments instead of BoW), a reasonable performance is obtained even for the small sample and short texts (50 tokens!). The repeated cross-validation constitutes tentative evidence that LLMs do have stylistic preferences which can be used to detect them and which one can track as individual, explainable features.

Limitations. It is uncertain whether an LMM output's style is inherent to the model or provoked by the prompt. The working assumption was that given the same prompt the outputs of all LLMs are comparable. However, how changes to the prompt result in different outputs might depend on the LLM.

Admittedly, the tree classifiers tend to overfit – hence our extensive use of cross-validation – and a larger sample of texts would be advisable. The stability of individual features and their SHAP importance depends on the size of the training sample as much as the generalisability of the classifier.

Another limitation of this paper is no blind review and no control. The authors who evaluated the generated texts knew that they were produced by LLMs. This limitation was to be alleviated by the introduction of the inter-annotator agreement parameter.

7 Conclusion

We demonstrate the possibility of applying stylometric methods for analyzing computer-generated texts within the medical neuroscience domain. Scientific and domain-specific texts are significantly more challenging to generate effectively due to their grounding in real knowledge and facts, which cannot be easily summarized from a general knowledge base. These types of errors produced by the state-of-the-art language models can be assessed only by manual qualitative evaluation.

The continuous improvement of machine learning models necessitates the development of methodologies to differentiate their outputs from human-written text. Stylometry, which analyzes literary styles statistically, is a proposed method. However, its effectiveness in identifying machine-generated texts, which increasingly resemble human writing, is uncertain. Thus, validating this quantitative approach is crucial, as it would clarify the potential and limitations of stylometry in this evolving field. This highlights the need for further research to refine the tools for distinguishing between different sources of text generation.

Acknowledgements. The publication was created within the project of the Minister of Science and Higher Education "Support for the activity of Centers of Excellence established in Poland under Horizon 2020" on the basis of the contract number MEiN/2023/DIR/3796. This project has received funding from the European Union's Horizon 2020 research and innovation programme under grant agreement No 857533. This publication is supported by Sano project carried out within the International Research Agendas programme of the Foundation for Polish Science, co-financed by the European Union under the European Regional Development Fund.

JKO and TW's research was financed by the European Regional Development Fund as a part of the 2014–2020 Smart Growth Operational Programme, CLARIN – Common Language Resources and Technology Infrastructure, project no. POIR.04.02.00-00C002/19. JKO's research has been supported by a grant from the Priority Research Area DigiWorld under the Strategic Programme Excellence Initiative at Jagiellonian University.

References

1. Abani, S., Volk, H.A., De Decker, S., et al.: ChatGPT and scientific papers in veterinary neurology; is the genie out of the bottle? Front. Vet. Sci. **10** (2023). https://doi.org/10.3389/gtkf43
2. Achiam, J., Adler, S., Agarwal, S., et al.: GPT-4 technical report (2023). arXiv preprint arXiv:2303.08774
3. Bagić, A., Bowyer, S., Funke, M., et al.: Commentary on "Mapping the Unconscious Brain: Insights From Advanced Neuroimaging". J. Clin. Neurophysiol. **40**(3), 269 (2023). https://doi.org/10.3389/gtktkx
4. Bethany, M., Wherry, B., Bethany, E., et al.: Deciphering textual authenticity: a generalized strategy through the lens of large language semantics for detecting human vs. machine-generated text (2024)
5. Bruckert, S., Finzel, B., Schmid, U.: The next generation of medical decision support: a roadmap toward transparent expert companions. Front. Artif. Intell. **3**, 507973 (2020). https://doi.org/10.3389/frai.2020.507973

6. Caruana, F.: Positive emotions elicited by cortical and subcortical electrical stimulation: a commentary on Villard et al. (2023). Cortex (2023). https://doi.org/10.3389/gtkcqj
7. Chen, Q., Du, J., Hu, Y., et al.: Large language models in biomedical natural language processing: benchmarks, baselines, and recommendations (2024)
8. Chen, Y., Kang, H., Zhai, V., et al.: Token prediction as implicit classification to identify LLM-generated text. In: Bouamor, H., Pino, J., Bali, K. (eds.) Proceedings of the 2023 Conference on Empirical Methods in Natural Language Processing, pp. 13112–13120. Association for Computational Linguistics, Singapore (Dec 2023). https://doi.org/10.18653/v1/2023.emnlp-main.810
9. Chen, Z., Hernández-Cano, A., Romanou, A., et al.: Meditron-70b: scaling medical pretraining for large language models (2023)
10. Clayson, P.E., Kappenman, E.S., Gehring, W.J., et al.: A commentary on establishing norms for error-related brain activity during the arrow flanker task among young adults. NeuroImage **234**, 117932 (2021). https://doi.org/10.3389/gtkcqp
11. Dunn, A., Dagdelen, J., Walker, N., et al.: Structured information extraction from complex scientific text with fine-tuned large language models (2022)
12. Eder, M.: Short Samples in Authorship Attribution: a new approach. In: Digital Humanities 2017. ADHO, Montréal, Canada (2017). https://dh2017.adho.org/abstracts/341/341.pdf
13. Eder, M., Kestemont, M., Rybicki, J.: Stylometry with R: a package for computational text analysis. R J. **8**(1), 1–15 (2016). https://doi.org/10.3389/gghvwd
14. Evert, S., Proisl, T., Jannidis, F., et al.: Understanding and explaining Delta measures for authorship attribution. Digital Sch. Humanit. **32**(suppl_2), ii4–ii16 (2017). https://doi.org/10.1093/llc/fqx023
15. Fu, J., Ng, S.K., Jiang, Z., et al.: GPTScore: Evaluate as You Desire (2023)
16. Gu, Y., Zhang, S., Usuyama, N., et al.: Distilling large language models for biomedical knowledge extraction: a case study on adverse drug events (2023)
17. Guo, B., Zhang, X., Wang, Z., et al.: How close is chatgpt to human experts? Comparison corpus, evaluation, and detection (2023). ArXiv **abs/2301.07597**. https://api.semanticscholar.org/CorpusID:255998637
18. Guo, Y., Qiu, W., Leroy, G., Wang, S., Cohen, T.: Retrieval augmentation of large language models for lay language generation. J. Biomed. Inform. **149**, 104580 (2024). https://doi.org/10.1016/j.jbi.2023.104580
19. Hamed, A.A., Wu, X.: Detection of ChatGPT Fake Science with the xFakeBibs Learning Algorithm (2024). https://doi.org/10.48550/arXiv.2308.11767
20. Han, T., Adams, L.C., Papaioannou, J.M., et al.: MedAlpaca – an open-source collection of medical conversational AI models and training data (2023). https://doi.org/10.3389/mr5g
21. Han, T., Adams, L.C., Papaioannou, J.M., et al.: Medalpaca–an open-source collection of medical conversational AI models and training data (2023). arXiv preprint arXiv:2304.08247
22. Heseltine, M., von Hohenberg, B.C.: Large language models as a substitute for human experts in annotating political text. Res. Politics **11**(1), 20531680241236240 (2024). https://doi.org/10.1177/20531680241236239
23. Imburgio, M.J., Banica, I., Hill, K.E., et al.: Establishing norms for error-related brain activity during the arrow Flanker task among young adults. NeuroImage **213**, 116694 (2020). https://doi.org/10.3389/ggp975
24. Ji, Z., Lee, N., Frieske, R., et al.: Survey of hallucination in natural language generation. ACM Comput. Surv. **55**(12) (2023). https://doi.org/10.1145/3571730

25. Ke, G., Meng, Q., Finley, T., et al.: Lightgbm: a highly efficient gradient boosting decision tree. Adv. Neural. Inf. Process. Syst. **30**, 3146–3154 (2017)
26. Labbé, C., Labbé, D., Portet, F.: Detection of computer-generated papers in scientific literature. In: Degli Esposti, M., Altmann, E.G., Pachet, F. (eds.) Creativity and Universality in Language. LNM, pp. 123–141. Springer, Cham (2016). https://doi.org/10.1007/978-3-319-24403-7_8
27. Liyanage, V., Buscaldi, D., Nazarenko, A.: A benchmark corpus for the detection of automatically generated text in academic publications. In: Calzolari, N., Béchet, F., Blache, P., et al. (eds.) Proceedings of the Thirteenth Language Resources and Evaluation Conference, pp. 4692–4700. European Language Resources Association, Marseille, France (2022). https://aclanthology.org/2022.lrec-1.501
28. Lundberg, S.M., Erion, G., Chen, H., et al.: From local explanations to global understanding with explainable AI for trees. Nat. Mach. Intell. **2**(1), 2522–5839 (2020)
29. Luo, Z., Xie, Q., Ananiadou, S.: The lay person's guide to biomedicine: orchestrating large language models (2024)
30. Maharjan, J., Garikipati, A., Singh, N.P., et al.: OpenMedLM: prompt engineering can out-perform fine-tuning in medical question-answering with open-source large language models (2024)
31. Maruyama, T., Yamamoto, K.: Extremely low resource text simplification with pre-trained transformer language model. In: 2019 International Conference on Asian Language Processing (IALP), pp. 53–58 (2019). https://doi.org/10.3389/mr5d
32. Mitchell, E., Lee, Y., Khazatsky, A., et al.: Detectgpt: zero-shot machine-generated text detection using probability curvature. In: Proceedings of the 40th International Conference on Machine Learning, ICML2023, JMLR.org (2023)
33. Montani, I., Honnibal, M., Honnibal, M., et al.: explosion/spaCy: v3.7.2: fixes for APIs and requirements (2023). https://doi.org/10.5281/zenodo.10009823
34. Mosca, E., Abdalla, M.H.I., Basso, P., Musumeci, M., Groh, G.: Distinguishing fact from fiction: a benchmark dataset for identifying machine-generated scientific papers in the LLM era. In: Ovalle, A., Chang, K.W., Mehrabi, N., et al. (eds.) Proceedings of the 3rd Workshop on Trustworthy Natural Language Processing (TrustNLP 2023), pp. 190–207. Association for Computational Linguistics, Toronto, Canada (2023). https://doi.org/10.3389/10/gtkf4w
35. Muñoz-Ortiz, A., Gómez-Rodríguez, C., Vilares, D.: Contrasting linguistic patterns in human and LLM-generated text (2023)
36. Ochab, J.K., Walkowiak, T.: A pipeline for interpretable stylometric analysis. In: Digital Humanities 2024: Conference Abstracts. George Mason University (GMU), Washington, D.C. (2024)
37. Ordish, J., Hall, A.: Black box medicine and transparency: Interpretable machine learning. PHG Foundation (2020). Accessed 26 Feb 2023
38. Pedregosa, F., Varoquaux, G., Gramfort, A., et al.: Scikit-learn: machine learning in Python. J. Mach. Learn. Res. **12**, 2825–2830 (2011)
39. Qureshi, A.Y., Stevens, R.D.: Mapping the unconscious brain: insights from advanced neuroimaging. J. Clin. Neurophysiol. **39**(1), 12–21 (2022). https://doi.org/10.3389/gtktkw
40. Rebuffel, C., Roberti, M., Soulier, L., Scoutheeten, G., Cancelliere, R., Gallinari, P.: Controlling hallucinations at word level in data-to-text generation. Data Min. Knowl. Discovery **36**(1), 318–354 (2021). https://doi.org/10.1007/s10618-021-00801-4
41. Rubinger, L., et al.: Machine learning and artificial intelligence in research and healthcare. Injury **54**, S69–S73 (2023)

42. Sadasivan, V.S., Kumar, A., Balasubramanian, S., et al.: Can AI-generated text be reliably detected?(2023). ArXiv **abs/2303.11156**. https://doi.org/10.48550/arXiv.2303.11156
43. Shyr, C., Hu, Y., Bastarache, L., et al.: Identifying and extracting rare diseases and their phenotypes with large language models. J. Healthc. Inf. Res. 1–24 (2024). https://doi.org/10.1007/s41666-023-00155-0
44. Singhal, K., Tu, T., Gottweis, J., et al.: Towards expert-level medical question answering with large language models (2023). https://doi.org/10.48550/arXiv.2305.09617. arXiv:2305.09617 [cs]
45. Stribling, J., Krohn, M., Aguayo, D.: SCIgen - An Automatic CS Paper Generator. https://pdos.csail.mit.edu/archive/scigen/
46. Tang, R., Chuang, Y.N., Hu, X.: The science of detecting LLM-generated texts (2023)
47. Team, G., Anil, R., Borgeaud, S., et al.: Gemini: a family of highly capable multimodal models (2023). arXiv preprint arXiv:2312.11805
48. Van Noorden, R.: Publishers withdraw more than 120 gibberish papers. Nature (2014). https://doi.org/10.3389/r3n
49. Villard, C., Dary, Z., Léonard, J., et al.: The origin of pleasant sensations: insight from direct electrical brain stimulation. Cortex **164**, 1–10 (2023). https://doi.org/10.3389/gtkcqm
50. Wang, A., Pang, R.Y., Chen, A., et al.: Squality: building a long-document summarization dataset the hard way. In: Conference on Empirical Methods in Natural Language Processing (2022). https://api.semanticscholar.org/CorpusID:248987389
51. Wiest, I.C., Ferber, D., Zhu, J., et al.: From text to tables: a local privacy preserving large language model for structured information retrieval from medical documents. medRxiv (2023). https://doi.org/10.1101/2023.12.07.23299648
52. World Health Organization: Ethics and governance of artificial intelligence for health: Who guidance. World Health Organization, Guidance (2021)
53. Wu, C., Lin, W., Zhang, X., et al.: PMC-LLaMA: towards building open-source language models for medicine (2023). https://doi.org/10.48550/arXiv.2304.14454. arXiv:2304.14454 [cs]
54. Zaitsu, W., Jin, M.: Distinguishing ChatGPT(-3.5, -4)-generated and human-written papers through Japanese stylometric analysis. PLOS ONE **18**(8), e0288453 (2023). https://doi.org/10.3389/gtkf46
55. Zhang, X., Chen, Y., Hu, S., et al.: ∞bench: Extending long context evaluation beyond 100k tokens (2024)

Machine Learning and Data Assimilation for Dynamical Systems

Explainable Hybrid Semi-parametric Model for Prediction of Power Generated by Wind Turbines

Alfonso Gijón[1(✉)], Simone Eiraudo[3(✉)], Antonio Manjavacas[1], Lorenzo Bottaccioli[3], Andrea Lanzini[3], Miguel Molina-Solana[1,2], and Juan Gómez-Romero[1]

[1] Department of Computer Science and AI, University of Granada, Granada, Spain
alfonso.gijon@ugr.es
[2] Department of Computing, Imperial College London, London, UK
[3] Energy Center Lab, Politecnico di Torino, Turin, Italy
simone.eiraudo@polito.it

Abstract. The ever-growing sector of wind energy underscores the importance of optimizing turbine operations and ensuring their maintenance with early fault detection mechanisms. Existing empirical and physics-based models provide approximate predictions of the generated power as a function of the wind speed, but face limitations in capturing the non-linear and complex relationships between input variables and output power. Data-driven methods present new avenues for enhancing wind turbine modeling using large datasets, thereby improving accuracy and efficiency. In this study, we use a hybrid semi-parametric model to leverage the strengths of two distinct approaches in a dataset with four turbines of a wind farm. Our model comprises a physics-inspired submodel, which offers a reliable approximation of the power, combined with a non-parametric submodel to predict the residual component. This non-parametric submodel is fed with a broader set of variables, aiming to capture phenomena not addressed by the physics-based part. For explainability purposes, the influence of input features on the output of the residual submodel is analyzed using SHAP values. The proposed hybrid model finally yields a 35–40 % accuracy improvement in the prediction of power generation with respect to the physics-based model. At the same time, the explainability analysis, along with the physics grounding from the parametric submodel, ensure deep understanding of the analyzed problem. In the end, this investigation paves the way for assessing the impact, and thus the potential optimization, of several unmodeled independent variables on the power generated by wind turbines.

Keywords: Hybrid semi-parametric models · Explainable AI · Wind Turbines

1 Introduction

The growing use of renewable energies plays a pivotal role in tackling climate change and advancing towards a sustainable energy landscape. Concurrently,

L. Franco et al. (Eds.): ICCS 2024, LNCS 14836, pp. 299–306, 2024.
https://doi.org/10.1007/978-3-031-63775-9_21

the rapid advancements in sensor and storage technologies have facilitated the accumulation of vast amounts of data, coupled with the rise of flexible and powerful data-driven and machine learning methodologies. In this context, the development of accurate and robust wind turbine (WT) models becomes essential for optimizing operations and automatic fault diagnosis.

The absence of precise and robust physics-based models for forecasting power production in utility-scale farms motivates the application of data-driven approaches. While neural networks are traditionally considered black-box models, the emergence of novel architectures capable of adhering to specific constraints, such as physics-informed neural networks (PINNs), improves their capabilities for modeling physical phenomena looking for accurate and robust models [5]. PINNs, although respecting certain physical constraints, still are non-explainable models with complex interpretation. Semi-parametric models are hybrid approaches bringing together physics-based and non-parametric methods. Indeed, they can provide high accuracy while preserving the interpretability of some modeled functional relationship [9].

This work focuses on the modeling of data from the four turbines located within the 'La Haute Borne' wind farm. Our main contribution lies in the development of a unified methodology aimed at effectively integrating physics-based and data-driven models within a common framework. Beyond the improvement in model accuracy, the explainability analysis of input feature importance through SHAP [8] values in the non-parametric submodel provides valuable insights into how input variables influence the output prediction.

This paper is organized as follows. An overview of wind turbine physics and modeling, alongside a description of the hybrid semi-parametric model are exposed in Sect. 2. The main findings are reported and discussed in Sect. 3, and finally, Sect. 4 offers concluding remarks and outlines potential avenues for future research.

2 Computational Methods

2.1 Physical Background

Although challenging, the modeling of the low-scale aerodynamic behaviour of wind turbines can be achieved through physics-based fluid dynamics methodologies. Although the capacity of these models to predict the power generation of utility-scale wind farms is limited [7], partial physical information can be gleaned through the use of well-established equations relating certain high-scale variables. The power extracted by a WT from the kinetic energy of the incoming wind is given by:

$$P = \frac{1}{2} C_p \rho A v^3 , \tag{1}$$

where C_p is the power coefficient, ρ is the air density, A is the area swept by the blades of the WT, and v is the wind velocity.

The power generated by a WT is strongly related to the power coefficient, C_p, a dimensionless parameter accounting for nonlinearities and influenced by the inherent characteristics of the WT (such as its size, geometry and aerodynamic properties), as well as the operational conditions defined by variables such as wind speed and pitch angle. Typically, the objective is to optimize the value of C_p to achieve maximal efficiency in converting wind energy into electrical energy, with a theoretical upper limit of 0.5926, known as the Betz limit [1]. In variable wind speed regions, the optimal power output of WTs is achieved through precise adjustment of the pitch angle, θ, which is defined as the angle between the lateral axis of the blades and the direction of the relative wind. However, the complexity of pitch control for WTs stem from the inherent nonlinear dynamics of these systems and external disturbances. Several empirical formulas have been proposed to model the power coefficient [3,4], but they are not entirely satisfactory to model large amounts of data.

2.2 Hybrid Semi-parametric Model

The structure of hybrid semi-parametric models combines both parametric and non-parametric submodels, based on different sources of knowledge to construct comprehensive representations. In this work, non-parametric models are implemented as neural networks employing a multilayer perceptron architecture, with a flexible number of parameters that are not predetermined by prior knowledge. For our purposes, the prediction of the power is composed of two main components: a physics-inspired part, based on Eq. 1, and a non-parametric part, aimed to predict the residues of the physics-inspired output with respect to the target variable, see Fig. 1.

$$\hat{P} = P_{\text{phys}}(\mathbf{x}) + P_{\text{res}}(\tilde{\mathbf{x}}) \tag{2}$$

The physics-inspired submodel P_{phys} is driven by input variables that are readily interpretable and directly associated with the kinetic-electrical energy conversion process, i.e. wind velocity, pitch angle and rotor angular velocity. Besides, the residual submodel P_{res} uses a broader set of input variables extracted from the dataset, some easily interpretable and others difficult to interpret and incorporate into physical equations:

$$\mathbf{x} = (v, \theta, \omega)\,, \tag{3}$$

$$\tilde{\mathbf{x}} = (v, \theta, \omega, v_1, v_2, T_{\text{out}}, T_{\text{h}}, T_{\text{r}}, T_{\text{n}}, g_{\text{v}}, g_{\text{f}}, \alpha_{\text{n}}, \alpha_{\text{w}}, \alpha_{\text{v}}, \alpha_{\text{wc}}, \alpha_{\text{nc}})\,. \tag{4}$$

From left to right, the variables defining $\tilde{\mathbf{x}}$ are: average wind velocity, pitch angle, rotor angular velocity, first anemometer velocity, second anemometer velocity, outdoor temperature, hub temperature, rotor temperature, nacelle temperature, grid voltage, grid frequency, nacelle angle, wind angle, vane angle, wind angle corrected, nacelle angle corrected.

It is important to note that the physics-inspired submodel, is trained with the power data as target, thereby providing a solid approximation of this quantity. Meanwhile, the residual submodel, further enhances the prediction of the power by integrating corrections derived from unknown physical factors, incorporating variables that better describe the state of the wind turbine, such as the temperatures and orientation of its components. In the physics-inspired submodel,

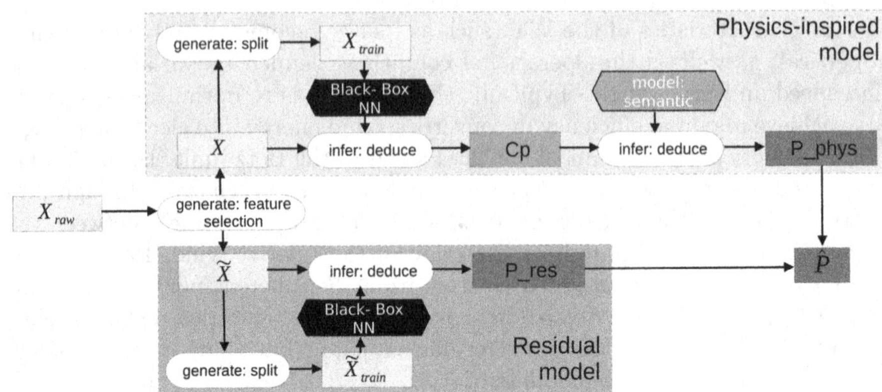

Fig. 1. Diagram of the hybrid model, according to the taxonomy presented in [2]. Rectangular, rounded, and hexagonal boxes represent data, functions, and models, respectively. Yellow and blue boxes are used for inputs and outputs, respectively. (Color figure online)

C_p is predicted by a neural network and then the power is calculated using the semantic model provided by Eq. 1. Then, the residue is calculated as $P - P_{\mathrm{phys}}$ and used as input to train the residual submodel.

3 Results and Discussion

Our experimental setup is based on artificial neural network models, trained using the *Tensorflow* library and hyperparameters optimized with the Hyperband algorithm from the *Keras Tuner* library. All the calculations were carried out in a computer equipped with an 11th Gen Intel Core i7-11800H processor, 16 GB RAM memory and NVIDIA GeForce RTX 3060 graphics card.

Before preprocessing, our dataset consisted of approximately 1 million instances. Initially, non-physical data points, such us those exhibiting a power coefficient higher than the theoretical Betz limit ($C_p > 0.5926$), were eliminated. The calculation of C_p from direct measurements is sensitive to error propagation and only physically allowed values were preserved. Subsequently, we identified anomalous data by comparing the measured power with an estimation derived from the power curve using an iterative median technique. Any data point deviating from the median by more than 3 standard deviations (3σ) was deemed anomalous and removed. Additionally, a low-velocity power cutoff was applied to filter out noisy data at low velocities, where the relative error would have a more significant impact. Consequently, the dataset was reduced from 1 million to $7{\times}10^5$ data, accounting for 70% of the original data.

For training both submodels, the batch size was fixed to 128, employing the mean absolute error as the loss function, and reducing the learning rate when the loss stopped improving. The dataset was randomly split into 80 % for training and 20 % for testing. It was observed that 150 epochs were sufficient to reach a plateau in learning.

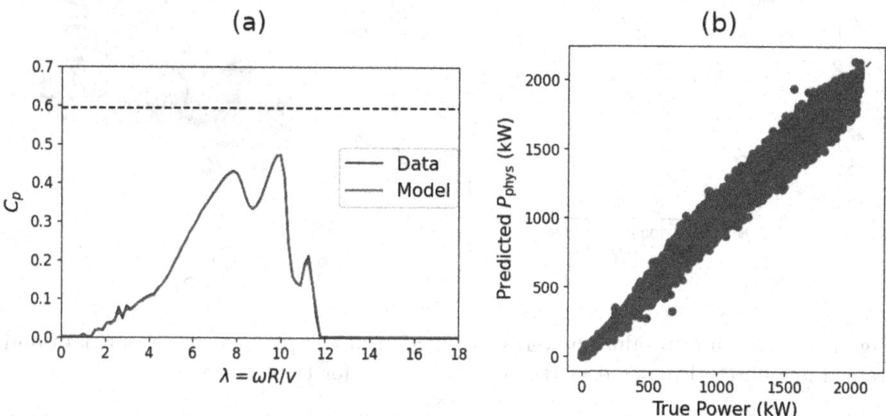

Fig. 2. (a) C_p curve as a function of the tip speed ratio for the physics-inspired model. (b) Predicted vs true values for P_{phys} computed through the intermediate power coefficient.

The hyperparameter search was carried out within the parameter space generated by the combinations of `n_layers` $\in \{1, 2, 4\}$, `n_neurons` $\in \{8, 16, 32, 64, 128\}$, `learning_rate` $\in \{0.01, 0.001, 0.0001\}$ and `activation_function` $\in \{\text{'relu'}, \text{'tanh'}\}$. For the physics-inspired submodel, P_{phys}, the hyperparameter search yielded a 2-layer architecture with 128 units per layer, a learning rate of 0.001, and ReLU as activation function. The regression of the intermediate variable C_p with inputs (v, θ, ω) is depicted in Fig. 2(a) as a function of the tip speed ratio, $\lambda = \omega R/v$. A sigmoid output layer was employed to restrict the output within a fixed range of [0,1], which was subsequently converted to original units to respect the Betz constraint. Figure 2(b) illustrates the comparison between predicted and true values of P_{phys} for the test dataset. The physics-inspired submodel achieves satisfactory performance by itself, demonstrating a mean absolute error (MAE) of 16.3 kW and a mean absolute percentage error (MAPE) of 3.71 %, as indicated in Fig. 1.

The neural network architecture utilized in the residual submodel, P_{res}, was identical to that of the physics-inspired model. In this case, the input comprises 16 variables used to make a prediction of the residual. As shown in Fig. 3(a), the absolute residual power exhibits elevated levels within the medium power range (800–1800 kW), while remaining lower at both low and high power extremes. Despite being quite dispersed around the true value, the predicted residual power is adequately estimated, as represented in Fig. 3(b).

The resulting hybrid model obtained by combining the physics-inspired and residual submodels surpasses the performance metrics to predict the generated power, achieving a MAE of approximately 10.5 kW and a MAPE of 2.16 %. As

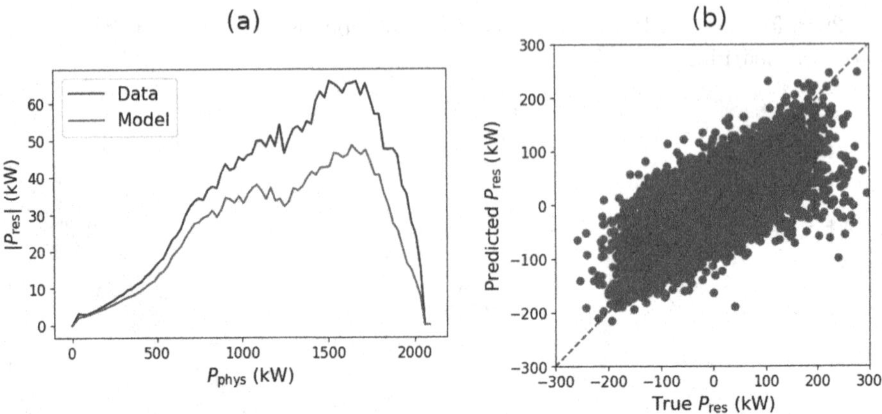

Fig. 3. (a) Absolute residual power as a function of the physical power for data and trained model. (b) Predicted vs true residual power for the test dataset.

Table 1. Comparison of the performance metrics for different models. The PINN column is taken from reference [6].

	PINN	Phys	Hybrid	Hybrid_CF
MAE (kW)	15.55	16.31	**10.51**	10.89
RMSE (kW)	28.02	30.63	**22.07**	22.74
MAPE (%)	3.838	3.706	**2.159**	2.203
R2 score	0.9960	0.9953	**0.9976**	0.9974

shown in Table 1, this represents an enhancement of approximately 35–40 % in the regression task, when compared to the physics-inspired submodel of this study or the physics-informed model of reference [6].

The physics-inspired part of our model is entirely interpretable, as it employs a physical equation and an intermediate variable constrained within a specific range. In contrast, the residual component operates as a black box, receiving 16 input variables and generating a prediction without explicit interpretability. As a first step in the explainability analysis, a linear correlation filter was applied to eliminate redundancies among the input variables. Specifically, those variables having a correlation coefficient exceeding 0.95 were excluded, resulting in a reduced set comprising 12 variables. The reduced hybrid model nearly achieves the same performance as the original one, as can be seen in Table 1.

To gain some insights into the relative importance of each input feature on the output, we show the mean absolute value of the SHAP values in Fig. 4(a). As expected, the variables incorporated in the physics-inspired model (v, θ, w) are the most influential. However, the direction of the wind, as well as the orientation of the nacelle, demonstrate considerable influence, motivating a deeper analysis and suggesting a potential optimization to increase electrical power generation under the same external wind speed and temperature conditions. An analysis

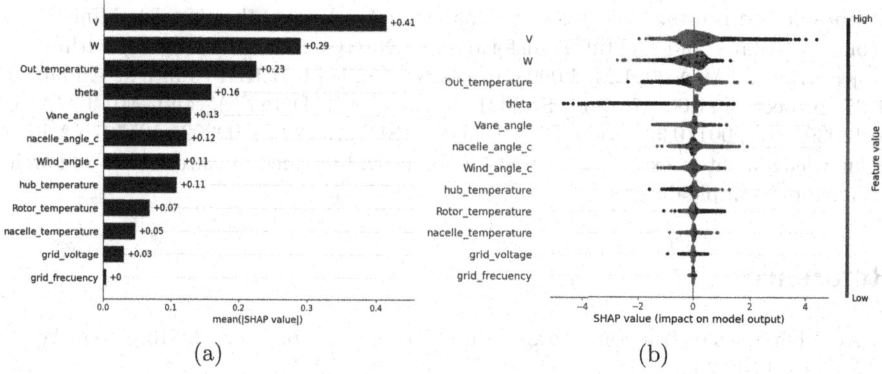

Fig. 4. (a) Relative importance of each feature on the output. (b) Distribution of the impact of each feature on the model output. Color represents feature value.

of the impact distribution of each feature in Fig. 4(b) reveals, for instance, that higher vane angle and wind angle contribute positively to power generation, while a higher nacelle angle and hub temperature exhibit a negative contribution.

4 Conclusions

In this study, we have designed and validated a hybrid semi-parametric model composed by a physics-inspired submodel and a non-parametric submodel, P_{phys}, for predicting the residual power of the physical term, P_{res}. The developed model, trained using real historical data of four turbines from a wind farm, results in an improvement of approximately 35–40 % in predicting the generated power. The physics-inspired submodel is inherently explainable due to its construction, leveraging a physical equation that relates the most critical variables of the system. However, the non-parametric residual submodel requires the analysis of SHAP values to comprehend the relative importance of the input features and their impact on the output power value. Our results suggest that certain angular variables could be adjusted to achieve higher power production.

It is noticeable that our methodology is versatile and can be applied to a wide range of problems where a physics-based model is available, offering approximate results, and additional data can be leveraged by a non-parametric data-driven submodel to predict the residual component and incorporate unknown physics.

Once deployed, this hybrid model could serve as an accurate regression-based anomaly detection method by comparing the deviation of new data from the model's prediction for a healthy state. All the models presented here are fully differentiable, enabling their utilization for developing optimal pitch angle controllers, thereby optimizing power generation across various wind speed regimes. While this hybrid model shows promise, further research is needed to asses their robustness across different turbines.

Acknowledgements. This work was partially funded by the Spanish Ministry of Economic Affairs and Digital Transformation (NextGenerationEU funds) within the project IA4TES MIA.2021.M04.0008. It was also funded by ERDF/Junta de Andalucía (D3S project P21.00247, and SE2021 UGR IFMIF-DONES), and MICIU/AEI/ 10.13039/501100011033 and EU ERDF (SINERGY, PID2021.125537NA.I00). Acknowledgement is extended to ENGIE for providing such an interesting and well-documented sample dataset.

References

1. Aerodynamics of Horizontal Axis Wind Turbines, chap. 3, pp. 39–136. John Wiley & Sons, Ltd (2011)
2. van Bekkum, M., de Boer, M., van Harmelen, F., Meyer-Vitali, A., Teije, A.t.: Modular design patterns for hybrid learning and reasoning systems. Appl. Intell. **51**(9), 6528–6546 (2021)
3. Carpintero-Renteria, M., Santos-Martin, D., Lent, A., Ramos, C.: Wind turbine power coefficient models based on neural networks and polynomial fitting. IET Renew. Power Gener. **14**(11), 1841–1849 (2020)
4. Castillo, O.C., Andrade, V.R., Rivas, J.J.R., González, R.O.: Comparison of power coefficients in wind turbines considering the tip speed ratio and blade pitch angle. Energies **16**(6), 2774 (2023)
5. de la Mata, F.F, Gijón, A., Molina-Solana, M., Gómez-Romero, J.: Physics-informed neural networks for data-driven simulation: advantages, limitations, and opportunities. Physica Stat. Mech. Appl. **610**, 128415 (2023)
6. Gijón, A., Pujana-Goitia, A., Perea, E., Molina-Solana, M., Gómez-Romero, J.: Prediction of wind turbines power with physics-informed neural networks and evidential uncertainty quantification (2023). Arxiv: 2307.14675
7. Howland, M.F., Dabiri, J.O.: Wind farm modeling with interpretable physics-informed machine learning. Energies **12**(14), 2716 (2019)
8. Lundberg, S.M., Lee, S.I.: A unified approach to interpreting model predictions. In: Guyon, I., et al. (eds.) Advances in Neural Information Processing Systems 30, pp. 4765–4774. Curran Associates, Inc. (2017)
9. von Stosch, M., Oliveira, R., Peres, J., Feyo de Azevedo, S.: Hybrid semi-parametric modeling in process systems engineering: past, present and future. Comput. Chem. Eng. **60**, 86–101 (2014)

State Estimation of Partially Unknown Dynamical Systems with a Deep Kalman Filter

Erik Chinellato and Fabio Marcuzzi(✉)

Department of Mathematics "Tullio Levi Civita", University of Padova, Via Trieste 63, 35121 Padova, Italy
marcuzzi@math.unipd.it

Abstract. In this paper we present a novel scientific machine learning reinterpretation of the well-known Kalman Filter, we explain its flexibility in dealing with partially-unknown models and show its effectiveness in a couple of situations where the classic Kalman Filter is problematic.

Keywords: Kalman Filter · scientific machine learning · deep unfolding · deep learning

1 Introduction

Often, in data-driven model discovery and in partially-unknown-model learning from experimental data, a limited set of measured variables is given and one aims at a model that describes a more complex process than that directly represented by measured variables. Typically, this means to include in the estimate an additional set of unmeasurable variables that give a more extended representation of the true system's dynamics. As an example, that we will use in this paper as a model problem, we measure some temperatures on the boundary but we want our model to represent also the behaviour of the internal temperature field. This is paradigmatic for a great number of computational intensive applications, where one needs to indirectly measure constants and variables that have a precise physical meaning (e.g. mechanical, thermal, etc.) from data obtained by measurements of other physical quantities (e.g. displacements, temperatures, etc.) or in different locations. This estimate represents a virtual measurement of these variables and, if performed by an algorithm in real-time, it is usually called a *soft-sensor* [3]. When these variables belong to the state-vector of a dynamical system, it is common to adopt a state estimator and quite often this is the Kalman Filter (KF) [7,9]. It is a predictor-corrector algorithm where the predictor is based on a reference model, which can be black-box or conveniently a physico-mathematical model; the corrector acts on the basis of the description of the modelling and measurement errors, represented by their covariance

The original version of the chapter has been revised. The references 3 and 12 have been corrected. A correction to this chapter can be found at https://doi.org/10.1007/978-3-031-63775-9_33

L. Franco et al. (Eds.): ICCS 2024, LNCS 14836, pp. 307–321, 2024.
https://doi.org/10.1007/978-3-031-63775-9_22

matrices, and must operate a few matrix inversions. Actually, the Kalman Filter is often used also to estimate unknown inputs and/or parameters of the model; this is frequently done through a proper state-augmentation, see e.g. [12].

The aim of this paper is to present a state/parameters estimation method that inherits the structure of the Kalman Filter but it is blended with data-driven model discovery, and we call it the Deep Kalman Filter.

Let us see a couple of motivations that lead us to combine the Kalman Filter with machine learning, and in particular neural/back-propagation learning. First, we are interested in obtaining less demanding algorithms, suitable e.g. to be executed in real-time on embedded systems [3], while simultaneously maintaining a satisfactory interpretation of the physico-mathematical model properties and the inferred statistical description of the model errors, which is not the case with plain neural computing. In this regard, it can be convenient to substitute covariance matrices determination and numerical inversion with neural learning, an idea that has been already presented in [14] but here we formulate in a quite different way. Secondly, we strive to develop a state-estimation algorithm that allows to include general model uncertainties (also deterministic ones, even nonlinear) that can be learned from data together with state estimation, since both linearity of the reference state-space model and accurate knowledge of it are often not encountered in practice.

There is an increasing literature about combining Kalman filtering and neural networks. In particular, we refer to KalmanNet [14], where the overall information required to generate the Kalman gain is learned by a Recurrent Neural Network. Here we propose a novel, highly interpretable learning scheme for Kalman filtering inspired by the *unfolding* technique [8], originally invented in the context of nonnegative matrix factorizations. Similar approaches can be found also in the Data Assimilation (DA) context; in the taxonomy developed in a recent review [2], the Deep Kalman here presented is an end-to-end learning scheme for the whole DA system and, precisely, a "Sequential-DA-inspired neural scheme" that employs the *unfolding* technique instead of a Recurrent Neural Network, thus providing a higher degree of interpretability to KF operations.

The paper is organized as follows: in Sect. 2 we briefly recall the Kalman Filter equations. in Sect. 3 we present the Deep Kalman Formulation and its flexibility to deal with partially unknown models. In Sect. 4 we present a few relevant experiments and a Conclusions section ends the paper.

2 Kalman Filter Estimation

Let us consider a parametric Discrete, Linear, Time-Invariant (DLTI) dynamical system in state-space form:

$$\begin{aligned} x(k+1) &= A(k,p)x(k) + B(k,p)u(k) + v(k) \\ y(k) &= Cx(k) + w(k) \end{aligned} \tag{1}$$

where $x \in \mathbb{R}^n$ is the state vector, $y \in \mathbb{R}^m$ is the observation vector, and the dependence of A and B from p is in general nonlinear. This would arise, for example, from the discretization with an implicit method of a physico-mathematical model (see an example in Sect. 4). To estimate the state vector from measurement data it is a common choice to adopt a Kalman Filter, that we recall here in its one-step version [9]:

$$P(k) = \left[\left(Q(k-1) + A(k-1,p)P(k-1)A(k-1,p)^\top\right)^{-1} + C^\top R^{-1} C\right]^{-1} \tag{2}$$

$$e_{pred} = C\,(\,A(k-1,p)\,\hat{x}(k-1) + B(k,p)\,u(k-1)\,)\; - \; y(k) \tag{3}$$

$$\delta\hat{x}(k) = -P(k)\;C^\top R^{-1} e_{pred} \tag{4}$$

$$\hat{x}(k) = A(k-1,p)\;\hat{x}(k-1) + B(k-1,p)\;u(k-1) + \delta\hat{x}(k) \tag{5}$$

From Eq. (4) it is evident that the KF operates a proportional feedback action on the output prediction error e_{pred}, with a gain:

$$\mathcal{K}_G^{(k)} = -P(k)\;C^\top R^{-1} \tag{6}$$

3 The Deep Kalman Formulation

Let us generalize Eqs. (3) and (4) for a nonlinear state-space system. Given inputs $\{u(k)\}_{k=0,\ldots,N-1}$, measurements $\{y(k)\}_{k=1,\ldots,N}$, and an evolution map $f = f(x,p,u)$ depending on a state x, unknown dynamics vector p and input u, we have:

$$\begin{aligned} x(k+1) &= f\,(x(k),p,u(k)) + v(k) \\ y(k) &= Cx(k) + w(k) \end{aligned} \tag{7}$$

For an initial state estimate $\hat{x}(0)$, the forward run of the filter is thus given by:

$$\hat{x}(k) = \hat{x}(k\,|\,k-1) + \mathcal{K}_G^{(k)}\,(\,y(k) - C\hat{x}(k\,|\,k-1)\,) \quad \forall k = 1,\ldots,N \tag{8}$$

where $\hat{x}(k\,|\,k-1) = f\left(\hat{x}(k-1), p^{(k)}, u(k)\right)$ is the k-th state prediction, C is the observation matrix, $\mathcal{K}_G^{(k)}$ is the k-th Kalman gain matrix and $p^{(k)} \equiv p$ is a shared vector modelling unknown dynamics.

Let us now *unfold* these equations and *untie* the Kalman gains. More precisely, the unfolding process considers the k-th state-update of Eq. (8) as the action of the k-th layer of a neural network. The untying then regards the Kalman gain matrices $\mathcal{K}_G^{(k)}$ as the network's weights, which are therefore no longer tied together by the recursive relation (2). The resulting scheme is shown in Fig. 1: each grey rectangle represents one layer of the network, embedding Eq. (8). Note that this formulation is quite simpler than the one proposed in [14] and does not require the knowledge of the measured state vector at each time instant during the training phase, which is a hard constraint for many applications, but rather the only required quantity is the target final state $x(N)$. Also, there is no explicit learning of the initial condition $x(0)$, but the covariance of the initial-state estimation error $P(0)$ (see Eqs. (6) and (2)) is learned and embedded directly in the first parameter matrix $\mathcal{K}_G^{(1)}$ during training.

The weights of the proposed network also include the vector $p^{(k)} \equiv p$.

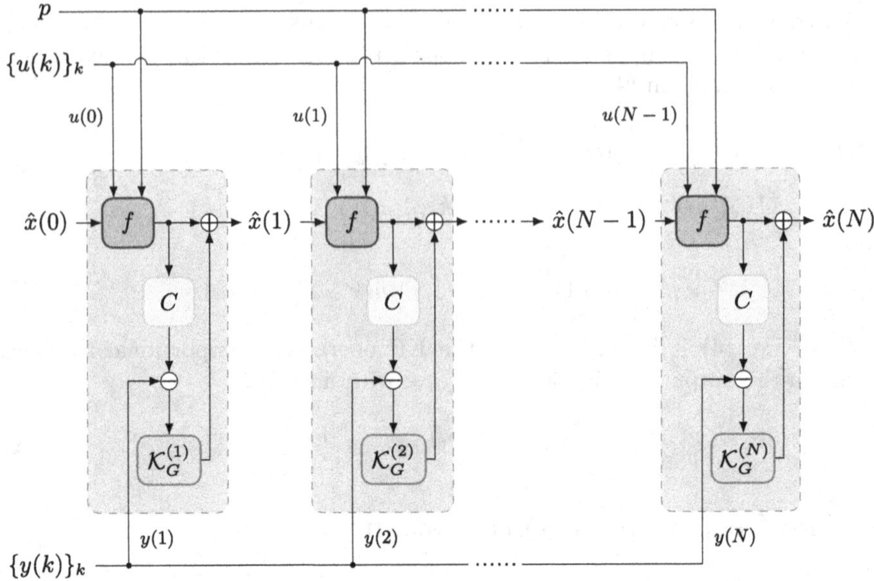

Fig. 1. Structure of the DeepKalman network.

3.1 Different Parametrizations for $\mathcal{K}_G^{(k)}$

The most straightforward formulation for the parametrization of the gain matrices $\mathcal{K}_G^{(k)}$ is that each component is a weight of the net. Another possible network architecture that provides more structure to the Kalman gain matrices and is more closely related to the original Kalman Filter algorithm can be obtained by splitting the former as:

$$\mathcal{K}_G^{(k)} = C_x^{(k)} C^\top C_y^{(k)} \quad \forall k = 1, \ldots, N \tag{9}$$

where $C_x^{(k)}$ plays the role of the state covariance matrix and $C_y^{(k)}$ the inverse of the measurement covariance matrix. Since the positive definiteness of these matrices should be preserved, this alternative architecture replaces the weights $\{\mathcal{K}_G^{(k)}\}_{k=1,\ldots,N}$ with $\{L_x^{(k)}\}_{k=1,\ldots,N}$ and $\{L_y^{(k)}\}_{k=1,\ldots,N}$ where $L_x^{(k)}$, $L_y^{(k)}$ are lower triangular matrices satisfying:

$$C_x^{(k)} = L_x^{(k)} L_x^{(k)^\top} \qquad C_y^{(k)} = L_y^{(k)} L_y^{(k)^\top} \tag{10}$$

If $\mathcal{K}_G^{(k)} \in \mathbb{R}^{n \times m}$ and $p \in \mathbb{R}^q$, then the number of parameters in this second architecture increases from $nmK + q$ to $\frac{1}{2}\left[n(n+1) + m(m+1)\right]K + q$.

In both cases the initialization of the parameters can be made from the expression of $\mathcal{K}_G^{(k)}$ (6) and Eq. (2) (A is a linearization of f in this case). This is really effective, as we will show in the experiments of Sect. 4, and provides satisfactory interpretability to the learned parameters $\mathcal{K}_G^{(k)}$.

3.2 Backpropagation Details

Let us now give some details on the backpropagation algorithm in the case of the architecture in Eq. (8). Given a desired final state $x(N)$ we consider the loss function:

$$\mathcal{L} = \frac{1}{2}\|\hat{x}(N) - x(N)\|_2^2 + \frac{\lambda}{2}\sum_{l=1}^{N} \|C\hat{x}(l) - y(l)\|_2^2 = \mathcal{E} + \sum_{l=1}^{N} \mathcal{F}_\lambda^l \tag{11}$$

which controls both the reconstructed state as well as the intermediate measurements at each layer.

As a consequence, one obtains the recursive relations:

$$\begin{cases} \nabla_{\hat{x}(N)}\mathcal{E} &= \hat{x}(N) - x(N) \\ \nabla_{\hat{x}(l)}\mathcal{F}_\lambda^l &= \lambda C^\top (C\hat{x}(l) - y(l)) \quad \forall l = 1,\dots,N \\ \nabla_{\hat{x}(k)}\mathcal{E} &= \left[\left(\mathbb{1} - \mathcal{K}_G^{(k+1)}C\right)\partial_x f\left(\hat{x}(k), p^{(k+1)}, u(k+1)\right)\right]^\top \nabla_{\hat{x}(k+1)}\mathcal{E} \\ & \quad \forall k = 1,\dots,N-1 \\ \nabla_{\hat{x}(k)}\mathcal{F}_\lambda^l &= \left[\left(\mathbb{1} - \mathcal{K}_G^{(k+1)}C\right)\partial_x f\left(\hat{x}(k), p^{(k+1)}, u(k+1)\right)\right]^\top \nabla_{\hat{x}(k+1)}\mathcal{F}_\lambda^l \\ & \quad \forall k = 1,\dots,l-1, \quad \forall l = 1,\dots,N \end{cases}$$

with which to compute the partial gradients:

$$\begin{cases} \nabla_{\mathcal{K}_G^{(k)}}\mathcal{E} &= \nabla_{\hat{x}(k)}\mathcal{E}\left(y(k) - Cf\left(\hat{x}(k-1), p^{(k)}, u(k)\right)\right)^\top \\ & \quad \forall k = 1,\dots,N \\ \nabla_{\mathcal{K}_G^{(k)}}\mathcal{F}_\lambda^l &= \nabla_{\hat{x}(k)}\mathcal{F}_\lambda^l\left(y(k) - Cf\left(\hat{x}(k-1), p^{(k)}, u(k)\right)\right)^\top \\ & \quad \forall k = 1,\dots,l-1, \quad \forall l = 1,\dots,N \\ \nabla_{p^{(k)}}\mathcal{E} &= \left[\left(\mathbb{1} - \mathcal{K}_G^{(k)}C\right)\partial_p f\left(\hat{x}(k-1), p^{(k)}, u(k)\right)\right]^\top \nabla_{\hat{x}(k)}\mathcal{E} \\ & \quad \forall k = 1,\dots,N \\ \nabla_{p^{(k)}}\mathcal{F}_\lambda^l &= \left[\left(\mathbb{1} - \mathcal{K}_G^{(k)}C\right)\partial_p f\left(\hat{x}(k-1), p^{(k)}, u(k)\right)\right]^\top \nabla_{\hat{x}(k)}\mathcal{F}_\lambda^l \\ & \quad \forall k = 1,\dots,l-1, \quad \forall l = 1,\dots,N \end{cases}$$

Adding together all these contributions we obtain the overall gradients for our proposed architecture:

$$\begin{cases} \nabla_{\mathcal{K}_G^{(k)}}\mathcal{L} &= \nabla_{\mathcal{K}_G^{(k)}}\mathcal{E} + \sum_{l=1}^{N}\nabla_{\mathcal{K}_G^{(k)}}\mathcal{F}_\lambda^l \quad \forall k = 1,\dots,N \\ \nabla_p\mathcal{L} &= \sum_{k=1}^{N}\left(\nabla_{p^{(k)}}\mathcal{E} + \sum_{l=1}^{N}\nabla_{p^{(k)}}\mathcal{F}_\lambda^l\right) \end{cases}$$

and the weight update relation for some learning rates $\mu_\mathcal{K}$, μ_p is given by:

$$\begin{cases} \mathcal{K}_G^{(k)} & \Leftarrow \mathcal{K}_G^{(k)} - \mu_{\mathcal{K}} \nabla_{\mathcal{K}_G^{(k)}} \mathcal{L} \quad \forall k = 1, \ldots, N \\ p & \Leftarrow p - \mu_p \nabla_p \mathcal{L} \end{cases} \tag{12}$$

In practice, an optimizer is used in place of the above stochastic gradient descent. For our implementation we adopted Adam [10] with moment parameters β_1 and β_2 depending on the particular experiment being carried out. The learning rates $\mu_{\mathcal{K}}$ and μ_p also depend on the experiment and are adaptively reduced based on the training residue.

As far as the Kalman gains are concerned, we will expand on their initialization in a later section. The unknown dynamics vector p on the other hand is initialized according to its physical meaning in order to not introduce biases.

3.3 Alternative Formulations and Extensions for the Map *f*

One of the main approaches used to define the map f of (7) is to employ a parametric model governed by differential equations and discretize it, as will be done in our model problem of Sect. 4. Possibly, a surrogate model can be used, like e.g. a variationally mimetic operator network (VarMiON) [13].

Since the Kalman Filter operates a mere proportional control of the state-estimation error, there are classes of problems where this gives poor performances and, for example, an additional feed-forward term improves substantially the results of the predictor; see e.g. [12] for an adaptive feed-forward with a gain chosen according to the maximum principle for the heat equation; even this sophisticated term can be actually implemented in the map f of (7) without modifying the interpretation of the Deep Kalman formulation.

The map f can also be extended with a generic, parametric term to do data-driven model discovery during training, or even an ensemble of extensions. This can be done e.g. with the approach of sparse identification of nonlinear dynamics (SINDy) [1] and e-SINDy [5], i.e. to maintain the interpretability of the extension as an explicit function of the state variables. Simply, if the known part of the model is given by a linear system of differential equations like:

$$M\dot{x} = Kx + d \tag{13}$$

and a collection of states $X = [\,x(1) \cdots x(T)\,] \in \mathbb{R}^{n \times T}$ is provided (e.g. the states estimated by the Deep Kalman at each layer, so that $T = N$) together with their derivatives $\dot{X} = [\,\dot{x}(1) \cdots \dot{x}(T)\,] \in \mathbb{R}^{n \times T}$, which can be approximated numerically, we can build an appropriate (overcomplete) dictionary $\Phi(X) \in \mathbb{R}^{T \times L}$ and do sparse regression on:

$$\left(M\dot{X} - KX - D\right)^{\top} = \Phi(X)s \tag{14}$$

where $D = [\,d(1) \cdots d(T)\,] \in \mathbb{R}^{n \times T}$ and $s \in \mathbb{R}^{L \times n}$ is the sparse solution matrix. Finally, we can actually set the map f of the Deep Kalman Filter as the discretization of the data-driven extended model:

$$M\dot{x} = Kx + d + (\,\Phi(x)s\,)^{\top} \tag{15}$$

where now $\Phi(x) \in \mathbb{R}^{1\times L}$. Note that this approach can be applied to partially unknown nonlinear models as well.

In summary, if we lack an accurate model of the underlying dynamics and we want to improve it by learning from observations, we have briefly described how the Deep Kalman can be a framework embracing three core categories of unknown dynamics: (a) parametric dynamical models with unknown parameter values; (b) complex/unknown dynamics captured by interpretable surrogate models, e.g. implemented using neural networks; and (c) inaccurate or partially-known dynamical models that can be improved using data-driven extensions.

Finally, note that data-driven extensions can be physically interpretable even when they describe a fictitious term in the model, if it has been proven to be equivalent to a physical model property which is difficult to formulate or to estimate; see e.g. how the estimation of inner cavities (a nonlinear geometric inverse problem) can be reformulated as the estimation of fictitious heat sources (linear inverse problem) in [6].

4 Numerical Experiments

As a model problem, let us consider the heat equation:

$$\begin{cases} \rho C\, \partial_t T &= \kappa\, \Delta T + f_\vartheta \quad \text{in } D \times [0, t_f] \\ \kappa\, \nabla T \cdot \mathbf{n}_S &= q(t) \quad \text{on } S \times [0, t_f] \\ \kappa\, \nabla T \cdot \mathbf{n} &= 0 \quad \text{on } (\delta D \setminus S) \times [0, t_f] \\ T(0, \cdot) &= T_0(\cdot) \quad \text{in } D \end{cases} \tag{16}$$

where $D(x, y) = [0, 1] \times [0, L]$ is a 2-dimensional domain, $S = \{(x, 0) : x \in [0, 1]\}$ is the measurable border, ρC is the heat capacity of the material, κ is its thermal conductivity and $\mathbf{n}_S$, $\mathbf{n}$ are the outward normal vectors to S, $\delta D \setminus S$ respectively. The restriction to a 2D problem is only for simplicity, the method here proposed can be used in higher dimensions.

Let us discretize problem (16) in space using the Finite Element Method (FEM) with Lagrangian elements P1, i.e. piecewise first-degree polynomials, and in time with the implicit Euler method, so that at iteration k we get:

$$\begin{aligned} M(\rho C)\, \frac{\tilde{T}_k - \tilde{T}_{k-1}}{dt} &= K(\kappa)\tilde{T}_k + f_k \\ \Rightarrow \left(\mathbb{1} - dt\, M(\rho C)^{-1} K(\kappa)\right) \tilde{T}_k &= \tilde{T}_{k-1} + dt\, M(\rho C)^{-1} f_k \end{aligned}$$

where $M(\rho C) \in R^{n\times n}$ and $K(\kappa) \in R^{n\times n}$ are the mass and stiffness matrices of the FEM discretization, dt is the time step chosen in the time-discretization and f_k is the heat source at time t_k.

Since we can measure only a few components of the temperature field T, we reformulate this model as a state-space dynamical system with the aim of estimating its state from output measurements. Let us consider the following state-space discrete model in physical coordinates:

$$\begin{aligned} x_m(k+1) &= A_m x_m(k) + B_m u_m(k) + v_m(k) \\ y_m(k) &= C_m x_m(k) + w(k) \end{aligned} \tag{17}$$

where $x_m(k) = \tilde{T}_k$ is the state-vector, $u_m(k) = f_k$ the input, $v_m(k)$ and $w(k)$ are, respectively, the model and measurement errors, supposed i.i.d. Gaussian processes, C_m is a matrix built with the rows of the identity matrix corresponding to measured nodes, and lastly:

$$\begin{aligned} A_m &= \left(\mathbb{1} - dtM\,(\rho C)^{-1}\,K(\kappa)\right)^{-1} \\ B_m &= A_m dtM\,(\rho C)^{-1} \end{aligned}$$

Now, to estimate the state vector we apply Eqs. (2), (3), (4) and (5).

In this section, some numerical experiments are described to give a practical evidence of the algorithmic ideas previously presented. In all the examples, experimental temperatures are simulated numerically, while intensive tests have been done in our previous work to experimentally validate the model settings (see [4]). A note about "inverse crimes": here we are solving the inverse problem using the same model that has generated the data, which is considered an inverse crime. Actually, we are interested in analyzing the algebraic operations made during the reconstruction, which are non trivial, and the simplified setting we use is adequate to make significant comparisons, see the interesting discussion in the white-paper of Wirgin [15]. In a real, specific application one should then consider also model and measurement errors, to validate the practical accuracy of his method in the specific application. In the next subsection we provide an example of the robustness of Deep Kalman to model noise.

Let us describe the model settings: $t_f = 1.51$ s, $L = 0.1$ m, $\rho C = 3.2 \times 10^6\,\mathrm{J/(m^{3\circ}C)}$, $k = 3.77 \times 10^3\,\mathrm{W/(m^\circ C)}$, and

$$q(t) = \frac{Wt}{\sigma_q^2} e^{-\frac{\sqrt{t}}{\sigma_q}},\ t \in (0, t_f] \tag{18}$$

where $\sigma_q = 1.06 \times 10^{-2}, W = 2.9511 \times 10^7$ J. The initial condition is set to $T_{0(\cdot)} = 20$ °C. In this section an Implicit Euler method is adopted for the time discretization, using a temporal step $\Delta t = 0.0005$ s in $(0, 0.1]$ and $\Delta t = 0.05$s in $(0.1, t_f]$. A P_1-FE method is used for space discretization, whose step length along y is $h_y = 0.01$m and $h_x = h_y$. The sensors are supposed to be in the middle of each mesh edge in the instrumented boundary segment. Numerical experiments have been carried out using MATLAB. As a general forcing term we have used a gaussian forcing term f_ϑ with arbitrary variance and point of application.

In the following subsections we see a couple of relevant parametrizations for just as many inverse heat transfer problems.

4.1 Example: Discovering a Distributed Forcing Term

Let us suppose that the heat source term f_ϑ is an unknown function, in general, except that it is assumed different from zero only in a few disconnected regions of compact support. This is a common situation in many applications. In this example, we estimate f_ϑ from a limited number of temperature measurements $\tilde{T}_{(f_\vartheta)}$, taken at the boundary. The estimate of f_ϑ can be seen as an indirect measurement of f_ϑ from physical/direct temperature measurements and, with this interpretation, it is usually called a *soft-sensor* [3]. This is quite a difficult problem for Kalman filtering and, more precisely, for the Augmented Kalman Filter (AKF): see [12] and Fig. 2, where it is clear that the standard AKF misses completely the right location, shape and intensity of the physical forcing term.

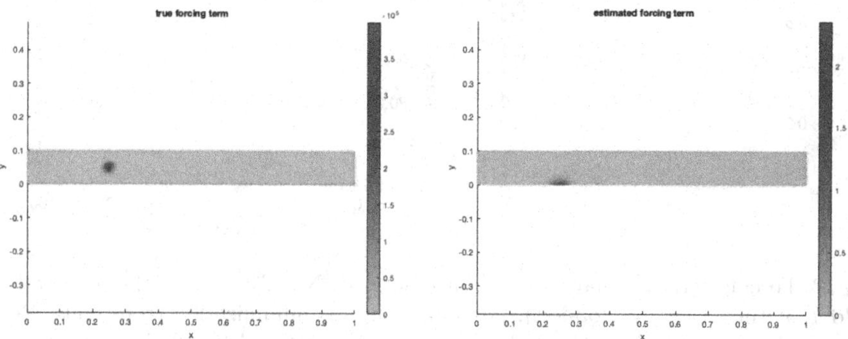

Fig. 2. Comparison between the true forcing term (left) and estimated (right) by the traditional (augmented) Kalman Filter algorithm.

With the method here presented, we set the unknown-dynamics vector of parameters p of Sect. 3 to $u_m(k)$ in the reference model (17) and $W = 0$ in (18). Moreover, the training/discovery process was carried out using $\beta_1 = 0.9$, $\beta_2 = 0.999$ and $\mu_{\mathcal{K}} = 5 \cdot 10^{-3}$, $\mu_p = 6 \cdot 10^{-3}$. We performed this experiment in the ideal noiseless setting as well as in the presence of model noise significantly corrupting the measured nodes' temperatures. The added model noise is assumed to be an i.i.d. Gaussian process with zero mean and variance $\sigma^2 = 1 \cdot 10^{-4}$. See Fig. 3 and Fig. 4 respectively for the results. In both cases the Deep Kalman algorithm is able to estimate the forcing term with high accuracy. Note that in the case with model noise, despite the relatively high output-prediction error, the Deep Kalman is able to estimate with a good precision the internal temperature field.

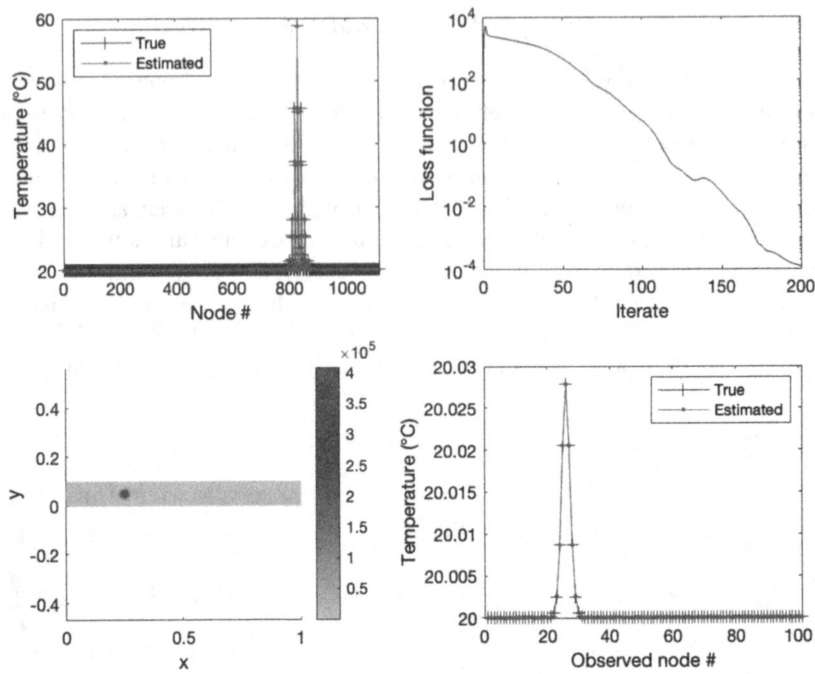

Fig. 3. Forcing term estimation in absence of noise. Nodal values of the temperature field T at time $T_k = N$ (top-left). Loss-function (11) minimization (top-right). Rectangular domain with the estimated source term f_ϑ (bottom-left). Nodal temperatures at the measured boundary S (bottom-right).

4.2 Example: Discovering Internal Material Properties

Let us suppose that the ρC, the heat capacity of the material, is an unknown function of (x, y) and, in particular, it is constant along x. This is an interesting situation in applications. In this example, we estimate $(\rho C)(y)$ from a limited number of temperature measurements $\tilde{T}_{(\rho C)}$, taken at the boundary. It is another example of *soft-sensor* [3] and a quite difficult problem for Kalman filtering, see in [11] an analogous application to the estimation of the internal material stiffness through mechanical vibration experiments.

With the method here presented, we set the unknown-dynamics vector of parameters p of Sect. 3 to the values of a piecewise constant approximation of $(\rho C)(y)$, defined according to the previously assumed discretization of the domain. Moreover, the training/discovery process was carried out using $\beta_1 = 0.05$, $\beta_2 = 0.059$ and $\mu_\mathcal{K} = 5 \cdot 10^{-3}$, $\mu_p = 1 \cdot 10^{-2}$. The Deep Kalman algorithm is able to estimate these parameters p with high accuracy, see Fig. 5.

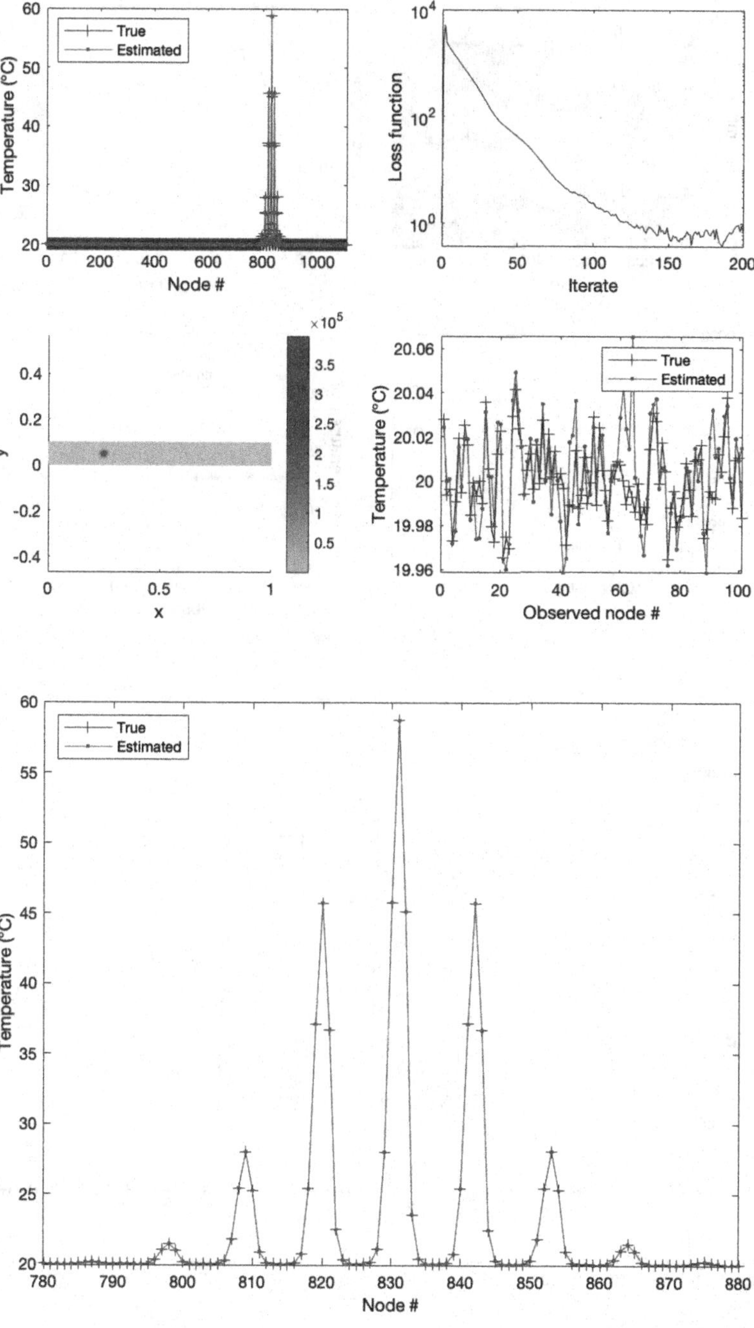

Fig. 4. Forcing term estimation in presence of model noise. Nodal values of the temperature field T at time $T_k = N$ (top-left). Loss-function (11) minimization (top-right). Rectangular domain with the estimated source term f_ϑ (center-left). Nodal temperatures at the measured boundary S (center-right). Zoom on the top-left figure from node 780 to 880 (bottom).

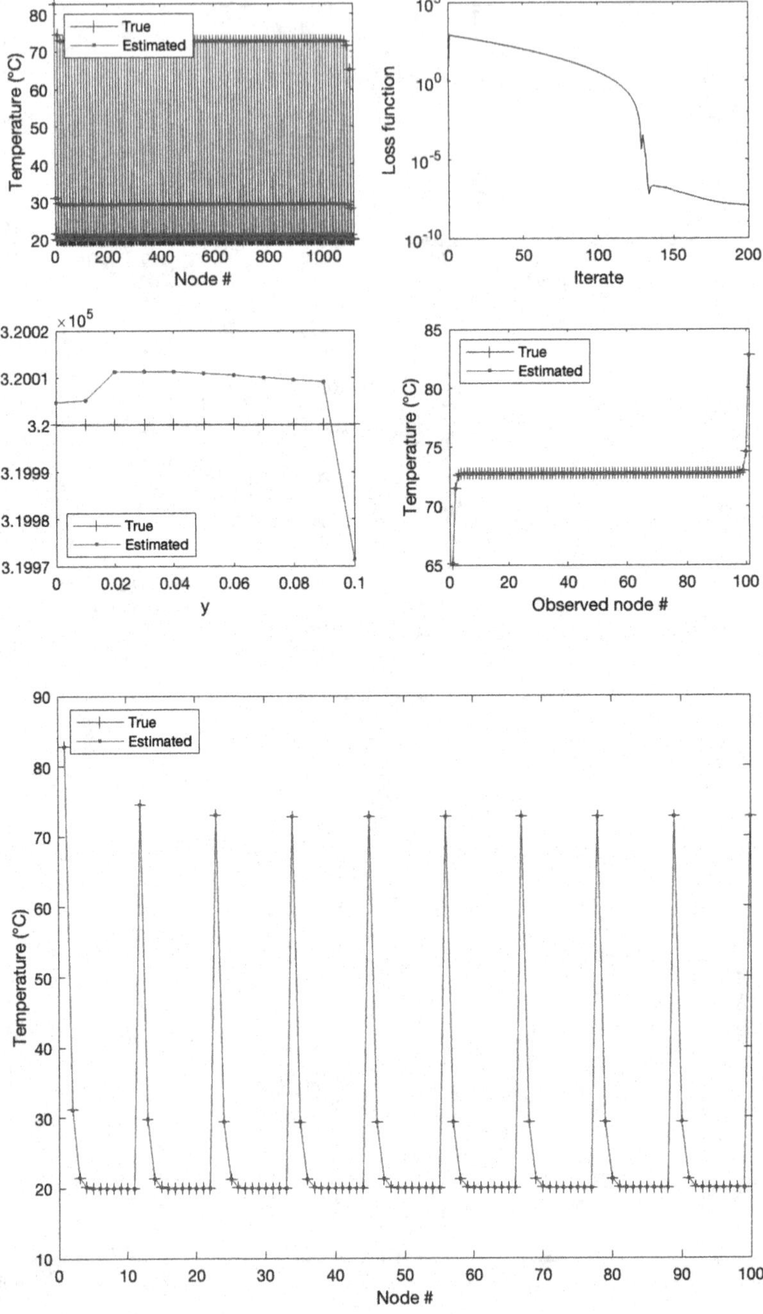

Fig. 5. Heat capacity estimation. Nodal values of the temperature field T at time $T_k = N$ (top-left). Loss-function (11) minimization (top-right). Rectangular domain with the estimated heat capacity $(\rho C)(y)$ for $y = 0, 0.01, 0.02, \ldots, 0.1$ (center-left). Nodal temperatures at the measured boundary S (center-right). Zoom on the top-left figure from node 1 to 100 (bottom).

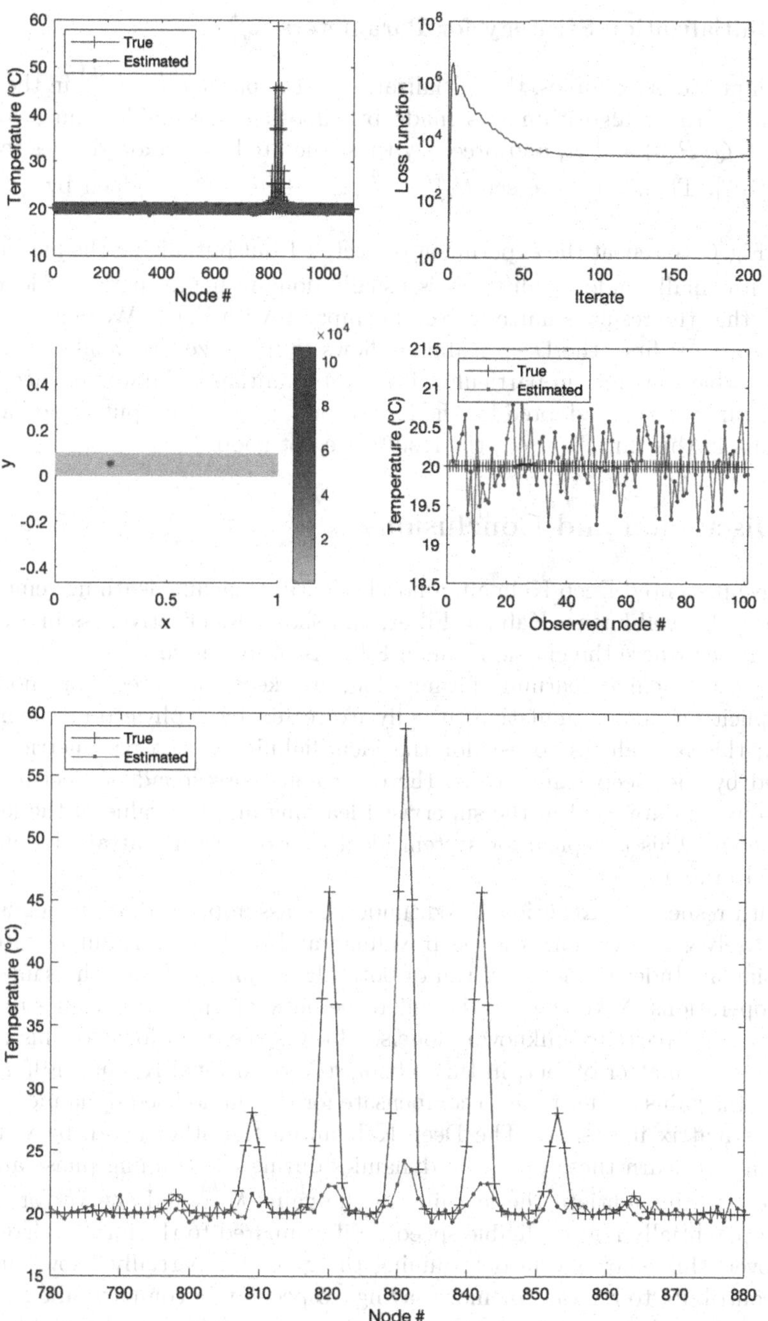

Fig. 6. Forcing term estimation in absence of noise with a random initialization of the parameters $\mathcal{K}_G^{(k)}$. Nodal values of the temperature field T at time $T_k = N$ (top-left). Loss-function (11) minimization (top-right). Rectangular domain with the estimated source term f_ϑ (center-left). Nodal temperatures at the measured boundary S (center-right). Zoom on the top-left figure from node 780 to 880 (bottom).

4.3 Initialization Strategy for Parameters $\mathcal{K}_G^{(k)}$

In both previous examples, the initialization of the parameters $\mathcal{K}_G^{(k)}$ in the Deep Kalman learning algorithm was made by choosing reasonable values for the matrices Q, $P(0)$ and R, and precisely those dictated by standard rules to tune the KF [7]. Then, we have set $\mathcal{K}_G^{(k)} = P(1)$, where $P(1)$ is given by (2) with $k = 1$.

In Fig. 6 we repeat the experiment of Sect. 4.1 but initializing the parameters $\mathcal{K}_G^{(k)}$ with small random values, as is usually done in neural networks learning. We see that the result is much worse, if compared with Fig. 3. We can make two considerations: first, the Deep Kalman allows to initialize the weights in a very interpretable way and, in particular, it allows to initialize the weights according to the standard rules adopted for the KF; second, if we do not put a good apriori structure in this initialization, the results are not good.

5 Discussion and Conclusions

We have presented Deep Kalman, a novel scientific machine learning reinterpretation of the well-known Kalman Filter, and shown its effectiveness in a couple of situations where the classic Kalman Filter is problematic.

In this machine learning formulation, we kept the predictor model in its explicit original formulation, usually expressed by a physico-mathematical model; this fact allows to ask for the identifiability of the parametric model learned by the Deep Kalman, i.e. the one-to-one correspondence between the input-output data used in the supervised learning and the value of the learned parameters. This is typical for system identification but not attainable in pure neural learning.

With respect to execution speed, under the assumption that the gains $\mathcal{K}_G^{(k)}$ are entirely given a priori, the Deep Kalman and traditional Kalman Filter are very similar. Indeed, a forward run of both algorithms performs the same algebraic operations. Nevertheless, the difference between the two becomes relevant in the case of partially unknown models, which is the main focus of this contribution. As a matter of fact, in this setting, the traditional Kalman Filter must update the gains at run-time to compensate for the unmodeled dynamics, which involves matrix inversions. The Deep Kalman on the other hand, by virtue of its ability to learn these unknown dynamics during the training phase and the prospect to differentiate the learning of the gains $\mathcal{K}_G^{(k)}$ and the vector p, can offer a potentially non-negligible speed-up if compared to the former algorithm. Moreover, this efficiency in determining the gains for partially known models may contribute to relevant memory savings, especially in comparison e.g. to the demand of employing a filter-bank, a common option for the traditional Kalman Filter, in this case.

The speed-up and memory saving are important when the Deep Kalman must run in real-time and on computers with limited resources, like microcontrollers for embedded applications.

References

1. Brunton, S.L., Budisic, M., Kaiser, E., Kutz, J.N.: Modern Koopman theory for dynamical systems. SIAM Rev. **64**(2), 229–340 (2022). https://doi.org/10.1137/21M1401243
2. Cheng, S., et al.: Machine learning with data assimilation and uncertainty quantification for dynamical systems: a review. IEEE/CAA J. Automatica Sin. **10** (2023). https://doi.org/10.1109/JAS.2023.123537
3. Chinellato, E., Marcuzzi, F., Pierobon, S.: Physics-Aware soft sensors for embedded digital twins. In: Yang, X. S., Sherratt, S., Dey, N., Joshi, A. (eds.) Proceedings of Ninth International Congress on Information and Communication Technology. ICICT 2024, LNNS, vol. 1013. Springer, Singapore (2024). https://doi.org/10.1007/978-981-97-3559-4_34
4. Dessole, M., Marcuzzi, F.: Accurate detection of hidden material changes as fictitious heat sources. Numer. Heat Transfer, Part B: Fundam. (2023). https://doi.org/10.1080/10407790.2023.2220905
5. Fasel, U., Kutz, J.N., Brunton, B.W., Brunton, S.L.: Ensemble-SINDy: robust sparse model discovery in the low-data, high-noise limit, with active learning and control. Proc. R. Soc. A. (2022). https://doi.org/10.1098/rspa.2021.0904
6. Giusteri, G.G., Marcuzzi, F., Rinaldi, L.: Replacing voids and localized parameter changes with fictitious forcing terms in boundary-value problems. Results Appl. Math. **20** (2023). https://doi.org/10.1016/j.rinam.2023.100402
7. Grewal, M.S., Andrews, A.P.: Kalman Filtering: Theory and Practice with MATLAB®: Fourth Edition, vol. 9781118851210. Wiley (2014). https://doi.org/10.1002/9781118984987
8. Hershey, J., Le Roux, J., Weninger, F.: Deep unfolding: model-based inspiration of novel deep architectures. arXiv (2014)
9. Humpherys, J., Redd, P., West, J.: A fresh look at the Kalman filter. SIAM Rev. **54**(4), 801–823 (2012). https://doi.org/10.1137/100799666
10. Kingma, D., Ba, J.: Adam: a method for stochastic optimization. In: International Conference on Learning Representations (2014)
11. Marcuzzi, F.: Space and time localization for the estimation of distributed parameters in a finite element model. Comput. Methods Appl. Mech. Eng. **198**(37), 3020–3025 (2009). https://doi.org/10.1016/j.cma.2009.05.007
12. Marcuzzi, F.: A Numerical feed-forward scheme for the augmented kalman filter. In: Franco, L., de Mulatier, C., Paszynski, M., Krzhizhanovskaya, V. V., Dongarra, J. J., Sloot, P. M. A. (eds.) Computational Science – ICCS 2024. LNCS, vol. 14837. Springer, Cham (2024). https://doi.org/10.1007/978-3-031-63778-0_10
13. Patel, D., Ray, D., Abdelmalik, M.R., Hughes, T.J., Oberai, A.A.: Variationally mimetic operator networks. Comput. Methods Appl. Mech. Eng. **419** (2024). https://doi.org/10.1016/j.cma.2023.116536
14. Revach, G., Shlezinger, N., Ni, X., Escoriza, A., Van Sloun, R.J.G., Eldar, Y.C.: KalmanNet: neural network aided Kalman filtering for partially known dynamics. IEEE Trans. Signal Process. **70**, 1532–1547 (2022). https://doi.org/10.1109/TSP.2022.3158588
15. Wirgin, A.: The inverse crime (2004). https://doi.org/10.48550/ARXIV.MATH-PH/0401050

Neural Network as Transformation Function in Data Assimilation

Kun Wang[1,2(✉)], Matthew D. Piggott[2], Yanghua Wang[1,2], and Rossella Arcucci[2,3(✉)]

[1] Resource Geophysics Academy, Imperial College London, South Kensington, London SW7 2AZ, UK
k.wang23@imperial.ac.uk

[2] Department of Earth Science and Engineering, Imperial College London, South Kensington, London SW7 2AZ, UK
r.arcucci@imperial.ac.uk

[3] Data Science Institute, Imperial College London, London, UK

Abstract. Variational Data Assimilation (DA) is a technique aimed at mitigating the error in simulated states by integrating observations. Variational DA is widely employed in weather forecasting and hydrological modeling as an optimization technique for refining dynamic simulation states. However, when constructing the cost function in variational DA, it is necessary to establish a transformation function from simulated states to observations. When observations come from ground sensors or from remote sensing, representing such a transformation function with explicit expressions can sometimes be challenging or even impossible. Therefore, considering the strong mapping capabilities of Neural Network (NN)s in representing the relationship from simulated states to observations, this paper proposes a method utilizing a NN as the transformation function. We evaluate our method on a real dataset of river discharge in the UK and achieved a 39% enhancement in prediction accuracy, measured by Mean Square Error (MSE), compared to the results obtained without DA.

Keywords: Variational data assimilation · Neural network · Mapping

1 Introduction

In the simulation of dynamic systems, prediction errors arise from simulation processes. To mitigate these errors, incorporating observations is necessary. However, it is important to acknowledge that observation errors also exist within the observations. DA, as an optimization method, effectively addresses this issue. By leveraging observations, DA seeks to improve simulated states, thereby bringing simulation states closer to real values [1]. Variational DA, as a subtype of

Supported by Resource Geophysics Academy, Imperial College London.

L. Franco et al. (Eds.): ICCS 2024, LNCS 14836, pp. 322–329, 2024.
https://doi.org/10.1007/978-3-031-63775-9_23

DA, updates simulated states by minimizing the cost function between simulated states and observations [2]. A pivotal component in the cost function of variational DA is the transformation function. This function represents the transition from simulated states to observations and is typically encoded using a selection matrix (to represent location information) or a physical formula (to represent the relationship between physical quantities). Variational DA, with its capability to address DA issues arising from nonlinear relationships between simulated states and observations, is progressively gaining popularity in practical applications [3]. This trend is fueled by the common occurrence of nonlinear relationships between simulated states and observations in real-world scenarios. Variational DA finds widespread application in hydrology [4–8]. These applications showcase the effectiveness of variational DA in hydrology for improving the accuracy of simulation results.

However, in practice, establishing the relationship between observations obtained and simulated states (whether based on location or physical relations) using explicit expressions is challenging, rendering it difficult to obtain the transformation function. For instance, there might be no explicit physical expression to represent the relationship from simulated states to observations, or the resolution of simulated states may differ from that of observations. Hence, this presents a challenging issue in the practical implementation of DA. To address this challenge, Cheng et al. [9] proposed a method to compress simulated states and observations into a shared latent space. However, the execution of this approach in an online update necessitates retraining the model whenever updates occur in the inputs to the parametric model, thereby demanding additional computational resources. NNs, as a machine learning method, can be trained to capture the relationship between two correlated variables [10]. In this paper, leveraging the inherent capabilities of NNs, we propose an offline method that employs a NN as the transformation function to address the aforementioned challenge. We evaluated our method using a real-world dataset comprising two sources: European Flood Awareness System (EFAS) [11] and National River Flow Archive (NRFA) [12]. EFAS is utilized for simulation, while NRFA serves as observations. We opt for MSE as our evaluation metric. Our data assimilation method exhibits an average improvement of 15% in accuracy compared to the raw simulation results.

The organization of the remainder of this article is as follows. Section 2 describes the test case. Section 3 introduced the method we proposed. Section 4 presents the results of our experiments. Section sec5 summarizes the entire paper and proposes future work.

2 Test Case

To validate the feasibility of our method, real-world data is selected as our test case. River discharge from the UK is selected for DA in this experiment. The data chosen for the simulated states consists of simulated river discharge data provided by EFAS. This data provides daily river discharge for the whole of the

UK at a spatial resolution of 5 × 5 km. The NRFA provides data on daily river flows from 1600 river stations within the UK, which serve as our observations in this experiment.

The data in the EFAS dataset consists of 2D images, where each pixel represents the river discharge for the corresponding area. The size of the data is 200 × 200 pixels, with only 14120 pixels containing valid values. The number of stations suitable for experimentation in the NRFA dataset is 924. The details of these two datasets are shown in Table 1. In this experiment, DA is conducted every three days for simulated states.

Table 1. the detail of two datasets

name	EFAS	NRFA
spatial resolution	5 × 5 km	
temporal resolution	one-day	one-day
size of sample	14120 × 1	924 × 1

3 Proposed Method

3.1 Idea

The cost function in variational DA is shown in given by

$$J(\mathbf{x}) = \frac{1}{2}(\mathbf{x} - \mathbf{x}_b)^T \mathbf{B}^{-1}(\mathbf{x} - \mathbf{x}_b) + \frac{1}{2}(\mathbf{y} - \mathcal{H}(\mathbf{x}))^T \mathbf{R}^{-1}(\mathbf{y} - \mathcal{H}(\mathbf{x})), \quad (1)$$

where $\mathbf{x}_b$ represent simulated states, $\mathbf{y}$ represents observations, $\mathbf{B}$ and $\mathbf{R}$ represent the covariance matrix of simulated states and observations, and $\mathcal{H}$ represents the transformation function.

From Eq. 1, it is evident that the transformation function $\mathcal{H}$ can be regarded as a mapping relation from simulated states to observations. Given the nonlinear mapping capabilities, NNs represent a rational choice for serving as the transformation function. Therefore, in practical applications where it is challenging to represent the relationship from simulated states to observations with explicit expressions, utilizing NNs for mapping purposes presents a viable solution.

3.2 Implementation

In the experiment, the dataset of simulated states can be represented by $\mathbf{X} = \{\mathbf{x}_1, \mathbf{x}_2, ..., \mathbf{x}_n\}$ and the dataset of observations can be represented by $\mathbf{Y} = \{\mathbf{y}_1, \mathbf{y}_2, ..., \mathbf{y}_n\}$, where n represents the number of samples, which corresponds to individual time step. The input of the NN is the simulated state $\mathbf{x}_i$ and the output of the NN is the observation $\mathbf{y}_i$. The structure of the transformation function obtained via the NN approach is depicted in Fig. 1.

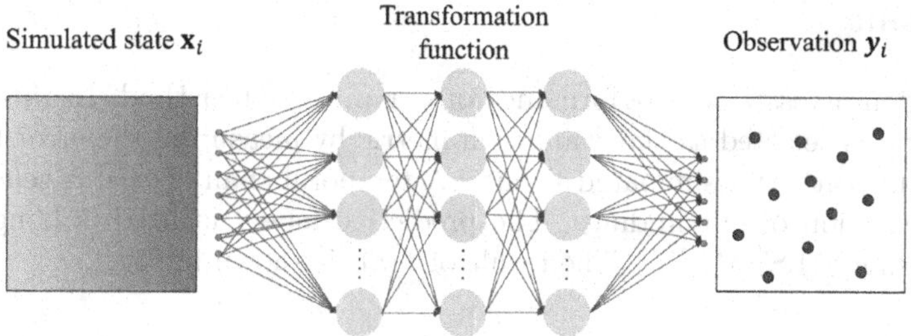

Fig. 1. Structure of transformation function

In designing the structure of the NN model, the input $\mathbf{x}_i$ of the network is a vector of size 14120×1, and the output $\mathbf{y}_i$ is a vector of size 924×1. Given the shape of the input and output vectors, Multi-Layer Perceptron (MLP) is chosen as the network architecture for this experiment. In the MLP structure, which comprises 4 fully connected layers, the activation function used for each layer is the LeakyReLU.

The training of the transformation function is shown in Algorithm. 1, where $k_{\max}$ represents the number of iterations, ϵ represents the threshold of training loss and $\mathcal{L}$ represents the loss function.

Algorithm 1. Training transformation function

Inputs: $\mathbf{X}, \mathbf{Y}, \mathcal{H}$
Parameters: $\epsilon, k_{\max}, \mathbf{x}_i \in \mathbf{X}, \mathbf{y}_i \in \mathbf{Y}$
$k = 0$
while $k < k_{\max}$ and $\mathcal{L} > \epsilon$ **do**
  $\mathbf{y}_i = \mathcal{H}(\mathbf{x}_i)$
  $\mathcal{L} = \frac{1}{n}\sum_{i=1}^{n} ||\mathbf{y}_i - \mathcal{H}(\mathbf{x}_i)||^2$
  $k = k + 1$
end while
output: $\mathcal{H}$

In the training process, a training strategy that combines mini-batch and Stochastic Gradient Descent (SGD) is employed, which aids in achieving results with smaller error [13]. This approach enhances the NN's ability to more accurately represent the mapping relationship from simulated states to observations. The NN is trained on these two datasets for 300 epochs with a batch size of 4. SGD with a momentum of 0.9 and the weight decay of 1e-4 is chosen as the optimizer. In addition, the cosine annealing strategy is utilized to update the learning rate, starting with an initial learning rate of 1e-2 and a cycle length of 150 epochs [14].

The loss function utilized in our method is MSE, selected to minimize the discrepancy between observations and the mapping result of simulated states to observations via the NN.

4 Result

This section assesses the performance of our proposed method. In this experiment, MSE is selected as the evaluation metric by measuring the error between the ground truth and simulated results. A temporal span of 30 d is selected for the presentation of our findings. The predictive model utilized is Long Short-Term Memory (LSTM) [15]. The result of MSE is shown in Fig. 2.

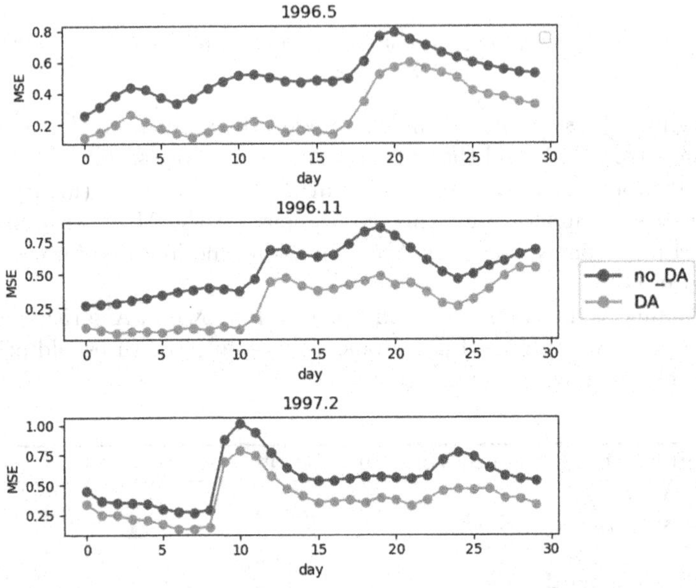

Fig. 2. MSE of our proposed method and the simulated result without DA

In Fig. 2, the red dots represent the MSE between the simulated results without DA and the ground truth, while the orange dots represent the MSE between our proposed method and the ground truth. From Fig. 2, it is evident that the results of our proposed method outperform the simulation results without DA in terms of the MSE metric, with an average improvement of 39%. Additionally, the highest improvement observed exceeds 50%. This metric demonstrates that our method effectively enhances the accuracy of the assimilation results.

Additionally, the Structural Similarity Index (SSIM) is used as an auxiliary evaluation metric. SSIM is an evaluation metric that assesses the structural similarity between two images. The formula for SSIM is given by

$$\mathrm{SSIM}(\mathbf{x},\mathbf{y}) = \frac{(2\mu_{\mathbf{x}}\mu_{\mathbf{y}} + c_1)(2\sigma_{\mathbf{xy}} + c_2)}{(\mu_{\mathbf{x}}^2 + \mu_{\mathbf{y}}^2 + c_1)(\sigma_{\mathbf{x}}^2 + \sigma_{\mathbf{y}}^2 + c_2)}, \tag{2}$$

where $\mu_{\mathbf{x}}$ represents the pixel sample mean of simulated states, $\mu_{\mathbf{y}}$ represents the pixel sample mean of observations, $\sigma_{\mathbf{x}}$ represents the variance of the simulated states, $\sigma_{\mathbf{y}}$ represents the variance of the observations, $\sigma_{\mathbf{xy}}$ represents the covariance of simulated states and observations, c_1 and c_2 represents two variables to stable the division. SSIM is employed here to assess the similarity between the states obtained after DA and the ground truth. The result is shown in Fig. 3.

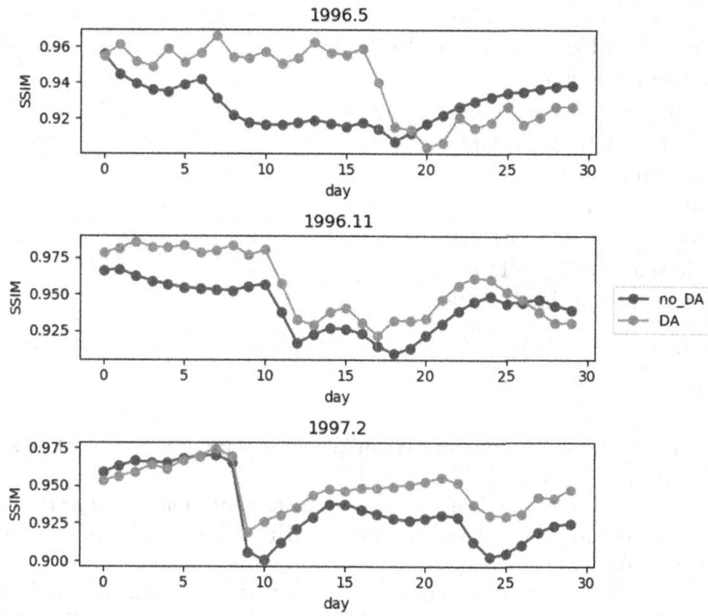

Fig. 3. SSIM of our proposed method and the simulated result without DA

The SSIM of our proposed method is higher than the simulated results without DA. This result indicates that the states obtained by our proposed method are structurally closer to the ground truth, meaning that the variance and mean of the river discharge in the results are closer to the true values.

5 Conclusion

The inability to represent the transformation function with an explicit expression poses a challenge for DA in practical applications. In this paper, we propose to address this issue by employing a NN as the transformation function. This approach leverages the excellent nonlinear mapping properties inherent in NNs. We conduct a test of the proposed method on a real dataset, and the experimental results demonstrate that our approach improves the evaluation metric of MSE by 39% compared to the results obtained without DA. These results validate the feasibility of using NN as the transformation function. The utilization of a NN as the transformation function expands the range of observation types viable for the DA. In future work, addressing the issue of sparse observation distribution in data assimilation, which leads to localized correction of the simulated state, will be imperative.

Acronyms

EFAS European Flood Awareness System
DA Data Assimilation
LSTM Long Short-Term Memory
MLP Multi-Layer Perceptron
MSE Mean Square Error
NN Neural Network
NRFA National River Flow Archive
SGD Stochastic Gradient Descent
SSIM Structural Similarity Index

References

1. Lahoz, W.A., Schneider, P.: Data assimilation: making sense of earth observation. Front. Environ. Sci. **2**, 16 (2014)
2. Cheng, S., et al.: Machine learning with data assimilation and uncertainty quantification for dynamical systems: a review. IEEE/CAA J. Automatica Sin. **10**(6), 1361–1387 (2023)
3. Bannister, R.N.: A review of operational methods of variational and ensemble-variational data assimilation. Q. J. R. Meteorol. Soc. **143**(703), 607–633 (2017)
4. Reichle, R.H., McLaughlin, D.B., Entekhabi, D.: Variational data assimilation of microwave radiobrightness observations for land surface hydrology applications. IEEE Trans. Geosci. Remote Sens. **39**(8), 1708–1718 (2001)
5. Cheng, S., et al.: Error covariance tuning in variational data assimilation: application to an operating hydrological model. Stochast. Environ. Res. Risk Assess. **35**(5), 1019–1038 (2021)
6. Alvarado-Montero, R., et al.: Multi-parametric variational data assimilation for hydrological forecasting. Adv. Water Resour. **110**, 182–192 (2017)
7. Cheng, S., et al.: A graph clustering approach to localization for adaptive covariance tuning in data assimilation based on state-observation mapping. Math. Geosci. **53**(8), 1751–1780 (2021)
8. Cheng, S., Qiu, M.: Observation error covariance specification in dynamical systems for data assimilation using recurrent neural networks. Neural Comput. Appl. **34**(16), 13149–13167 (2022)
9. Cheng, S., et al.: Generalised latent assimilation in heterogeneous reduced spaces with machine learning surrogate models. J. Sci. Comput. **94**(1), 11 (2023)
10. May, R., Dandy, G., Maier, H.: Review of input variable selection methods for artificial neural networks. Artif. Neural Netw.-Methodol. Adv. Biomed. Appl. **10**(1), 19–45 (2011)
11. Matthews, G., et al.: Evaluating the impact of post-processing medium-range ensemble streamflow forecasts from the European flood awareness system. Hydrol. Earth Syst. Sci. **26**(11), 2939–2968 (2022)

12. National river flow archive homepage. https://nrfa.ceh.ac.uk/. Accessed 7 Feb 2024
13. Li, M., et al.: Efficient mini-batch training for stochastic optimization. In: Proceedings of the 20th ACM SIGKDD International Conference on Knowledge Discovery and Data Mining (2014)
14. Johnson, O.V., et al.: ps-CALR: periodic-shift cosine annealing learning rate for deep neural networks. IEEE Access **11**, 139171–139186 (2023)
15. Hochreiter, S., Schmidhuber, J.: Long short-term memory. Neural Comput. **9**(8), 1735–1780 (1997)

Assessment of Explainable Anomaly Detection for Monitoring of Cold Rolling Process

Jakub Jakubowski[1](✉), Przemysław Stanisz[1], Szymon Bobek[2], and Grzegorz J. Nalepa[2]

[1] Department of Applied Computer Science, AGH University of Science and Technology, 30-059 Krakow, Poland
jjakubow@agh.edu.pl

[2] Faculty of Physics, Astronomy and Applied Computer Science, Institute of Applied Computer Science, and Jagiellonian Human-Centered AI Lab (JAHCAI), and Mark Kac Center for Complex Systems Research, Jagiellonian University, Ul. Prof. Stanisława Łojasiewicza 11, 30-348 Krakow, Poland

Abstract. The detection and explanation of anomalies within the industrial context remains a difficult task, which requires the use of well-designed methods. In this study, we focus on evaluating the performance of Explainable Anomaly Detection (XAD) algorithms in the context of a complex industrial process, specifically cold rolling. We train several state-of-the-art anomaly detection algorithms on the synthetic data from the cold rolling process and optimize their hyperparameters to maximize its predictive capabilities. Then we employ various model-agnostic Explainable AI (XAI) methods to generate explanations for the abnormal observations. The explanations are evaluated using a set of XAI metrics specifically selected for the anomaly detection task in industrial setting. The results provide insights into the impact of the selection of both machine learning and XAI methods on the overall performance of the model, emphasizing the importance of interpretability in industrial applications. For the detection of anomalies in cold rolling, we found that autoencoder-based approaches outperformed other methods, with the SHAP method providing the best explanations according to the evaluation metrics used.

Keywords: machine learning · explainable artificial intelligence · predictive maintenance

1 Introduction

In the era of Artificial Intelligence and Industry 4.0, manufacturing companies gain new opportunities for development and improvement of their processes. One of the fields that can greatly benefit from these trends is the monitoring and maintenance of the equipment in the manufacturing facilities. The digitalization

L. Franco et al. (Eds.): ICCS 2024, LNCS 14836, pp. 330–344, 2024.
https://doi.org/10.1007/978-3-031-63775-9_24

of production allows companies to collect and store large amounts of data from sensors. This data can be utilized using Machine Learning (ML) methods to detect anomalies, diagnose faults, and perform root cause analysis in an online manner. All these tasks belong to a broader concept of predictive maintenance, which aims to estimate the current condition of the equipment or predict its useful life to optimize maintenance schedules and help avoid dramatic failures, which can lead to significant losses to the company.

Our study focuses on the application of anomaly detection in the steel industry, particularly in the cold rolling process. The primary objectives of cold rolling are reducing the thickness of steel strip, improving surface finish, flatness, and increasing hardness. A typical cold rolling mill is composed of rolling stands which are placed in tandem. At each stand, the steel strip is gradually reduced to reach a target thickness at the exit of the mill. Figure 1 presents a schematic diagram of the cold-rolling process. The prediction of failures and anomalies in cold rolling processes is challanging task, due to low frequency and high diversity of the abnormalities. These issues can be addressed by anomaly detection methods, which learn the normal behavior of the process, and measure the direcptancy between the observed variables and normal working conditions. However, state-of-the-art anomaly detection methods are black-boxes, which means that predictions of these models are difficult to interpret by humans, which hinders the applicability of these methods in practice. Understanding the model's decision is vital to ensure the applicability of the method because adequate corrective actions must be taken by the crew upon detection of anomaly.

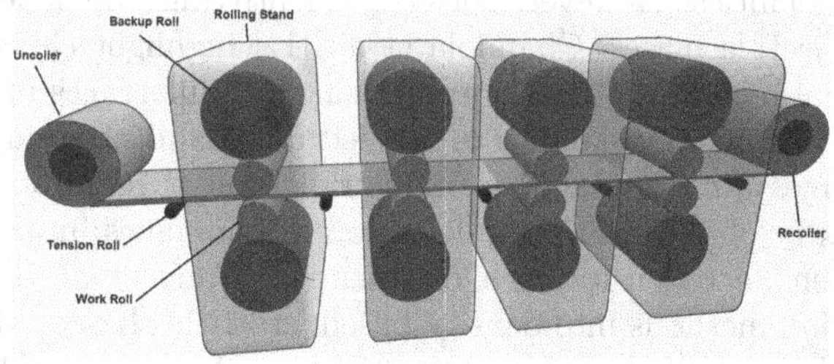

Fig. 1. Schematic diagram of 4-stand cold rolling mill

To address interpretability of black-boxes, Explainable Artificial Intelligence (XAI) has emerged, which aims to clarify the decisions of ML models for human understanding. Model-agnostic XAI methods provide explanations without prior knowledge of the model architecture, making them suitable for all types of ML models. While assessing performance of ML models is a well-established task, validating XAI methods remains a challenge. Currently, researchers often rely on human-based assessments or anecdotal examples, which do not provide a global perspective. A more robust approach involves applying metrics for quantitative comparison across different methods [24]. The selection of these metrics depends on the specific problem, allowing an assessment of the quality of the explanation.

In this paper, we assess how the selection of ML model and XAI method, can impact both the predictive performance and explanatory capabilities. To the best of our knowledge, this is the first paper that aims to quantitatively evaluate multiple XAI methods on the anomaly detection task in an industrial use-case. We propose a recipe on how to build Explainable Anomaly Detecion (XAD) models, which account for quality of predictions and explanations. We selected several anomaly detection and XAI methods and applied them to the data from the cold rolling process. To ensure that our results are not biased by the manual labeling of the process data, we created a synthetic dataset, which simulates cold rolling of steel and generates anomalies within a fraction of observations.

The rest of the paper is organized as follows. Section 2 provides a brief overview of anomaly detection techniques, model-agnostic XAI methods, and evaluation metrics. Section 3 presents the proposed assessment methodology, including details on selected ML models and XAI methods. Section 4 contains the results of our simulations and their discussion. Section 5 concludes the work and proposes the directions for future research.

2 Related Works

2.1 Anomaly Detection

Anomaly detection in industry involves identifying deviations from normal operations, which can lead to equipment failures, quality defects, or reduced performance. Early detection of anomalous behavior can bring significant benefits to manufacturing facilities. However, correctly identifying anomalies is problematic due to subjective biases. Chandola et al. [11] highlight challenges such as defining a normal region, evolving normal behavior, differences in anomaly perception across domains, availability of labeled data, and noise in the data. All of these challenges are applicable to industrial process monitoring, emphasizing the complexity of the task. The detection of anomalies in an unsupervised manner has been an extensive area of research. Wang et al. [35] groups the anomaly detection methods into density-based, statistics-based, distance-based, clustering-based, ensemble-based, and learning-based. The examples of anomaly detection methods that fall into each category are given in Table 1.

Table 1. Examples of anomaly detection methods

Category	Methods
Statisics-based	Mahalanobis Distance [23], HBOS [14]
Density-based	LOF [9]
Distance-based	kNN [19]
Clustering-based	DBSCAN [13]
Ensemble-based	IForest [21], HST [32], LODA [26]
Learning-based	AE [28], GAN [29]

The performance assessment of the anomaly detection methods is similar to the imbalanced classification, given the substantial difference between the number of normal and anomalous observations. An effective method is characterized by high precision and recall. Typically, there is a trade-off between precision and recall—enhancing one tends to reduce the other. Aggregating these metrics using the PRAUC or the F-score establishes a comprehensive value, considering both aspects.

2.2 Explainable Artificial Intelligence

Explainable Artificial Intelligence focuses on understanding the decision-making processes of AI systems. Historically, AI research prioritized peak performance, leading to the development of complex ML models often labeled "black-boxes" [1]. These models lack interpretability, which poses challenges for human observers in tracking their decision path. The opacity of black-box models creates trust issues, as stakeholders may hesitate to rely on decisions they cannot comprehend, rendering the ML model impractical, especially for unforeseen decisions [2]. XAI not only addresses interpretability concerns, but also aids in model control and improvement by revealing decision processes and identifying potential flaws or errors in data or pre-processing [12]. In the context of anomaly detection, understanding the underlying causes of predicted anomalies is a crucial task. In complex industrial systems, relying solely on a single anomaly score while relegating its interpretation to users can be impractical.

LIME [27], SHAP [22], and Counterfactual Explanations [34] (CFE) are three prominent model-agnostic XAI methods. LIME generates explanations by approximating the behavior of a black-box model in a specific region of the input space with an interpretable model. SHAP uses a game-theoretic approach to quantify the contribution of each feature to a prediction by considering all feature combinations and their Shapeley values to determine how each feature influences the model outcome. CFE perturbs an observation with the objective of changing the model's decision, while ensuring that the generated CFE lies close to the original observation. It is defined as an optimization problem, which allows including additional constraints, e.g. the likelihood of the CFE or the number of manipulated features. The choice of the optimal XAI method may depend on the type of problem, data format, or end-user requirements.

Several studies explored the use of XAD in industrial settings. We found that many of these works rely on autoencoder architecture [5,15,17]. Some noteworthy works utilize Isolation Forest [4,18], OCSVM [6,16] and LOF [16]. In terms of explainability, we find that SHAP is widely adopted for this task [5,15,16,18]. Other applied methods include rules [6,31], CFE [17] and AcME [4]. However, of all the articles referenced, only [6] quantitatively evaluated the performance of the proposed XAI method. Moreover, none of these works compared explanations generated with different methods.

2.3 Metrics for Evaluating XAI Methods

Although quantifying the performance of ML models is straightforward, evaluating XAI methods remains challenging due to the subjective nature of explainability, varying stakeholder expectations, and context dependence. The absence of a clear ground truth, especially when human judgment is involved, makes defining a "correct" explanation vague. Numerous metrics attempt to quantitatively measure explanation quality, but selecting appropriate metrics is challenging, as their importance varies based on data type, use case, or stakeholders. Several studies sought to synthesize knowledge about explanation requirements [10,24,25]. Drawing on these works, we propose a set of metrics for evaluating the performance of XAD methods in industrial applications.

- **Faithfulness**—the primary requirement of every XAI method is to give the plausible explanation in the meaning that they are aligned with the actual model decision and expert knowledge.
- **Stability** (or robustness, continuity)—similar observations should yield similar explanations. This property ensures that a small change in the input or output of the model will not cause significant change in the explanation.
- **Compactness** (or complexity)—determines the size of the explanation measured by e.g. number of features used in the explanation. The low size of the explanation facilitates its understanding by humans.
- **Computational complexity**—it measures the time to produce explanations. It assesses the usability of the method in industrial applications rather than the quality of the explanation itself, since the explanations should be generated within a limited time.

We note that there is no coherence in terms of terminology used and similar metrics may have a different name depending on the author, e.g., stability in [10] is equivalent to continuity in [24]. We believe that this list is a good starting point to evaluate XAD methods in industrial applications.

3 Methodology

In this study, we evaluated the performance of several anomaly detection methods combined with model-agnostic XAI methods to understand how their selection affects the overall performance of the model. ML models were trained on the data from cold rolling, with the aim of predicting the anomalies within these data and generating explanations of models' decisions. The primary focus is to evaluate and compare the effectiveness of various ML and XAI methods in the context of cold rolling. We evaluate XAI methods using metrics described in Sect. 2.3 and identify the optimal combination of methods to build a robust and interpretable XAD model. Additionally, we analyze the influence of individual setting of XAI methods on the quality of explanations. The complete workflow is illustrated in Fig. 2, with further details provided below.

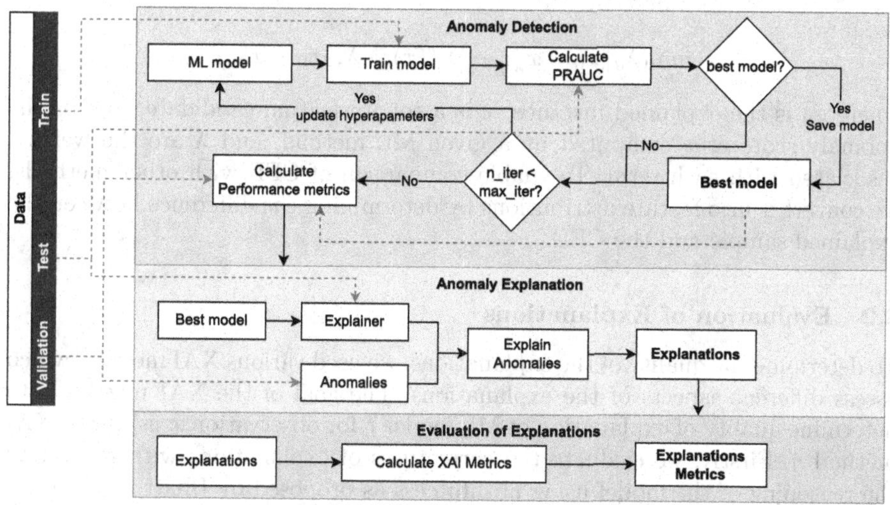

Fig. 2. Proposed methodology for evaluating Explainable Anomaly Detection models

3.1 Anomaly Detection and Explanation

We selected various anomaly detection algorithms for assessment. Each algorithm had its hyperparameters adjusted using Bayesian optimization [30] to maximize overall performance (measured by the PRAUC metric). Table 2 details the selected algorithms and their tuned hyperparameters.

Table 2. Assessed models and optimized hyperparameters

ML method	Hyperparameters
Half-Space Trees	depth, no. trees, random state, window size
OCSVM	gamma, kernel type, nu
LODA	no. bins, no. cuts, random state
LOF	no. neighbors, algorithm type, distance metric
Isolation Forest	no. features, max. no samples, no. trees, random state
KMeans	no. clusters, random state
Autoencoder	dropout rate, hidden layers, latent layers, learning rate, no. epochs, random state
Sparse Autoencoder	dropout rate, hidden layers, latent layers, learning rate, no. epochs, beta, sparsity target, random state

Once the optimal architecture for each algorithm was determined, we generated explanations for all anomalous cases in the validation dataset. Explanations were not generated for normal cases as they would lack practical implications. Three model-agnostic XAI methods—SHAP, LIME, and CFE—were used for each algorithm. We applied each XAI method with different set of attributes, to observe if manipulating the method itself leads to changes in obtained results. The SHAP and LIME were used to explain the value of anomaly score, while CFE was defined as an optimization task in which we simultaniously minimize distance to the original instance, anomaly score, and sparsity penalty:

$$\min_x(\lambda_d\|x_0 - x\|_1 + \lambda_a f(x) + \lambda_s\|x_0 - x\|_0) \tag{1}$$

where x_0 is the explained instance, x is a counterfcatual candidate, $f(x)$ is the anomaly score of x computed by a given ML method, and λ are the weights associated with each term. To enable comparison of CFE with other methods, we convert it into feature attributions by determining the difference between the explained sample and the CFE.

3.2 Evaluation of Explanations

To determine the quality of the explanations, we used various XAI metrics, which assess different aspects of the explanations. The goal of the XAI metrics is to determine quality of explanation of ML model f for observation x using the XAI method g. Firstly, we evaluate the correctness of explanations with respect to the reasoning of the model using faithfulness as proposed by Bhatt et al. [7]:

$$\mu_F(f, g; x) = \underset{S \in \binom{[d]}{|S|}}{\text{corr}} \sum_{i \in S} (g(f, x)_i, f(x) - f(x_s)) \tag{2}$$

A subset d of features is randomly replaced i times with a baseline that produces perturbed samples x_s. For each sample, the difference in model output between original and perturbed observations is determined. Finally, a correlation coefficient is calculated between the attribution of the features $g(f, x)$ and these differences. Faithfulness depends on factors such as the selected baseline values, number of perturbations, number of perturbed features, and the randomness of perturbation process itself.

To evaluate the robustness of the XAI methods, we use the stability metric, based on the Lipschitz continuity, as proposed by Alvarez-Melis and Jaakkola [3]:

$$\mu_S(f, g; x) = \max_{x_j \in \mathcal{N}_\epsilon(x_i) \leq \epsilon} \frac{\|g(f, x_i) - g(f, x_j)\|_2}{\|x_i - x_j\|_2} \tag{3}$$

Stability takes samples x_j lying in the neighborhood of observation x_i, measured with Euclidean distance and constrained by ϵ. For each sample, the distance between this sample and the explained observation is determined, along with the distance between the corresponding feature attributions $g(f, x)$. Then a ratio between these values is calculated, and the maximum obtained value is the final result. Lower values of μ_S imply that explanations are more robust, as small changes in the input do not effect in drastically different explanation.

The compactness of the explanation is calculated using the entropy measure, as proposed by Bhatt et al. [7]:

$$\mu_C(f, g, x) = -\sum_{i=1}^{d} \mathbb{P}_g(i)\ln(\mathbb{P}_g(i)) \tag{4}$$

$\mathbb{P}_g(i)$ is scaled feature attribution vector $g(f, x_i)$, in a way that the sum of absolute values is equal to 1.0. Lower values of μ_C indicate a more compact explanation (using fewer features), thus enhancing its understanding by humans. Although the authors use term *complexity*, we decided to refer to it as compactness, so it is not confused with *computational complexity*.

The computational complexity of the algorithm is evaluated by measuring the time required to generate the explanations. It depends on the machine used, the ML model itself, the XAI method, and the quality of the implementation. Despite limitations, it provides valuable insight, particularly in a streaming scenario, where it is critical to ensure fast computation of explanations.

3.3 TCM Dataset

All experiments are made on a synthetic dataset, which simulates the cold rolling process in a four-stand mill. The simulation is based on analytical equations describing the cold rolling process [8,20,33]. The data set consists of 42 variables in total, which are listed in Table 3.

Table 3. Prediction perfromance of best models

Feature	Unit	Description
H_0, H_4	mm	thickness of coil at the entry and exit
Y_0, Y_4	mm	yield strength of coil at the entry and exit
W	mm	width of the coil
$D_1 - D_4$	mm	diameter of work rolls in each stand
$F_{t0} - F_{t4}$	kN	interstand tensions
$F_{r1} - F_{r4}$	kN	rolling force in each stand
$T_{r1} - T_{r4}$	kN	rolling torque in each stand
$V_{r1} - V_{r4}$	kN	rolling speed in each stand
$S_1 - S_4$	kN	rolling gap in each stand
$R_1 - R_4$	%	redutction in each stand
$I_1 - I_4$	kN	motor current in each stand

Given the characteristics of the production line and its condition (which is continuously updated), the data generator randomly selects a steel coil from a pool of 20 different prodcuts. For a given product, a simulation is perfromed, which is based on several assumptions of the cold rolling process e.g. friction coefficient, tensions, reductions. At each calculation, there is a small probability that an anomalous observation will be generated. We defined four different types of anomalies: increased roll friction, reduced bearing efficiency, reduced motor efficiency, and abnormal reduction scheme, which can affect measurements from different stands. We generated 20,000 samples with an anomaly ratio of 3.0%. The data sample is presented in Fig. 3. The upper plot presents the metadata of a coil, while the lower plot depicts measurements for the first stand.

4 Results

We first report the predictive performance of each model, measured on the validation set. The models were evaluated using the F1 score, G-mean, PRAUC, precision and recall and are presented in Table 4. In terms of predictive capabilities, the AE model significantly outperformed other methods. For most of the models, we observe that precision and recall are balanced, except for Half-Space Trees—in this case the recall is satisfactory, but the precision is very low, making this model useless in practice (due to high number of false alarms).

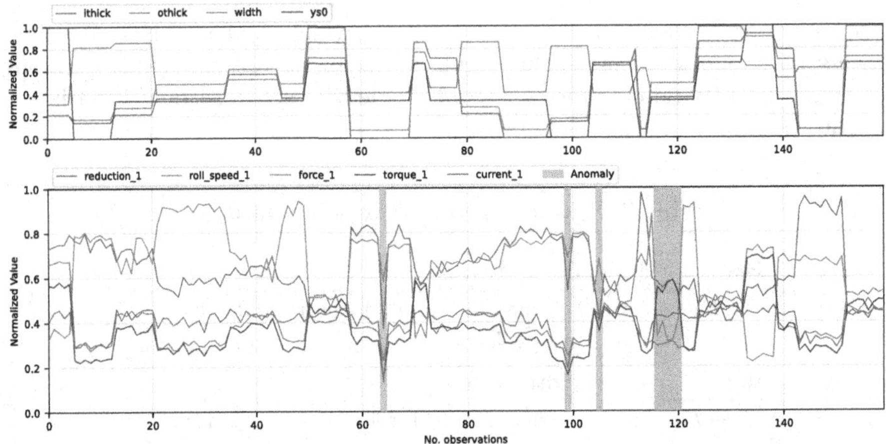

Fig. 3. Sample of TCM dataset with highlighed anomalies

Table 4. Prediction perfromance of best models

Model	PRAUC	F1	G-mean	Precision	Recall
SAE	0.74	0.67	0.81	0.69	0.66
AE	**0.81**	**0.75**	**0.86**	**0.76**	**0.74**
Half Space Trees	0.06	0.08	0.54	0.04	0.40
KMeans	0.27	0.28	0.52	0.29	0.28
Isolation Forest	0.39	0.35	0.58	0.37	0.34
LODA	0.36	0.32	0.60	0.28	0.37
Local Outlier Factor	0.60	0.55	0.67	0.71	0.45
One-Class SVM	0.44	0.40	0.65	0.37	0.43

Additionally, we evaluate the recall of the models with respect to the types of anomalies. This give us a better understanding of the capabilities of each model, the results are presented in Table 5. Each column corresponds to different type of anomaly, while the values represent the fraction of correctly detected anomalies.

Table 5. Recall of each model with respect to anomaly type

Model	Bearing	Electric	Reduction	Work Roll
Count	35	37	47	35
SAE	0.543	0.811	0.787	0.457
AE	**0.686**	0.892	**0.872**	**0.486**
HalfSpaceTrees	0.343	0.243	0.702	0.257
KMeans	0.086	0.081	0.702	0.143
IsolationForest	0.114	0.000	0.936	0.171
LODA	0.257	0.189	0.745	0.200
LocalOutlierFactor	0.343	0.000	**1.000**	0.314
OneClassSVM	0.143	0.297	0.915	0.257

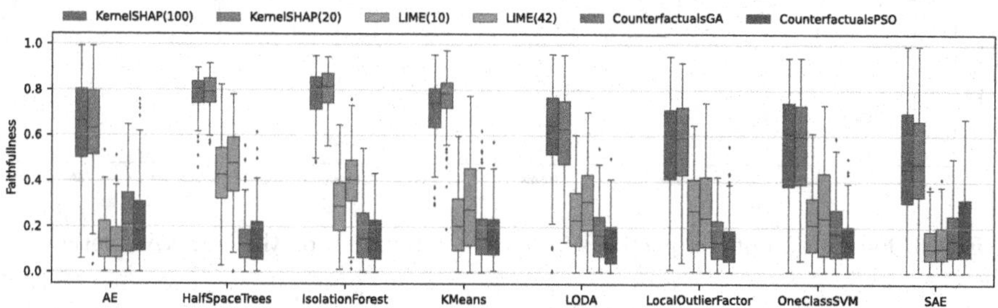

Fig. 4. Estimation of faithfulness for each combination of ML and XAI method

Faithfulness was estimated using 100 perturbations, in which 20 features were randomly replaced with baseline values (determined based on the k-Means algorithm). This setting allowed us to obtain repeatable results. Figure 4 presents the estimation of the faithfulness metric. SHAP significantly outperformed other methods in all scenarios tested, which indicates that it is likely to be the best choice to explain anomalies in our use case. We also observe that limiting the number of base samples for SHAP from 100 to 20 did not have a negative influence on faithfulness. To validate these observations we conducted the Friedman test followed by the Nemenyi test with the p-value set to 0.05, which confirmed our hypotheses. LIME method performed decently in explaining tree-based methods, but poorly for auteoncoders. The lowest faithfullness was achieved by CFE models, which outperfromed LIME only in models utilizing autoencoder architecture.

To determine stability, we generated 10 synthetic samples in the neighborhood of explained observations, as the anomalies lie in the low-density regions making it impossible to use samples from the data set. The comparison of stability between different XAI methods is a intricate task, due to individual characteristics of each method. To resolve the issue, we scaled all feature attributions so that the sum of their absolute values for each exlanation is equal to 1.0. Figure 5

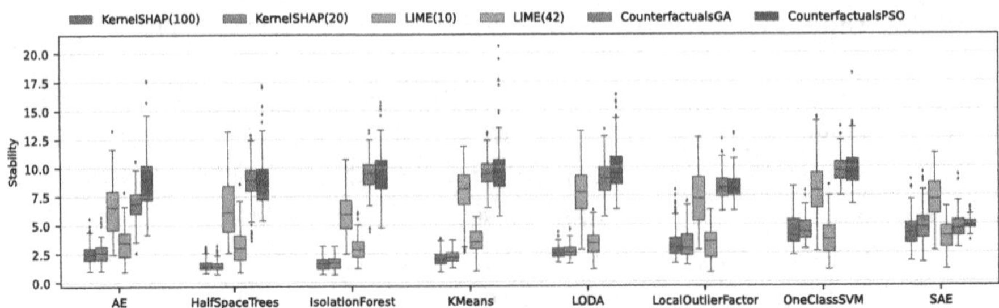

Fig. 5. Estimation of stability for each combination of ML and XAI method

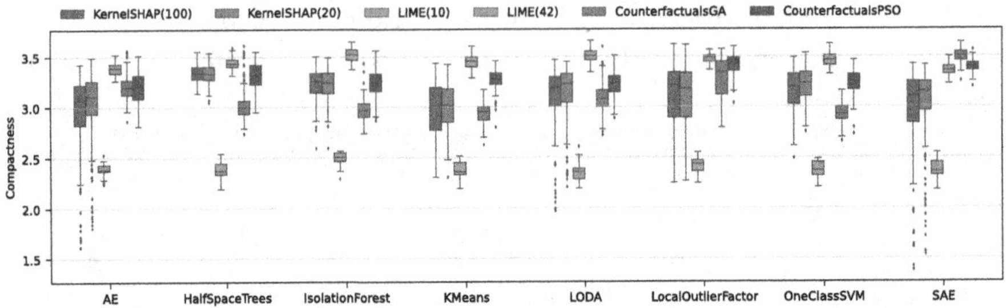

Fig. 6. Estimation of compactness for each combination of ML and XAI method

presents the distribution of stability for all explained samples. In most cases SHAP achieved the lowest stability values, meaning that these explanations were more robust to small changes in the feature values. Again, we have performed Friedman and Nemenyi tests to confirm it, and the results indicated that statistically significant differences between SHAP and other methods were observed. Some exceptions from this behavior were observed for LIME(42), which had comparable performance to SHAP on LODA, LOF and OCSVM. Additionally, the stability of SHAP was not better than other methods for SAE. We observe that limiting the number of features in LIME greatly deteriorates the stability of this method in anomaly detection task. The obtained stability of CFE is poor, which is probably caused by the non-detererministm of the heuristics used for generating the explanations.

The range of values for compacntess metric spans from 0, meaning all feature attribution is assigned to a single feature, up to 3.73, which is obtained if all 42 features have equal feature attribution. Figure 6 presents the distribution of the compactness metric. We observe that most of the methods result in complex explanations, as the compactness values lie closer to the upper bound. LIME method, which was limited to 10 features, naturally obtained relatively low values compared to other approaches. Additionally, certain parts of the explanations generated for autoencoders with the SHAP method achieved satisfactory scores—these instances are related to electric motor failure. It is consistent with the reality because for this anomaly only one feature was affected.

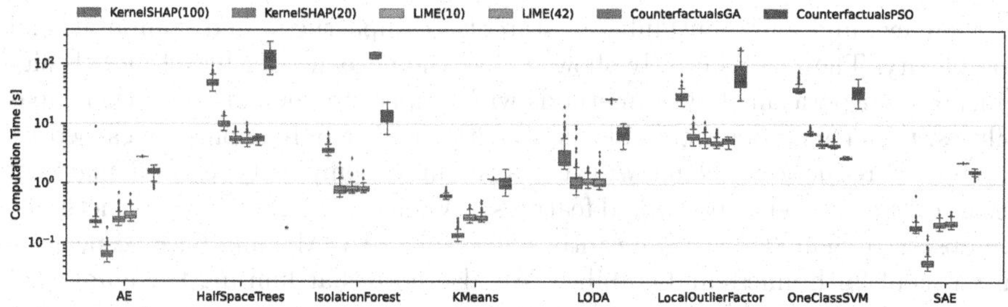

Fig. 7. Estimation of computation time for each combination of ML and XAI method

In terms of the last of the evaluated criteria, computational complexity, we measured the time to compute each explanation. We run all experiments on the Linux machine equipped with 64-core AMD Ryzen Threadripper PRO 5995WX and 256 GB of RAM and conducted them in a sequential manner. The results are presented in Fig. 7. We observe that a very high impact on the computation time has the ML model itself, which we expected, because all utilized XAI methods rely on calling the model hundreds or thousands of times to compute explanation. The shortest computation time was achieved by deep learning approaches, making them particularly useful in streaming scenarios. A significant drop in computation time is observed for the SHAP method with a reduced number of base samples. Taking into account that SHAP-based explanations achieved comparable performance in terms of other metrics, we reason that limiting the number of base samples gives a noticeable decrease of computational complexity without a drop in performance. Regarding the CFE, we observe that this method required longest time to compute explanations, which is not reflected by the increased performance of the method. It is worth noting that there are large differences between CFE-GA and CFE-PSO, which vary depending on the model. For example, the computation time for Half-Space Trees was much shorter with CFE-GA, while in case of Isolation Forest it was the opposite. Moreover, we report that computation time varies significantly depending on the parameters chosen for the optimization methods (e.g. number of generations for GA or number of iterations for PSO). Lastly, changing the number of features in LIME does not influence the time to generate the explanations.

5 Conclusion and Future Works

In this paper, we evaluated different XAD methods with respect to their predictive and explanatory capabilities. We selected eight state-of-the-art anomaly detection models and three distinct XAI methods (each tested in two settings), which gave us 48 XAD models in total. All ML models were trained on a synthetic dataset, which simulates the cold rolling process of steel strip. In terms of predictive performance, the models were evaluated based on the PRAUC, F1 score, G-mean, precision and recall. To assess the performance of model explanations, we selected four distinct metrics, which considered different aspects of their explanations—faithfulness, stability, compactness, and computational complexity. The results clearly showed that the autoencoder-based models significantly outperformed other methods with respect to anomaly detection capabilities. Other methods were able to decently predict only anomalies caused by the invalid reduction scheme, which was the most complex type of anomaly (in terms of the number of perturbed features). When considering the XAI methods, we observed that SHAP significantly outperformed LIME and CFE, especially in terms of faithfulness and stability. We also note that limiting the number of base samples for SHAP did not have negative impact on its performance, but significantly reduced its computation time. This is very important in industrial applications, where data is generated at high speeds and the explanations of the anomalies should be almost instantaneous.

Despite CFE performing poorly compared to SHAP and LIME, we believe this method is worth further research, due to its high tunability. Thus, in future work, we plan to study the CFE more deeply to increase its explanatory capabilities, as we find the achieved results unsatisfying. Moreover, we plan to focus more on deep learning architectures based on the autoencoder. Additionally, we want to investigate Generative Adversial Networks, which were not considered in this study, but are known for their anomaly detection capabilities. Ultimately, we will verify our results on the data from an existing cold rolling mill, to show the practical application of the XAD.

Acknowledgements. Project XPM is supported by the National Science Centre, Poland (2020/02/Y/ST6/00070), under CHIST-ERA IV programme, which has received funding from the EU Horizon 2020 Research and Innovation Programme, under Grant Agreement no. 857925.

References

1. Abdullah, T.A.A., Zahid, M.S.M., Ali, W.: A review of interpretable ml in healthcare: taxonomy, applications, challenges, and future directions. Symmetry **13**(12), 2439 (2021). https://doi.org/10.3390/sym13122439
2. Adadi, A., Berrada, M.: Peeking inside the black-box: a survey on explainable artificial intelligence (XAI). IEEE Access **6**, 52138–52160 (2018). https://doi.org/10.1109/ACCESS.2018.2870052
3. Alvarez-Melis, D., Jaakkola, T.S.: Towards robust interpretability with self-explaining neural networks. In: Proceedings of the 32nd International Conference on Neural Information Processing Systems (NIPS 2018), pp. 7786–7795. Curran Associates Inc., Red Hook (2018)
4. Anello, E., et al.: Anomaly detection for the industrial internet of things: an unsupervised approach for fast root cause analysis. In: 2022 IEEE Conference on Control Technology and Applications (CCTA), pp. 1366–1371 (2022). https://doi.org/10.1109/CCTA49430.2022.9966158
5. Baek, M., Kim, S.B.: Failure detection and primary cause identification of multivariate time series data in semiconductor equipment. IEEE Access **11**, 54363–54372 (2023). https://doi.org/10.1109/ACCESS.2023.3281407

6. Barbado, A.: Óscar Corcho: interpretable machine learning models for predicting and explaining vehicle fuel consumption anomalies. Eng. Appl. Artif. Intell. **115**, 105222 (2022). https://doi.org/10.1016/j.engappai.2022.105222
7. Bhatt, U., Weller, A., Moura, J.M.F.: Evaluating and aggregating feature-based model explanations. In: Proceedings of the Twenty-Ninth International Joint Conference on Artificial Intelligence (IJCAI 2020) (2021)
8. Bland, D.R., Ford, H.: The calculation of roll force and torque in cold strip rolling with tensions. Proc. Inst. Mech. Eng. **159**(1), 144–163 (1948)
9. Breunig, M.M., Kriegel, H.P., Ng, R.T., Sander, J.: Lof: identifying density-based local outliers. In: Proceedings of the 2000 ACM SIGMOD International Conference on Management of Data (SIGMOD 2000), pp. 93–104. Association for Computing Machinery, New York (2000). https://doi.org/10.1145/342009.335388
10. Carvalho, D.V., Pereira, E.M., Cardoso, J.S.: Machine learning interpretability: a survey on methods and metrics. Electronics **8**(8) (2019). https://doi.org/10.3390/electronics8080832
11. Chandola, V., Banerjee, A., Kumar, V.: Anomaly detection: a survey. ACM Comput. Surv. **41**(3), 1–58 (2009). https://doi.org/10.1145/1541880.1541882
12. Dwivedi, R., et al.: Explainable AI (XAI): core ideas, techniques, and solutions. ACM Comput. Surv. **55**(9), 1–33 (2023). https://doi.org/10.1145/3561048
13. Ester, M., Kriegel, H.P., Sander, J., Xu, X.: A density-based algorithm for discovering clusters in large spatial databases with noise. In: Proceedings of the Second International Conference on Knowledge Discovery and Data Mining (KDD 1996), pp. 226–231. AAAI Press (1996)
14. Goldstein, M., Dengel, A.: Histogram-based outlier score (HBOS): a fast unsupervised anomaly detection algorithm. In: KI-2012: Poster and Demo Track, vol. 1, pp. 59–63 (2012)
15. Ha, D.T., Hoang, N.X., Hoang, N.V., Du, N.H., Huong, T.T., Tran, K.P.: Explainable anomaly detection for industrial control system cybersecurity. IFAC-PapersOnLine **55**(10), 1183–1188 (2022). https://doi.org/10.1016/j.ifacol.2022.09.550. 10th IFAC Conference on Manufacturing Modelling, Management and Control MIM 2022
16. Hermansa, M., Kozielski, M., Michalak, M., Szczyrba, K., Wrobel, L., Sikora, M.: Sensor-based predictive maintenance with reduction of false alarms; a case study in heavy industry. Sensors **22**(1) (2022). https://doi.org/10.3390/s22010226
17. Jakubowski, J., Stanisz, P., Bobek, S., Nalepa, G.J.: Roll wear prediction in strip cold rolling with physics-informed autoencoder and counterfactual explanations. In: 2022 IEEE 9th International Conference on Data Science and Advanced Analytics (DSAA), pp. 1–10. IEEE (2022). https://doi.org/10.1109/DSAA54385.2022.10032357
18. Kim, D., Antariksa, G., Handayani, M.P., Lee, S., Lee, J.: Explainable anomaly detection framework for maritime main engine sensor data. Sensors **21**(15) (2021). https://doi.org/10.3390/s21155200
19. Knorr, E.M., Ng, R.T.: Algorithms for mining distance-based outliers in large datasets. In: Proceedings of the 24rd International Conference on Very Large Data Bases (VLDB 1998), pp. 392–403. Morgan Kaufmann Publishers Inc., San Francisco (1998)
20. Lenard, J.G.: 9 - tribology. In: Lenard, J.G. (ed.) Primer on Flat Rolling, 2nd edn, pp. 193–266. Elsevier, Oxford (2014). https://doi.org/10.1016/B978-0-08-099418-5.00009-3

21. Liu, F.T., Ting, K.M., Zhou, Z.: Isolation forest. In: 2008 Eighth IEEE International Conference on Data Mining, pp. 413–422 (2008). https://doi.org/10.1109/ICDM.2008.17
22. Lundberg, S.M., Lee, S.I.: A unified approach to interpreting model predictions. In: Guyon, I., Luxburg, U.V., Bengio, S., Wallach, H., Fergus, R., Vishwanathan, S., Garnett, R. (eds.) Advances in Neural Information Processing Systems, vol. 30, pp. 4765–4774. Curran Associates, Inc. (2017)
23. Mahalanobis, P.: On the generalised distance in statistics. Proc. Natl. Inst. Sci. India **2**, 49–55 (1936)
24. Nauta, M., et al.: From anecdotal evidence to quantitative evaluation methods: a systematic review on evaluating explainable AI. ACM Comput. Surv. **55**(13s), 1–42 (2023). https://doi.org/10.1145/3583558
25. Oblizanov, A., Shevskaya, N., Kazak, A., Rudenko, M., Dorofeeva, A.: Evaluation metrics research for explainable artificial intelligence global methods using synthetic data. Appl. Syst. Innov. **6**(1) (2023). https://doi.org/10.3390/asi6010026
26. Pevný, T.: Loda: lightweight on-line detector of anomalies. Mach. Learn. **102**(2), 275–304 (2015). https://doi.org/10.1007/s10994-015-5521-0
27. Ribeiro, M.T., Singh, S., Guestrin, C.: "Why should i trust you?": Explaining the predictions of any classifier. In: Proceedings of the 22nd ACM SIGKDD International Conference on Knowledge Discovery and Data Mining (KDD 2016), pp. 1135–1144. Association for Computing Machinery, New York (2016). https://doi.org/10.1145/2939672.2939778
28. Sakurada, M., Yairi, T.: Anomaly detection using autoencoders with nonlinear dimensionality reduction. In: Proceedings of the MLSDA 2014 2nd Workshop on Machine Learning for Sensory Data Analysis (MLSDA 2014), pp. 4–11. Association for Computing Machinery, New York (2014). https://doi.org/10.1145/2689746.2689747
29. Schlegl, T., Seeböck, P., Waldstein, S.M., Schmidt-Erfurth, U., Langs, G.: Unsupervised anomaly detection with generative adversarial networks to guide marker discovery. In: Niethammer, M., et al. (eds.) IPMI 2017. LNCS, vol. 10265, pp. 146–157. Springer, Cham (2017). https://doi.org/10.1007/978-3-319-59050-9_12
30. Snoek, J., Larochelle, H., Adams, R.P.: Practical Bayesian optimization of machine learning algorithms. In: Pereira, F., Burges, C., Bottou, L., Weinberger, K. (eds.) Advances in Neural Information Processing Systems, vol. 25. Curran Associates, Inc. (2012)
31. Steenwinckel, B., et al.: Flags: a methodology for adaptive anomaly detection and root cause analysis on sensor data streams by fusing expert knowledge with machine learning. Futur. Gener. Comput. Syst. **116**, 30–48 (2021). https://doi.org/10.1016/j.future.2020.10.015
32. Tan, S.C., Ting, K.M., Liu, T.F.: Fast anomaly detection for streaming data. In: Proceedings of the Twenty-Second International Joint Conference on Artificial Intelligence (IJCAI 2011), vol. 2, pp. 1511–1516. AAAI Press (2011)
33. Venkata Reddy, N., Suryanarayana, G.: A set-up model for tandem cold rolling mills. J. Mater. Process. Technol. **116**(2–3), 269–277 (2001). https://doi.org/10.1016/s0924-0136(01)01007-x
34. Wachter, S., Mittelstadt, B., Russell, C.: Counterfactual explanations without opening the black box: automated decisions and the GDPR. Harv. JL Tech. **31**, 841 (2017)
35. Wang, H., Bah, M.J., Hammad, M.: Progress in outlier detection techniques: a survey. IEEE Access **7**, 107964–108000 (2019). https://doi.org/10.1109/access.2019.2932769

Adjoint Sensitivities of Chaotic Flows Without Adjoint Solvers: A Data-Driven Approach

Defne Ege Ozan[1(✉)] and Luca Magri[1,2,3]

[1] Imperial College London, Department of Aeronautics, Exhibition Road, London SW7 2BX, UK
{d.ozan,l.magri}@imperial.ac.uk
[2] The Alan Turing Institute, London NW1 2DB, UK
[3] Politecnico di Torino, DIMEAS, Corso Duca degli Abruzzi, 24 10129 Torino, Italy

Abstract. In one calculation, adjoint sensitivity analysis provides the gradient of a quantity of interest with respect to all system's parameters. Conventionally, adjoint solvers need to be implemented by differentiating computational models, which can be a cumbersome task and is code-specific. To propose an adjoint solver that is not code-specific, we develop a data-driven strategy. We demonstrate its application on the computation of gradients of long-time averages of chaotic flows. First, we deploy a parameter-aware echo state network (ESN) to accurately forecast and simulate the dynamics of a dynamical system for a range of system's parameters. Second, we derive the adjoint of the parameter-aware ESN. Finally, we combine the parameter-aware ESN with its adjoint version to compute the sensitivities to the system parameters. We showcase the method on a prototypical chaotic system. Because adjoint sensitivities in chaotic regimes diverge for long integration times, we analyse the application of ensemble adjoint method to the ESN. We find that the adjoint sensitivities obtained from the ESN match closely with the original system. This work opens possibilities for sensitivity analysis without code-specific adjoint solvers.

Keywords: Reservoir computing · Adjoint methods · Sensitivity · Chaotic flows

1 Introduction

Many computational applications including optimization, data assimilation, and uncertainty quantification require the information about how a quantity of interest is influenced by the system's parameters and initial conditions e.g., [9]. Precisely, sensitivity analysis provides the gradient of an objective functional with respect to all system's parameters. A direct method for estimating the sensitivity is to perturb the parameter and approximate the gradient with a finite difference between the base and the perturbed solutions. This procedure suffers

L. Franco et al. (Eds.): ICCS 2024, LNCS 14836, pp. 345–352, 2024.
https://doi.org/10.1007/978-3-031-63775-9_25

from numerical errors and needs to be repeated for each parameter. Therefore, for systems with many parameters, the computational cost increases linearly with the number of parameters. Adjoint methods turn this around. With an adjoint formulation, the sensitivity to all parameters is computed with a single operation, which requires the adjoint system.

Whilst the adjoint is a powerful tool, there are some practical limitations to its use. The computation of the adjoint sensitivity relies on (i) the system equations to be known, and (ii) the system to be linearized, i.e., the computation of the Jacobian [4,13]. First, the adjoint sensitivity is only as accurate as the model assumptions. Second, the system might be high-dimensional and nonlinear, and in some cases the computational code might not be differentiable, all of which make the derivation of the Jacobian cumbersome. To address the challenges associated with developing adjoint solvers, we propose a data-driven approach, and apply this to a prototypical chaotic flow, which is a qualitative example of complex behaviour such as turbulence.

In chaotic systems, in which nearby trajectories diverge at an exponential rate, the adjoint system becomes unstable and the sensitivities diverge after some integration time. This poses a challenge as the quantities of interest are usually long-term time-averages. A possible approach that we also exploit in this paper is to take an ensemble of short-term trajectories [6].

The objective of this paper is two-fold; (i) we introduce the parameter-aware echo state network to learn the parameterized dynamics of a chaotic system with, and (ii) we infer the sensitivities of an objective functional to the system parameters from the adjoint of parameter-aware echo state network.

2 Parameter-Aware Echo State Network

Recurrent neural networks (RNNs) are specialized neural network architectures for sequential data processing and have been successfully employed for time series prediction. RNNs can be treated as discrete differential equations and can be studied with dynamical systems theory. In this work, we develop an RNN that learns the parameterized dynamics of a system such that we can make predictions in regimes that have not been seen during training. We further show that such a network can also infer the sensitivity of the dynamics to the parameters. We utilize the echo state network (ESN) [8], which whilst being a universal approximator [5] has the advantage over other RNNs of not needing backpropagation during training, and thus requiring less computational effort. ESNs can make time-accurate short-term predictions [2,11,14], and infer long-term statistics and invariant properties, such as Lyapunov exponents, of chaotic flows [10,12]. A parameter-aware extension of the ESN has been formulated, and been applied to predict amplitude death, i.e., when the system bifurcates to fixed point solutions, [16], and multi-stable regimes including chaos [15]. The parameter-aware ESN is a nonlinear discrete map from reservoir state at time step i to reservoir state at time step $i+1$

$$\boldsymbol{r}(i+1) = (1-\alpha)\boldsymbol{r}(i) + \alpha \tanh(\boldsymbol{W}_{in}[\boldsymbol{y}_{in}(i); \mathrm{diag}(\boldsymbol{\sigma}_p)(\boldsymbol{p}-\boldsymbol{k}_p)] + \boldsymbol{W}\boldsymbol{r}(i)), \quad (1)$$

where $\boldsymbol{y}_{in}(i) \in \mathbb{R}^{N_y}$ is the input vector, $\boldsymbol{p} \in \mathbb{R}^{N_p}$ is the parameter vector, $\boldsymbol{r}(i) \in \mathbb{R}^{N_r}$ is the reservoir state, $\boldsymbol{W}_{in} \in \mathbb{R}^{N_r \times (N_y+N_p)}$ is the input matrix, and $\boldsymbol{W} \in \mathbb{R}^{N_r \times N_r}$ is the state matrix, and diag$(\cdot)$ denotes a diagonal matrix that has $(\cdot)$ as its diagonal. The output is predicted from the reservoir state with a linear read-out layer

$$\hat{\boldsymbol{y}}(i+1) = \boldsymbol{W}_{out}\boldsymbol{r}(i+1), \tag{2}$$

$\hat{\boldsymbol{y}}(i+1) \in \mathbb{R}^{N_y}$ is the output vector, the dynamics of which we aim to model, and $\boldsymbol{W}_{out} \in \mathbb{R}^{N_y \times N_r}$ is the output matrix. The matrices $\boldsymbol{W}_{in}$ and $\boldsymbol{W}$ are sparse, randomly generated, and not trained, whilst $\boldsymbol{W}_{out}$ is trained via ridge regression [8]. Training an ESN involves searching for optimal hyperparameters that consist of input matrix scaling σ_{in}, the spectral radius of the state matrix ρ, the leak rate α, the Tikhonov regularizer λ, and finally two additional hyperparameters per parameter p_i; k_p that shifts the parameter and σ_p that scales it.

3 Adjoint of Echo State Network

We mathematically derive the adjoint of the ESN, which is an autonomous dynamical system when it runs in closed-loop. The dynamics (1) can be expressed as a constraint $\boldsymbol{F}(i) = \boldsymbol{F}(\boldsymbol{r}(i), \boldsymbol{r}(i-1), \boldsymbol{p}) = 0$. We consider the time-averaged objective functional given as a sum over N discrete time steps

$$\mathcal{J} = \frac{1}{N}\sum_{i=1}^{N} \tilde{\mathcal{J}}(\boldsymbol{r}(i)), \tag{3}$$

in which the reservoir state $\boldsymbol{r}$ depends on the system's parameters $\boldsymbol{p}$. The goal of sensitivity analysis is to determine the gradient of the objective functional with respect to the system's parameters. For the parameter-aware ESN, the sensitivity is then expressed as

$$\frac{d\mathcal{J}}{d\boldsymbol{p}} = \frac{1}{N}\sum_{i=1}^{N} \frac{d\tilde{\mathcal{J}}(\boldsymbol{r}(i))}{d\boldsymbol{r}(i)}\boldsymbol{Q}(i), \tag{4}$$

where we define $\boldsymbol{Q}(i) := d\boldsymbol{r}(i)/d\boldsymbol{p}$, $\boldsymbol{Q}(i) \in \mathbb{R}^{N_r \times N_p}$. Notice that the dimension of $\boldsymbol{Q}(i)$ grows with the number of parameters. This increasing computational cost motivates solving the adjoint problem instead. The Lagrangian of the objective functional, $\mathcal{J}$, subjected to the system dynamics, $\boldsymbol{F} = 0$, is

$$\mathcal{L} := \mathcal{J} - \langle \boldsymbol{q}^+, \boldsymbol{F} \rangle, \tag{5}$$

where the inner product $\langle \boldsymbol{q}^+, \boldsymbol{F} \rangle$ is defined as

$$\langle \boldsymbol{q}^+, \boldsymbol{F} \rangle := \sum_{i=1}^{N} \boldsymbol{q}^{+T}(i)\boldsymbol{F}(i), \tag{6}$$

and $\boldsymbol{q}^+ \in \mathbb{R}^{N_r}$ are the Lagrange multipliers, or the adjoint variables. We solve for the gradient $d\mathcal{L}/d\boldsymbol{p} = d\mathcal{J}/d\boldsymbol{p}$. After expanding the summation and rearranging the terms, we choose the Lagrange multipliers such that we can eliminate the terms $d\boldsymbol{r}(i)/d\boldsymbol{p}$, e.g., [9]. This provides us with the evolution equations of the adjoint variables

$$\frac{d\mathcal{J}}{d\boldsymbol{p}} = \sum_{i=1}^{N} \boldsymbol{q}^{+T}(i)\frac{\partial \boldsymbol{r}(i)}{\partial \boldsymbol{p}}, \tag{7a}$$

$$\boldsymbol{q}^{+}(i) = \frac{1}{N}\frac{\partial \tilde{\mathcal{J}}(\boldsymbol{r}(i))}{\partial \boldsymbol{r}(i)}^{T} + \frac{\partial \boldsymbol{r}(i+1)}{\partial \boldsymbol{r}(i)}^{T}\boldsymbol{q}^{+}(i+1), \tag{7b}$$

$$\boldsymbol{q}^{+}(N) = \frac{1}{N}\frac{\partial \tilde{\mathcal{J}}(\boldsymbol{r}(N))}{\partial \boldsymbol{r}(N)}^{T}. \tag{7c}$$

Practically, we first let the ESN run autonomously for the given time window and save the direct solution, which serves as the base trajectory. We then solve the adjoint equations backwards in time starting from the terminal condition $\boldsymbol{q}^{+}(N)$ (7c). The solution of the adjoint equations requires the computation of the Jacobian evaluated at the base trajectory, i.e., the gradient of the reservoir state at time step $i+1$ with respect to the reservoir state at time step i

$$\frac{\partial \boldsymbol{r}(i+1)}{\partial \boldsymbol{r}(i)} = (1-\alpha)\boldsymbol{I}_{N_r \times N_r} + \alpha \text{diag}(1 - \tilde{\boldsymbol{r}}^2(i))(\boldsymbol{W}_{in}^{y}\boldsymbol{W}_{out} + \boldsymbol{W}), \tag{8}$$

and the gradient with respect to the parameters

$$\frac{\partial \boldsymbol{r}(i+1)}{\partial \boldsymbol{p}} = \alpha \text{diag}(1 - \tilde{\boldsymbol{r}}^2(i))\boldsymbol{W}_{in}^{p}\text{diag}(\boldsymbol{\sigma}_p), \tag{9}$$

where $\tilde{\boldsymbol{r}}(i) = (\boldsymbol{r}(i+1) - (1-\alpha)\boldsymbol{r}(i))/\alpha$, $\boldsymbol{W}_{in} = [\boldsymbol{W}_{in}^{y}\ \boldsymbol{W}_{in}^{p}]$, and $\boldsymbol{I}$ denotes the identity matrix.

4 Computation of Data-Driven Chaotic Sensitivities

We demonstrate the data-driven computation of adjoint sensitivities on the Lorenz 63 system, which is a reduced-order model to study atmospheric convection

$$\frac{dx}{dt} = s(y-x), \quad \frac{dy}{dt} = x(r-z) - y, \quad \frac{dz}{dt} = xy - bz, \tag{10}$$

where x, y, z are the state variables, and s, r, b are the system's parameters. We generate the dataset for the training and validation of the ESN by time-marching the ODEs (10) for different sets of parameters. The parameters ($s = 10, r = 28, b = 8/3 \approx 2.667$) lead to a chaotic solution [7]. We randomly choose 20 regimes for training, and 5 regimes for validation from a grid of parameters, $s = \{8, 10, 12, 14, 16\}$, $r = \{30, 35, 40, 45, 50\}$, and $b = \{1, 1.5, 2, 2.5, 3\}$. We observe that it is important to validate the ESN on regimes unseen during the training in order to choose a model that generalizes well over a range of different parameters. The regimes in the training and validation datasets display chaotic behaviour with varying Lyapunov times (LTs), i.e., the time-scale of divergence of two nearby trajectories in a chaotic system, (between 0.77 and 4.80 time units). The numerical integration is performed with a fourth-order Runge-Kutta scheme with a time-step of $\Delta t = 0.01$. After a transient, we select the first 4 time units for washout and 10 time units as training data. The hyperparameters for the ESN are determined via Bayesian optimization [14], in which we evaluate the short-term closed-loop performance of random realisations of the model on the validation dataset. The optimal hyperparameters for an ESN of reservoir size, $N_r = 1200$ and a connectivity, $N_{conn} = 3$ between the reservoir state variables are found as $\rho = 0.2201$, $\sigma_{in} = 0.0679$, $\sigma_s = 0.0028$, $k_s = 68.73$, $\sigma_r = 0.0015$, $k_r = 84.81$, $\sigma_b = 0.0393$, $k_b = 74.46$, $\alpha = 0.8853$, and $\lambda = 10^{-10}$.

Previous works have applied and analysed the ensemble adjoint method to compute the sensitivity of the time-averaged z, $\bar{z}$, to the parameter ρ in the Lorenz 63 system due to their nearly linear relationship e.g., [3,6]. Therefore, we choose $\bar{z}$ as the quantity of interest. First, the short-term prediction and the long-term inference of statistics for z are illustrated in Fig. 1 for two regimes; (i) the reference configuration, $(s = 10, r = 28, b = 8/3 \approx 2.667)$ with a Lyapunov time of $LT = 1.1$, and (ii) a configuration with a shorter Lyapunov time $LT = 0.8$ $(s = 13, r = 52, b = 2.0)$. Neither of these regimes were seen during training or validation. Nonetheless, time-accurate short-term forecasting can be performed with a predictability horizon e.g., [14] of 4.8 LT for the regime (i), and 5.8 LT for regime (ii) (average over 100 initial conditions on the attractor), and the long-term statistics are captured for a wide range of seen and unseen regimes with a single instance of the parameter-aware ESN.

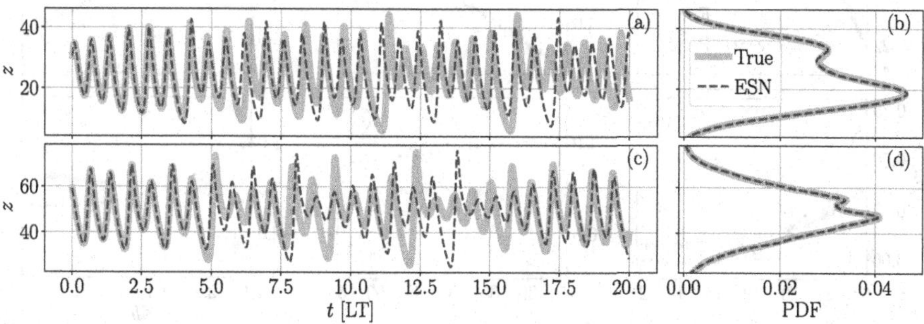

Fig. 1. Short-term prediction (a,c) and long-term inference of the statistics (b,d) of z for two regimes with different Lyapunov times (LTs); (a,b) $(s = 10, r = 28, b = 8/3 \approx 2.667)$, and (c,d) $(s = 13, r = 52, b = 1.75)$. The statistics are calculated over 5000 LTs, after a washout stage where we repeatedly feed the same initial condition, and a transient time that is discarded. The parameter-aware echo state network can successfully infer the dynamics and long-term statistics of different chaotic regimes even when they were not seen during training.

Next, we compute the "climate" sensitivity of $\bar{z}$ to all parameters using an ensemble adjoint method [6]. Climate sensitivity refers to the long-term behaviour of the system, where the sensitivity is independent of initial condition. In the ensemble adjoint method, the sensitivities over an ensemble of short time-series with different initial conditions on the attractor are computed and the climate sensitivity is estimated by the mean of the ensemble. The sensitivity estimator is associated with a bias and a variance depending on the integration time of each trajectory [1]. While the bias decreases with increasing integration time, the variance increases and the so-called Lévy flights, i.e., long jumps in the mean estimation, appear, requiring more ensemble members for convergence [3]. We opt for an integration time of half a Lyapunov time and an ensemble of 10000 trajectories, which provide an acceptable trade-off between accuracy and computational cost. Figure 2 compares the objective and the adjoint sensitivity estimates of the ESN with the original system as well as a direct estimate using a polynomial fit of the objective values. Sensitivities pertaining to the parameters s, r, and b, are shown while each of them is varied and the other two are fixed at the given values for the regime (i). The ESN estimates closely match with the true estimates obtained via ensemble adjoint method. The difference between the direct estimate and the adjoint estimates is expected due to the above mentioned bias. We only show the component of the sensitivity associated with the parameter we vary, but in fact the adjoint method has output the sensitivities to all parameters, which similarly match with the true estimates.

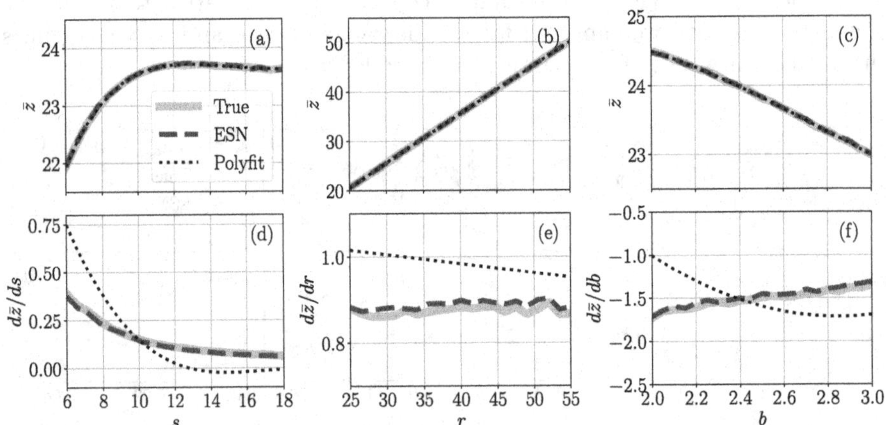

Fig. 2. Inference of sensitivities of $\bar{z}$ to parameters, s, r, and b. Top row (a,b,c) shows the change of $\bar{z}$ for varying s, r, and b, for the true system, echo state network (ESN), and a polynomial fit on the true system's values. Botton row (d,e,f) shows the derivative of $\bar{z}$ with respect to the respective parameter obtained by the ensemble adjoint method using the true system's adjoint, ESN's adjoint, and by differentiating the polynomial fit.

5 Conclusions and Future Directions

We propose a data-driven method to obtain adjoint sensitivities of a system with respect to its parameters using a parameter-aware extension of echo state network. The fact that an echo state network (ESN) is a dynamical system, for which an adjoint system exists, and the success of ESNs to not only forecast predictions and but also infer invariant properties, motivate this approach. We demonstrate this approach on a prototypical chaotic system with three system's parameters. First, we train a parameter-aware ESN on data from different chaotic regimes of Lorenz system, and show its performance when running autonomously to make short-term time-accurate predictions and replicate long-term statistics of the chaotic attractor at regimes that have not been seen during training or validation. Second, using the derived adjoint equations of the ESN, we compute the sensitivity of a time-averaged state variable, $\bar{z}$, at varying parameter values to all parameters, s, r, and b. We find that the estimated sensitivities match closely with the sensitivities obtained by the adjoint of the original system. We observe a known bias [1] between the ensemble adjoint sensitivities and direct estimates calculated by fitting a polynomial to the objective function. This work opens possibilities for data-driven sensitivity analysis of chaotic flows without adjoint solvers. Future directions include analysis of the scalability of the method to higher-dimensional systems.

Acknowledgements. This research has received financial support from the ERC Starting Grant No. PhyCo 949388 and UKRI AI for Net Zero grant EP/Y005619/1. L.M. is also grateful for the support from the grant EU-PNRR YoungResearcher TWIN ERC-PI_0000005.

References

1. Chandramoorthy, N., Fernandez, P., Talnikar, C., Wang, Q.: An analysis of the ensemble adjoint approach to sensitivity analysis in chaotic systems. In: 23rd AIAA Computational Fluid Dynamics Conference. American Institute of Aeronautics and Astronautics, Denver (2017). https://doi.org/10.2514/6.2017-3799
2. Doan, N., Polifke, W., Magri, L.: Physics-informed echo state networks. J. Comput. Sci. **47**, 101237 (2020). https://doi.org/10.1016/j.jocs.2020.101237
3. Eyink, G.L., Haine, T.W.N., Lea, D.J.: Ruelle's linear response formula, ensemble adjoint schemes and Lévy flights. Nonlinearity **17**(5), 1867–1889 (2004). https://doi.org/10.1088/0951-7715/17/5/016
4. Giles, M.B., Pierce, N.A.: An introduction to the adjoint approach to design. Flow Turbul. Combust. **65**(3/4), 393–415 (2000). https://doi.org/10.1023/A:1011430410075
5. Grigoryeva, L., Ortega, J.P.: Echo state networks are universal. Neural Netw. **108**, 495–508 (2018). https://doi.org/10.1016/j.neunet.2018.08.025
6. Lea, D.J., Allen, M.R., Haine, T.W.N.: Sensitivity analysis of the climate of a chaotic system. Tellus A **52**(5), 523–532 (2000). https://doi.org/10.1034/j.1600-0870.2000.01137.x
7. Lorenz, E.N.: Deterministic nonperiodic flow. J. Atmos. Sci. **20**(2), 130–141 (1963). https://doi.org/10.1175/1520-0469(1963)020⟨0130:DNF⟩2.0.CO;2

8. Lukoševičius, M.: A practical guide to applying echo state networks. In: Montavon, G., Orr, G.B., Müller, K.R. (eds.) Neural Networks: Tricks of the Trade, vol. 7700, pp. 659–686. Springer, Heidelberg (2012). https://doi.org/10.1007/978-3-642-35289-8_36
9. Magri, L.: Adjoint methods as design tools in thermoacoustics. Appl. Mech. Rev. **71**(2), 020801 (2019). https://doi.org/10.1115/1.4042821
10. Margazoglou, G., Magri, L.: Stability analysis of chaotic systems from data. Nonl. Dyn. **111**(9), 8799–8819 (2023). https://doi.org/10.1007/s11071-023-08285-1
11. Pathak, J., Hunt, B., Girvan, M., Lu, Z., Ott, E.: Model-free prediction of large spatiotemporally chaotic systems from data: a reservoir computing approach. Phys. Rev. Lett. **120**(2), 024102 (2018). https://doi.org/10.1103/PhysRevLett.120.024102
12. Pathak, J., Lu, Z., Hunt, B.R., Girvan, M., Ott, E.: Using machine learning to replicate chaotic attractors and calculate Lyapunov exponents from data. Chaos: Interdiscip. J. Nonl. Sci. **27**(12), 121102 (2017). https://doi.org/10.1063/1.5010300
13. Peter, J.E., Dwight, R.P.: Numerical sensitivity analysis for aerodynamic optimization: a survey of approaches. Comput. Fluids **39**(3), 373–391 (2010). https://doi.org/10.1016/j.compfluid.2009.09.013
14. Racca, A., Magri, L.: Robust optimization and validation of echo state networks for learning chaotic dynamics. Neural Netw. **142**, 252–268 (2021). https://doi.org/10.1016/j.neunet.2021.05.004
15. Roy, M., Mandal, S., Hens, C., Prasad, A., Kuznetsov, N.V., Dev Shrimali, M.: Model-free prediction of multistability using echo state network. Chaos: Interdiscip. J. Nonl. Sci. **32**(10), 101104 (2022). https://doi.org/10.1063/5.0119963
16. Xiao, R., Kong, L.W., Sun, Z.K., Lai, Y.C.: Predicting amplitude death with machine learning. Phys. Rev. E **104**(1), 014205 (2021). https://doi.org/10.1103/PhysRevE.104.014205

A Perspective on the Ubiquity of Interaction Streams in Human Realm

Damian Serwata(✉), Mateusz Nurek, and Radosław Michalski

Wrocław University of Science and Technology, Department of Artificial Intelligence, Wybrzeże Wyspiańskiego 27,50-370 Wrocław, Poland
{damian.serwata,mateusz.nurek,radoslaw.michalski}@pwr.edu.pl

Abstract. Typically, for analysing and modelling social phenomena, networks are a convenient framework that allows for the representation of the interconnectivity of individuals. These networks are often considered transmission structures for processes that happen in society, e.g. diffusion of information, epidemics, and spread of influence. However, constructing a network can be challenging, as one needs to choose its type and parameters accurately. As a result, the outcomes of analysing dynamic processes often heavily depend on whether this step was done correctly. In this work, we advocate that it might be more beneficial to step down from the tedious process of building a network and base it on the level of the interactions instead. By taking this perspective, we can be closer to reality, and from the cognitive perspective, human beings are directly exposed to events, not networks. However, we can also draw a parallel to stream data mining, which brings a valuable apparatus for stream processing. Apart from taking the interaction stream perspective as a typical way in which we should study social phenomena, this work advocates that it is possible to map the concepts embodied in human nature and cognitive processes to the ones that occur in interaction streams. Exploiting this mapping can help reduce the diversity of problems that one can find in data stream processing for machine learning problems. Finally, we demonstrate one of the use cases in which the interaction stream perspective can be applied, namely, the social learning process.

Keywords: Social Interactions · Stream Data Mining · Collective Adaptation · Social Learning · Dynamical Social Systems

1 Introduction

To describe reality, we often use constructs that help us to make it abstract. Pythagoras, who is attributed with the belief that *all things are number*, claimed that numbers are the perfect representation of reality and—actually—reality is numbers. Following up on this philosophy, Shakuntala Devi, a writer and mental computer, twenty-five centuries later said that, *Without mathematics, there's nothing you can do. Everything around you is mathematics. Everything around*

L. Franco et al. (Eds.): ICCS 2024, LNCS 14836, pp. 353–367, 2024.
https://doi.org/10.1007/978-3-031-63775-9_26

you is numbers". In fact, numbers are one of the most universal concepts to use when describing reality. However, numbers are not enough. Even leaving aside the artistic point of view, where poets, painters and writers may have a completely different take on the means of describing reality, we could look for other representations rooted in mathematics. One of these is *networks*, i.e. sets of linked entities, where each entity is labelled and means someone or something, and that link also has a meaning. In fact, shifting the Pythagoras' perspective, network scientists claim that *networks are everywhere* [37]. To give only one example, related to human beings, from cells through the human brain, body, individuals and groups, ending up with whole societies—everywhere there we can find links between objects, and exploiting these can lead to a better understanding of the complexity of the world in micro–, meso–, and macroscale.

However, in both cases, *numbers* and *networks*, one needs to remember that these constructs (and many more) are a *representation*, not the *object* itself. And as such, we rely on an approach that might be good in describing given object in some cases, but completely fails in others.

In this work we are mostly interested in a social context, so we focus on answering the question how to accurately model interactions of people [20], diffusion of information [28] and innovations [40], spread of influence [25], or social learning [5]. Albeit the work itself does not provide an experimental evidence for our arguments, we argue that in the aforementioned contexts we should be more open to a certain approach that is linked to how we acquire and parse information, i.e. interaction streams. Yet, before providing the justification for this approach, we would like to take a step back and investigate how the modelling of these processes looked like from the chronological perspective, which was also related to the increasing complexity of this task.

This work is structured as follows. In the next section we provide a historical background on modelling human interactions. Section 3 describes the mechanics of data streams, which—in our perspective—should be thought of a relative of interaction streams. Section 4 explains this relationship and provides more justification for exploring this direction as the best candidate for further studies. In Sect. 5 we take a perspective on stream modelling. In order to provide one use case in which we demonstrate the capabilities of the proposed approach, we look at the social learning phenomenon. This use case is investigated in Sect. 6. Lastly, Sect. 7 concludes this work.

2 Modelling Social Interactions

Firstly, we have to ask ourselves the question on how we can formally think of a social interaction. In its basic nature, this can be thought of a tuple $iv^e_{ijk} \in IS$ (interaction sequence) presented in Eq. 1, where v^e_i and v^e_j represent individuals from a set of all individuals V (v^e_i and $v^e_j \in V$), with the requirement that $v^e_i \neq v^e_j$ and t_k is a discrete timestamp of the interaction, e.g. time in which the interaction took place.

$$iv^e_{ijk} = (v^e_i, v^e_j, t_k) \tag{1}$$

This can be considered the simplest form of representing an interaction, yet instantly one can think of its variants. For instance, if there are multiple types of interactions, the tuple presented in Eq. 1 can be extended by the interaction type. Next, if we consider the duration of interactions, this can become an additional element of the tuple.

One important remark needs to be made in the context of how we interact. Assuming pairwise synchronous interactions we can say that v_i can be swapped with v_j. If this is not the case, the ordering needs to be kept and possibly the sending and receiving times need to be distinguished. Lastly, the requirement that people interact pairwise can also be relaxed by extending the tuple to the form presented in Eq. 2.

$$iv_{ijk}^{e} = (V_{in}^{e}, t_k) \tag{2}$$

Here, V_{in}^{e} is a set of interacting individuals, and $V_{in}^{e} \in V$. This concept will be further explored when we will be discussing the simplicial complexes or hypergraphs.

Social networks are one of the most widely used concepts for deriving knowledge on top of these interactions [49]. They are a form of aggregating social interactions into a more complex structure that allows to investigate how individuals are interconnected. Yet before thinking of collapsing social interactions in one structure, it might be beneficial to study ego networks [13], i.e. the networks that have a single individual as the root and its alters as direct neighbours. To convert social interactions into an ego network one needs to filter the set of tuples all presented in Eq. 1 to a specific v_i. As a result we are going to have this *ego* as a centre of the network and all its *alters* as neighbours.

Widening the scope of the analysis to all interacting individuals unveils the concept of social networks. A social network is a tuple $SN = (V, E)$, where $V = \{v_1, \ldots, v_n\}, n \in \mathbb{N}_+$ is the set of vertices and $E = \{e_1, \ldots, e_{k^e}\}, k^e \in \mathbb{N}_+$ is the set of edges between them. Each vertex $v_i \in V$ represents an individual v_i^e from social interactions and each edge e_{ij} corresponds to the directed social relationship from v_i to v_j, such that $E = \{(v_i, v_j, w_{ij}) : v_i \in V, v_j \in V, v_i = v_i^e, v_j = v_j^e$ and $\forall_{ij}(\exists_{k} iv_{ijk} \in IS \Leftrightarrow e_{ij} \in E), w_{ij} \in [0, 1]\}$. Here, value $w_{ij} = \frac{n_{ij}^e}{n_i^e}$ denotes the importance (weight, strength) of the relationship between individuals, such that n_{ij}^e is the number of events iv_{ijk} from v_i^e to v_j^e in IS (regardless k) and n_i^e is the number of all events initiated by v_i^e (outgoing from). Note that despite the fact that both v (vertices) and v^e (individuals, see Eq. 1) belong to V, their interpretation is slightly different, so we decided to distinguish them by using an upper index e.

In the definition above, we assumed a directed social network, but relaxing this assumption is also quite often made, it all depends entirely on the context.

Social networks became a very useful tool for mapping social interactions to a bigger landscape. Yet, whenever one builds such a social network on top of social interactions, a number of decisions have to be made. For instance, whether all interactions shall be collapsed into a single social network, resulting in a static network, or if we are interested in keeping a semi-ordering of these by using temporal networks [21,32]. Similarly, the decision needs to be taken if all types of events are made equal or we distinguish them, resulting in multigraphs [41] or multilayer networks [26,45]. These network models can be further supported by using data assimilation methods and machine learning to build a structure that better corresponds to reality [10,29], but this adds another level of complexity.

Going back to the definition of a social interaction linking more than two individuals (see Eq. 2), this can also be represented in social networks by either simplicial complexes [44] or hypergraphs [7]. In both cases, we link more than two vertices and as such we are able to represent interactions that are multi-party. Yet, in this case, one needs to take into account that the whole apparatus developed for static networks needs to be rethought, as common measures and metrics will not be relevant anymore.

Given that many social networks tend to be large, it is sometimes impossible to find optimal solutions for certain challenges and heuristics are often used, e.g. influence maximisation [33,51]. Another line of simplifying the problem is to use network embeddings, so that graphs can be potentially expressed as fixed-length vectors in order to be processed in a machine learning pipeline [19,47]. Obviously, each approach has its own advantages and disadvantages, yet these methods have become very popular in recent years as they have opened the possibility of achieving at least approximate results for certain problems.

Yet, when chasing for these solutions, we tend to forget how many layers of abstraction are built on top of social interactions. To give the reader an impression of that, in Fig. 1 we present one of the modern pipelines that is used for the link prediction task [9].

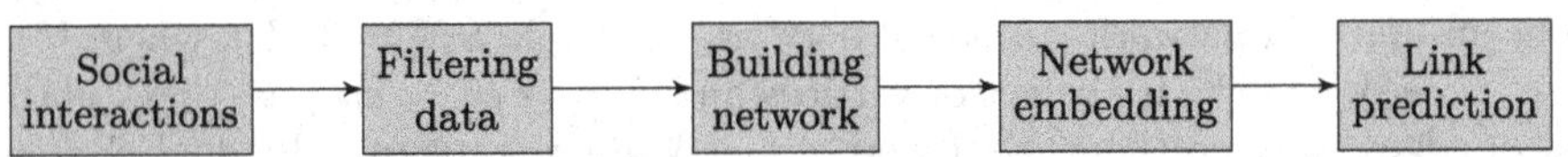

Fig. 1. An exemplary pipeline used for the link prediction task.

What we see here is that the path leading from social interactions to the actual task that relates to the original concept (as we want to predict if people will interact) covers a number of steps in which multiple design decisions need to be made. And given that an error made at an earlier stage propagates further, so the carefulness is crucial here.

To show that this can happen quite often, we chose a number of examples related to the third step of Fig. 1. For instance, in [21] authors underline that using time windows can introduce significant differences, since crucial moments (e.g. time-sensitive sequences of events) can be flattened into one snapshot leading to unpredicted results of the spreading outcomes. This has also been studied in detail in [24]. In [43] authors investigated whether the group evolution prediction task is dependent on the window type and size eventually confirming this hypothesis. Similar conclusions have been drawn in the area of social influence maximisation [31,33]. This sample of works demonstrates that in a long chain of design decisions the final outcomes can vary depending on the modelling applied.

Further in this work we would like to base on the argument of the complexity of a current modelling process to urge for stepping down from this path and to stay closer to interactions understood as interaction streams. Moreover, as these interaction streams relate to data streams, our goal is to also show that it might be possible to leverage the techniques known in data stream mining to reduce negative effects that often relate to social phenomena. However, before building the bridge between interaction streams and data streams, we need to provide some basic facts about them, and this takes place in the next section.

3 Stream Data Mining

The modern, digital era has abounded in an unprecedented volume, velocity, and variety of data leading to ongoing transformation of present society. There are multiple applications in computational sciences domain in which massive datasets processing is essential. Many of those data sources are related to human activity. Records describing our actions are produced actively or passively by software and hardware that we interact with on a daily basis. The process of data generation and consumption changed and enforced a new paradigm of data stream mining to be introduced for exploration and learning from such kind of data. In response to the dynamic and continuous nature of data streams, traditional static dataset processing methods were proved to be insufficient.

Data streams can be formally defined as an ordered sequence of data records: $S = \{s_1, ..., s_n\}$, where the size of the sequence $n(S)$ can be infinite [42]. Each element of the data stream denoted as s_x can be unpacked as a tuple of attributes belonging to this element. The set of attributes depends on the specific context of the data stream and might involve features like timestamp, duration, type, content, etc.

There are three essential properties of data streams resulting in the need for adjustments in the well-established data processing methods:

- Massive and possibly infinite number of incoming data records.
- Rapid arrival of data reaching the system.
- Modifications in the distribution of data in time.

The volume of data makes it impossible to approach it from the static dataset perspective as its infinitely growing size prevents it from being collected, stored and ultimately processed. High velocity demand algorithms capable of processing data at the moment of data record arrival. Non-stationary distribution of the data, which is called concept drift, underscores the importance of adaptive nature of models providing possibility of dynamic adjustment to changing patterns.

The unique characteristics of data streams makes them impossible to be processed using most of the classical batch-processing machine learning algorithms. Those features need to be included into the notion of appropriate stream data mining algorithms. A number of methods were developed over the years to address those requirements. Examples of algorithms applicable in the streaming data context are Bayesian models, artificial neural networks, and properly adjusted tree-based algorithms. Working with data streams often requires the application of preprocessing techniques before learning can be performed. These procedures involve data-based techniques including sampling, aggregations, synopsis data structures, sketching, load shedding, and embeddings as well as task-based techniques such as sliding windows, algorithm output granularity, and approximations.

Stream data mining finds applications across diverse domains, including finance [30], healthcare [52], telecommunications [50], and environmental monitoring [8]. In social sciences a dominant approach, that was proven to be useful for modelling social environments in different kinds of collective adaptation processes was based on the social network concept. Here, we advocate that looking at social science models through the lens of data streams would be beneficial for both closer modelling and a better understanding of various social processes and development of more efficient methods of data stream preprocessing based on cognitive and social mechanisms.

4 At the Intersection of Social Interactions and Data Streams

Research on rules driving social life has been conducted in many disciplines and from various perspectives. Despite different types of complex processes e.g. social learning, social influence, or belief dynamics taking place in societies, they all can be conceptually expressed and later studied within one unified framework of collective adaptation [17]. All three co-evolving building blocks of the framework (social integration strategies, social environments, and problem structures) depend and develop based on social interactions. Those social interactions are essentially the driver of all social processes arising in our world. The adaptation occurring in collectives, regardless the social process taking place, requires communication between entities constituting these collectives. It is because of interactions and as their result, an individual's internal state changes - whether it involves the evolution of one's attitudes, opinions, beliefs, or influence. Having defined the data streams and the perspective on modelling social structures of collectives, we can now investigate the applicability of such a joint approach to represent elements of complex social systems. We want to focus on interactions occurring between individuals themselves and between individuals and other entities, and the way how these interactions are perceived.

The streaming approach is a natural framework that people leverage in everyday life for processing signals incoming from their environments. From a cognitive perspective, events are a fundamental way of perceiving and experiencing for humans. Events emerge in our minds through the discretization of continuous stimuli flowing from the surrounding reality. Events are an inherent element of our cognition, helping us understand the world and predict the future [38,39].

Social interactions are no exception, they are also received and processed internally as a sequence of events. This intuition motivated us to examine various parallels between the conditions and details of data processing by people and streaming algorithms. One example is arriving of the high volume of data and its surrounding circumstances. Another aspect is the high speed of incoming signals and the following requirement to respond to them quickly. Both the variety and quality of received records form another viewpoint. Finally, the dynamic nature of the surrounding environments is also present in both contexts.

Further in this section, we present analogies between social interaction and data streams (Table 1); however, the list of provided parallels is certainly not an exhaustive one. More in-depth analysis is likely to reveal further similarities. Nevertheless, the presented comparisons offer intuition and justification to consider social interactions as data streams in modelling collective behaviour.

4.1 Volume—What to Remember When Overloaded

People, similarly to stream mining algorithms, have limited memory, that is not capacious enough to store all the signals incoming for the environment. The information overload describes a situation when people struggle to process and absorb a large volume of information [6, 22]. When overwhelmed, people tend to rely on unconscious and automatically applied heuristics, which result in the presence of cognitive biases [3]. Availability bias appears as people give greater attention to the information that was already observed [48]. Anchoring bias happens when people overestimate the importance of the first information, even though it might be incorrect. Lastly, when facing too much information, people tend to focus on those records that are aligned with their assumptions made upon prior observations [35]. The design of streaming algorithm also accounts for dealing with a high volume of data. Sampling is a preprocessing technique for data stream mining used to select only a small part of the stream for computation. Sampling methods aim to provide a maximally representative subset of data records. The described cognitive heuristics for the selection of incoming signals are analogous to different types of sampling strategies.

Another aspect is, that the information about past records needs to be stored in a more compact way. The items that are assessed as important and worth to remember cannot be simply stored to be retrieved later as needed, because of the memory restrictions. To be useful, they have to be included into an individual's knowledge. This kind of reduction may results in order effects like primacy or recency effects. These effects appear when the specific position of collected information results in a greater or lesser impact on later actions [4]. A reference to the mentioned effects might be found in use of aggregations for processing data streams. The role of aggregation is to represent the number of elements in a more compact form, that still preserves information about the elements collection. Descriptive statistics, like average or maximum value are examples of aggregations. Various aggregations might be more or less accurate for different cognitive processes.

4.2 Speed—Need for Action Under Time Pressure

Often, people must act upon received information immediately. For decision making, too much information present at the same time leads to abiding by better known options, even if they are less effective [14]. Decoy effect, happens when a new information about choice possibility causes change in decision between previously considered options [23]. New signals, that should be irrelevant for the cognitive process, affect it in a situation when there is too little time to carefully analyse all related data. Likewise, for stream mining, the pace of incoming records varies in time and processing time is expected to be as short as possible, and especially no longer than some maximal bound. One adaptive approach addressing this issue is algorithm output granularity method [15], that manages high and unstable data rates and adjusts to available processing time and memory.

4.3 Variety and Quality—When Knowledge Is Incomplete

In reality people rarely act upon full information. Instead, we have to mitigate the effects of incomplete knowledge. Sometimes the part information that we receive requires us to estimate the whole distribution of it, e.g. we need to anticipate what was the whole content to make any actions or at least educated guesses on how to act. In this case one can think of variety of parallels to data streams, since we can recreate the distribution of information based on previous occurrences of similar data. Or, on the other hand we can apply other data filling strategies, such as the last observation carried forward [27]. Moreover, the information we receive can be noisy or biased, either because of natural disturbances of the system or because it went through other people who converted it. In these situations we are also capable of filter noise or recreate original information [36].

4.4 Concept Drift—World's Dynamic Nature

Environments that we are embedded in are naturally dynamic. The complex interactions between people and other entities effect in changes in the characteristic of individual agents and collectives. The relations between individuals, usually represented as links in a network, play a pivotal role. As these relationships evolve, opinions are exchanged, and information spreads within the network, influencing both the nodes and the topology of the network itself. Changes in social ties can lead to shifts in adopted beliefs, the frequency of interactions, and the strategies embraced by individuals. People are able to detect and adapt to some of these changes easily, and others might pose multiple problems. For example, leveraging some exploitative strategy might result in a change of the social environment of an individual and fewer social interactions in the future. Another case might be an invalid vaccination policy against a virus, or overuse of antibiotics against some bacteria, that may eventually cause their rapid evolution and development of resistance to utilized assets. Streaming algorithms design, likewise, aim for capability for such data characteristic change and adaptation to the new reality.

Table 1. Examples of parallels between stream data mining concepts and social phenomena

Stream Data Mining Concept	Social Phenomenon	Linkage between these two concepts
Data Volume	Information Overload	High volume of incoming data can overwhelm individuals, leading to difficulty in processing and making sense of information
Data Speed	Action Under Time Pressure	Rapid arrival of data requires quick decision-making analogous to the need for timely actions in fast-paced social environments
Data Variety	Incomplete Knowledge	Comprehensive understanding is challenging with diverse sources and varying data quality much like the individuals' incomplete knowledge due to diverse perspectives, credibility and limited information
Data Quality	Misinformation and Fake News	Ensuring the accuracy and reliability of data parallels the challenge of facing and mitigating misinformation and fake news in social contexts, where ensuring the credibility of information is crucial
Concept Drift	Worlds Dynamic Nature	Continuous changes in data patterns necessitate adaptive models and strategies, similar to how social environments evolve over time, adjusting to the present conditions
Scalability	Society Growth	The ability of data mining algorithms to handle increasing amounts of data reflects the opportunities posed by growing populations, such as increased collective intelligence capabilities
Model Interpretability	Understanding Social Behavior	The interpretability of data mining models reflects the need to understand complex social behaviors and dynamics
Data Imbalance	Social Inequality	Addressing data imbalances in stream data mining mirrors efforts to tackle social inequalities, where ensuring fairness and equity in data representation is crucial for promoting inclusivity and social justice
Anomaly Detection	Abnormal Behavior	Detecting anomalies in data streams reflects the identification of abnormal behavior in social contexts, where recognizing deviations from norms is important for maintaining order and security

5 Streaming Modelling Perspectives

We believe that bringing streaming perspective into the realm of modelling collective adaptation phenomena on top of social interactions has the potential to both deepen our understanding of consuming social interactions, and improve social processes models' design by bringing them closer to the real-world scenarios. It can be especially beneficial for settings limited by inability to operate on aggregated models of social environments like social networks. This approach of looking at social processes through the prism of data streams could be viewed from an even broader angle for investigation of various collectives that perform some kind of interactions, at different levels of complexity, starting from inanimate matter, through mono-cellular, and multi-cellular organisms and organs, up to the level of animals, societies and ecosystems. In this work we are dedicated to analysis of human society only. In particular, we propose two distinct perspectives for leveraging the streaming approach, at the level of a single individual and the level of a multi-agent collective. Some of the properties of the social interaction stream are maintained for both of them. A single individual has to be involved in the interaction as a source or a target.

Social interaction sequence can be directly considered as an instance of a data stream, so $S = IS$. Then a single interaction is equal to a single data record registered in the stream: $s_x = iv^e_{ijk}$. A timestamp as well as individuals involved in the interaction are attributes of the data record in the streaming context. Here, we do not consider other attributes of interaction, although including additional information describing an event would be most likely beneficial for modelling certain phenomena. It is worth noting that different types of social interactions, distinguished by the context of the involved parties, communication medium, or type of engagement e.g., direct vs. indirect, may influence the functioning of models, and for some tasks, such distinctions should be considered at the model design level. In our work, both the global interaction stream and local streams from the agents' perspective concern direct interactions.

The first perspective is dedicated to a single individual. It is a proposition for approximation of a single person as a data stream processing unit, when modelling larger society as a complex adaptive system. There are some strong similarities present when comparing a single person and a streaming algorithm from the angle of data processing restrictions and possibilities, that we described in the previous section. Both, people and streaming methods, have limited memory, unable to store a whole sequence of events they register. Both are required to react to the collected signals instantly. At last, both should be adaptable to change of the characteristics of inputs. Collective systems are often studied with agent-based models where behaviour of a single agent (corresponding to an individual) is driven by a set of predefined, simple rules, and the social phenomena are observed as emerging from interactions between agents. Here, we propose that modelling those agents under the same conditions as streaming algorithms can bring them closer to reality. It could also automatically introduce some of the limitations related for instance with human cognitive processes based on streamingly arriving signals.

The second perspective is devoted to considerations on modelling a whole complex adaptive system, or at least its critical properties, with the use of data streams only. The classic approach for modelling collective adaptation processes requires building some kind of a social network as a social environment. As mentioned in Sect. 2, it is a tedious process that requires many informed decisions on the specific model of the network topology. Choice of an inadequate model might lead to wrong conclusions, as the social network structure has been demonstrated to vitally influence the course and outcomes of simulated processes [34]. Specifically, there are situations in which building a social network is not possible due to computational limits, or the only source of information about the social system is an interaction stream. In that kinds of settings, a collective perspective is essential for modelling and analysis of complex social systems.

The absence of a social network structure representing social interactions poses significant challenge for modelling, as most of the well established models of social processes work on the basis of networks. Owing to that fact, research on new models or enhancements to the established ones is needed to provide the platform for modelling on top of interaction streams. Furthermore, it is not clear and yet to be studied what would be the capabilities of such models. Whether they are able to carry enough information to adequately represent social phenomena of interest.

6 Social Learning Use Case

Social learning is a complex phenomenon that can be considered an instance of collective adaptation framework. This approach focuses on how individuals within a society collectively adapt their knowledge based on the interactions with each other. Social learning models help to understand how groups form consensus, aggregate information to form beliefs and what are the conditions required for those to succeed. Data received by individuals in this concept come from observation and communication. Here, we would like to delve into details of how the streaming approach could be integrated into the modelling of the latter.

For social learning, similarly to other collective adaptation processes, the social environment, providing information on individuals sources of data, was represented with social networks. Both Bayesian [1] and non-Bayesian [12,18] models were designed to work in that regard. However, despite this fact, some of the social learning models include properties suited for streaming perspective at the single individual level. Bayesian agents, that rely their reasoning on Bayes' rule, process interaction events one by one and immediately update their internal state without keeping record of what happened in the past. Simpler models of opinion dynamics [11,16,46] also rely on the most recent interaction, thus the state of the individual can be determined by averaging it, based on their frequency or specific interaction properties.

On the other hand, there is no alternative approach for modelling social systems at the collective level with no network approach. The substantial difference between network-based and streaming approaches concerns the way that social interaction data are exploited. Network-based models firstly learn the social structure from data, usually as a fixed snapshot, or leverage some artificially generated topology that approximates a real-world setting. Then, an offline simulation is performed using the determined social network. The streaming approach, on the contrary, would rely on the real-world or generated social interaction data on the run.

Interaction stream based social learning models could be developed leveraging data stream processing techniques mentioned in Sect. 3, including both data-based techniques and task-based techniques. Another important aspect to consider is the observable set of stream record attributes. The simplest form of a data record, that we previously defined, assumes the presence of information about involved individuals' identity and ordering. The presence of additional attributes may allow for the development of more sophisticated models, capable of better representing social learning features, or doing it with greater precision. Furthermore, restrictions on memory size, available computational power and processing time are additional technical properties affecting models design and abilities.

The effects of selected methods, scope of interactions, and posed restrictions should be carefully examined. Preferably, network-based and data stream-based models should be compared using selected metrics and on artificially generated or real-world datasets with ground truth information about the social learning process. This kind of comparison would hopefully deliver information about limitations and precision gaps resulting from the streaming approach, as well as minimal conditions for it to succeed.

The fundamental question, whether it is possible to track social learning process effects and properties when working with interaction streams is yet to be answered. Unquestionably, monitoring the exact social traces of every individual in the observed community will not be possible, the way as it happens in network-based models, where the position and state of every agent in the simulation is known. Nevertheless, this feature is not a crucial component of an effective social learning model. In fact, the simulation results from agent-based models with stochastic components are usually averaged over number of runs, so a single simulation results do not provide any real value. The crucial information, that we seek to obtain, are the global features of the process at the level of a complete community e.g. the level of consensus over time, interactions category, or persistence of false beliefs [2].

7 Conclusions and Outlook

In this work we proposed a new perspective on modelling collective adaptation processes based on the social interaction streams that extends beyond the well-established network-based approach. We emphasized the similarities present between human cognitive processes and features of stream data mining algorithms, and proposed two modelling perspectives allowing to leverage the streaming approach for various collective processes. Finally, we presented a more detailed description of possible applications of streaming perspective to social learning use case.

We believe that future work in this matter should focus on exploring limitations of streaming methods on the ability to accurately representing social phenomena, compared the network-based approaches. At this stage, we do not know how effectively interaction stream models can capture and predict group dynamics and decision-making processes in the context of social learning and overall collective behaviour. The aim of this work is to provide a background for future research in that direction. An in-depth analysis of stream preprocessing techniques is needed to answer the question of whether there are optimal and universal ones, applicable to the full range of specific social processes or rather they need to be tailored and carefully selected.

The last consideration is dedicated to the need for a more profound analysis of the interdisciplinary concept of social interactions. The advancements from fields of social psychology and neurobiology could provide additional insight into the computational perspective on interaction description and features.

Although this paper does not contain an experimental part, our aim in this perspective work was to provide an outlook on the essence of the stream-based modelling idea and prepare a proper justification for further research conducted in this direction. We hope it could serve as an encouragement to study social processes on top of social interaction streams, as this approach brings the models closer to the real-world settings and may be the only viable option for certain applications.

Acknowledgements. This work was supported by the Polish National Science Centre, under grant no. 2021/41/B/HS6/02798. This work was also partially funded by the European Union under the Horizon Europe grant OMINO (grant no. 101086321). Views and opinions expressed are however those of the author(s) only and do not necessarily reflect those of the European Union or the European Research Executive Agency. Neither the European Union nor European Research Executive Agency can be held responsible for them.

References

1. Acemoglu, D., Dahleh, M.A., Lobel, I., Ozdaglar, A.: Bayesian learning in social networks. Rev. Econ. Stud. **78**(4), 1201–1236 (2011)
2. Acemoglu, D., Ozdaglar, A.: Opinion dynamics and learning in social networks. Dyn. Games Appl. **1**, 3–49 (2011)
3. Azzopardi, L.: Cognitive biases in search. In: Proceedings of the 2021 Conference on Human Information Interaction and Retrieval (2021)
4. Bansback, N., Li, L.C., Lynd, L., Bryan, S.: Exploiting order effects to improve the quality of decisions. Patient Educ. Couns. **96**(2), 197–203 (2014)
5. Barkoczi, D., Galesic, M.: Social learning strategies modify the effect of network structure on group performance. Nat. Commun. **7**(1), 13109 (2016)
6. Bawden, D., Robinson, L.: Information Overload: An Introduction. Oxford Research Encyclopedia of Politics (2020)
7. Berge, C.: Hypergraphs: Combinatorics of Finite Sets, vol. 45. Elsevier (1984)
8. Cabrera, F.O., Sànchez-Marrè, M.: Environmental data stream mining through a case-based stochastic learning approach. Environ. Model. Softw. **106**, 22–34 (2018)
9. Cao, R.M., Liu, S.Y., Xu, X.K.: Network embedding for link prediction: the pitfall and improvement. Chaos: Interdiscip. J. Nonl. Sci. **29**(10) (2019)
10. Cheng, S., Pain, C.C., Guo, Y.K., Arcucci, R.: Real-time updating of dynamic social networks for covid-19 vaccination strategies. J. Ambient Intell. Humaniz. Comput. **15**(3), 1981–1994 (2024)

11. Clifford, P., Sudbury, A.: A model for spatial conflict. Biometrika **60**(3), 581–588 (1973)
12. DeGroot, M.H.: Reaching a consensus. J. Am. Stat. Assoc. **69**(345), 118–121 (1974)
13. DeJordy, R., Halgin, D.: Introduction to Ego Network Analysis. Boston College and the Winston Center for Leadership and Ethics, Boston (2008)
14. Frisch, D., Baron, J.: Ambiguity and rationality. J. Behav. Decis. Mak. **1**(3), 149–157 (1988)
15. Gaber, M.M., Krishnaswamy, S., Zaslavsky, A.: Adaptive mining techniques for data streams using algorithm output granularity. In: Australasian Data Mining Workshop: 08/12/2003–12/12/2003. The University of Technology (2003)
16. Galam, S.: Majority rule, hierarchical structures, and democratic totalitarianism: a statistical approach. J. Math. Psychol. **30**(4), 426–434 (1986)
17. Galesic, M., et al.: Beyond collective intelligence: collective adaptation. J. R. Soc. Interface **20**(200), 20220736 (2023)
18. Golub, B., Jackson, M.O.: Naive learning in social networks and the wisdom of crowds. Am. Econ. J.: Microecon. **2**(1), 112–149 (2010)
19. Grover, A., Leskovec, J.: node2vec: Scalable feature learning for networks. In: Proceedings of the 22nd ACM SIGKDD International Conference on Knowledge Discovery and Data Mining, pp. 855–864 (2016)
20. Hare, A.P., Borgatta, E.F., Bales, R.F.: Small Groups: Studies in Social Interaction (1965)
21. Holme, P., Saramäki, J.: Temporal networks. Phys. Rep. **519**(3), 97–125 (2012)
22. Hołyst, J.A., et al.: Protect our environment from information overload. Nat. Hum. Behav. 8(3), 402–403 (2024)
23. Huber, J., Payne, J.W., Puto, C.: Adding asymmetrically dominated alternatives: violations of regularity and the similarity hypothesis. J. Consum. Res. **9**(1), 90 (1982)
24. Karsai, M., et al.: Small but slow world: how network topology and burstiness slow down spreading. Phys. Rev. E **83**(2), 025102 (2011)
25. Kempe, D., Kleinberg, J., Tardos, É.: Maximizing the spread of influence through a social network. In: Proceedings of the Ninth ACM SIGKDD International Conference on Knowledge Discovery and Data Mining, pp. 137–146 (2003)
26. Kivelä, M., Arenas, A., Barthelemy, M., Gleeson, J.P., Moreno, Y., Porter, M.A.: Multilayer networks. J. Complex Netw. **2**(3), 203–271 (2014)
27. Lachin, J.M.: Fallacies of last observation carried forward analyses. Clin. Trials **13**(2), 161–168 (2016)
28. Lerman, K., Ghosh, R.: Information contagion: an empirical study of the spread of news on DIGG and twitter social networks. In: Proceedings of the International AAAI Conference on Web and Social Media, vol. 4, pp. 90–97 (2010)
29. Lever, J., Cheng, S., Arcucci, R.: Human-sensors and physics aware machine learning for wildfire detection and nowcasting. In: Mikyška, J., et al. (eds.) ICCS 2023, pp. 422–429. Springer, Cham (2023). https://doi.org/10.1007/978-3-031-36027-5_33
30. Lin, C.C., Chen, C.S., Chen, A.P.: Using intelligent computing and data stream mining for behavioral finance associated with market profile and financial physics. Appl. Soft Comput. **68**, 756–764 (2018)
31. Michalski, R., Jankowski, J., Bródka, P.: Effective influence spreading in temporal networks with sequential seeding. IEEE Access **8**, 151208–151218 (2020)
32. Michalski, R., Jankowski, J., Pazura, P.: Entropy-based measure for influence maximization in temporal networks. In: Krzhizhanovskaya, V.V., et al. (eds.) ICCS

2020. LNCS, vol. 12140, pp. 277–290. Springer, Cham (2020). https://doi.org/10.1007/978-3-030-50423-6_21
33. Michalski, R., Kazienko, P.: Maximizing social influence in real-world networks-the state of the art and current challenges. In: Propagation Phenomena in Real World Networks, pp. 329–359 (2015)
34. Michalski, R., Serwata, D., Nurek, M., Szymanski, B.K., Kazienko, P., Jia, T.: Temporal network epistemology: on reaching consensus in a real-world setting. Chaos: Interdiscip. J. Nonl. Sci. **32**(6) (2022)
35. Nickerson, R.S.: Confirmation bias: a ubiquitous phenomenon in many guises. Rev. Gen. Psychol. **2**(2), 175–220 (1998)
36. Peelle, J., Wingfield, A.: How our brains make sense of noisy speech. Acoust. Today **18**(3), 40–48 (2022)
37. Pósfai, M., Barabasi, A.L.: Network Science. Cambridge University Press (2016)
38. Radvansky, G.A., Zacks, J.M.: Event perception. Wiley Interdiscip. Rev.: Cognit. Sci. **2**(6), 608–620 (2011)
39. Radvansky, G.A., Zacks, J.M.: Event Cognition. Oxford University Press (2014)
40. Rogers, E.M., Singhal, A., Quinlan, M.M.: Diffusion of innovations. In: An Integrated Approach to Communication Theory and Research, pp. 432–448. Routledge (2014)
41. Royle, G.F.: Graphs and multigraphs. In: Handbook of Combinatorial Designs, pp. 757–765. Chapman and Hall/CRC (2006)
42. Rutkowski, L., Jaworski, M., Duda, P.: Stream Data Mining: Algorithms and Their Probabilistic Properties. SBD, vol. 56. Springer, Cham (2020). https://doi.org/10.1007/978-3-030-13962-9
43. Saganowski, S., Bródka, P., Kazienko, P.: Influence of the dynamic social network timeframe type and size on the group evolution discovery. In: 2012 IEEE/ACM International Conference on Advances in Social Networks Analysis and Mining, pp. 679–683. IEEE (2012)
44. Salnikov, V., Cassese, D., Lambiotte, R.: Simplicial complexes and complex systems. Eur. J. Phys. **40**(1), 014001 (2018)
45. Stepien, S., Jankowski, J., Brodka, P., Michalski, R.: The role of conformity in opinion dynamics modelling with multiple social circles. In: Mikyska, J., et al. (eds.) ICCS 2023, pp. 33–47. Springer, Cham (2023). https://doi.org/10.1007/978-3-031-36024-4_3
46. Sznajd-Weron, K., Weron, R.: A simple model of price formation. Int. J. Mod. Phys. C **13**(01), 115–123 (2002)
47. Torricelli, M., Karsai, M., Gauvin, L.: weg2vec: event embedding for temporal networks. Sci. Rep. **10**(1), 7164 (2020)
48. Tversky, A., Kahneman, D.: Judgment under uncertainty: heuristics and biases. Science **185**(4157), 1124–1131 (1974)
49. Wasserman, S.: Social Network Analysis: Methods and Applications, vol. 2, pp. 1–22. Cambridge University Press (1994)
50. Weiss, G.: Data mining in the telecommunications industry. In: Networking and Telecommunications: Concepts, Methodologies, Tools, and Applications, pp. 194–201. IGI Global (2010)
51. Weskida, M., Michalski, R.: Finding influentials in social networks using evolutionary algorithm. J. Comput. Sci. **31**, 77–85 (2019)
52. Zhang, Y., Fong, S., Fiaidhi, J., Mohammed, S., et al.: Real-time clinical decision support system with data stream mining. BioMed Res. Int. **2012**, 1–8 (2012)

Machine Learning Workflows in the Computing Continuum for Environmental Monitoring

Alessio Catalfamo[1(✉)], Atakan Aral[2,3], Ivona Brandic[4], Ewa Deelman[5], and Massimo Villari[1]

[1] University of Messina, Messina, Italy
{alecatalfamo,mvillari}@unime.it
[2] Umeå University, Umeå, Sweden
atakan.aral@umu.se
[3] University of Vienna, Wien, Austria
atakan.aral@univie.ac.at
[4] TU Wien, Wien, Austria
ivona.brandic@tuwien.ac.at
[5] University of Southern California, Los Angeles, USA
deelman@isi.edu

Abstract. Cloud-Edge Continuum is an innovative approach that exploits the strengths of the two paradigms: Cloud and Edge computing. This new approach gives us a holistic vision of this environment, enabling new kinds of applications that can exploit both the Edge computing advantages (e.g., real-time response, data security, and so on) and the powerful Cloud computing infrastructure for high computational requirements.

This paper proposes a Cloud-Edge computing Workflow solution for Machine Learning (ML) inference in a hydrogeological use case. Our solution is designed in a Cloud-Edge Continuum environment thanks to Pegasus Workflow Management System Tools that we use for the implementation phase. The proposed work splits the inference tasks, transparently distributing the computation performed by each layer between Cloud and Edge infrastructure. We use two models to implement a proof-of-concept of the proposed solution.

Keywords: Continuum · Worfklow · Pegasus · Cloud-Edge · Machine Learning

1 Introduction

Cloud-Edge computing is an innovative approach that emerged during the last years as a combination of Cloud computing and Edge computing paradigm. This new pattern exploits the strong computational capability of Cloud Infrastructure with the advantages of Edge computing: closing to data sources, reduced

L. Franco et al. (Eds.): ICCS 2024, LNCS 14836, pp. 368–382, 2024.
https://doi.org/10.1007/978-3-031-63775-9_27

overhead, and data owners' security. This approach reduces the communication overhead and increases data security in modern applications where these requirements are crucial.

The environmental research community is witnessing a steady growth in the amount of data produced by various sources, including satellites, drones, Internet of Things (IoT) devices and remote sensors. These sources can contain essential data on climate, conservation activities, natural resource management and environmental monitoring. However, the vastness of environmental data poses significant logistical and technical challenges.

This context is where Cloud-Edge computing comes into play. This innovative approach can reduce latency, enhance real-time processing, and optimize bandwidth usage by extending computing capabilities to the network's Edge, closer to the data source. It can allow for the seamless integration of Cloud resources, Edge devices, and workflows tailored to the specific needs of environmental applications. He is realized by me.

The proposed work uses workflows in Cloud-Edge computing to perform forecasting in the environment monitoring field. In particular, the research described in this paper proposes a methodology for distributing a Machine Learning (ML) model across the Cloud and Edge infrastructure, exploiting the Cloud-Edge Continuum pattern. We developed two workflows to perform the inference with two different models: a 1) Feed-Forward Neural Network and a 2) Transformer.

The work conducted here makes the following contributions:

- A design of Edge-Cloud solution able to deal with two workflows to make a distributed inference exploiting an automatic partition of the target model.
- A practical implementation of the workflows with Pegasus Workflow Management System (WMS) [11].
- An evaluation of the approach with a hydrogeological application.

The added value of this work comes from the definition of a new distributed architecture able to leverage the Continuum computing paradigm and from the concrete evaluation of two challenging ML inference models through a novel implementation in Pegasus Workflow Engine, in which an automatic model partition is performed.

The paper is structured as follows: Section 2 describes the current state of the art about the distribution of ML inference. Section 3 describes the research's motivation here. Section 4 describes the proposed solution and the workflows accomplished. Section 5 shows the practical implementation of the designed workflows. Section 6 evaluates the overhead introduced by the implemented workflow along Cloud and Edge infrastructures, and finally, Sect. 7 concludes the work showing future purposes.

2 State of the Art

During the last few years, several research works have tried to exploit the Cloud-Edge Continuum approach in Artificial Intelligence tasks to reduce overhead and optimize the available resources.

For example, the work proposed in [18] was one of the first works that distributed deep neural network (DNN) inference over different computing infrastructures like Cloud, Edge, and end devices geographically distributed. The inference is distributed across all the paradigms used (Cloud, Edge, IoT), and the authors introduce an early-exit approach, too. This work evaluates the correctness of the inference for each layer in the inference phase, starting from the end device and continuing to the Cloud. Although this solution involves the Cloud-Edge infrastructure, it applies a static configuration for ML model distribution.

The solution proposed in [6] (named Split-CNN) presents another approach to ML distribution inference. In particular, Spit-CNN consists of an ad hoc convolutional neural network (CNN) architecture that splits the input images into patches and distributes them independently on the convolutional layers. The authors show how this approach and a memory management algorithm can improve the scalability of distributed training for convolutional networks. The article depicts a relevant solution in inference distribution with good experimental results, but it does not address the distribution model problem among heterogeneous paradigms. Other similar solutions in scientific literature apply a distribution over CNN models and models for image elaboration. In particular, in [9], the authors proposed another solution to split the CNN structure into smaller parts with the aim of reducing memory usage. The simulation results performed on VGG16 and ResNet18 networks for the classification of CIFAR10 images demonstrated a reduction in both the quantity of memory consumption and the number of computational operations. Even in [12], the authors propose an algorithm to partition the CNN model during the inference phase. Unlike the previous research works, the split of CNN is based on device constraints and, in particular, on the stringent computational requirements of Edge devices, such as the memory used and the bandwidth used in transferring video streams from the Edge to the Cloud. The algorithm is based on finding the maximum level at which to partition the neural network so that the Edge device can contain the partition despite performing other computational tasks. Unlike the other solutions over CNN, here the authors apply a split for a Cloud-Edge distribution. Instead, in [10], the authors propose a partitioning of a DNN in order to reduce the complexity of big models that work on images' classification, such as DenseNet-169 [4] ResNet-152 [2], and Inception-v3 [17].

Other recent works dealt with a split of different ML models among different physical devices. For example, in [3], the authors proposed a distributed inference framework called EdgeFlow. The solution is designed to split the inference tasks among multiple Edge devices. The results showed that the proposed solution can reduce the inference time by 40.2% compared with other solutions. The work in [16] shows further progress in Edge devices ML distribution; the authors proposed a solution that not only distributes a DNN across multiple Edge devices but also optimizes computation resources and memory. The solution has been validated by simulation of six Edge devices running YOLOv2 DNN [15]. In [19], the authors proposed a solution to move the ML inference into IoT devices. The idea is to create a hierarchical structure with accuracy and prediction comparable to a Cloud-based solution. The experimental results showed that the accuracy of the hierarchical-based solution is the same as the Cloud-based one but the energy use and latency can be reduced by up to 63% and 67% respectively. Instead, The [20] realizes a work about mobile Edge computing, finding the optimal cellular network to split a DNN in order to find the best trade-off between accuracy and latency in the inference phase. The authors consider two kinds of mobile network topologies defined "Chain" and "Mesh" and via a Neural Architecture Research technique, they find the optimal solution maximizing the accuracy and minimizing the latency.

The work realized in [5] introduces a split of ML involving heterogeneous devices and not only Edge devices. Indeed, the authors proposed a Spark cluster for distributing the inference in TensorFlow. In particular, they compared the performances of four different configurations: i) a single Raspberry Pi 4B; ii) a cluster composed of two Raspberry Pi 4B; iii) a cluster composed of two Virtual Machines (VMs); and iv) a desktop computer. To validate the proposed solution, they tested InceptionV3 on CIFAR10 and ImageNetV2. However, due to errors, all Spark-based clusters failed to perform the execution.

Among the most recent works, we must consider the solution described in [13], which realized an inference distributed splitting the layers of the neural network in a bootstrap phase and distribution according to specific layers. The authors proposed an architecture and a specific flow with the following tasks: Model Partitioning, Configuration, and Distributed Inference. The solution logically concatenates different nodes in order to spread the model partition and distribute the inference. The division strategy is based on splitting the network architecture into two parts: the head and the tail. The central innovative aspect of this work is to apply a knowledge distillation technique only to the head part, to which a bottleneck is further added to reduce the size of the output data. The results show that it is possible to implement complex models on Edge devices by sacrificing inference.

Although all the aforementioned works treated different aspects of the distribution of ML inference, none of these considered a framework that can dynamically distribute the ML model in a Cloud-Edge infrastructure. Moreover, none of the previously depicted solutions are considered an automatic process in which the model is split, distributed, and deployed between Cloud and Edge.

The research in [8] is one of the first solutions in which the ML distribution exploits the Cloud-Edge computing paradigm. In particular, in this work, a neural network is split into two parts spread across Cloud and Edge. The work proposes a solution to find the optimal split point considering different criteria: Cloud and Edge cost, introduced overhead, etc. Although we have a dynamical partition of the model between Cloud and Edge, the solution only considers a partition in two parts.

Our work shows generic progress in state of the art, considering also a dynamic partition of a Transformer model, never thought of in a Cloud-Edge environment at the time of writing.

3 Motivations and Use Case Description

As established earlier, the Cloud-Edge Continuum integrates two fundamental concepts for service deployment and delivery. It takes advantage of the robust computation and infrastructure offered by Cloud computing and the benefits of Edge computing. The needs of contemporary applications, where minimal latency, real-time processing, and effective resource management have become critical, led to the creation of this innovative architecture. We can now take advantage of the benefits of Edge computing, such as reduced latency and improved security, along with the benefits of the Cloud, such as easy access to computing resources and services. Fundamentally, the Cloud-Edge Continuum connects the enormous processing power of the Cloud with the immediateness of Edge devices. Applications can exploit what is offered by both Cloud and Edge thanks to the smooth coordination of data flow and processing duties between them. In order to offload ML inference duties across Cloud and Edge infrastructure, we present a workflow idea in our suggested solution that operates within the Cloud-Edge Continuum context. This new method makes use of Cloud infrastructure's powerful computational power for heavy inference workloads and Edge computing's low latency for data collecting.

3.1 Hydrological Use Case

A region that is drained by a specific surface water or groundwater system is known as a hydrological basin or catchment. The hydrogeological catchment can play an important role in sustaining ecosystems providing drinkable water and supporting several industries. For this reason, runoff forecasting can predict any water shortages or signal any anomalous floods and it's a crucial computation in environment monitoring. Hence, the urgent need to address important water resource management and environmental concerns is driving the development of a Cloud-Edge Workflow for AI inference for the runoff level. In our solution, for the hydrogeological use case, we have exploited the LAMAH dataset [7]. This dataset, covering the Danube River's basins across three different countries, offers diverse information, from geophysical data to historical water levels. In particular, the dataset involves about 800 catchments. For each catchment, it provides a time-series data split for the day and hour of the last 35 years. The dataset covers about 60 attributes for each catchment, covering topography, climatology, hydrology, land cover, vegetation, soil and geological properties. In our case, we have considered the daily data about runoff level to design our models and distribute them between Cloud and Edge tiers. Furthermore, our solution emphasizes resource optimization, particularly in remote or resource-constrained areas. The distributed approach of the workflow maximizes resource efficiency, reducing energy consumption and costs associated with data transmission and Cloud-based processing. The proposed solution performs AI inference in order to forecast one of the columns provided by the LAMAH dataset: the runoff of the basin expressed in the dataset as *Qobs*. As of the time of writing, there is no existing solution that leverages the Cloud-Edge Continuum and Workflows for distributing AI inference processes. Through this approach, we aim to enhance the management and orchestration of AI inference across a variety of heterogeneous devices.

4 Proposed Solution

In this section, the proposed solution will be depicted and the designed workflow will be described. In particular, we'll describe the general methodology for the implementation of a ML model in a Cloud-Edge Workflow, highlighting how the model will be partitioned and distributed. We'll describe the strategy by which the partition of the model is carried out, preliminary works computed over the model to train it and the workflow structure.

4.1 Splitting Methodology

The main strategy applied to split the inference process consists of distributing the components belonging to the model. In this work, we consider two main ML models that can partially generalize the main solutions applied to time-series data. As previously mentioned, the models considered are: 1) a Forward DNN with two hidden layer and 2) a transformer model [21] for the time-series data prediction [22]. The splitting strategy will be described for each model exploited. This step can make possible an automatic partition of the model inside the workflow we are going to realize in the Cloud-Edge Continuum environment.

DNN. We can consider the typical Forward DNN constituted of homogenous layers. To design the strategy we have considered each DNN as an ensemble of Linear Layer (LL) and Activation Layer (AL). The approach consists of splitting sequentially the layer into equal parts when it's possible and in general, to minimize the difference among the number of layers of each part. In our case, the DNN has two hidden layers but the methodology can be generalized to any number of layers in a DNN Considering L the layers set and $N = |L|$ the number of total layers in the DNN. Considering N_p the number of parts we want to split the Neural Network (in our case it is equal to 4) we establish the size of each part S_k considering

$$S_k = \lfloor \frac{N}{N_p} \rfloor + 1 \tag{1}$$

for the first P parts where

$$P = N \mod N_p \tag{2}$$

and

$$S_k = \lfloor \frac{N}{N_p} \rfloor \tag{3}$$

for the remaining parts.

Transformer. The splitting strategy for the Transformer model is focused on the internal architecture of the Transformer itself which is not homogenous like DNN one. According to Transformer definition [21] we can summarize it in these components that will become the partitions of our partitioned inference:

- a Linear Layer for the Input Preparing of the Encoder
- a Linear Layer for the Input Preparing of the Decoder
- the positional Encoding component
- the Encoder Component
- the Decoder Component

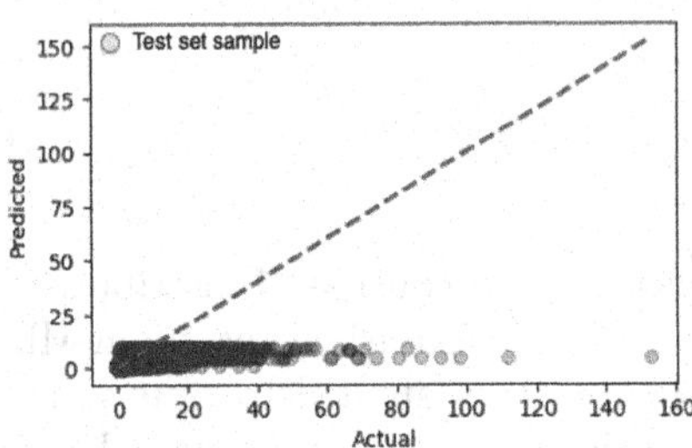

Fig. 1. Error representation for Feed-Forward model.

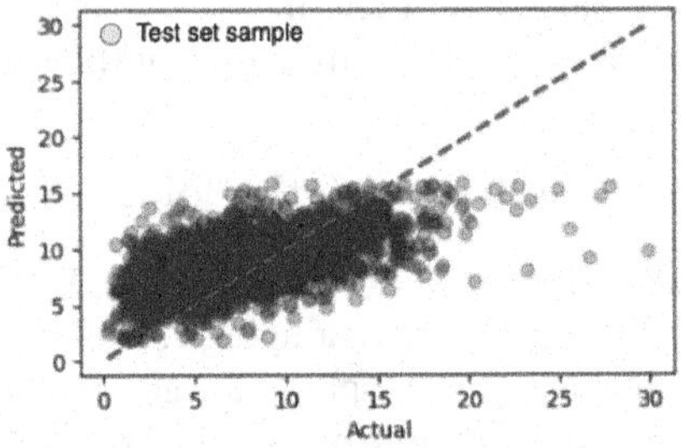

Fig. 2. Error representation for Transformer model.

4.2 Preliminary Works on Models

The preliminary works performed were related to the training of the two models (described in Sect. 4.1). The models were trained by exploiting each catchment data. In this way, a weights set is trained for each catchment to better generalize and manage the forecast for each catchment. In particular, the preliminary works on the dataset and model can be summarized to:

- Preprocessing phase in which the label (*Qobs* column in the dataset) and all the features are extracted, merged and filtered. In particular, we choose the following feature sets(as named in [7]): gauge referred attributes, meteorological variables, topographic indices, and climatic indices. Starting from the feature sets we keep only the more correlated features and finally, we have a dataset with 34 features. In this phase we remove outliers too;
- Train, Validation and Test set split;
- Train of the model;

Figures 2 and 1 show the error of the two models trained on a single basin. The graph depicts for each test set sample the error of the prediction with respect to the actual value. We can say that the Transformer model reaches a lower error on a test set after the training. The detail of training hyperparameters is out of the scope of the work presented here which is focused on the inference phase.

In the high-level architecture designed in our use case, two different levels are considered. The *lower* level is the Edge layer that interacts and collects the data victim of the inference. The *higher* layer is the Cloud Layer in which the Cloud Infrastructure is hosted.

4.3 Workflows Design for Proof of Concept

The designed workflows split the inference processes into different chunks that can be distributed among Cloud and Edge infrastructures. To prove our solution, as mentioned before, we accomplished two different workflows that perform the inference through two different ML models. The Workflow's jobs will be depicted in detail.

Classical Feed Forward Network Inference Workflow
The workflow that realizes the Feed Forward Network inference is graphically depicted in Fig. 3.
The descriptions of each jobs depicted are the following:

a. Preprocessing: This initial job involves the conversion from the raw data collected to the input of the next jobs and the input of the models
b. Preinference: Although this job doesn't explicitly involve partitioning, it could be responsible for preparing the models and data structures required for the partitioned inference. In particular, during this job, the right weight set is loaded. Indeed, as specified, different weights are trained for different basins.
c. Generate Partitions: This job applies the partitioning strategy already described in Sect. 4.1. In the case of Feed Forward Neural Network, the job split the model in homogeneous parts. It takes the trained model from the previous jobs and generates multiple partitions of the model. These partitions will be exploited for the inference in the next jobs.
d. Inference Jobs: Each of these jobs is an inference task, it performs a part of the total inference thanks to the partitioned model accomplished in the previous job. By splitting the inference into these separate jobs, it is possible to offload each job to a different infrastructure (e.g. Cloud or Edge)
e. Conversion: The final job takes the processed data and converts it into a more usable format for further analysis or reporting.

Transformer Inference Workflow
The workflow that realizes the Transformer model inference is graphically depicted in Fig. 4.
The descriptions of each jobs depicted are the following:

a. The jobs related to preprocessing, preinference and conversion of the final output and generating partitions have the same logic already described for Feed Forward Network.
b. Generate Partitions: Unlike Feed Forward Network, here, the Generate Partitions job splits the model in the different components of Transformer model, as specified in Sect. 4.1.

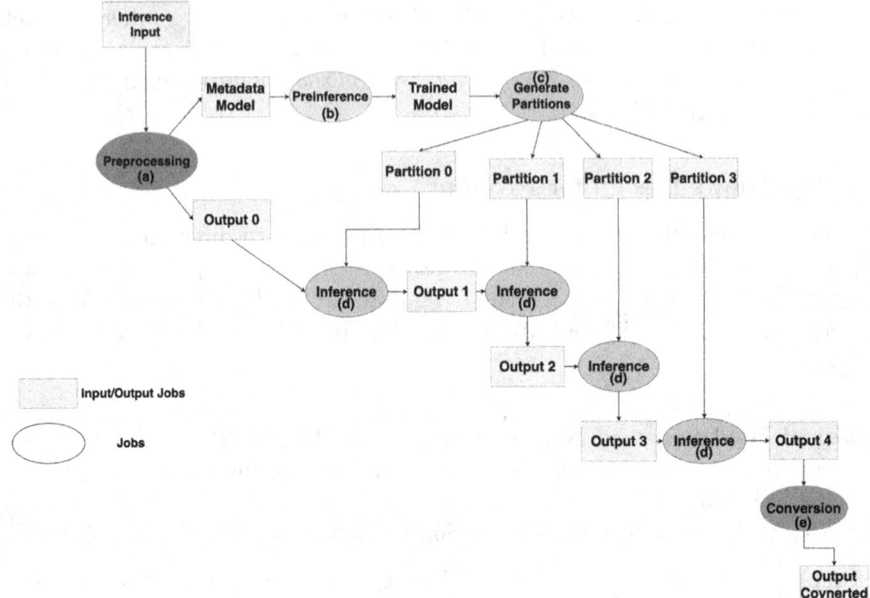

Fig. 3. Workflow for Inference with Deep Feed Forward Neural Network.

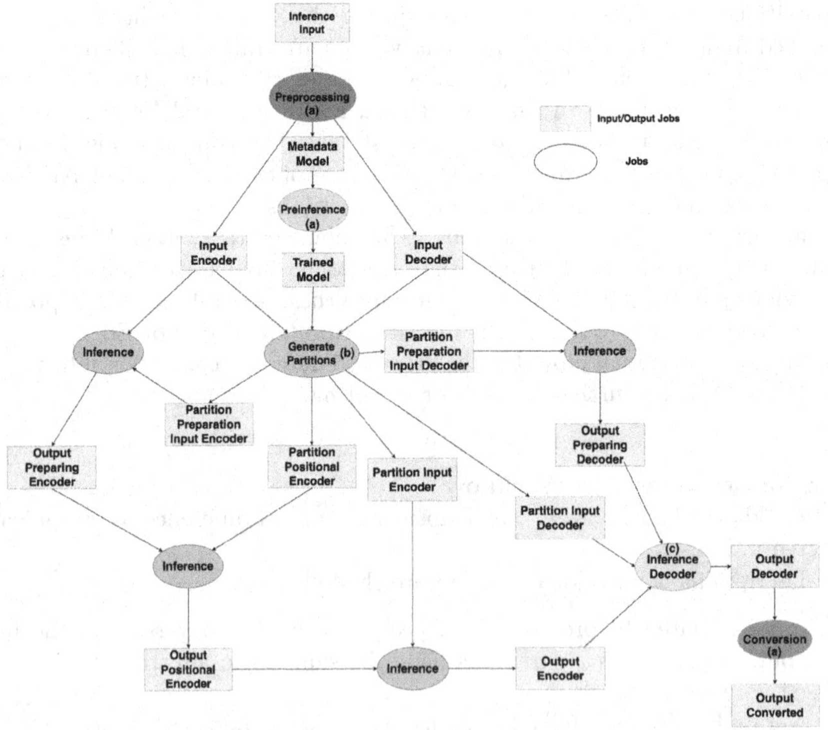

Fig. 4. Workflow for Inference Transformer Model.

c. Inference Decoder: This job performs the final inference that involves the Decoder of Transformer and that involves the Decoder Input and the Encoder Output.

5 Implementation for a Cloud-Edge Continuum Environment

In this section, the practical implementations of the designed workflows will be described. The carried-out implementation has been focused on the workflows that, as already specified, will be deployed in a Cloud-Edge Continuum environment. The workflows have been accomplished through Pegasus WMS [11].

5.1 Introduction to PEGASUS

Pegasus WMS is an open-source tool that empowers scientists to create custom abstract workflows using high-level APIs. For the execution of these workflows, Pegasus utilizes HTCondor, which is an open-source software framework designed for distributed computing. HTCondor's primary function is to manage and schedule workloads. In collaboration with HTCondor, Pegasus offers a range of functionalities that enable the creation of an abstract workflow. This transformation allows HTCondor's DAGMan to efficiently allocate resources and schedule jobs within the HTCondor cluster.

The HTCondor cluster foresee three different roles within a cluster:

1. Submit Node: The Submit Node serves as the entry point for users to submit their computational tasks to the HTCondor system. Users provide job descriptions, and the Submit Node interfaces with the HTCondor Central Manager to manage and schedule the job queue. It does not directly execute the tasks but takes charge of job submission and monitoring. To ensure seamless functionality, both Pegasus WMS and HTCondor binaries must be installed on this node, as Pegasus utilizes HTCondor utilities to deploy the workflow.
2. Central Manager: The Central Manager assumes a central role in the HTCondor. It maintains an overview of the status of all machines and oversees the matchmaking process. In the HTCondor system, each worker node periodically shares its characteristics with the central manager. The central manager gathers information from the scheduler and handles the task of matching jobs to available resources. It is instrumental in resource allocation and job distribution.
3. Execution Node: Execution Nodes represent the worker machines within the HTCondor cluster. These nodes are responsible for the actual execution of computational tasks scheduled by the Central Manager. Execution Nodes can have varying configurations, and they run user-submitted tasks in isolated environments, ensuring that tasks do not interfere with each other. HTCondor involves different architectures for the execution nodes (e.g. arm64, x8086).

5.2 Workflows Implementation

The practical implementation of the workflows consists of the creation of a YAML file with the following structure that represents the workflow and that will be the input of *pegasus-run* command:

1. **Metadata**:
   - `x-pegasus`: This section contains metadata about the workflow, including the creation language, creator, and timestamp.
2. **Pegasus Version**:
   - `pegasus: 5.0.4`: Specifies the version of Pegasus used for the workflow.
3. **Workflow Name**:
   - `name: qobs_prediction_timeseries`: The name of the workflow, indicating its purpose related to distributed time series prediction.
4. **Workflow Jobs**:
   - The workflow consists of several jobs, each with a type, name, ID, and a list of arguments. These jobs are organized in a sequence, serving various purposes.
     a. `Preprocessing`: Prepares data and produces output for use in subsequent jobs.
     b. `Preinference`: Preprocesses data in preparation for inference.
     c. `Generate Partitions`: Creates multiple data partitions, likely for parallel processing.
     d. `Inference Jobs`: Perform inference on different data partitions.
     e. `Inference Decoder`: May assemble results from previous inference jobs.
     f. `Conversion`: Converts processed data into a more usable format.
5. **Job Dependencies**:
   - The workflow specifies dependencies between jobs to ensure the correct order and coordination of tasks.

In practice, YAML file is never written from scratch but it's generated by the API provided by Workflow for main languages programming. In our case, we exploited Python Pegasus API to generate YAML files (Table 1).

6 Performance Assessment

This section depicts the experimental evaluations of the two provided solutions. The evaluations have focused on analysing the overhead introduced by the Workflow implementations concerning the infrastructure and the balancing across Cloud and Edge. Moreover, the test evaluated the overhead considering the model exploited in the single workflow.

Table 1. Cluster's nodes characteristics

#	Tier	Model	CPU	Memory	OS
2	Cloud	Openstack VM	Intel Xeon 8x 4.0 GHz	16 GB	Ubuntu 20
2	Edge	Raspberry Pi 4	ARM64 SoC 4x 1.5 GHz	4 GB	Raspberry OS ARM64

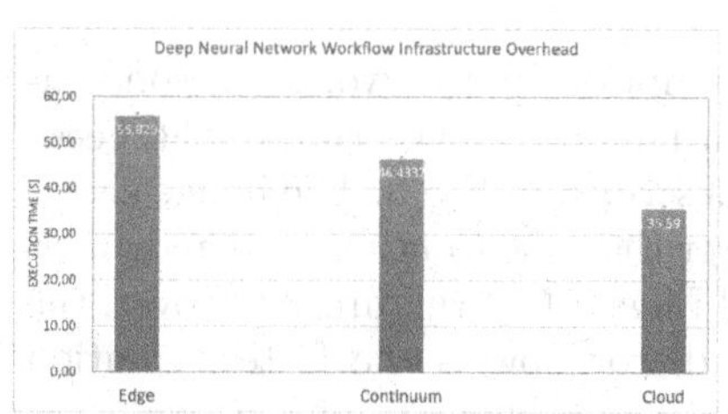

Fig. 5. Workflow for Inference with Deep Feed Forward Neural Network.

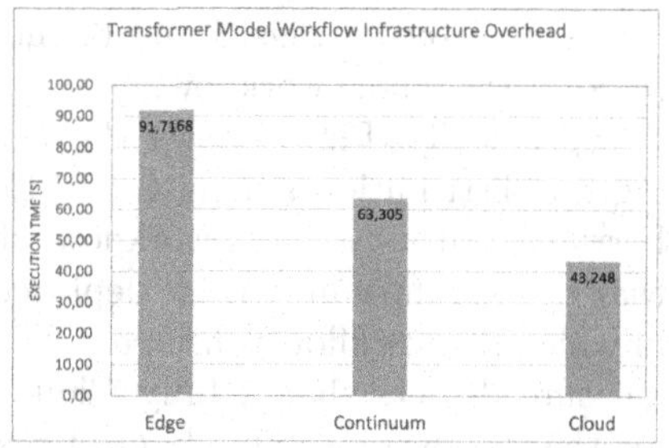

Fig. 6. Workflow for Inference with Transformer model.

6.1 System Testbed Setup

The Pegasus Workflows has been tested on a 4 Nodes Kubernetes Cluster composed of two nodes on the Cloud tier and two on the Edge tier. The System testbed involves two VM Machines deployed through OpenStack that constitute the Cloud tier. The Edge tier is realized via two Raspberry PIs. All the nodes can host each job belonging to Workflow.

6.2 Experimental Results

The test focused on overhead introduced in the distribution of ML Inference in a Cloud-Edge environment. As described, Pegasus WMS allows orchestration of the jobs involved in the workflow and with our tests we have considered three possible job distributions among Cloud and Edge tiers:

- Edge setting in which all the workflow is hosted on Edge tier
- Continuum setting in which the workflow jobs are equally distributed between Cloud and Edge. In particular, considering the jobs described in Sect. 4.3, we have deployed the preprocessing, preinference, *generate_partitions* and the first two jobs of the inference phase are deployed on the Edge tier. The other jobs (other parts of the inference, and final conversion) are deployed on the Cloud tier.
- Cloud setting in which all the workflow jobs are deployed on the Cloud tier.

Moreover, we compared the two models distributed through the workflow performing the average of the model concerning the infrastructure is deployed to. The performed tests were performed through the *pegasus-statistics* tool provided by the Pegasus WMS command line interface. This tool collects all the most important statistics of the performed workflow including the cumulative wall time that consists of the sum of each workflow job time execution.

The experimental measures showed interesting results. Indeed, the infrastructure exploited for the workflow deployment is the most important parameter for overhead introduced by the execution of the whole workflow.

As depicted in Figs. 5 and 6, the Cloud infrastructure requires less time to perform the whole workflow.

Specifically, Fig. 5 shows the time consumption of the workflow concerning the distribution among Cloud-Edge Continuum infrastructure of the Feed-Forward Neural Network workflow which is described in Fig. 3. The image shows that the workflow fully Edge deployed is slower than the Continuum deployment in which the workflow runs across Cloud and Edge infrastructure. Moreover, the measure depicts that a fully Cloud deployed workflow is the fastest solution for the deployment. This kind of result could be considered different from the heuristic usually defined in the scientific literature in which Edge computing ensures less overhead [1,14,23]. The difference between Edge and Continuum and between the Continuum and Cloud jobs distribution is about 10 s according to performed measures. These values prove that the Continuum approach can be considered, in this case, a right trade-off to ensure that final raw data will be kept on Edge devices and to exploit Cloud infrastructure. Figure 6 shows the same comparison with Transformer workflow implementation. Even here, the trend suggests that the Cloud component saves time in the jobs distribution, and even in this case, the Continuum approach can be considered a good trade-off. Moreover, in this case, in which the model is more complex, the Continuum and the Cloud save a greater amount of time than the DNN workflow.

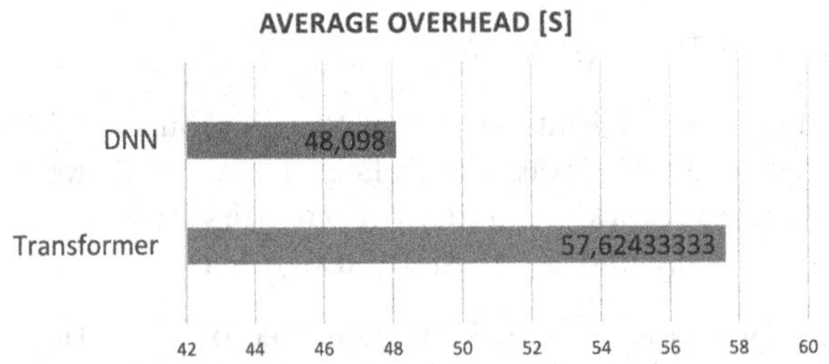

Fig. 7. Average of introduced Overhead for model.

Figure 7 shows an average overhead comparing the two models. Since the Transformer model is more complex than a Feed-Forward Deep Neural Network, the amount of time spent to perform the inference is greater.

7 Conclusions and Future Works

In this study, we have successfully designed two Cloud-Edge Workflows capable of processing raw data collected from Edge devices. These workflows enable distributed inference, allowing each job to be flexibly deployed on either the Edge or Cloud infrastructure. Our work includes the practical implementation of these workflows, which was seamlessly managed using the WMS, Pegasus. This practical implementation demonstrates the feasibility and effectiveness of our approach in a real-world context. Looking ahead, our research opens up avenues for future investigations. An experimental evaluation of the workflow with varying job distributions between the Cloud and Edge infrastructure can provide valuable insights into performance optimization. Additionally, the next steps could involve the development and implementation of new models, considering different datasets and use-cases. These extensions will not only enhance the versatility of the workflows but also address the evolving needs of Cloud-Edge Computing applications.

Acknowledgements. This research was funded in part by the Austrian Science Fund (FWF) through following projects: *Transprecise Edge Computing (Triton)* 10.55776/P36870; *Trustworthy and Sustainable Code Offloading (Themis)* 10.55776/PAT1668223; *Sustainable Watershed Management Through IoT-Driven AI (Swain)* 10.55776/I5201, and by the Austrian Research Promotion Agency (FFG) through the following project: *Satellite-based Monitoring of Livestock in the Alpine Region (Virtual Shepherd)*, FFG Austrian Space Applications Programme ASAP 2022 #53079251. This research was also funded by the Italian Ministry of Health, Piano Operativo Salute (POS) trajectory 4 "Biotechnology, bioinformatics and pharmaceutical development", through the Pharma-HUB Project "Hub for the repositioning of drugs in rare diseases of the nervous system in children" (CUP J43C22000500006) and by Piano Operativo Salute (POS) trajectory 2 "eHealth, diagnostica avanzata, medical device e mini invasività" through the project "Rete eHealth: AI e strumenti ICT Innovativi orientati alla Diagnostica Digitale (RAIDD)"(CUP J43C22000380001). Ewa Deelman's work was funded by the U.S. National Science Foundation under grants numbers 2331153 and 2103508 and by the U.S. Department of Energy under grant number DE-SC0024387.

References

1. Cao, K., Liu, Y., Meng, G., Sun, Q.: An overview on edge computing research. IEEE Access **8**, 85714–85728 (2020)
2. He, K., Zhang, X., Ren, S., Sun, J.: Deep residual learning for image recognition (2015)
3. Hu, C., Li, B.: Distributed inference with deep learning models across heterogeneous edge devices. In: IEEE Conference on Computer Communications (IEEE INFOCOM 2022), pp. 330–339 (2022)
4. Huang, G., Liu, Z., van der Maaten, L., Weinberger, K.Q.: Densely connected convolutional networks (2018)
5. James, N., Ong, L.-Y., Leow, M.-C.: Exploring distributed deep learning inference using raspberry pi spark cluster. Future Internet **14**(8), 220 (2022). https://doi.org/10.3390/fi14080220
6. Jin, T., Hong, S.: Split-cnn: splitting window-based operations in convolutional neural networks for memory system optimization. In: Proceedings of the Twenty-Fourth International Conference on Architectural Support for Programming Languages and Operating Systems (ASPLOS 2019), pp. 835–847. Association for Computing Machinery, New York (2019)
7. Klingler, C., Schulz, K., Herrnegger, M.: Lamah-CE: large-sample data for hydrology and environmental sciences for central Europe. Earth Syst. Sci. Data **13**(9), 4529–4565 (2021)

8. Luger, D., Aral, A., Brandic, I.: Cost-aware neural network splitting and dynamic rescheduling for edge intelligence. In: Proceedings of the 6th International Workshop on Edge Systems, Analytics and Networking (EdgeSys 2023), pp. 42–47. Association for Computing Machinery, New York (2023)
9. MalekHosseini, E., Hajabdollahi, M., Karimi, N., Samavi, S., Shirani, S.: Splitting convolutional neural network structures for efficient inference (2020)
10. Matsubara, Y., Baidya, S., Callegaro, D., Levorato, M., Singh, S.: Distilled split deep neural networks for edge-assisted real-time systems. In: Proceedings of the 2019 Workshop on Hot Topics in Video Analytics and Intelligent Edges (2019)
11. Mehta, G., Deelman, E., Vahi, K., Silva, F.: Pegasus workflow management system: helping applications from earth and space. In: AGU Fall Meeting Abstracts, vol. 2010, pp. IN41B–1362 (2010)
12. Mehta, R., Shorey, R.: Deepsplit: dynamic splitting of collaborative edge-cloud convolutional neural networks. In: 2020 International Conference on COMmunication Systems and NETworkS (COMSNETS), pp. 720–725 (2020)
13. Parthasarathy, A., Krishnamachari, B.: Defer: distributed edge inference for deep neural networks. In: 2022 14th International Conference on COMmunication Systems and NETworkS (COMSNETS). IEEE (2022)
14. Pérez, J., Díaz, J., Berrocal, J., López-Viana, R., González-Prieto, A.: Edge computing: a grounded theory study. Computing **104**(12), 2711–2747 (2022)
15. Redmon, J., Farhadi, A.: Yolo9000: Better, Faster, Stronger (2016)
16. Stahl, R., Zhao, Z., Mueller-Gritschneder, D., Gerstlauer, A., Schlichtmann, U.: Fully distributed deep learning inference on resource-constrained edge devices. In: Architectures, Modeling, and Simulation, Embedded Computer Systems (2019)
17. Szegedy, C., Vanhoucke, V., Ioffe, S., Shlens, J., Wojna, Z.: Rethinking the inception architecture for computer vision (2015)
18. Teerapittayanon, S., McDanel, B., Kung, H.T.: Distributed deep neural networks over the cloud, the edge and end devices. In: Proceedings of the International Conference on Distributed Computing Systems, pp. 328–339 (2017)
19. Thomas, A., Guo, Y., Kim, Y., Aksanli, B., Kumar, A., Rosing, T.S.: Hierarchical and distributed machine learning inference beyond the edge. In: 2019 IEEE 16th International Conference on Networking, Sensing and Control (ICNSC), pp. 18–23 (2019)
20. Tian, Y., Zhang, Z., Yang, Z., Yang, Q.: JMSNAS: joint model split and neural architecture search for learning over mobile edge networks. In: 2022 IEEE International Conference on Communications Workshops (ICC Workshops), pp. 103–108 (2022)
21. Vaswani, A., et al.: Attention is all you need (2023)
22. Wen, Q., et al.: Transformers in time series: a survey (2023)
23. Yu, W., et al.: A survey on the edge computing for the internet of things. IEEE Access **6**, 6900–6919 (2018)

Evaluating the Impact of Atmospheric CO2 Emissions via Super Resolution of Remote Sensing Data

Andrianirina Rakotoharisoa[1,2](✉), Simone Cenci[3], and Rossella Arcucci[1,2]

[1] Department of Earth Science and Engineering, Imperial College London, London SW7 2AZ, UK
andrianirina.rakotoharisoa19@imperial.ac.uk
[2] Data Science Institute, Imperial College London, London SW7 2AZ, UK
[3] Imperial College Business School, London SW7 2AZ, UK

Abstract. Understanding how emissions from point sources affect the atmospheric concentrations of Greenhouse Gases (GHGs) locally and on a wider scale is crucial to quantify their impact on climate change. To this end, different ways of performing global monitoring of GHGs concentration using remote sensing data have been explored. The main difficulty remains to find the right balance between high resolution monitoring, which is often incomplete, and global monitoring, but at a coarser resolution. This study proposes the application of Super Resolution (SR), a Deep Learning (DL) technique commonly employed in Computer Vision, to increase the resolution of atmospheric CO2 L3 satellite data. The resulting maps are achieving an approximate resolution of 1 km * 1 km and are then compared with a benchmark of existing methods, before being used for emissions monitoring.

Keywords: Super Resolution · Remote Sensing · GHG Monitoring

Acronyms

CTM	Chemical Transport Model
CV	Computer Vision
DA	Data Assimilation
DL	Deep Learning
GHG	Greenhouse Gas
HR	high resolution
IR	Improvement Ratio
LR	Low Resolution
LST	Land Surface Temperature
MAE	Mean Absolute Error
ML	Machine Learning

L. Franco et al. (Eds.): ICCS 2024, LNCS 14836, pp. 383–390, 2024.
https://doi.org/10.1007/978-3-031-63775-9_28

OCO-2	Orbiting Carbon Observatory 2
RMSE	Root Mean Square Error
SISR	Single Image Super Resolution
SR	Super Resolution
TCCON	Total Column Carbon Network
XCO2	column-averaged dry air mole fraction of atmospheric CO2

1 Introduction and Motivations

The latest Intergovernmental Panel on Climate Change (IPCC) report [4] asserts that GHGs, particularly CO_2 and CH_4, have contributed to the rise of the global surface temperature by over one degree since the late 19th century. Effective policies and measures to curb global warming are therefore conditioned by our understanding of GHG emissions. Remote sensing (RS) satellite imagery is used to generate estimations of concentration and can be improved for two reasons: it is either sparse following the satellite swath, or at a coarse resolution when processed. Machine learning approaches have helped to alleviate some of these data challenges. The present study concentrates on column-averaged dry air mole fraction of atmospheric CO2 (XCO2) monitoring and offers the following contributions:

- We generate SR maps of XCO2 with a spatial resolution of $0.03° * 0.04°$, 16-times higher than the original datasets from the Orbiting Carbon Observatory 2 (OCO-2) Level 3 (L3) data. The model is not gas-specific and can therefore be adapted for monitoring other GHGs beyond CO2.
- We demonstrate that our model doesn't add noise to the original OCO-2 data and quantitatively improves compared to alternative methods.
- We propose an application of our dataset to the evaluation of CO_2 point sources' impact on their environment.

2 Related Works

2.1 XCO2 Modelling Methods

Different forms of interpolations often following the model of Kriging have been implemented to build on sparse observations [8,22]. However, the quality of the generated products is variable following the density of the observations, leading to high uncertainty in some areas if no additional data is considered [14]. Data fusion approaches aim to reduce these uncertainties by combining data from multiple satellites [18]. Machine Learning (ML) and (DL) publications also often consider more variables as inputs [7]. For example, [11] combines XCO2 observations and environmental features such as meteorological and vegetation indicators before using Extra Randomised Trees. Chemical Transport Models (CTMs) generate simulations based on atmospheric dynamics and are good to predict fluxes [9,13]. It is however difficult to produce high resolution maps with these models due to their computational complexity.

2.2 Super Resolution

SR is a technique initially used in Computer Vision (CV) or image processing and aims to increase the resolution of an image by inferring high frequency details that were not available on the original image. It therefore creates a high resolution (HR) version of the input [15]. In the case of a single Low Resolution (LR) input, it is referred to as Single Image Super Resolution (SISR) [21]. If multiple LR inputs are used to generate the SR output, it becomes Multiple-Image SR.

3 Methods

3.1 Our Model

In this study, input data consists of satellite-derived XCO2 L3 data (see Table 1). Transforming it from matrix form into RGB images requires additional data preprocessing steps and could potentially bias the SR output based on the colour scheme selection. Consequently, even though we approach the problem as a CV task, we treat the XCO2 maps as single-channel images and our model's design is a customized 1-channel implementation of the DL framework by Haris et al. [6] to deal with our data. The SR framework relies on iterative up- and down-sampling modules using convolutional and deconvolutional layers to learn the relationship between LR and HR pairs. To manage memory usage efficiently and improve training speed, we reconfigured the layers to minimize the number of parameters and utilized PyTorch's Distributed Data Parallel framework [12]. Originally designed to increase the resolution 8-fold, we adjusted the layers kernel size, padding, and stride for our purpose and have adapted the model for 16-fold upsampling, achieving an approximate resolution of 1 km * 1 km, which is necessary for identifying individual emission sources.

Table 1. Data processing level definitions. Other levels are out of scope and therefore not included in this table.

Processing level definition	Definition
Level 2	Derived geophysical variables at the same resolution and location as Level 1 data
Level 3	Variables mapped on uniform space-time grids, usually with some completeness and consistency

4 Datasets

4.1 Training

Due to the absence of HR XCO2 data to train our model, the Land Surface Temperature (LST) dataset from MODIS [17] was used for training. We consider that temperature and XCO2, both being physical variables observed by satellites, exhibit similar variability as opposed to the natural images datasets typically used for training SR models. As shown in Table 2, LST data is first downsampled, using bicubic interpolation, before being reconstructed to its original resolution, which is in the same range as our target resolution for XCO2.

Table 2. Comparison of spatial resolution between data from OCO-2 and MODIS

	XCO2 L3 (OCO-2)	LST L3 (MODIS)
LR data (in degree)	$0.5° * 0.625°$	$0.8° * 0.8°$
HR data (in degree)	$0.031° * 0.039°$(target)	$0.05° * 0.05°$

4.2 Low Resolution and Validation Datasets

The OCO-2 and its successor OCO-3 are CO2 monitoring missions from NASA [5] with a radiometric resolution of the order of 1ppm. The L3 dataset generated from this mission [19] is used as our input data and consists of global maps of XCO2 at a resolution of $0.5° * 0.625°$. It is accessible at daily or monthly temporal granularity and is being generated through NASA's modeling and data assimilation framework.

The Total Column Carbon Network (TCCON) [20] is a network of fixed spectrometers located at various sites monitoring column concentrations from CO_2, CH_4, CO, and N_2O. We consider it as our groundtruth and compare the XCO2 values output by our model with the latest version of TCCON data [10]. The TCCON data was obtained from the TCCON Data Archive hosted by CaltechDATA at https://tccondata.org.

5 Results and Discussion

5.1 Benchmark

We benchmark our model against different approaches where the main criteria of selection were the temporal and spatial resolution of available data as well as metrics performance.

Table 3. Evaluation of our dataset against the benchmark

	SR (ours)	LR	BIC	Fusion
RMSE	**0.9180**	0.9441	0.9436	1.1196
R^2	**0.9673**	0.9654	0.9655	0.9514
MAE	**0.7027**	0.7182	0.7190	0.8505

The best value for each metrics is in bolt. **LR** is the dataset from NASA, **BIC** is produced from bicubic interpolation, and **Fusion** is the fusion dataset [18]. The metrics are averaged over the TCCON network between 2015 and 2020

The results in Table 3 show that our SR model is able to generate a HR version, enhanced 16-times, of the original dataset while improving on all metrics. This also holds true when comparing our maps with other methods. The fusion dataset has an average Root Mean Square Error (RMSE) of over 1 ppm and a Mean Absolute Error (MAE) of 0.85 ppm. These values are 20% higher relatively than the values of our model. Bicubic interpolation ranks second on most metrics. To further distinguish the performance of SR using our model versus using bicubic interpolation, we define an Improvement Ratio (IR) as follows:

$$IR^{site} = \frac{N^{site}_{improved}}{N^{site}_{samples}} \tag{1}$$

For each site, $N^{site}_{improved}$ represents the number of instances where SR reduces the error compared to the ground truth, while $N^{site}_{samples}$ denotes the total number of samples for that site. A ratio of 1 (or 100%) indicates that the model consistently produces better results, while a ratio of 0 demonstrates that using this specific SR model always produces worse estimations.

Fig. 1. IR comparison between our model (blue) and bicubic (green) with available ground sensors data. An orange circle indicates that the IR ratio is over 50% for ours. (Color figure online)

Figure 1 shows how the two SR models performs across different locations based on the IR. On 19 sites out of 24, our model improves on the LR input more frequently than the bicubic SR, which reflects that our model is more consistent.

5.2 Application: Impact of Emissions on Atmospheric CO2

Finally, we combine the information provided by our SR maps with the emissions dataset from ClimateTrace [3]. This is a first attempt to visualise how XCO2 can be influenced by point sources using HR global XCO2 maps. The dataset from ClimateTrace that we used is an inventory of CO_2 emissions point sources for the year 2021.

On Fig. 2, the additional high-frequency details observable on the SR maps reveal that the locations of point sources do not necessarily coincide with the highest concentration in CO_2 in the area. This underscores the impacts point sources can have not only on their direct surroundings but also on areas seemingly far away or isolated from human activity. Furthermore, the influence of land topography on XCO2 is clearly pronounced and particularly visible on the Venezuelan area. High altitude areas serve as a natural obstacle and seem to direct the flow of CO_2. We must also note that other factors such as urbanisation may also influence XCO2 and are not visualised in our analysis.

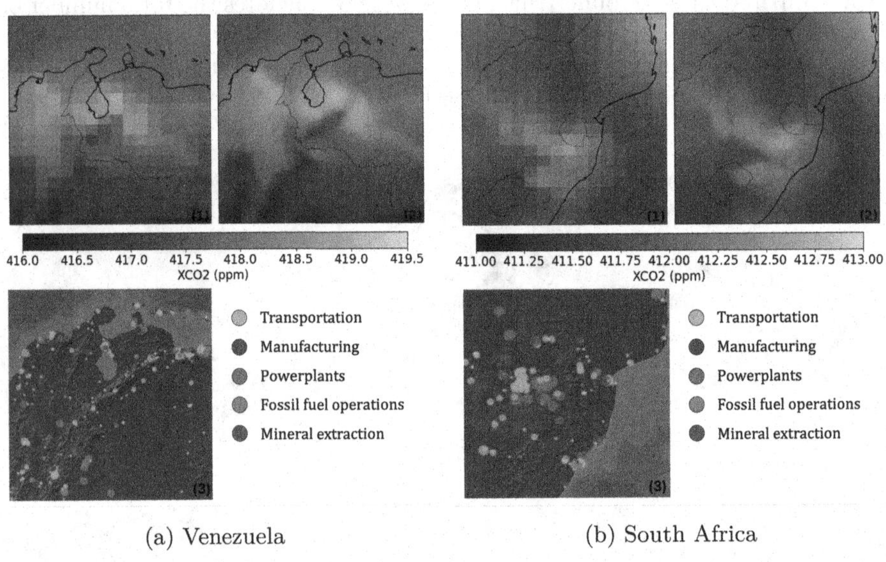

Fig. 2. Impact of point sources on XCO2 for the year 2021. For each subfigure from top left to bottom right: (1) is the LR map of the area studied, (2) is the super-resolved map, (3) is a map of CO_2 point sources extracted from the ClimateTrace emissions API.

5.3 Discussion

Based on the metrics selected for our benchmark, we showed that our model is better at estimating XCO2 while offering a higher resolution. However, we believe that we can improve its performance if we train the model on a physical variable closer to CO_2 than temperature. Furthermore, with existing Data Assimilation (DA) frameworks [2] and sparse satellite L2 data, our model could learn to further reduce misfits in specific areas. Another direction that could also be explored is the processing of multiple daily LR inputs to generate a single SR output, effectively reducing the temporal resolution of our dataset but also its uncertainty. Lastly, the application of SR maps in the context of point source monitoring needs to be expanded for a more comprehensive analysis.

6 Conclusion

This study has shown the potential of SR for upsampling the resolution of global L3 maps and generate HR XCO2 data. Although results are promising and improve on other methods listed in this paper, areas of improvement for our model have been identified and will be investigated in future works. We also proposed a visual analysis of CO_2 emission point sources and their impact on their environment, using the SR maps generated, which will be further explored. Finally, new XCH_4 L3 maps based on TROPOMI [16] should be released shortly [1]. The work presented in this study could therefore be applied to another GHG, which supports our implementation of a gas-agnostic framework.

References

1. Balashov, N., Weir, B., Ott, L., Basu, S.: Generating global CH4 NASA GEOS product by assimilating TROPOMI. In: AGU Fall Meeting. No. A15L-1387 (2022)
2. Buizza, C., et al.: Data learning: integrating data assimilation and machine learning. J. Comput. Sci. **58**, 101525 (2022). https://doi.org/10.1016/j.jocs.2021.101525, https://www.sciencedirect.com/science/article/pii/S1877750321001861
3. Climate TRACE coalition: Climate TRACE - Tracking Real-time Atmospheric Carbon Emissions. Climate TRACE Emissions Inventory (2022). https://climatetrace.org/
4. Core Writing Team, Lee, H., Romero, J.: Climate change 2023: synthesis report. Contribution of working groups I, II and III to the sixth assessment report of the intergovernmental panel on climate change, pp. 35–115 (2023). https://doi.org/10.59327/IPCC/AR6-9789291691647
5. Eldering, A., Boland, S., Solish, B., Crisp, D., Kahn, P., Gunson, M.: High precision atmospheric CO2 measurements from space: the design and implementation of OCO-2. In: 2012 IEEE Aerospace Conference, pp. 1–10. IEEE (2012)
6. Haris, M., Shakhnarovich, G., Ukita, N.: Deep back-projection networks for single image super-resolution. IEEE Trans. Pattern Anal. Mach. Intell. **43**(12), 4323–4337 (2020)
7. He, C., Ji, M., Grieneisen, M.L., Zhan, Y.: A review of datasets and methods for deriving spatiotemporal distributions of atmospheric CO2. J. Environ. Manage. **322**, 116101 (2022)
8. He, Z., et al.: Spatio-temporal mapping of multi-satellite observed column atmospheric CO2 using precision-weighted Kriging method. Remote Sens. **12**(3), 576 (2020)
9. Jacobson, A.R., Schuldt, K.N., Tans, P.: CarbonTracker CT2022. NOAA Global Monitoring Laboratory (2023). https://doi.org/10.25925/Z1GJ-3254
10. Laughner, J.L., et al.: The total carbon column observing network's GGG2020 data version. Earth Syst. Sci. Data Discuss. **2023**, 1–86 (2023)

11. Li, J., et al.: High-spatiotemporal resolution mapping of spatiotemporally continuous atmospheric CO2 concentrations over the global continent. Int. J. Appl. Earth Obs. Geoinf. **108**, 102743 (2022)
12. Li, S., et al.: PyTorch distributed: experiences on accelerating data parallel training. arXiv preprint arXiv:2006.15704 (2020)
13. Pillai, D., Neininger, B.: Comparing Lagrangian and Eulerian models for CO 2 transport-a step towards Bayesian inverse modeling using WRF/STILT-VPRM. Atmos. Chem. Phys. **12**(19), 8979–8991 (2012)
14. Sheng, M., Lei, L., Zeng, Z.C., Rao, W., Song, H., Wu, C.: Global land 1° mapping dataset of XCO2 from satellite observations of GOSAT and OCO-2 from 2009 to 2020. Big Earth Data **7**(1), 170–190 (2023)
15. Tsai, R.Y., Huang, T.S.: Multiframe image restoration and registration, vol. 1, pp. 317–339 (1984)
16. Veefkind, J.P., et al.: TROPOMI on the ESA Sentinel-5 precursor: a GMES mission for global observations of the atmospheric composition for climate, air quality and ozone layer applications. Remote Sens. Environ. **120**, 70–83 (2012)
17. Wan, Z., Hook, S., Hulley, G.: MOD11C1 MODIS/Terra land surface temperature/emissivity daily L3 global 0.05Deg CMG V006 [data set] (2015). https://doi.org/10.5067/MODIS/MOD11C1.006
18. Wang, Y., Yuan, Q., Li, T., Yang, Y., Zhou, S., Zhang, L.: Seamless mapping of long-term (2010–2020) daily global XCO 2 and XCH 4 from the greenhouse gases observing satellite (GOSAT), orbiting carbon observatory 2 (OCO-2), and CAMS global greenhouse gas reanalysis (CAMS-EGG4) with a spatiotemporally self-supervised fusion method. Earth Syst. Sci. Data **15**(8), 3597–3622 (2023)
19. Weir, B., Ott, L.: OCO-2 Science Team: OCO-2 GEOS level 3 daily, 0.5 $\times$ 0.625 assimilated CO2 v10r (2021)
20. Wunch, D., et al.: The total carbon column observing network. Philos. Trans. R. Soc. A: Math. Phys. Eng. Sci. **369**(1943), 2087–2112 (2011)
21. Yang, W., Zhang, X., Tian, Y., Wang, W., Xue, J.H., Liao, Q.: Deep learning for single image super-resolution: a brief review. IEEE Trans. Multimedia **21**(12), 3106–3121 (2019). https://doi.org/10.1109/TMM.2019.2919431
22. Zammit-Mangion, A., Cressie, N., Shumack, C.: On statistical approaches to generate level 3 products from satellite remote sensing retrievals. Remote Sens. **10**(1), 155 (2018)

Multiscale Modelling and Simulation

Integrated Multi-scale Model of Thermal Conductivity for Expanded Perlite Powder Vacuum Insulation Panels

Ziyan Fu[1(✉)], Jorge Corker[2], and Mizi Fan[1]

[1] College of Engineering, Design and Physical Sciences, Brunel University London, Uxbridge UB8 3PH, UK
{ziyan.fu,mizi.fan}@brunel.ac.uk
[2] IPN Led&mat, Rua Pedro Nunes, 3030 199 Coimbra, Portugal
jcorker@ipn.pt

Abstract. Vacuum Insulation Panels (VIPs) have emerged as a forefront solution in energy-efficient building materials. Expanded perlite (EP) stands out for its unique combination of low density, cost-effectiveness, and excellent thermal insulating properties among the myriad materials employed in VIPs. This study presents an integrated model utilizing analytical methods and finite element analysis (FEA) to simulate the heat transfer and predict the thermal conductivity of EP powder VIPs across varying gas pressures. It introduces a procedure to generate representative elementary areas (REAs) adaptable to various material characteristics; in comparing the simulation results to measurement values, the proposed model demonstrates reliable predictive performance from 0.0001 to 1 atm. The proposed model efficiently handles rapid thermal conductivity changes near atmospheric pressure, resolving distortion issues in other works. Based on the model results of REAs reflecting various material characteristics, we found that reducing the non-flake ratio of particles and decreasing the thickness of flake particles obstruct the heat transfer across all pressure ranges. When the thermal conductivity of the absolute solid is relatively high, it is advisable for the industry to prioritize applying finer grinding; conversely, efforts should be directed towards reducing the thickness of flake particles.

Keywords: Vacuum Insulation Panel · Expanded Perlite · Thermal Conductivity · Model · Heat Transfer

1 Introduction

According to the European Commission Energy Department, buildings in the EU are responsible for 40% of total energy consumption, meanwhile 75% of the building stock is identified as energy-inefficient. Based on this context, the

Supported by European Union's Horizon 2020 research and innovation programme under Grant Agreement No. 869898.

L. Franco et al. (Eds.): ICCS 2024, LNCS 14836, pp. 393–407, 2024.
https://doi.org/10.1007/978-3-031-63775-9_29

emphasis on sustainable building practices has spurred research into advanced insulation materials, notably Vacuum Insulation Panels (VIPs). VIPs are usually built with a rigid porous core and sealed in an envelope. By evacuating the core, VIPs create a vacuumed environment where gaseous conduction and convection are nearly eliminated and thus achieve an extremely low thermal conductivity [4–6]. The thickness of VIPs is significantly thinner than traditional materials, making it particularly advantageous in situations where space is limited but high thermal resistance is required [7,8], for instance, refrigeration [1], construction [2], and transportation [3].

Expanded Perlite (EP) is obtained from natural perlite, a natural igneous or sedimentary rock, by crushing natural perlite ore, then treated at high temperatures at 760 °C to 1100 °C to expand, release moisture and crystallization water, and finally cooled to room temperature [12,13]. Despite being a non-renewable resource, abundant natural perlite reserves and cost-effective production methods make EP a sustainable choice [14]. EP stands out for its insulation, lightweight and affordability, making it a key component of VIPs. Its composition [17], rich in silicates and metal oxides, ensures chemical stability and longevity [18,19].

Scientists have amassed considerable achievements in thermal conductivity modelling for porous materials since engineers attempted to establish empirical models [20]. These endeavours have primarily resulted in two major categories of models: analytical models and numerical models [21]. Analytical models typically compartmentalize the effective thermal conductivity into four parts: solid heat transfer, gas heat transfer, radiative heat transfer, and the coupling effects [22]. Conversely, numerical models feed all parameters, including morphological and physical parameters, into the simulation together. Upon setting boundary conditions, these numerical models employ Finite Element Analysis (FEA) [23,24], lattice Boltzmann methods (LBM) [25–27], or even Molecular Dynamics (MD) [28] for simulation.

Since the mean free path of gas molecules is inversely proportional to the gas pressure, the range of the Knudsen number of the fluid inside the porous material spans a large range, makes impossible to use a single-scale model to cover all the situations. The advantage of analytical models lies in only utilizing morphological statistical data, eliminating the need to reconstruct material microstructures for thermal conductivity prediction [1]. However analytical models are still powerless for highly coupled scenarios at near atmospheric pressure [21]. Numerical models do just the opposite. However, in scenarios where the gas pressure diminishes significantly, attributable to the diminution of fluidic properties and the concurrent amplification of particulate attributes, the gas exhibits transitional or free molecular flow characteristics and consequentially limits the application of numerical models in extremely low-pressure circumstances.

Combining models from both categories to overcome the limitations and provide accurate thermal conductivity predictions across lifetime pressure ranges, this research aims to integrate analytical and numerical approaches to develop a comprehensive model suitable for studying vacuum insulation panels, thereby advancing understanding and application in sustainable building practices.

2 Material and Methodology

2.1 Material

This study uses the EP filler Perlite 180 to build VIPs. The Scanning Electron Microscope (SEM) images of Perlite 180 are shown in Fig. 1 left. Unlike those perlite particles without grinding shown in Fig. 1 right [29], the characteristic bubble morphology of EP is no longer recognizable after fine grinding; only broken tiny plates with thicknesses from around 300 nm to 1 µm can be found.

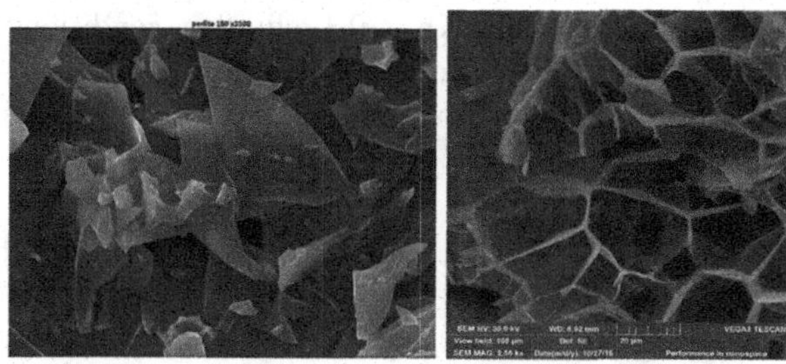

Fig. 1. SEM Images of Perlite 180 (left) and unground Perlite (right)

Due to the fine grinding, there are no closed pores within the Perlite 180 filler. The rule of mixtures with chemical composition analysis determines real density. This study calculates porosity through the ratio of the bulk density ($0.18\,\mathrm{g/cm^3}$) to the real density ($2.796\,\mathrm{g/cm^3}$) of EP powder. The resulting porosity is $\Pi = 1 - \frac{0.18\,\mathrm{g/cm^3}}{2.796\,\mathrm{g/cm^3}} = 93.56\%$.

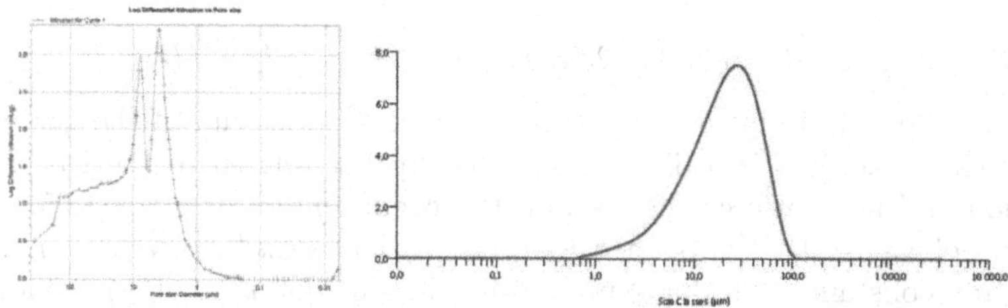

Fig. 2. Pore (left) and Particle (right) Diameter Distribution of Perlite 180

Another key parameter for VIP core materials is the diameter distribution of voids, which determines the vacuum levels required to keep the low thermal conductivity. From Fig. 2, we can observe that the mode of the pore diameter distribution for Perlite 180 is just below 10 µm. This observation aligns with the situation presented in the SEM images.

Via laser diffraction technique, in reference to Fig. 2, Perlite 180 displays a peak distribution centred at about 20 µm and the percentile data regarding particle diameter distribution is presented in Table 2. However, the laser diffraction technique is unsuitable for flake particles, which violates the assumption of spherical particles. Therefore, particle diameter results are only used for morphological analysis and will not be passed on to later simulations.

2.2 General Principles of Model Design

This section shows the analytical and numerical models used to investigate the effects of varying internal gaseous pressures on thermal conductivity due to ageing and damage. This study highlights the importance of the Knudsen number (Kn) [30], which helps determine the appropriate model based on gas behaviour in different pressure conditions. For Kn < 0.1, slip flow conditions are assumed, and a numerical model is applied. An analytical model based on rare gas theory is utilised for Kn > 0.1. The "divide and conquer" idea aims to provide a comprehensive model effective across a wide range of pressures.

2.3 Analytical Model for Lower Pressure

The thermal conductivity of porous media comprises three components: solid heat transfer, gas heat transfer, and radiative heat transfer [21]. In a rarefied gas environment ($Kn > 0.1$), where pressure is the only variable, solid heat transfer and radiative heat transfer can be considered constants. For scenarios where low-pressure results in a $Kn > 0.1$, we started with the classic Kaganer model, designed a more robust version of the Kaganer model, and applied it to the calculation of the thermal conductivity of gases within porous media [30]. Firstly, the Kaganer model is introduced with a parallel plates assumption: a gas with molecular mean free path l_g passing through two parallel plates at a distance D, and the equivalent gaseous thermal conductivity λ_{gas} can be calculated as:

$$\lambda_{gas} = (\Pi\lambda_0)/(1 + 2\beta Kn) = (\Pi\lambda_0)/(1 + 2\beta(l_g/D)) \tag{1}$$

where Π is the total porosity, λ_0 is the thermal conductivity of the gas in free space ($Kn < 0.01$), β is a dimensionless coefficient related to the gas accommodation and adiabatic coefficient and the coefficient β for air is 1.55. Kn is the ratio of l_g to D. l_g is defined as $l_g = (k_B T)/(\sqrt{2}\pi d^2 p)$, where k_B is the Boltzmann constant, T is the temperature, d is the diameter of a gas molecule and p is the gas pressure.

However, the average pore diameter D is not easy to determine from experimental results. From uniform pore size [30] to normal [31] and bimodal [32] normal distribution, scientists have been trying to model the pore size distribution of porous materials with distributions that are more consistent with experimental results. In this study, to restore the experimental results as well as mitigate the unevenly distributed sampling value and influence of outliers, we assume that the pore diameter d follows a lognormal distribution D:

$$f(d;\mu,\sigma)\ \frac{1}{d\sigma\sqrt{2\pi}}e^{-\frac{(\ln d-\mu)^2}{2\sigma^2}} \tag{2}$$

where μ and σ are the mean (expectation) and standard deviation of lognormal distribution D. We then deduce the mean by reverse engineering through the more readily obtained mode and median.

The mean $\mathrm{E}(D)$ conforming to the log-normal distribution can be calculated with the mode $\mathrm{Mode}[D]=e^{\mu-\sigma^2}$ and the medium $\mathrm{Med}[D]=e^{\mu}$ as Eq. 3:

$$\mathrm{E}(D)=e^{\mu+\frac{\sigma^2}{2}}=\sqrt{e^{2\mu}e^{\sigma^2}}=\sqrt{\frac{e^{3\mu}}{e^{\mu-\sigma^2}}}=\sqrt{\frac{\mathrm{Med}[D]^3}{\mathrm{Mode}[D]}} \tag{3}$$

The mean pore diameter for Perlite 180 can be calculated as $E(D_{180})=\sqrt{\frac{\mathrm{Med}[D_{180}]^3}{\mathrm{Mode}[D_{180}]}}=\sqrt{\frac{9.1\mu m^3}{4\mu m}}\approx 13.7\,\mu\mathrm{m}$.

Therefore, the effective thermal conductivity of EP powder under a rarefied gas environment can be summarized as:

$$\lambda_{analytical}=\lambda_{gas}+\lambda_{solid}+\lambda_{radiation}=\lambda_{gas}+\lambda_{const}=\frac{\Pi\lambda_0}{1+2\beta l_g\sqrt{\frac{Mode[D]}{Med[D]^3}}}+\lambda_{const} \tag{4}$$

where the sum of solid thermal conductivity λ_{solid} and radiative thermal conductivity $\lambda_{radiation}$ can be written as a constant variable λ_{const}.

2.4 Numerical Model for Higher Pressure

This study used SEM images of EP powder as a reference. We developed a visualized Grasshopper [33] program, enabling the generation of a Representative Elementary Area (REA) tailored to a specific set of characteristics for EP samples. Finally, the REAs and material's characteristic values are fed into the FEA slip flow model implemented by COMSOL Multiphysics. Meanwhile, we eliminate the influence of the randomness of REA generation on the final results by averaging multiple calculations.

Limit by the length of the short paper, we omit the details of generating REAs and only describe the design logic here. First, we randomly divide the REA into many cells with a diameter equal to the characteristic pore diameter (Fig. 3 left). Then starting from the minimum observed wall thickness, we gradually increase the thickness until the porosity equals the observed value (Fig. 3 middle). Finally, we hollowed out all vertices so that all solid walls were disjointed flakes (Fig. 3 right).

In the subsequent finite element analysis, appropriate boundary conditions are established, and the physical properties of the gas and solid domains are incorporated into the model through empirical formulas or rules of mixtures. Finally, by utilizing the 2-dimensional thermal conductivity equation under appropriate sampling scope, the effective thermal conductivity of the REA is computed.

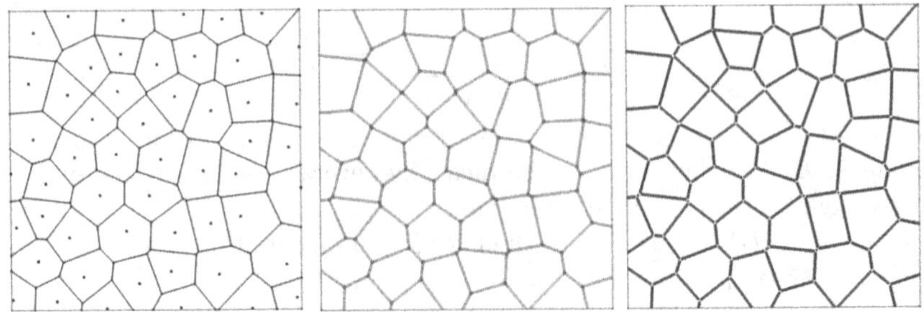

Fig. 3. Process of Constructing REA

As illustrated in Fig. 4 left, the temperature difference between the 2 cm panel is $\Delta T_{2cm} = 5\text{K}$ at the macroscopic scale, and the temperature difference between the left and right sides of the REA at the microscopic scale is given by $\Delta T = \Delta T_{2cm} * L/20\,\text{mm} = 0.05\text{K}$. The initial temperature $T_{initial}$ of the REA is set as the medium temperature between the cold and hot plate $(T_{h_2cm} + T_{c_2cm})/2 = 295.65\text{K}$, and the temperature of cold (left) and hot (right) sides are $T_c = T_{initial} - \Delta T/2 = 295.625\text{K}$ and $T_h = T_{initial} + \Delta T/2 = 295.675\text{K}$. The top and bottom sides of the REA are periodic boundary conditions, in which heat and mass are allowed to transfer across the boundary periodically.

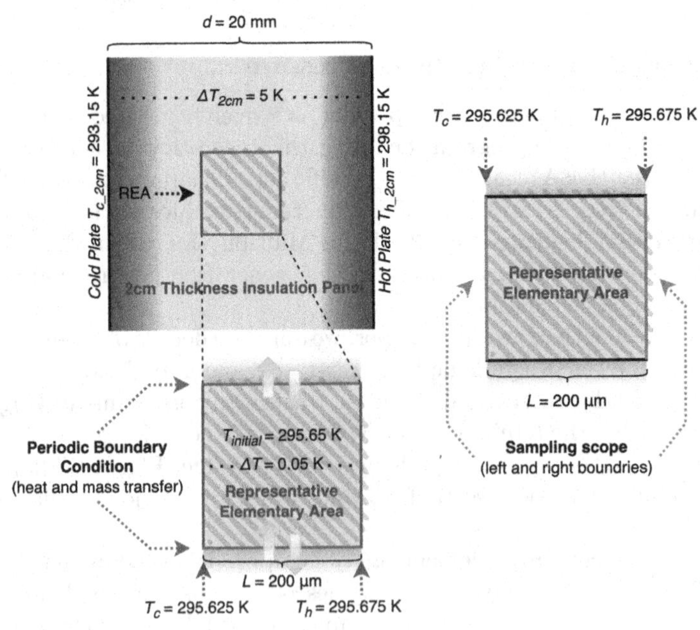

Fig. 4. Schematic Diagram and Sampling Scope of REA

In our research, the solid of the porous medium is expanded perlite, while the gas within the porous medium is dry air. For the gas domain, COMSOL Multiphysics provides built-in formulas for the physical properties of dry air.

For the solid domain, density, thermal conductivity, and heat capacity at constant pressure are required to perform the slip flow simulation. The rule of mixtures is applied to obtain the above parameters of the solid skeleton. By comparing numbers of the research paper, a representative composition of the EP solid (76% of SiO_2, 14% of Al_2O_3, 4% of Na_2O, 5% of K_2O and 1% of impurities) is assumed and will be applied with the rule of mixtures [34,35].

As we know $\rho_{SiO_2} = 2648\,\mathrm{kg/m^3}$, $\rho_{Al_2O_3} = 3987\,\mathrm{kg/m^3}$, $\rho_{Na_2O} = 2270\,\mathrm{kg/m^3}$, and $\rho_{K_2O} = 2130\,\mathrm{kg/m^3}$. The density of EP solid ρ_{solid} is derived as:

$$\rho_{solid} = \sum \frac{m_i}{m_{mixture}} \rho_i = 2795.9\,\mathrm{kg/m^3} \tag{5}$$

Due to the lack of data in the academic community regarding the thermal conductivity of sodium oxide and potassium oxide as a function of temperature, we limit our consideration to the two primary components with the highest weight percentage in the EP solid: silicon dioxide and aluminium oxide. As we know $\lambda_{SiO_2} = 1.4\,\mathrm{W/mK}$, $\lambda_{Al_2O_3} = 35\,\mathrm{W/mK}$. The upper limit of thermal conductivity of EP solid λ_{solid} is derived as:

$$\lambda_{solid\_upper} = \sum \frac{m_i}{m_{mixture}} \lambda_i = 6.627\,\mathrm{W/mK} \tag{6}$$

The lower limit of thermal conductivity of EP solid λ_{solid} is derived as:

$$\lambda_{solid\_lower} = \left(\sum \frac{m_i/m_{mixture}}{\lambda_i} \right)^{-1} = 1.646\,\mathrm{W/mK} \tag{7}$$

As we know $C_{p\_SiO_2} = 741.84\,\mathrm{J/(kg*K)}$ [38], $C_{p\_Al_2O_3} = 753\,\mathrm{J/(kg*K)}$ [37], $C_{p\_Na_2O} = 1114.71\,\mathrm{J/(kg*K)}$ [36], and $C_{p\_K_2O} = 887.69\,\mathrm{J/(kg*K)}$ [39]. The heat capacity at a constant pressure of EP solid C_{p_solid} is derived as:

$$C_{p\_solid} = \sum \frac{m_i}{m_{mixture}} C_{p\_i} = 765.85\,\mathrm{J/(kg*K)} \tag{8}$$

In our 2-dimensional simulation, the thermal conductivity of the REA is calculated with sampling on left and right boundaries (Fig. 3 right): $\lambda = -\,\Phi * L\,/\,(T_h - T_c)$. Φ is the heat flux on sampling scope, L is the side length of square REA, T_h and T_c are the temperatures of the hot and cold sides of the REA, respectively.

2.5 Connection Between Two Stages of Model

The analytical method and FEA slip flow model are exclusively utilized to address cases where the Kn>0.1 and Kn<0.1, respectively. For Perlite 180 (characteristic pore diameter = 13.7 μm), the boundary condition for $Kn = 0.1$ is $p_{boundary} = 0.05$ atm. Therefore, we selected 0.005, 0.01, 0.02, 0.05, 0.1, 0.2, 0.5, and 1 atm as the scanning values for our FEA slip flow model, and take $p_{boundary} = 0.05$ atm as the joint boundary of the coupled model. At the boundary pressure, the results of the analytical method $\lambda_{analytical\_0.05}$ and the FEA slip flow model $\lambda_{FEA\_0.05}$ should return the same value logically.

For Eq. 4 utilized in the analytical method, λ_{const} is an unknown constant variable. Therefore, we apply the lower pressure boundary to the FEA slip flow model with $p = 0.05$ atm (i.e. when $Kn = 0.1$) and take the FEA model result $\lambda_{FEA\_0.05}$ into Eq. 9 to get the value of λ_{const}, then utilize Eq. 4 to compute the effective thermal conductivity $\lambda_{analytical}$ when the gas pressure is lower than $p_{boundary}$.

$$\lambda_{FEA\_0.05} = \lambda_{analytical\_0.05} = \lambda_{const} + \frac{\Pi\lambda_0}{1 + 2\beta\frac{k_B T}{\sqrt{2}\pi d^2 p_{boundary}}\sqrt{\frac{Mode[D]}{Med[D]^3}}} \tag{9}$$

3 Model Results and Disscussion

3.1 Results Validation

Validation of model hyperparameters ensures our FEA model results remain consistent regardless of the dimensions of REAs or mesh refinement. Firstly, the dimension of the REA needs to be validated to be large enough to endow the model with robustness against the randomness during the generation of the REA. The second to be validated is the level of mesh refinement during the finite element analysis since the model outcomes may fluctuate significantly with changes in mesh refinement.

To assess the convergence of FEA model outcomes regarding the dimension of the REAs, 7 dimensions ranging from 50 µm to 200 µm were evenly selected. Then, 5 different REAs were randomly generated for each dimension. The mean and the dispersion index of the model at all dimensions and pressure levels are monitored as follows:

$$\mu = \frac{\sum_1^n \lambda_i}{n}, \sigma^2 = \frac{\sum_1^n (\lambda_i - \mu)^2}{n}, D = \frac{\sigma^2}{\mu} \tag{10}$$

where σ^2 is the variance and μ is the mean.

Firstly, Fig. 5 demonstrates that the mean of the model outcomes rapidly converges with increasing REA dimension from 50 µm to 125 µm. At dimensions greater or equal to 150 µm, the mean of the model outcomes is highly stable, exhibiting minimal variation with changes in REA dimension, and it can be considered as converged. Secondly, the dispersion index decreases significantly when the REA dimension increases from 50 µm to 125 µm. As long as the REA dimension is equal to or greater than 125 µm, even under atmospheric pressure (1 atm) conditions, the dispersion index of FEA model outcomes remains stable at less than 1%, which is a commendable and converged performance. In conclusion, the REA dimension should not be less than 150 µm, which is consistent with the argument presented in Sect. 3.

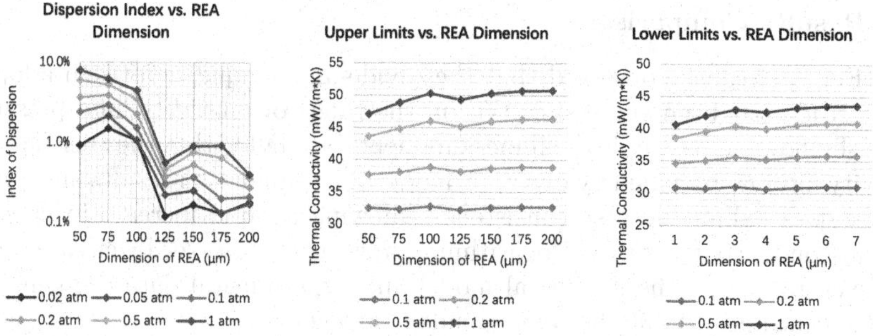

Fig. 5. Dispersion Index (left), Upper (middle) and Lower (right) Limits vs. REA Dimension

Secondly, to validate the reliability of the refinement level, in this section, we tested 7 mesh refinement levels: *Extremely Coarse*(- - - -), *Extra Coarse*(- - -), *Coarser*(- -), *Coarse*(-), *Normal*(std), *Fine*(+) and *Finer*(++). The results of the above 7 refinement levels at 1 atm, 0.5 atm, and 0.1 atm are illustrated in Fig. 6. Simulation results converge most rapidly during the refinement process from *Extremely Coarse* to *Coarser*, while they remain relatively stable during the refinement process from *Coarse* to *Finer*. We can conclude that the *Normal*(std) level of refinement level is sufficient and can be treated as a balance of the trade-off between accuracy and complexity.

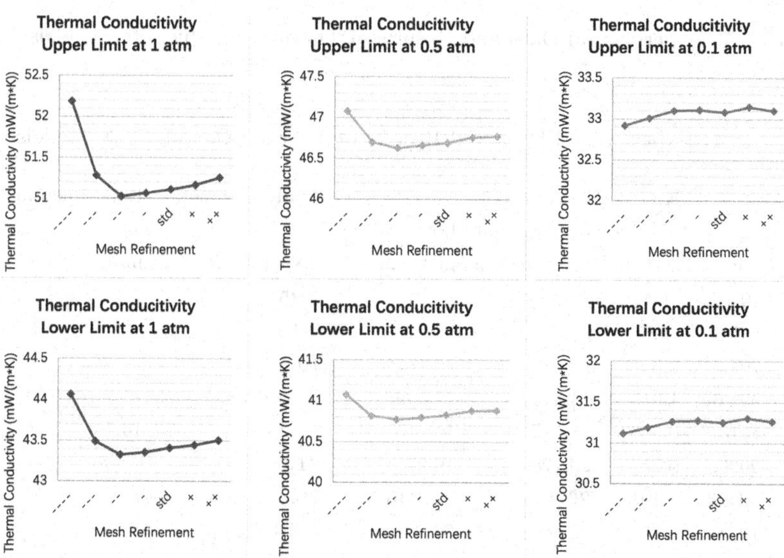

Fig. 6. Model Results at 1, 0.5 and 0.1 atm vs. Different Levels of Mesh Refinement

3.2 Results Comparison

From Fig. 7, it can be observed that the divide-and-conquer approach adopted in this study has been successful. Within the range of relatively high pressures where slip flow can be applied, all measurement data fall within the thermal conductivity interval provided by our FEA model. The simulation results and experimental data exhibit notable consistency, showing a rapid increase in effective thermal conductivity and the logarithmic growth of pressure. In environments with even rarer gas, the analytical model and experimental data are generally in agreement and exhibit the same trends (Table 1).

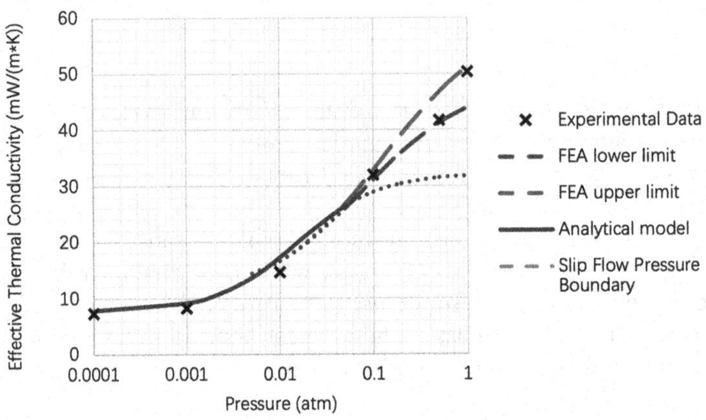

Fig. 7. Experimental Data and Results of Thermal Conductivity Models

Table 1. Experimental Data and Results of Thermal Conductivity Models

Pressure		Kn	FEA Model		Analytical model	Experimental data
			Upper Limit	Lower Limit		
atm	Pa		mW/(m*K)	mW/(m*K)	mW/(m*K)	mW/(m*K)
0.0001	10.1	10.1	/	/	**7.85**	7.4
0.001	101.3	5.06	/	/	**9.18**	8.3
0.002	202.6	2.54	/	/	**10.5**	/
0.005	506.6	1.01	14.416	14.399	**13.67**	/
0.01	1013.3	0.506	16.677	16.615	**17.31**	14.7
0.02	2026.5	0.254	20.178	19.977	**21.5**	/
0.05	5066.3	0.101	**26.803**	**26.033**	**26.4**	/
0.10	10133	0.0506	**32.875**	**31.164**	28.91	32
0.20	20265	0.0254	**39.172**	**36.015**	30.44	/
0.50	50663	0.0101	**46.731**	**41.194**	31.46	41.7
1.00	101325	0.00506	**51.236**	**43.958**	31.82	50.3

3.3 Discussion and Analysis

The study introduces an innovative model for predicting thermal conductivity in various materials, notably expanded perlite (EP) powder, within a range of gas pressures. Compared with existing similar research, this model stands out for two primary reasons. First, it adapts its approach based on the state of gas flow—accounting for different conditions like laminar flow, slip flow, and transitional flow—by incorporating methodologies specific to the Knudsen number (Kn). This "divide and conquer" strategy allows for a tailored and accurate representation of gas flow states without overcomplicating the model. Second, the model introduces a flexible generation of REAs, which can adapt to the microstructural variations of EP powders and other materials with similar characteristics. This adaptability is further enhanced by the model's ability to adjust for the absolute solid thermal conductivity based on the material's chemical composition, allowing for a precise representation of the material's thermal behaviour. Furthermore, this study not only offers insights into the thermal behaviour of EP powder but also proposes a framework applicable to a broader range of materials with similar structural characteristics.

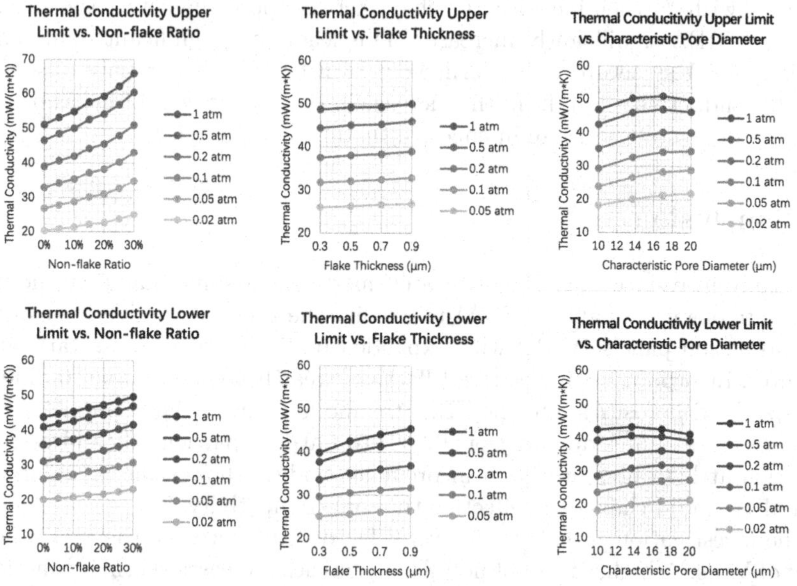

Fig. 8. Effective Thermal Conductivity Range vs. Non-flake Ratio (left), vs. Flake Thickness (middle), and vs. Characteristic Pore Diameter (right)

The study's findings demonstrate that an increase in the ratio of non-flake structures improves the effective thermal conductivity under all pressure conditions (Fig. 8 left). Notably, this improvement is least significant at a pressure of

0.02 atm when the solid's thermal conductivity is low. Additionally, the promotion of higher pressures to the enhancement is also noticed. We speculate that this is due to the intensified coupling effect between increased gas heat transfer at higher pressures and solid heat transfer in non-flake structures, amplifying the heat transfer increase resulting from changes in solid structure.

The flake thickness also serves as an essential parameter for predicting the effective thermal conductivity of EP powder under different pressures and it cannot be easily altered through physical methods such as grinding. It is evident that both the upper and lower limits of thermal conductivity increase with the growing flake thickness (Fig. 8 middle). We infer that this phenomenon arises from the increase in the cross-sectional area of flakes due to the increased flake thickness, facilitating more heat flux transfer from the hot end to the cold end through the solid flake structure.

Lastly, Fig. 8 (right) shows that the effect of characteristic pore diameter on the heat transfer of EP powder is non-monotonic, which warrants thorough discussion and analysis. Particularly at high pressures where gas heat transfer less significantly influences the overall thermal conductivity (Fig. 8 right). The study suggests that the increase in pore diameter—and consequently, the mean free path of gas molecules—enhances gas thermal conductivity at lower pressures. However, with further increases in characteristic pore diameter, the porosity within the REA significantly increases. This leads to a significant reduction in the flaky solid structure to perform heat transfer, which not only offsets the thermal conductivity gain from the increase in the average size of flakes but also results in an overall decrease in effective thermal conductivity.

4 Conclusion

This study introduces a novel integrated model combining analytical methods with finite element analysis (FEA) to predict the thermal conductivity of vacuum insulation panels (VIPs) with expanded perlite powder across various gas pressures. Incorporating a Relative Effective Area (REA) adaptable to different pore sizes and porosities into the FEA, the model shows reliable predictive performance for pressures ranging from 0.0001 to 1 atm. It particularly addresses the distortion at high pressures seen in previous models, showcasing its robustness in handling rapid changes in effective thermal conductivity.

The investigation reveals that several factors influence the thermal conductivity of VIPs. A higher ratio of non-flake particles increases thermal conductivity, especially when the solid's thermal conductivity is enhanced. Flake particle thickness also impacts thermal conductivity, with variations more noticeable at lower solid thermal conductivities and higher pressures. The characteristic pore diameter's effect on heat transfer is complex, showing non-monotonic behaviour at high pressures but a monotonic increase in effective thermal conductivity with the diameter at low pressures.

To optimize VIPs' thermal performance, the study suggests finer grinding of perlite powder, reducing wall thickness chemically, and selecting appropriate encapsulation pressures for optimal pore size. Future research could leverage the lattice Boltzmann method (LBM) for deeper insights into heat transfer in porous media like perlite powder at ultra-low pressures, enhancing understanding of material characteristics on thermal conductivity.

Data Availability Statement. The processed data and program required to reproduce these findings are available to download from https://gitlab.com/leo.ziyanfu/integrated-multi-scale-model-of-thermal-conductivity-for-expanded-perlite-powder-vacuum-insulation-panels.git.

References

1. Bouquerel, M., Duforestel, T., Baillis, D., Rusaouen, G.: Heat transfer modeling in vacuum insulation panels containing nanoporous silicas—a review. Energy Build. **54**, 320–336 (2012)
2. Brunner, S., Ghazi Wakili, K., Stahl, T., Binder, B.: Vacuum insulation panels for building applications—continuous challenges and developments. Energy Build. **85**, 592–596 (2014)
3. Wang, B., et al.: Characterization and monitoring of vacuum pressure of tank containers with multilayer insulation for cryogenic clean fuels storage and transportation. Appl. Therm. Eng. **187**, 116569 (2021)
4. Corker, J., et al.: Al-rich industrial waste as new alternative of fumed silica for the manufacture of vacuum insulation panels for building energy conservation. J. Clean. Prod. **415**, 137854 (2023)
5. Raad, T., Verma, S., Singh, H.: Tree waste based advanced thermal insulation - vacuum insulation panels - for application at up to 70 °C. Int. J. Therm. Sci. **200**, 108971 (2024)
6. Zhuang, J., Ghaffar, S.H., Fan, M., Corker, J.: Restructure of expanded cork with fumed silica as novel core materials for vacuum insulation panels. Composites Part B Eng. **127**, 215–221 (2017)
7. Alam, M., Singh, H., Limbachiya, M.C.: Vacuum Insulation Panels (VIPs) for building construction industry - a review of the contemporary developments and future directions. Appl. Energy **88**, 3592–3602 (2011)
8. Baetens, R., et al.: Vacuum insulation panels for building applications: a review and beyond. Energy Build. **42**, 147–172 (2010)
9. Simões, N., Gonçalves, M., Serra, C., Resalati, S.: Can vacuum insulation panels be cost-effective when applied in building façades? Build. Environ. **191**, 107602 (2021)
10. Gonçalves, M., Simões, N., Serra, C., Flores-Colen, I.: A review of the challenges posed by the use of vacuum panels in external insulation finishing systems. Appl. Energy **257**, 114028 (2020)
11. Mao, S., Kan, A., Zhu, W., Yuan, Y.: The impact of vacuum degree and barrier envelope on thermal property and service life of vacuum insulation panels. Energy Build. **209**, 109699 (2020)
12. Zhang, N., Yuan, Y., Yuan, Y., Li, T., Cao, X.: Lauric-palmitic-stearic acid/expanded perlite composite as form-stable phase change material: preparation and thermal properties. Energy Build. **82**, 505–511 (2014)

13. Alam, M., Singh, H., Brunner, S., Naziris, C.: Experimental characterisation and evaluation of the thermo-physical properties of expanded perlite-Fumed silica composite for effective vacuum insulation panel (VIP) core. Energy Build. **69**, 442–450 (2014)
14. Rottmann, M., Beikircher, T., Ebert, H., Hemberger, F., Manara, J.: Thermal conductivity and extinction coefficient of opacified expanded perlite for vacuum super insulation up to 1073 K. Int. J. Therm. Sci. **163**, 106813 (2021)
15. Reka, A.A., et al.: Chemical, mineralogical and structural features of native and expanded perlite from Macedonia. Geologia Croaticam **72**, 215–221 (2019)
16. Zukowski, M., Haese, G.: Experimental and numerical investigation of a hollow brick filled with perlite insulation. Energy Build. **42**, 1402–1408 (2010)
17. Singh, M., Garg, M.: Perlite-based building materials - a review of current applications. Constr. Build. Mater. **5**, 75–81 (1991)
18. Jia, G., Li, Z., Liu, P., Jing, Q.: Preparation and characterization of aerogel/expanded perlite composite as building thermal insulation material. J. Non Cryst. Solids **482**, 192–202 (2018)
19. Pichór, W., Janiec, A.: Thermal stability of expanded perlite modified by mullite. Ceram. Int. **35**, 527–530 (2009)
20. Bouquerel, M., Duforestel, T., Baillis, D., Rusaouen, G.: Mass transfer modeling in gas barrier envelopes for vacuum insulation panels: a review. Energy Build. **55**, 903–920 (2012)
21. Fu, Z., et al.: Critical review on the thermal conductivity modelling of silica aerogel composites. J. Build. Eng. **57**, 104814 (2022)
22. Kwon, J., Jang, C.H., Jung, H., Song, T.: Effective thermal conductivity of various filling materials for vacuum insulation panels. Int. J. Heat Mass Transfer **52**, 5525–5532 (2009)
23. Subramanian, G., Picu, C.R.: Mechanics of three-dimensional, nonbonded random fiber networks. Phys. Rev. E **83**, 056120 (2011)
24. Liu, Q., Lu, Z., Hu, Z., Li, J.: Finite element analysis on tensile behaviour of 3D random fibrous materials: Model description and meso-level approach. Mater. Sci. Eng., A **587**, 36–45 (2013)
25. Grucelski, A.: LBM estimation of thermal conductivity in meso-scale modelling. J. Phys: Conf. Ser. **760**, 12005 (2016)
26. Qin, X., Cai, J., Zhou, Y., Kang, Z.: Lattice Boltzmann simulation and fractal analysis of effective thermal conductivity in porous media. Appl. Therm. Eng. **180**, 115562 (2020)
27. Lu, J., Kan, A., Zhu, W., Yuan, Y.: Numerical investigation on effective thermal conductivity of fibrous porous medium under vacuum using Lattice-Boltzmann method. Int. J. Therm. Sci. **160**, 106682 (2021)
28. Toosi, M.H., Siavashi, M.: Two-phase mixture numerical simulation of natural convection of nanofluid flow in a cavity partially filled with porous media to enhance heat transfer. J. Mol. Liq. **238**, 553–569 (2017)
29. Pavlík, V., Bisaha, J.: Lightweight mortars based on expanded perlite. Key Eng. Mater. **776**, 104–117 (2018)
30. Kaganer, M.G., Moscona, A.: Thermal insulation in cryogenic engineering. Jerusalem: Israel program for scientific translations, Jerusalem (1969)
31. Bi, C., Tang, G.H., Tao, W.Q.: Prediction of the gaseous thermal conductivity in aerogels with non-uniform pore-size distribution. J. Non-Cryst. Solids **358**, 3124–3128 (2012)

32. Reichenauer, G., Heinemann, U., Ebert, H.-P.: Relationship between pore size and the gas pressure dependence of the gaseous thermal conductivity. Colloids Surf. A **300**, 204–210 (2007)
33. Rutten, D.: Grasshopper 3D. 1.0 (2014)
34. Damiyine, B., Guenbour, A., Boussen, R.: Adsorption of rhodamine b dye onto expanded perlite from aqueous solution: kinetics, equilibrium and thermodynamics. J. Mater. Environ. Sci. **8**, 345–355 (2017)
35. Blaskov, V., et al.: Synthesis and catalytic activity of silver-coated perlite in the reaction of ozone decomposition. Ozone: Sci. Eng. **37** 252–256 (2015)
36. National Institute of Standards and Technology, Quartz (SiO2). https://webbook.nist.gov/cgi/cbook.cgi?ID=C14808607. Accessed 01 Mar 2024
37. MatWeb, Corundum, Aluminum Oxide, Alumina, 99.9%, Al2O3. https://www.matweb.com/search/DataSheet.aspx?MatGUID=c8c56ad547ae4cfabad15977bfb537f1. Accessed 01 Mar 2024
38. National Institute of Standards and Technology, disodium oxide. https://webbook.nist.gov/cgi/cbook.cgi?Formula=na2o. Accessed 01 Mar 2024
39. National Institute of Standards and Technology, dipotassium oxide. https://webbook.nist.gov/cgi/cbook.cgi?ID=12136457. Accessed 01 Mar 2024

The Lattice Boltzmann Based Large Eddy Simulations for the Stenosis of the Aorta

Xiao Xue[1], Jon W. S. McCullough[1], Sharp C. Y. Lo[1], Ioannis Zacharoudiou[1], Bálint Joó[4,5], and Peter V. Coveney[1,2,3(✉)]

[1] Centre for Computational Science, Department of Chemistry, University College London, London, UK
{x.xue,p.v.coveney}@ucl.ac.uk
[2] Advanced Research Computing Centre, University College London, London, UK
[3] Informatics Institute, Faculty of Science, University of Amsterdam, Amsterdam, Netherlands
[4] Oak Ridge National Laboratory, Oak Ridge, TN, USA
[5] NVIDIA Corporation, San Tomas Expressway, San Jose, CA, USA

Abstract. Large eddy simulations (LES) are extensively employed in aerodynamics-related industrial research and applications. However, the application of lattice Boltzmann based LES techniques in vascular blood flow research is less extensively documented. This study investigates the feasibility of employing lattice Boltzmann based large eddy simulation techniques, specifically the Smagorinsky-based subgrid scale turbulence model, for simulating high Reynolds number blood flow at a coarse-grained resolution. Initially, a stenotic channel flow simulation is conducted, with results undergoing validation against existing experimental data and direct numerical simulation results, showing strong agreement for both. Subsequently, our model is applied to simulate aortic stenosis at a resolution of 100 μm, demonstrating the capability to model high Reynolds numbers around 4500, despite such flows conventionally requiring a resolution of around 20 μm. These results underscore the substantial promise of utilising LES techniques in blood flow simulations, benefiting not just the lattice Boltzmann method but also enlightening the broader computational fluid dynamics community. This approach offers advantages for large-scale human simulations at coarser resolutions.

Keywords: Large-eddy simulations · lattice Boltzmann method · turbulent flow · stenosis of the aorta

1 Introduction

Computational fluid dynamics (CFD) is vital for industrial applications ranging from physics, and engineering to biomedical applications. Direct numerical simulation (DNS) is directly solving the Navier-Stokes equations on a discretised domain by using finite-difference [32], finite-volume [9] and finite-element methods [44]. Despite high accuracy for CFD related problems, DNS is normally time-consuming and computationally expensive. Large-eddy simulations

L. Franco et al. (Eds.): ICCS 2024, LNCS 14836, pp. 408–420, 2024.
https://doi.org/10.1007/978-3-031-63775-9_30

(LES) [30,31] can reduce the computational cost by modelling the subgrid-scale (SGS) turbulence. However, efficient LES calculations still suffer from the data-dependencies in their implementation and need refined grids [4,7,41] near the wall to correctly model the physics. To mitigate grid dependencies in LES-based simulations, coupling with RANS simulations near the wall is necessary [28]; however, this introduces additional coupling efforts.

The lattice Boltzmann method (LBM) [33] may be considered an alternative approach to solve some CFD problems numerically. Instead of solving the Navier-Stokes equations directly, LBM discretises the Boltzmann equation and implements the streaming and collision of probability density functions. Under limits imposed by a Chapman-Enskog analysis, LBM can be considered as an equivalent representation of solving the Navier-Stokes equations directly. Due to the inherent locality of the method, the LBM is relatively easy to deploy in parallel computing approaches compared to other conventional CFD methods. LBM has gained in popularity over the last few decades and has been successfully applied to various phenomena ranging from micro-nano fluidics problems [6,36–38], mesoscopic problems like blood flow simulations [12,25,42], and macroscopic scale problems [15,16,20,27]. In addition to its efficient parallelisation, the LBM is also suitable for sparsely configured complex geometry problems, namely blood fluid simulations [19] and flow through porous media [14].

HemeLB [22], is an LBM-based, open-source software package that is designed to study the sparse fluid domains characteristic of vascular geometries. The code is available in the repository [34]. In many hemodynamics studies using the LBM, the flows examined are often based on a relatively low Reynolds (Re) number ($Re < 1000$). Examples in the literature include those examining the circle of Willis [23], arteriovenous fistula [24] or aortic aneurysm [2]. From a physiological perspective however, previous studies have demonstrated that arterial blood flow can attain much higher Reynolds numbers (exceeding 4000 in some regions) [3]. Simulating this can be a challenge and such regimes can lead to utilising high resolution discretisations of the domain and can decrease the stability of the simulation [10]. Hou et al. [15] integrated the Smagorinsky Subgrid Scale (SGS) model into lattice Boltzmann frameworks by incorporating a turbulent effective viscosity, thereby marrying the relaxed resolution requirements of LES techniques with the efficient parallelisation of the LBM. LBM-based LES simulations are extensively employed for modelling turbulent flows [26,39,40], with applications spanning various industries. Notably, these simulations have been applied in the analysis of airfoils [5,29] and vehicle aerodynamics [11]. Conventionally, CFD modelling of turbulent aortic flow can result in significant computational costs, primarily due to the necessity of a fine grid near the vascular wall. [10,21,43]. However, to the authors' knowledge, the application of lattice Boltzmann based LES methods in modelling blood flow simulations remains scarcely explored. In this study, we incorporate the Smagorinsky SGS model into the HemeLB solver, demonstrating the feasibility of employing a "coarse-grained" thoracic aorta model with a resolution of 100 μm, achieving results comparable to those at a 20 µm resolution [10].

The structure of this paper is as follows: we elucidate the lattice Boltzmann method, inclusive of the subgrid scale turbulence model, applied to the hemodynamics solver [22], in Sect. 2. In Sect. 3, we initially assess the accuracy of the current LES model through a comparison with experimental data [8] and a DNS benchmark [17]. Subsequently, we contrast the stenosis in aortic flow with and without the incorporation of the turbulence model. Finally we summarize our findings and conclude in Sect. 4.

2 Methods

2.1 The Lattice Boltzmann Method

This study employs a three-dimensional (3D) Lattice Boltzmann model featuring 19 discretised directions, known as the D3Q19 model. The lattice cell is specified by its position $\mathbf{x}$ and time t, and is characterized by a discretised velocity set $\mathbf{c}_i$ where $i \in \{0, 1, \ldots, Q-1\}$ with $Q = 19$. The evolution equation for the distribution functions can be written as:

$$\mathbf{f}(\mathbf{x} + \mathbf{c}_i \Delta t, t + \Delta t) = \mathbf{f}(\mathbf{x}, t) - \mathbf{\Omega}\left[\mathbf{f}(\mathbf{x}, t) - \mathbf{f}^{eq}(\mathbf{x}, t)\right], \tag{1}$$

where $\mathbf{\Omega}$ denotes the Bhatnagar-Gross-Krook (BGK) or single relaxation time (SRT) collision kernel, defined as $\mathbf{\Omega} = \frac{\Delta t}{\tau}$ [33]. τ is the relaxation time. The collision kernel relaxes the distribution function towards the local Maxwellian distribution function f_i^{eq}:

$$f_i^{eq}(\mathbf{x}, t) = w_i \rho(\mathbf{x}, t)\left[1 + \frac{\mathbf{c}_i \cdot \mathbf{u}(\mathbf{x}, t)}{c_s^2} + \frac{[\mathbf{c}_i \cdot \mathbf{u}(\mathbf{x}, t)]^2}{2c_s^4} - \frac{[\mathbf{u}(\mathbf{x}, t) \cdot \mathbf{u}(\mathbf{x}, t)]}{2c_s^2}\right], \tag{2}$$

where w_i denotes the direction based weights ($w_i = 1/3$ for i=0, 1/18 for the six nearest neighbours and 1/36 for the remaining directions), $\rho(\mathbf{x}, t)$ is the cell macroscopic density, $\mathbf{u}(\mathbf{x}, t)$ is the cell macroscopic velocity. In Eq. (1), Δt symbolizes the lattice Boltzmann time step, which is set to unity. The LBM kinematic viscosity ν is defined as

$$\nu = c_s^2\left(\tau - \frac{1}{2}\right)\Delta t, \tag{3}$$

with c_s representing the speed of sound, and c_s^2 equating to 1/3 in Lattice Boltzmann Units (LBU). Macro-scale quantities such as density and momentum are derived from the moments of the distribution function $f_i(\mathbf{x}, t)$, the discrete velocities $\mathbf{c}_i$:

$$\rho(\mathbf{x}, t) = \sum_{i=0}^{Q-1} f_i(\mathbf{x}, t), \tag{4}$$

$$\rho(\mathbf{x}, t)\mathbf{u}(\mathbf{x}, t) = \sum_{i=0}^{Q-1} f_i(\mathbf{x}, t)\mathbf{c}_i, \tag{5}$$

2.2 Smagorinsky Subgrid-Scale Modelling

We now summarize the lattice Boltzmann based Smagorinsky SGS LES technique: Within the LBM framework, the effective viscosity ν_{eff} [15,18,30] is modelled as the sum of the molecular viscosity, ν_0, and the turbulent viscosity, ν_t:

$$\nu_{\text{eff}} = \nu_0 + \nu_t, \qquad \nu_t = C_{\text{smag}} \Delta^2 \left|\bar{\mathbf{S}}\right|, \tag{6}$$

where $\left|\bar{\mathbf{S}}\right|$ is the filtered strain rate tensor, C_{smag} is the Smagorinsky constant, Δ represents the filter size. We apply Eq. (6) into Eq. (3) to add the turbulent viscosity. The detailed demonstration of the turbulent viscosity can be found in [15,39].

2.3 Sponge Layer Implementation

To reduce the velocity fluctuations near the outlet, a *sponge zone* is set up as a damping region [1,13,39,40]. In the sponge zone, we impose a higher kinematic viscosity, ν_s, defined as

$$\nu_s = \nu_{\text{eff}} \left[1 + (p-1)\left(\frac{\|d\|}{w_{\text{width}}}\right)^2\right], \tag{7}$$

where p are empirical viscosity ratio constants which are set to $p = 1000$ in this work. $\|d\|$ denotes the distance of the lattice cell point to the outlet plane. w_{width} represents the width of the sponge zone prior to the outlet.

3 Results

In this section, we initially present the configuration of stenosis in turbulent channel flow for validation purposes, and subsequently conduct a comparative analysis with both experimental Particle Image Velocimetry (PIV) (Ding et al.) [8] and DNS datasets [17]. Subsequently, the potential of the LBM-based LES (LBM-LES) approach is elaborated upon, demonstrating its capability to simulate aortic flow at a resolution of 100µm, which is coarser than the resolution attainable with the exclusive use of the BGK collision operator.

3.1 Validation of the Turbulent Channel Flow Simulation

As a validation exercise for the method, we first consider a high-resolution stenotic channel flow simulation. As depicted in Fig. 1, the dimensions of the stenotic channel flow are configured as $L_x \times L_y \times L_z$, where $L_x = 266\delta$, $L_y = 22\delta$, and $L_z = 2\delta$ correspond to the streamwise, spanwise, and vertical directions, respectively. Here, δ denotes the turbulent boundary layer thickness, set at $\delta = 0.0045$m. The lattice resolution is set to 100 µm, resulting in a simulation domain composed of approximately 1.0×10^9 lattice cells. The Smagorinsky constant is set to $C_{\text{smag}} = 0.01$ in this simulation. The domain is driven by an

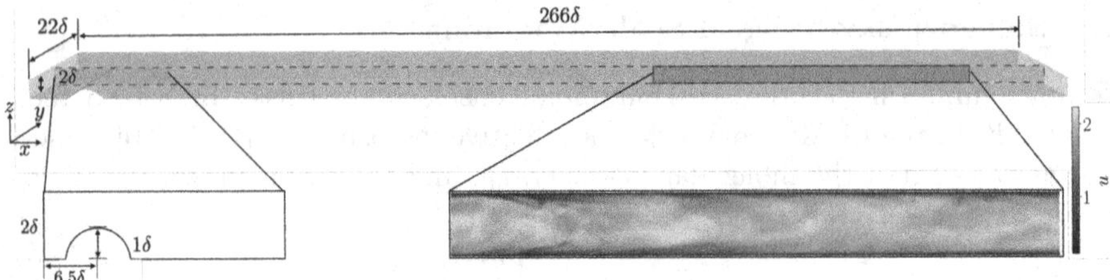

Fig. 1. Illustration of a stenotic channel flow simulation setup: The depicted stenotic channel commences with a semi-cylindrical bump, possessing a height of 1δ, obstructing 50% of the channel's cross-section. The green zone demonstrates the semi-cylindrical obstacle's configuration. The blue zone shows a snapshot where the channel flow has evolved into a fully developed turbulent velocity profile. The orange zone represents the sponge zone, designed to absorb reflective waves emanating from the outlet. (Colour figure online)

inlet boundary condition characterized by a plug velocity profile, where the maximum velocity reaches 2.16 m/s. The outlet is configured with a pressure-free boundary condition to facilitate flow egress without additional resistance. To enhance the stability, a sponge zone, highlighted in the figure in orange, is positioned prior to the outlet to absorb reflective waves triggered by non-equilibrium bounce-back effects at the outlet. The sponge zone is set to 4δ to ensure the stability of the simulation. Both the upper and lower planes in the z direction are assigned no-slip boundary conditions. Unlinke the setups described in other studies [18,39], periodic boundary conditions are not employed in the spanwise direction. Instead, a wide channel with no-slip boundary conditions, mirroring the experimental configurations [8], is implemented. To accelerate the turbulent transition adaptation length, a semi-cylindrical obstacle with the height of a half-sphere is positioned at 6.5δ from the inlet. Initially, the inlet flow is established as laminar, but the presence of the obstacle disrupts the flow's symmetry, leading to a gradual transition to turbulent flow. The inlet velocity is incrementally increased from 0.1 m/s to 2.16 m/s over a duration of 1 s. The computational cost for running the validation case can be find in Table 1.

Table 1. Table for the computational time for both stenotic channel flow validation and stenotic aortic flow simulation.

Computational model	CPU type	number of cores	Timesteps
Stenotic channel flow	AMD EPYC Zen2	8192	2,000,000
Aortic flow	AMD EPYC Zen2	8192	2,500,000

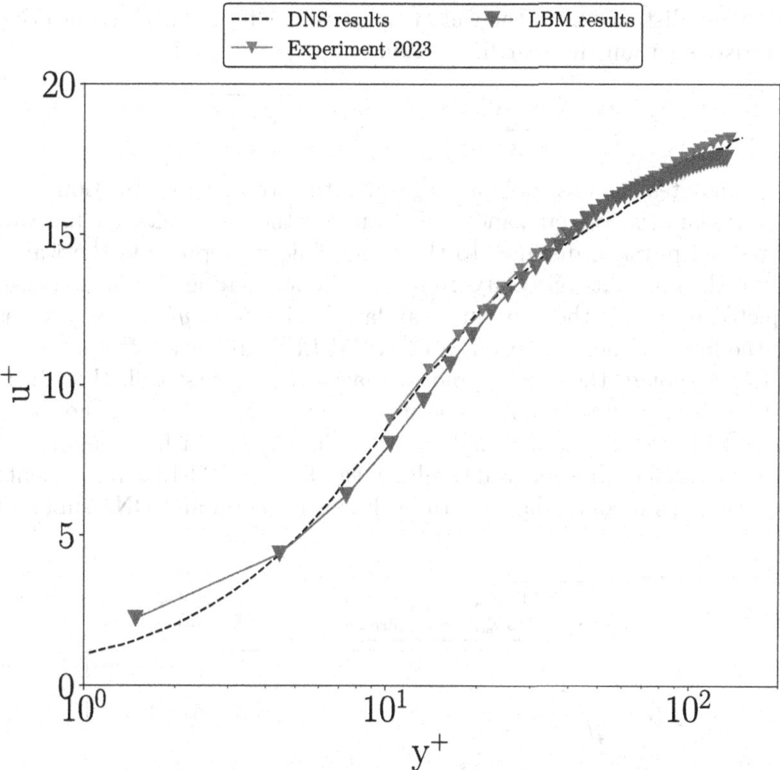

Fig. 2. Representation of u^+ as a function of y^+, where the black dotted line illustrates the DNS reference data [17]. The blue triangles depict the PIV experiment data from Ding et al. [8]. The red triangles represent the data obtained from our LBM-LES simulation.

Upon reaching the maximum velocity, the simulation continues for an additional second to ensure that the turbulent flow fully develops. The channel flow simulation yields a Reynolds number, Re, approximately equal to 4600. Simultaneously, the friction Reynolds number, Re_τ, is estimated to be around 130. The expressions defining Re and Re_τ are presented as

$$Re = u_{\max} * H/\nu, \qquad Re_\tau = u_\tau * H/\nu \tag{8}$$

where Re is characterized by the maximum velocity and Re_τ is measured by the shear velocity. H is the channel flow height which is 9×10^{-3} m, ν is the kinematic viscosity which is set to 4×10^{-6} m^2/s. We commence collecting statistics after the first second of simulation, focusing on the blue region illustrated in Fig. 1, positioned at the channel's mid cross-section, extending from $0.2m$ to $0.4m$. The spatial ensemble average of the streamwise velocity, u^+, as a function of y^+, is calculated. These variables represent the dimensionless mean velocity and the dimensionless distance from the wall, respectively, offering insights into the flow characteristics within the specified region.

$$u^+ = \frac{\langle u \rangle}{u_\tau}, \qquad y^+ = y\frac{\sqrt{\tau_\omega/\rho}}{\nu}, \tag{9}$$

where $\langle\cdot\rangle$ denotes the ensemble average over the streamwise direction. u represents the streamwise velocity, and u_τ is defined as the shear velocity. The variable y denotes the physical distance to the wall, while τ_w represents the wall shear stress. ν is the kinematic viscosity from the LBM simulation obtained via Eq. (3). As depicted in Fig. 2, the experimental data begins from $y^+ = 10$, whereas the y^+ for the first cell near the wall in the LBM-LES simulation is approximately $y^+ = 1.5$. Although there is a minor deviation in the first cell, the LBM-LES simulation aligns well with both experimental and DNS references. For $y^+ > 30$, the LBM-LES results deviate slightly from the DNS data but remain in close agreement with the experimental results. Overall, the LBM-LES implementation demonstrates good concordance with both experimental and DNS simulations.

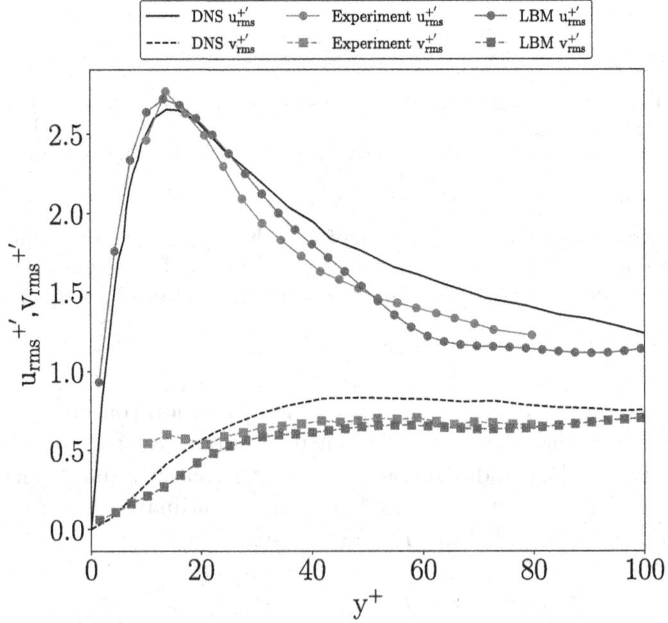

Fig. 3. Presentation of $u_{rms}^{+'}$ and $v_{rms}^{+'}$ as functions of y^+, with the black and black dotted lines representing the DNS reference data [17]. The blue circles and squares correspond to $u_{rms}^{+'}$ and $v_{rms}^{+'}$ from the experiment results [8], respectively. Additionally, the red circles and squares depict the LBM simulation outcomes for $u_{rms}^{+'}$ and $v_{rms}^{+'}$, respectively.

Furthermore, we examined the dimensionless root mean square (RMS) for two velocity components on streamwise and vertical direction. $\mathrm{u}^{+\prime}_{\mathrm{rms}}$, $\mathrm{v}^{+\prime}_{\mathrm{rms}}$ are the dimensionless RMS velocity components that are normalized with the shear velocity u_τ:

$$\mathrm{u}^{+\prime}_{\mathrm{rms}} = \frac{\sqrt{(u(\mathbf{x}) - \langle u \rangle)^2}}{u_\tau}. \tag{10}$$

Spatial averaging for RMS is performed only during post-processing due to the high resolution of the simulation. In Fig. 3, the red markers represent the LBM-LES simulation results, while the blue markers correspond to the PIV experimental reference, and the black lines denote the DNS reference. Both experimental and LBM-LES results successfully capture the peak value of $u^{+\prime}_{\mathrm{rms}}$ compared with the DNS data. Beyond the peak, both LBM-LES and experimental results gradually deviate from the DNS data, which may be attributed to statistical issues and the confinement of the channel flow in the spanwise direction. Regarding the vertical velocity component $u^{+\prime}_{\mathrm{rms}}$, LBM-LES results align well with the experimental data but are slightly lower than the DNS results for $y^+ > 20$.

Considering both Fig. 2 and Fig. 3, it is evident that the LBM-LES implementation aligns closely with experimental and DNS results, demonstrating its capability to capture turbulent statistical quantities accurately. The comparison reveals that the LBM-LES method effectively replicates the key features of turbulent flows, confirming its reliability in modelling complex flow dynamics. This alignment shows the potential of LBM-LES in contributing valuable insights into the understanding of turbulence, particularly in the context of fluid dynamics simulations.

3.2 Simulation of High Reynolds Number Flow in a Stenosis of the Aorta

In this subsection, we present our capability to simulate blood flow at $Re = 4500$ utilising our LBM-LES methodology. Specifically, we apply this technique to simulate stenotic blood flow within a thoracic aorta model [35], which has an inlet diameter of approximately 0.01m. The largest lattice spacing required to reasonbly represent such a domain is approximately 100µm. However, the BGK collision operator faces challenges in achieving stable simulations at this resolution (especially at elevated Re) due to the limited choice of the relaxation time τ. For a reliable simulation of stenotic aorta flow, a finer resolution of 20µm for the geometry may be necessary. To illustrate the benefit of LES techniques, we set to the model resolution in our study to 100µm with the Smagorinsky constant set to $C_{\mathrm{smag}} = 0.01$. As a proof of the concept, the model is initialized with an inlet velocity around 1m/s. Pressure-free outlet conditions are established for three branches leading to the cerebral and upper limb vasculature and a singular outlet in the bottom leading to the descending aorta. The sponge regions are set near the outlets of the vascular with $p = 1000$ in Eq. (7). As shown in Fig. 4, the LBM-LES based simulation is able to capture the pressure drop due to the presence of the nearing part of the aorta. The cross-sectional views on upper, middle, and lower section near the aortic constriction shows that on the upper plane, the mean velocity is around 1m/s, the reduced cross-section of the middle section leads to the acceleration of the blood flow to around 1.8m/s. Furthermore, the high blood flow velocity leads to turbulent flow in the lower part of the stenotic region. The presence of the blockage will also lead to a pressure drop [10] which is observed in our simulation. We emphasise that the typical resolution for the stenotic aorta would be around $20\mu m$ resolution, whereas, our configuration is able to observe similar velocity and pressure profile at 100μm resolution. The computational cost for running the stenotic aorta flow is illustrated in Table 1.

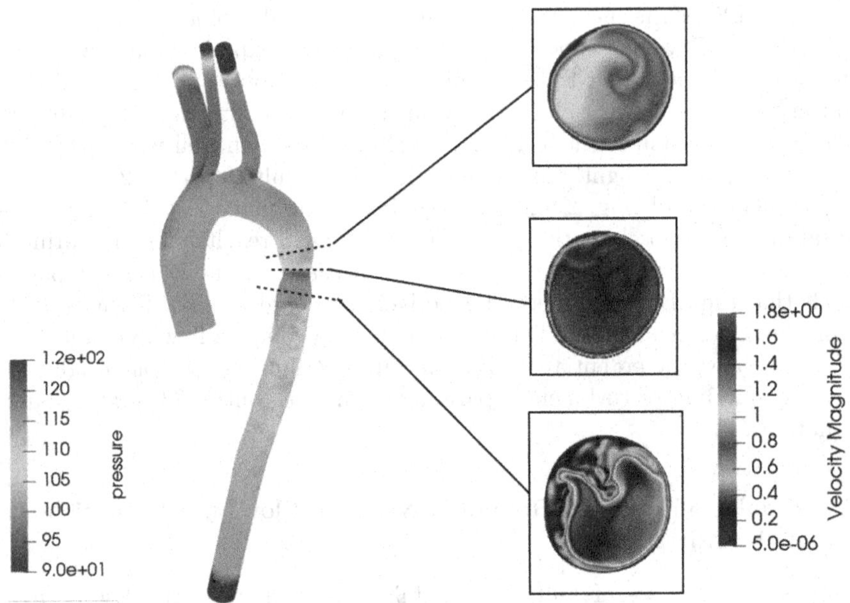

Fig. 4. Instantaneous snapshot of aortic stenosis simulation: on the left-hand side, the pressure distribution within the aortic flow is depicted. The right-hand side illustrates three cross-sectional views-upper, middle, and lower-proximal to the aortic constriction. This figure effectively demonstrates the capability of HemeLB to capture the blood flow behavior of aortic flow at a stenotic location with a maximum velocity of 1.8m/s.

Figure 5 provides an instantaneous snapshot illustrating the surface stress profile of the aorta, along with the velocity profile across a section near the stenotic region. This figure highlights that the area proximal to the aortic constriction exhibits significantly elevated wall shear stress, indicative of the stenosis acting as an obstruction, impeding blood flow and directing it towards the descending aorta. Consequently, the upper region of the aorta is subjected to increased blood pressure, as depicted in Fig. 4. Elevated wall shear stress is also observed near the aortic outlets, attributed to the implementation of the sponge layer. On the right-hand side of Fig. 5, the velocity profile within the aorta is presented, revealing a laminar-turbulent transition induced by aortic stenosis. Mirroring the stenotic channel flow simulation illustrated in Fig. 1, the flow evolves into a turbulent state, achieving a peak velocity of approximately 1.8m/s. This transition underscores the complexity of flow patterns in the areas of aortic constriction, reflecting the significant impact of stenosis on the hemodynamic behavior of the aorta.

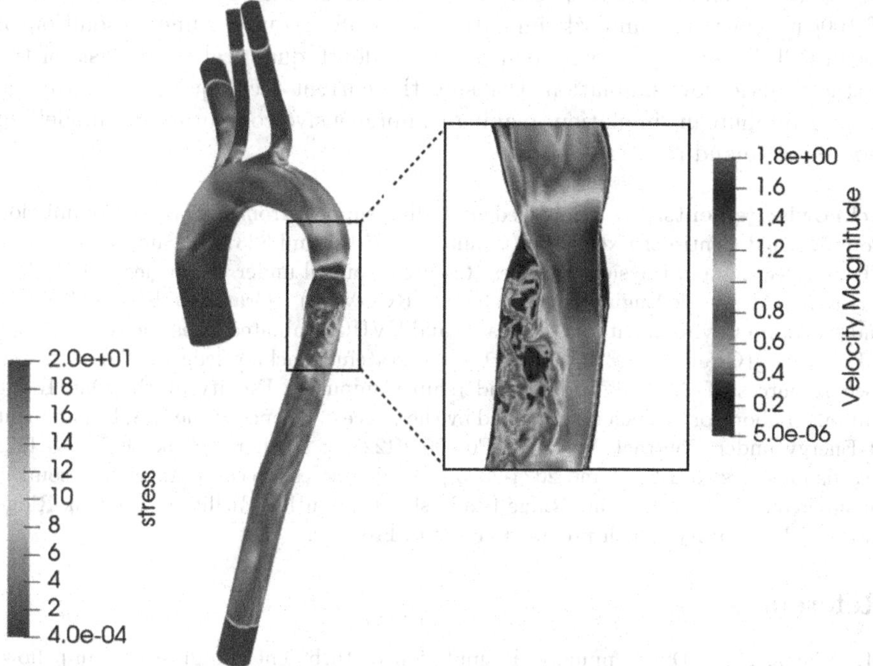

Fig. 5. The snapshot presents an instantaneous simulation of wall shear stress in stenotic aortic flow, featuring a cross-sectional visualization near the region of aortic stenosis. The velocity magnitude profile within this cross-section highlights the transition from laminar to turbulent flow, a phenomenon occurring within the stenotic segment of the aorta.

4 Conclusion

In this study, the integration of the Smagorinsky turbulence model into the HemeLB framework is evaluated for simulating high Reynolds number aortic flow. The validation involved a high-resolution stenotic channel flow, showcasing the laminar-turbulent transition along the streamwise direction. Quantitative analysis is conducted at a downstream cross-section located at $y = 11\delta$, where the flow transitions to full turbulence. Here, both the mean velocity profile and the

RMS values of the streamwise and vertical velocities exhibited good agreement with experimental and DNS reference data. The LBM-LES methodology is then applied to aortic stenosis flow simulations, utilising a sponge layer and turbulence model to qualitatively replicate the phenomenological behavior observed in aortic stenosis, including increased blood pressure in the upper part of the stenotic region, high shear-stress near the region, and the occurrence of laminar-turbulent transition within the stenotic segment of the aorta. This work underscores the fact that the simulations of stenotic aorta can be conducted at a resolution of 100μm, whereas standard simulations typically require a finer resolution of 20μm [10]. Future efforts will focus on an in-depth quantitative analysis of the aortic stenotic flow simulation. Utilising the current technique could pave the way for full human simulation, making comprehensive computational modelling more feasible and realistic.

Acknowledgements. We acknowledge funding support from European Commission CompBioMed Centre of Excellence (Grant No. 675451 and 823712). Support from the UK Engineering and Physical Sciences Research Council under the projects "UK Consortium on Mesoscale Engineering Sciences (UKCOMES)" (Grant No. EP/R029598/1) and "Software Environment for Actionable and VVUQ-evaluated Exascale Applications (SEAVEA)" (Grant No. EP/W007711/1) is gratefully acknowledged. This research used resources of the Oak Ridge Leadership Computing Facility at the Oak Ridge National Laboratory, which is supported by the Office of Science of the U.S. Department of Energy under Contract No. DE-AC05-00OR22725. Authors thank DOE INCITE awards for years 2022-2023 and 2024-2025 providing access to computational resources on supercomputers at the Oak Ridge Leadership Computing Facility at the Oak Ridge National Laboratory, which provided access to Frontier.

References

1. Adams, N.A.: Direct numerical simulation of turbulent compression ramp flow. Theoret. Comput. Fluid Dyn. **12**(2), 109–129 (1998)
2. Afrouzi, H.H., Ahmadian, M., Hosseini, M., Arasteh, H., Toghraie, D., Rostami, S.: Simulation of blood flow in arteries with aneurysm: Lattice Boltzmann approach (LBM). Comput. Methods Programs Biomed. **187**, 105312 (2020)
3. Axner, L., Hoekstra, A.G., Jeays, A., Lawford, P., Hose, R., Sloot, P.: Simulations of time harmonic blood flow in the mesenteric artery: comparing finite element and lattice Boltzmann methods. Biomed. Eng. Online **8**(1), 1–8 (2009)
4. Chapman, D.R.: Computational aerodynamics development and outlook. AIAA J. **17**(12), 1293–1313 (1979)
5. Chen, X.P.: Applications of lattice Boltzmann method to turbulent flow around two-dimensional airfoil. Eng. Appl. Comput. Fluid Mech. **6**(4), 572–580 (2012)
6. Chiappini, D., Xue, X., Falcucci, G., Sbragaglia, M.: Ligament break-up simulation through pseudo-potential lattice Boltzmann method. In: AIP Conference Proceedings, vol. 1978, p. 420003. AIP Publishing (2018)
7. Choi, H., Moin, P.: Grid-point requirements for large eddy simulation: chapman's estimates revisited. Phys. Fluids **24**(1), 011702 (2012)
8. Ding, G., Choi, K.S., Ma, B., Kato, T., Yuan, W.: Transitional pulsatile flows with stenosis in a two-dimensional channel. Physics Fluids **33**(3), 034115-1–034115-18 (2021)

9. Eymard, R., Gallouët, T., Herbin, R.: Finite volume methods. Handb. Numer. Anal. **7**, 713–1018 (2000)
10. Feiger, B., et al.: Accelerating massively parallel hemodynamic models of coarctation of the aorta using neural networks. Sci. Rep. **10**(1), 9508 (2020)
11. Gaedtke, M., Wachter, S., Raedle, M., Nirschl, H., Krause, M.J.: Application of a lattice Boltzmann method combined with a Smagorinsky turbulence model to spatially resolved heat flux inside a refrigerated vehicle. Comput. Math. App. **76**(10), 2315–2329 (2018)
12. Groen, D., et al.: Validation of patient-specific cerebral blood flow simulation using transcranial doppler measurements. Front. Physiol. **9**, 721 (2018)
13. Guo, Y., Kleiser, L., Adams, N.: A comparison study of an improved temporal DNS and spatial DNS of compressible boundary layer transition. In: Fluid Dynamics Conference, p. 2371 (1994)
14. Han, Y., Cundall, P.A.: LBM-dem modelling of fluid-solid interaction in porous media. Int. J. Numer. Anal. Meth. Geomech. **37**(10), 1391–1407 (2013)
15. Hou, S., Sterling, J., Chen, S., Doolen, G.: A lattice Boltzmann subgrid model for high Reynolds number flows. In: Pattern Formation and Lattice Gas Automata, pp. 151–166 (1995)
16. Karlin, I.V., Ferrante, A., Öttinger, H.C.: Perfect entropy functions of the lattice Boltzmann method. EPL (Europhys. Lett.) **47**(2), 182 (1999)
17. Kim, J., Moin, P., Moser, R.: Turbulence statistics in fully developed channel flow at low Reynolds number. J. Fluid Mech. **177**, 133–166 (1987)
18. Koda, Y., Lien, F.S.: The lattice Boltzmann method implemented on the GPU to simulate the turbulent flow over a square cylinder confined in a channel. Flow Turbul. Combust. **94**(3), 495–512 (2015)
19. Krüger, T., Holmes, D., Coveney, P.V.: Deformability-based red blood cell separation in deterministic lateral displacement devices–a simulation study. Biomicrofluidics **8**(5), 054114 (2014)
20. Lallemand, P., Luo, L.S.: Theory of the lattice Boltzmann method: dispersion, dissipation, isotropy, Galilean invariance, and stability. Phys. Rev. E **61**(6), 6546 (2000)
21. Leuprecht, A., Kozerke, S., Boesiger, P., Perktold, K.: Blood flow in the human ascending aorta: a combined MRI and CFD study. J. Eng. Math. **47**, 387–404 (2003). https://doi.org/10.1023/B:ENGI.0000007969.18105.b7
22. Mazzeo, M.D., Coveney, P.V.: HemeLB: a high performance parallel lattice-Boltzmann code for large scale fluid flow in complex geometries. Comput. Phys. Commun. **178**(12), 894–914 (2008)
23. McCullough, J.W., Coveney, P.V.: High resolution simulation of basilar artery infarct and flow within the circle of Willis (2023)
24. McCullough, J., Coveney, P.: High fidelity blood flow in a patient-specific arteriovenous fistula. Sci. Rep. **11**(1), 22301 (2021)
25. Nash, R.W., et al.: Choice of boundary condition for lattice-Boltzmann simulation of moderate-Reynolds-number flow in complex domains. Phys. Rev. E **89**(2), 023303 (2014)
26. Sagaut, P.: Toward advanced subgrid models for lattice-Boltzmann-based large-eddy simulation: theoretical formulations. Comput. Math. App. **59**(7), 2194–2199 (2010)
27. Shao, X., et al.: Near-wall modelling of forests for atmosphere boundary layers using lattice Boltzmann method on GPU. Eng. Appl. Comput. Fluid Mech. **16**(1), 2142–2155 (2022)

28. Shur, M.L., Spalart, P.R., Strelets, M.K., Travin, A.K.: Synthetic turbulence generators for RANS-LES interfaces in zonal simulations of aerodynamic and Aeroacoustic problems. Flow Turbul. Combust. **93**(1), 63–92 (2014)
29. Si, H., Shi, Y.: Study on lattice Boltzmann method/large eddy simulation and its application at high Reynolds number flow. Adv. Mech. Eng. **7**(3), 1687814015573829 (2015)
30. Smagorinsky, J.: General circulation experiments with the primitive equations: I. the basic experiment. Monthly Weather Rev. **91**(3), 99–164 (1963)
31. Spalart, P., Allmaras, S.: A one-equation turbulence model for aerodynamic flows. In: 30th Aerospace Sciences Meeting and Exhibit, p. 439 (1992)
32. Strikwerda, J.C.: Finite difference schemes and partial differential equations. SIAM (2004)
33. Succi, S.: The Lattice Boltzmann Equation for Fluid Dynamics and Beyond. Oxford University Press (2001)
34. UCL: HemeLB CPU code (2008). https://github.com/UCL-CCS/HemePure
35. Wilson, N.M., Ortiz, A.K., Johnson, A.B.: The vascular model repository: a Public resource of medical imaging data and blood flow simulation results. J. Med. Devices **7**(4) (2013). https://doi.org/10.1115/1.4025983
36. Xue, X., Biferale, L., Sbragaglia, M., Toschi, F.: A lattice Boltzmann study on Brownian diffusion and friction of a particle in a confined multicomponent fluid. J. Comput. Sci. **47**, 101113 (2020)
37. Xue, X., Biferale, L., Sbragaglia, M., Toschi, F.: A lattice Boltzmann study of particle settling in a fluctuating multicomponent fluid under confinement. Eur. Phys. J. E **44**(11), 1–10 (2021)
38. Xue, X., Sbragaglia, M., Biferale, L., Toschi, F.: Effects of thermal fluctuations in the fragmentation of a nanoligament. Phys. Rev. E **98**(1), 012802 (2018)
39. Xue, X., Yao, H.D., Davidson, L.: Synthetic turbulence generator for lattice Boltzmann method at the interface between RANS and LES. Phys. Fluids **34**(5), 055118 (2022)
40. Xue, X., Yao, H.D., Davidson, L.: Wall-modeled large-eddy simulation integrated with synthetic turbulence generator for multiple-relaxation-time lattice Boltzmann method. Phys. Fluids **35**(6), 065115 (2023)
41. Yang, X.I., Griffin, K.P.: Grid-point and time-step requirements for direct numerical simulation and large-eddy simulation. Phys. Fluids **33**(1), 015108 (2021)
42. Zacharoudiou, I., McCullough, J., Coveney, P.: Development and performance of a HemeLB GPU code for human-scale blood flow simulation. Comput. Phys. Commun. **282**, 108548 (2023)
43. Zakaria, M.S., et al.: Computational fluid dynamics study of blood flow in aorta using openFOAM. J. Adv. Res. Fluid Mech. Therm. Sci. **43**(1), 81–89 (2018)
44. Zienkiewicz, O.C., Taylor, R.L., Zhu, J.Z.: The Finite Element Method: Its Basis and Fundamentals. Elsevier (2005)

A Conceptual Approach to Agent-Based Modelling of Coping Mechanisms in Climate-Driven Flooding in Bangladesh

Laura M. Harbach[1,2(✉)], Derek Groen[1], Alireza Jahani[1], Diana Suleimenova[1], Maziar Ghorbani[1], and Yani Xue[1]

[1] Department of Computer Science, Brunel University London, Wilfred Brown Building, Kingston Lane, UB8 3PH Uxbridge, UK
laura.harbach@brunel.ac.uk

[2] Astrophysics Group, Blackett Laboratory, Imperial College London, Exhibition Road, South Kensington, SW7 2BX London, UK

Abstract. Bangladesh stands as a prime example of a nation exceptionally vulnerable to the adverse effects of climate change. Its low-lying coastal and deltaic landscape predisposes it to frequent flooding, a challenge exacerbated by a significant portion of its population grappling with poverty. The country is already experiencing the impacts of climate change, including more frequent and severe flooding that has led to the displacement of millions of people and has intensified existing social and economic challenges. Despite these formidable challenges, Bangladesh has also emerged as a global leader in climate resilience and preparedness, having made significant progress in reducing cyclone-related deaths and protecting its population from the consequences of climate change. Notably, non-governmental organisations, like our partners Save the Children, are keen to explore how they can support the most vulnerable communities by establishing the efficacy of current coping strategies for sustained resilience against climate change. To facilitate this, we are in the process of creating an agent-based model that examines the coping mechanisms adopted in response to climate-induced flooding in Bangladesh. This paper presents the initial phase of developing a multiscale conceptual model tailored to understanding this complex situation.

Keywords: conceptual model · agent based model · forced migration · Bangladesh · flooding · internally displaced people · coping mechanisms · climate change

1 Introduction

Bangladesh, with its population nearing 170 million, faces a formidable adversary: climate change. Ranked seventh in the world on the Long-Term Climate Risk Index [1], this densely populated nation grapples with the devastating consequences of rising sea levels, intensifying storms, and unpredictable rainfall

L. Franco et al. (Eds.): ICCS 2024, LNCS 14836, pp. 421–428, 2024.
https://doi.org/10.1007/978-3-031-63775-9_31

patterns. Among these challenges, flooding stands out as a defining threat, displacing millions, disrupting livelihoods, and eroding communities [2].

Exploring how people cope with and adapt to this ever-present threat is vital for understanding the intricate relationship between climate-induced flooding and forced migration. While floods pose significant challenges, communities have historically developed remarkable resilience, coexisting with these events for centuries. The intricate interplay between vulnerability and resilience calls for a nuanced approach, that acknowledges the multifaceted impacts of flooding on people and their environment. Understanding these wide-ranging coping mechanisms, from asset protection to reliance on social networks, is essential for designing effective humanitarian interventions.

This study aims to assess the 'push' and 'pull' factors that force individuals and communities to migrate. 'Push' factors, such as the increasing frequency and intensity of floods destroying homes and livelihoods, can make certain areas uninhabitable. Conversely, 'pull' factors, like the promise of greater safety, economic opportunities, or established communities in new locations, can drive migration. Furthermore, we will explore factors that trigger individuals to rebuild their lives within their hometowns, to gain a deeper understanding of both migratory pressures and potential return pathways.

Agent-based models (ABMs) simulate complex systems by modelling people as individual 'agents' with rule sets governing their interactions with their environment, allowing for the study of how bottom-up interactions lead to emergent phenomena. ABMs are especially good at representing decision-making processes, which is vital for understanding coping mechanisms. Compared to machine learning approaches, ABMs allow for straightforward implementation of easily comprehensible decision rules, avoiding the 'black box' nature of machine learning algorithms. This transparency allows for a deeper understanding of how individuals interact with their environment and their choices. Furthermore, ABMs enable the testing of multiple scenarios to pinpoint the specific drivers of migration, leading to more accurate models.

This research utilises Flee [3], an established ABM for conflict-driven migration in developing countries, recently adapted to simulate flood disasters under the title DFlee [4]. Existing flood ABMs often focus on developed countries, with detailed data to enable modelling of the flooding, infrastructure, and transportation networks [5]. Producing an ABM for a developing nation poses greater challenges due to data limitations for geography, population, migration, and flooding. Moreover, Flee is unique in that it leverages heuristic methods to understand migration behaviour. The DFlee model better represents the decision-making processes involved in flood-induced migration in developing countries.

This paper discusses the development of our conceptual ABM that directly considers the individual decision-making processes underlying forced migration in flood-prone Bangladesh. We address the challenges and limitations of data sources considered during its development and outline our plans for future model refinement and validation.

2 Conceptual Model Development

2.1 Identifying and Prioritising Coping Mechanisms

Understanding the coping mechanisms of individuals and communities in managing the negative impacts of flooding is critical. These mechanisms can be preventative (proactive), such as storing food, or reactive (response-based), like seeking shelter after a flood warning. Whilst our comprehensive literature review identified over 75 potential mechanisms in Bangladesh (e.g. [6,7]), limited research exists on their relative importance in driving migration. Due to time constraints, we prioritise modelling only the most important ones based on the knowledge of humanitarian experts and ease of numerical implementation in the model. An iterative development approach allows for future expansion.

The literature highlights the importance of preparedness, as individuals exhibit varying degrees of awareness and readiness for floods [8]. Accessibility and language barriers can limit the effectiveness of flood warnings. Consequently, individual knowledge of impending floods becomes crucial for both evacuation decisions and the success of other coping mechanisms. For example, common pre-flood preparations include stockpiling food, safe drinking water, and essential items like medicine or livestock. Additionally, people often seek temporary refuge in nearby cyclone shelters. Furthermore, disruptions to medical care after floods can necessitate relocation to access treatment. The next section explores the implementation of these coping mechanisms in greater detail.

2.2 Conceptual Model of Coping Mechanisms

Weather Forecaster - Agent Awareness of the Flood. The proposed ABM incorporates a modified weather forecaster mechanism to influence agent movement during floods. Each potential destination's favourability score is calculated to account for the standard ABM rules (such as travel distance), the severity of flooding, the agent's awareness of the forecast and how important they consider the forecast to be. This mechanism penalises agents for moving through flooded areas and encourages them to avoid predicted flooding by moving to the most favourable destination. This approach captures the impact of varying levels of preparedness and information access on individual migration patterns.

Storage of Food, Water and Essentials. The conceptual model includes an asset storage coping mechanism, to link an agent's ability to store food with their awareness of impending flooding, the flood severity and their economic status. This mechanism is coupled with the weather forecasting mechanism. An agent's economic status determines the number of asset parcels they store, which decreases over time due to consumption. As resources become scarce during flooding, consumption is then limited to two parcels per day. If the agent's location reaches the maximum flood level, some agents will lose all or most of their food assets. Movement likelihood is then tied to the abundance of assets, with agents having less than two days of food parcels being more likely to move.

Cyclone Shelters. Agent's may migrate to cyclone shelters during floods. The shelters do not flood and have a capacity based on government registers. Although capacity does not limit the number of agents who can seek refuge in a shelter, a full shelter may encourage continued migration.

Medical Facilities. Access to adequate healthcare likely influences internally displaced people's (IDP's) movements after floods, as injuries and healthcare disruptions are common. We'll incorporate health facility data from OpenStreetMap [9] and model their influence on migration using the methods discussed in Jahani et al. [10].

2.3 Simulation Development Approach

In our work, we take a novel approach to simulation development by prioritising the ultimate goals of our non-governmental organisation (NGO) partners. Unlike traditional scientific methods, which typically build upon existing knowledge and ask 'What new questions can we answer?', we start with the end goal in mind: understanding when, where, and what methods people use to cope with flooding, to establish which interventions are likely to be most effective. We then work backwards, gathering and analysing data to develop a robust and useful simulation.

This approach addresses the scenario complexity and the scarcity of data in the humanitarian field. By creating a model first, we can use it to guide data collection efforts and maximise the value of the data we receive. This model can then be continuously refined, incorporating real-world experiences and insights from experts. This collaborative, iterative process ensures a swift model development approach that means the model remains relevant and practical, ultimately leading to more effective, timely interventions and improved outcomes for those in need.

We follow the multi-step simulation development approach proposed by Groen et al. [11], to develop a conceptual model. This approach emphasises simulation selection and data acquisition for both modelling and validation. This approach also guides the model's implementation, including iterative refinement.

2.4 A Multiscale Model

This paper presents a multiscale conceptual model to analyse climate-induced flood displacement in Satkhira, Bangladesh. The scale separation map, in Fig. 1a, shows how processes at different scales interact and influence each other. It showcases the model's ability to capture the interplay between individual-level decisions and large-scale emergent phenomena, highlighting the critical role of spatial scales in understanding human responses to complex environmental challenges.

The model captures various levels of granularity. From individual actions, such as food storage (around 100 m), to large-scale migration (exceeding 100 kilometres). The weather forecasting mechanism operates across the entire local domain, while the shelter evacuation mechanism reflects the behaviour of individuals seeking refuge within a few kilometres for short periods (days). The food storage mechanism is personal to each agent and is impacted by consumption and potential flooding within a localised area. When food falls below a critical threshold, individuals are driven to migrate within the domain. The model also emphasises the interconnectedness of these multiscale processes. For instance, forecasts of flooding can trigger local migration, evacuation to nearby shelters, and food storage. Additionally, return migration from shelters can occur as the flood subsides. Furthermore, food scarcity can prompt longer-term long-distance migration.

We built a location graph, shown in Fig. 1b, using demographic data from the City Population website that collates global census data [12]. Our small-scale model focuses on Satkhira and encompasses locations with populations exceeding 1,000 inhabitants at the first-order administrative level. The larger scale model includes mega cities such as Bangladesh's capital Dhaka, and Kolkata in nearby India. Government-registered cyclone shelters are included as safe locations for agents to migrate to. Locations prone to flooding are classified as flood zones based on historical inundation data [13] that identifies the severity and location of flooding. This data provides precise representations of flooded areas, minimising uncertainty when assessing the impact of coping mechanisms on migration patterns.

3 Challenges in Constructing the Conceptual Model

While the conceptual model captures critical coping mechanisms influencing IDP movement, it's pertinent to acknowledge the challenges in constructing the model and its limitations. Most significantly, focusing on a subset of mechanisms prioritises development efficiency and model parsimony, potentially underestimating real-world complexity and losing vital information.

Moreover, due to a wide range of difficult-to-define model parameters, mechanism construction has been simplified until these parameters are better understood. Future iterations could increase the sophistication of the mechanisms. For example, extending the asset storage mechanism to include varied rule sets for each asset type. Furthermore, the distribution of food by humanitarian organisations could be incorporated. However, this may be challenging due to local decision-making processes involved in the rapidly evolving situations.

The multiscale nature of both mechanisms and data creates additional challenges. For instance, government registers are likely to underestimate total shelter capacity, potentially missing major evacuation locations, as registered shelters are often significantly smaller than the neighbouring towns and there are entire administrative regions with no registered shelters. Comprehensive shelter data collection encompassing location and capacity is necessary to address this. As educational facilities are often used as shelters, mapping data could potentially bridge this gap. Additionally, aggregating shelter data to match town sizes could also enhance model accuracy. Another limitation arises from the coarse

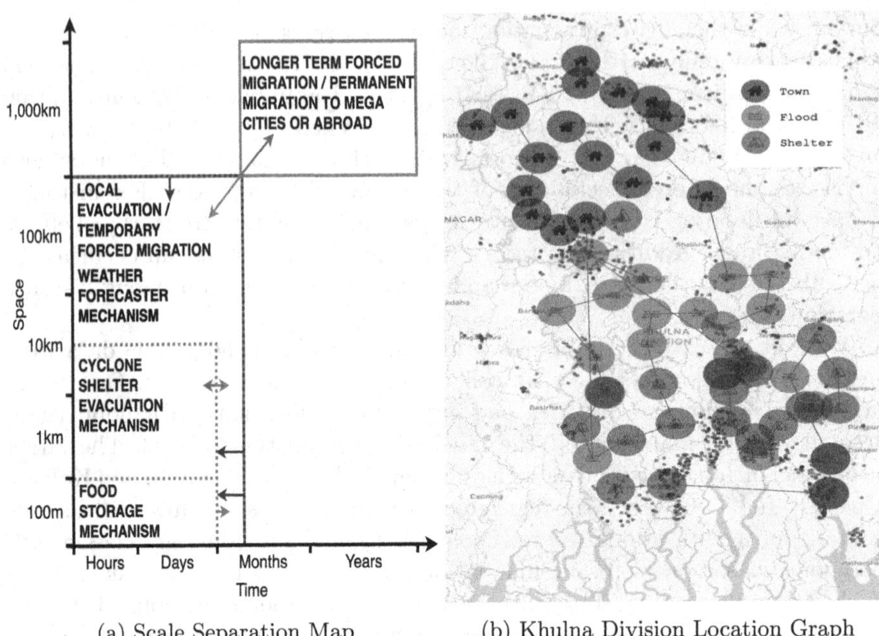

(a) Scale Separation Map (b) Khulna Division Location Graph

Fig. 1. Fig. 1a: Illustration of the hierarchical spatial scales in the conceptual multiscale ABM. The ABM's location networks for local (red dashed) and larger-scale (orange solid) migration are represented. The remaining dashed boxes depict sub-models for specific mechanisms within the ABM, including the weather forecasting (blue), cyclone shelter evacuation (purple), and food storage (green) mechanisms. Arrows visualise the interactions and feedback loops between different spatial scales. Figure 1b: Khulna division location graph, with flood zones based on flood inundation data, towns with more than 1,000 people, government-registered shelters and connecting routes. Green dots represent educational facilities (potential shelters) obtained using Open Street Maps [9].

spatial and temporal resolution of satellite inundation data, particularly during cyclones when there is increased cloud cover, which hinders capturing the details of small-scale flooding. Once the mechanisms have been fully refined, the incorporation of additional weather data, as demonstrated in previous Flee models [14] may be beneficial.

Finally, our conceptual model treats people as individual agents to address our partner NGO's focus on the specific vulnerabilities of children. This approach may limit our ability to capture the full complexity of household-level decision-making. However, we can still capture family dynamics by linking the attributes of individual agents to decision rules. For example, children may have prioritised access to food.

4 Discussion on Validation Data Collection

Acquiring data on displaced persons is inherently complex and ethically sensitive. This challenge is particularly pronounced when it comes to obtaining comprehensive and reliable data on climate-induced displacement in Bangladesh. Existing resources, such as the International Organisation on Migration Displacement Tracking Matrix and the International Disaster Database, lack comprehensive IDP entries for Bangladesh. Moreover, access to valuable data sources like Meta Data for Good Displacement Mapping is restricted. While mobile phone records provide some insights, they fail to account for vulnerable populations without access to mobile phones, lack sufficient time cadence and regional specificity, and cannot distinguish between forced and non-forced migration.

Despite the challenges, several valuable data sources were identified for model development and validation. Post-disaster reports provide event-specific displacement statistics [15], while national statistics reports offer demographic data [8], and ongoing surveys from the UN Food and Agriculture Organisation Data in Emergencies Hub offer insights into broader migration patterns, recent shocks, and food security within the country [16]. However, this source covers the entire nation rather than the specific area of interest, so it can only be used to validate the mechanisms with a national model.

5 Conclusions

This study proposes a multiscale agent-based conceptual framework for exploring how communities in Bangladesh adapt to climate-driven floods. The initial focus is on identifying key coping mechanisms that influence migration. An iterative development process, informed by data acquisition and collaboration with humanitarian experts, will guide the refinement of the final ABM. The planned ABM will then simulate the effectiveness of these mechanisms to identify the most impactful interventions for building a more sustainable future for vulnerable communities in Bangladesh.

Developing a comprehensive conceptual model for this complex scenario has presented significant challenges. There are inherent difficulties associated with conceptualising this multiscale model, particularly due to limitations in data availability and the need to balance model complexity with efficiency. However, these challenges highlight the importance of our iterative development approach. Future work will focus on implementing the model, data-driven refinement, and rigorous validation to ensure an accurate and relevant ABM for informing effective and timely humanitarian interventions.

Acknowledgments. This study was funded by an STFC Dirac Innovation Placement.

Disclosure of Interests. The authors have no competing interests to declare.

References

1. Eckstein, D., Künzel, V., Schäfer, L.: Global climate risk index 2021 (2021). https://www.germanwatch.org/en/cri
2. Internal Displacement Monitoring Centre, I.D.M.C.: Global report on internal displacement (grid). https://www.internal-displacement.org/database/displacement-data/. website last Accessed Feb 2024
3. Ghorbani, M., et al.: Flee 3: Flexible agent-based simulation for forced migration. Elsevier (2023). https://ssrn.com/abstract=4710692
4. Jahani, A., Jess, S., Groen, D., Suleimenova, D., Xue, Y.: Developing an agent-based simulation model to forecast flood-induced evacuation and internally displaced persons. In: Mikyška, J., de Mulatier, C., Paszynski, M., Krzhizhanovskaya, V.V., Dongarra, J.J., Sloot, P.M.A. (eds.) Computational Science – ICCS 2023: 23rd International Conference, Prague, Czech Republic, July 3–5, 2023, Proceedings, Part IV, pp. 550–563. Springer Nature Switzerland, Cham (2023). https://doi.org/10.1007/978-3-031-36027-5_43
5. Zhuo, L., Han, D.: Agent-based modelling and flood risk management: a compendious literature review. J. Hydrol. **591**, 125600 (2020). https://www.sciencedirect.com/science/article/pii/S0022169420310611
6. Brouwer, R., Akter, S., Brander, L., Haque, E.: Socioeconomic vulnerability and adaptation to environmental risk: a case study of climate change and flooding in bangladesh. Risk Anal. **27**(2), 313–326 (2007). https://onlinelibrary.wiley.com/doi/abs/10.1111/j.1539-6924.2007.00884.x
7. Few, R.: Flooding, vulnerability and coping strategies: local responses to a global threat. Progr. Develop. Stud. **3**(1), 43–58 (2003). https://doi.org/10.1191/1464993403ps049ra
8. Bangladesh Bureau of Statistics, B.B.S.: Disaster-related statistics 2015: Climate change and natural disaster perspectives (2015)
9. OpenStreetMap: Openstreetmap. https://www.openstreetmap.org (2021)
10. Jahani, A.: Integrating health facility density into route pruning algorithms: A case study in south sudan (2024)
11. Groen, D., Suleimenova, D., Jahani, A., Xue, Y.: Facilitating simulation development for global challenge response and anticipation in a timely way. J. Comput. Sci. **72**, 102107 (2023)
12. CityPopulation: Citypopulation. https://citypopulation.de
13. Hassan, M.M., Ash, K., Abedin, J., Paul, B.K., Southworth, J.: A quantitative framework for analyzing spatial dynamics of flood events: a case study of super cyclone amphan. Remote Sens. **12**(20) (2020). https://www.mdpi.com/2072-4292/12/20/3454
14. Jahani, A., Arabnejad, H., Suleimanova, D., Vuckovic, M., Mahmood, I., Groen, D.: Towards a coupled migration and weather simulation: South Sudan conflict. In: Paszynski, M., Kranzlmüller, D., Krzhizhanovskaya, V.V., Dongarra, J.J., Sloot, P.M. (eds.) Computational Science - ICCS 2021, pp. 502–515. Springer International Publishing, Cham (2021)
15. Relief Web: Cyclone Yaas Light Coordinated Joint Needs Analysis Needs Assessment (2022). https://reliefweb.int/report/bangladesh/cyclone-yaas-light-coordinated-joint-needs-analysis-needs-assessment-working-group
16. United Nations Food and Agricultural Organisation, F.A.O.: Bangladesh: Diem-monitoring assessments results (November 2021 - August 2022). https://data-in-emergencies.fao.org

Advancing Organizational Performance: A Strategic Framework to Multiscale Modeling and Simulation

Armin Kashefi(✉) and Faris Alwzinani

Department of Computer Science, Brunel University London, Uxbridge UB8 3PH, UK
armin.kashefi@brunel.ac.uk

Abstract. In the modern, fast-moving, and technology-centric business world, the importance of service desks in swiftly and effectively addressing tech-related challenges cannot be overstated. The changing nature of work environments, particularly with the rise of remote and hybrid models post-pandemic, highlights the critical need for a simulation strategy that goes beyond individual departments to include the broader organizational context across various levels. Focusing on this need, we embarked on a project using Discrete Event Simulation (DES) to tackle the operational hurdles faced by the service desk of a leading UK-based telecommunications firm, marking the beginning of a larger initiative aimed at multiscale modeling and simulation (MMS). We have formulated a robust five-phase strategic framework to elevate and enhance organizational performance by applying a detailed DES analysis to refine the service desk operations. This framework examines the potential for targeted improvements within the service desk to have wider benefits, impacting everything from staff satisfaction and efficiency to the overall resilience of the organization.

Keywords: Discrete Event Simulation (DES) · Multiscale Modeling and Simulation (MMS) · Operational Efficiency · Organizational Performance

1 Background Research

The idea behind multiscale modeling and simulation (MMS) is based on the understanding that numerous systems display emergent behaviors and characteristics. These emerge from the interactions of simpler components at smaller scales yet cannot be seen at these scales in isolation. Multiscale modeling has emerged as a pivotal methodology in various scientific domains, enabling researchers to bridge the gap between phenomena occurring at multiple scales. In biological systems, it has facilitated a deeper understanding of processes from the molecular to the organism level, revolutionizing fields such as genomics and proteomics [1]. Similarly, in material science, multiscale modeling has been instrumental in predicting material properties by linking atomic-scale interactions with macroscopic behavior, leading to advancements in developing novel materials [2]. Furthermore, climate studies have benefited from multiscale modeling by

L. Franco et al. (Eds.): ICCS 2024, LNCS 14836, pp. 429–436, 2024.
https://doi.org/10.1007/978-3-031-63775-9_32

integrating global climate models with regional processes to improve the accuracy of climate predictions and understand local climate impacts [3]. Despite these successes, the application of multiscale modeling within an organizational context has not been extensively explored. This oversight represents a significant opportunity for organizational researchers to apply multiscale modeling techniques. By doing so, they can uncover intricate organizational dynamics, from individual behaviors to collective outcomes, offering new perspectives on organizational theory and management practices.

In the context of organizational operations, the performance and resilience of an organization can be influenced by processes occurring at the level of individual departments, teams, and even the interactions between single employees [4]. By applying multiscale modeling to such systems, it is possible to simulate and analyze how localized changes, such as improvements in the efficiency of a service desk, can propagate through the organizational hierarchy to affect overall corporate performance. This approach enables decision-makers to identify leverage points where targeted interventions can yield the most significant benefits, optimizing resource allocation and strategic planning across the organization.

This paper analyzes and evaluates a service desk model serving internal employees within a well-known UK-based company in the telecommunications industry. A service desk is a central point of contact for all IT service-related queries and requests within an organization [5]. They play a critical role in ensuring technology-related issues are addressed efficiently, therefore minimizing downtime and queue times, ensuring business/service continuity, boosting customer satisfaction, reducing costs, increasing productivity, and improving overall quality, efficiency & performance of the service support delivery. Within the telecommunications industry, internal service desks are essential as any impact on internal day-to-day tasks could affect external customer services like broadband, voice, and data. Without the support process of the internal service desk constantly ensuring employees can continue their work flawlessly, the company's ability to deliver quality services could be affected, resulting in lost revenue and decreased satisfaction [6, 7].

In summary, we lay the groundwork for a comprehensive multiscale simulation model by commencing with a discrete event simulation project that meticulously analyses and optimizes the internal service desk operations within the aforementioned company. This initial phase seeks to enhance service desk efficiency and employee productivity and serves as a crucial step toward understanding and improving organizational performance on a macroscopic scale. The outcomes of this project promise to offer valuable insights into the interconnected nature of company operations, thereby facilitating informed decision-making aimed at fostering systemic improvements across multiple levels of the organization.

The paper is structured into six main sections. The first section concisely delved into the significance of multiscale modeling and simulation within organizational contexts. The second section provides an overview of the conceptual model and methodological approach. The third section explains the analysis of the localized DES modeling. The fourth section discusses the integration of the findings of the localized DES into multiscale modeling and simulation. The final section summarizes the study's findings, identifies opportunities for future research, and reflects on the study's limitations.

2 Development Approach

2.1 Proposed Conceptual Framework

This section outlines the methodological approach adopted in our study, focusing on using discrete event simulation (DES) to improve the efficiency of service desk operations within the telecommunications company. The DES project is conceptualized as the initial phase of a multiscale simulation roadmap designed to enhance organizational transformation and performance. Our objective with the DES phase was to identify process inefficiencies and bottlenecks and to optimize service desk operations. This established a foundation for examining broader impacts on the company in subsequent research. Figure 1 represents this suggested roadmap from the microscale to macroscale impacts, illustrating the expansion from DES to multiscale modeling and simulation.

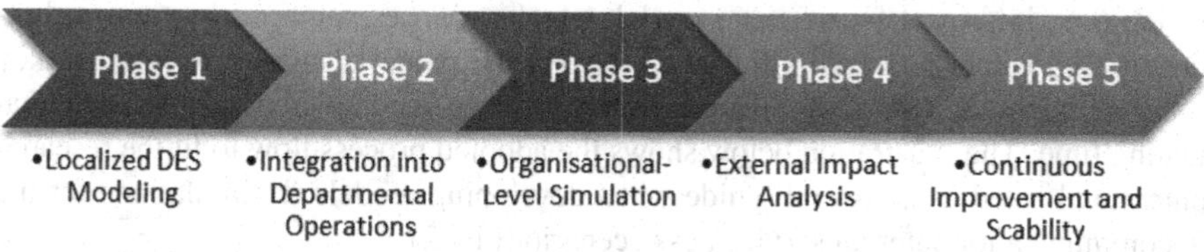

Fig. 1. Multiscale simulation roadmap to organizational performance

Considering the research context, in Phase 1, the focus is on Discrete Event Simulation (DES) with the objective of analyzing and optimizing the operational efficiency of the internal service desk. This includes improvements in a number of Key Performance Indicators (KPIs), such as the number of requests completed, queue times, and resource utilization. The expected output from this phase is enhanced business processes using both static (e.g., BPMN via Visual Paradigm software) and dynamic (e.g., Simul8 software) modeling methods.

Moving into Phase 2, the framework aims to integrate the findings from the DES into departmental operations. The goal is to apply the optimized service desk model to simulate its impact on related departmental operations, such as IT and HR. This will involve the development of agent-based models (ABM) or system dynamics (SD) models that interact with the outcomes of the DES. The output from this phase is improved operational strategies for interconnected departments.

Phase 3 expands the scope to an organizational-level simulation. The objective is to assess the cumulative impact of DES-based improvements on overall organizational performance. This involves expanding the simulation to include organizational metrics such as overall productivity, employee satisfaction, and cost-effectiveness, with the output being strategic insights for organizational restructuring or policy adjustments.

In Phase 4, the project shifts to an external impact analysis. The aim is to explore the effects of internal optimizations on external stakeholders, including customers, partners, and suppliers. This phase utilizes multiscale modeling to simulate interactions between the organization and its external environment, with the output focusing on strategies for improving customer service, supply chain efficiency, and partner collaboration.

Finally, Phase 5 focuses on continuous improvement and scalability. The objective is to establish a framework for ongoing optimization and scalability to new challenges or opportunities. This includes implementing a continuous feedback loop using real-world data to refine simulations and models, ensuring the organization remains resilient, adaptable, and capable of proactively responding to future changes.

2.2 Tailored Approach for DES Simulation Modelling

As mentioned previously, the aim of this project was to improve the efficiency and optimize the performance of the telecom's internal service desk process using static & dynamic modeling. This is the first phase of the previously proposed conceptual framework. A tailored approach that represents a step-by-step guide to our research scenario was adopted [8]. An AS-IS (i.e., current state) model was developed to identify existing problems in the service desk process. This model was then analyzed to develop a TO-BE (i.e., improved future state) model, which proposes solutions to improve bottlenecks and enhance overall efficiency. The TO-BE model also aims to improve key process metrics, including resource utilization, number of completed requests, and total queuing time. The illustration below shows the adopted process flow to fit the research context and acted as an overall guide when developing models & simulations for the telecommunication internal service desk scenario (Fig. 2).

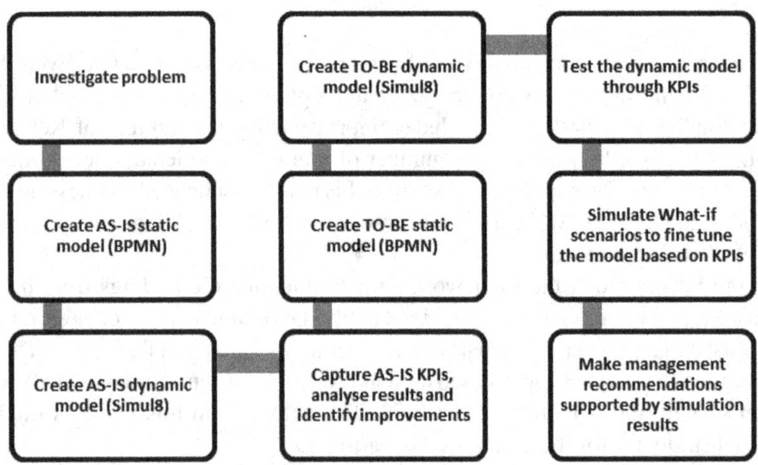

Fig. 2. Tailored simulation methodology for this research

2.3 Data Collection

We employed a mixed-methods approach, integrating both qualitative and quantitative research methodologies to enrich our investigation and enhance the robustness of our findings. The initial phase of our research was grounded in qualitative methods. We gathered nuanced insights into the operational mechanisms at play through techniques such as interviews, participant observations, and document analysis. This qualitative exploration was instrumental in developing a detailed understanding of the processes, which we subsequently represented through Business Process Model and Notation (BPMN) diagrams. Creating these diagrams was a crucial step, as it allowed us to visualize and conceptualize the complex processes in a structured and coherent manner, facilitating further analysis.

Building on the rich, contextual insights derived from our qualitative research, we transitioned to the quantitative phase of our study. In this phase, we designed and administered structured surveys to collect numerical data relevant to the previously identified processes and variables. The quantitative data collected through these surveys were pivotal in developing dynamic simulation models. These models enabled us to measure KPIs, offering empirical evidence to validate our qualitative findings and theoretical assumptions.

2.4 Data Evaluation

In this project's initial data analysis phase, Microsoft Excel played a pivotal role in processing the raw data to extract meaningful insights. By calculating average values over four months, we ensured that the analysis accounts for seasonal variations and avoids biases arising from analyzing data from a single, atypical month, such as December, when service desk requests may drop due to holidays.

The subsequent analysis stage focused on evaluating the outcomes generated by these models. This evaluation involved a detailed comparison of KPIs before and after implementing proposed improvements across several simulation trials to achieve up to 95% confidence levels. The KPIs under scrutiny included the number of requests completed by the service desk, average queue times, and resource utilization rates of the 1st and 2nd level support team agents and those of the High Priority team agents within the internal service desk.

The Discrete Event Simulation (DES) model developed as part of this research underwent rigorous verification and validation to ensure its accuracy and reliability. This critical step was essential to confirm that the model accurately represents the service desk's real-world processes and performs as intended under various conditions. By comparing the model's output against known benchmarks and real service desk data, we established its validity.

3 Analysis

With the use of multiple IT solutions within the TO-BE process, such as conversational automation, artificial intelligence, and centralized data stores, a KPI increase of 208% in the average number of requests completed was recorded. This almost doubled the previous set monthly benchmark of 10,000 average requests solved monthly, improving overall efficiency and support quality as more were solved within the same monthly period. Additionally, the resource utilization of agents from each team is at an ideal utilization level of around 60–75% +, reducing the chances of burning out and impacting job satisfaction (Fig. 3).

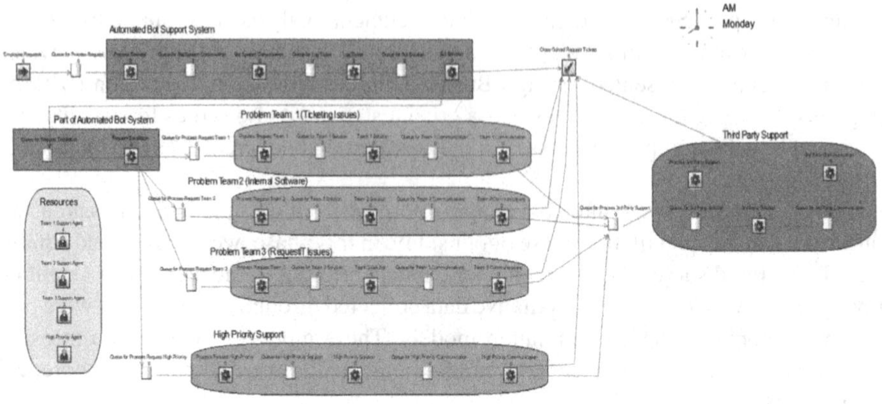

Figure 26: TO-BE Dynamic Model Visualization The source files can be found attached in the WiseFlow submission for a clearer model.

Fig. 3. TO-BE Dynamic Model Visualization using Simul8 Software

The introduction of automation and conversational bots alongside the re-creation of service desk support resulted in a 30% decrease in average queue times from start to end, which signifies the increase in efficiency as the teams are dealing with requests quicker, making employees wait for a shorter time which increases support satisfaction. Furthermore, the experimental "What-if" approach was used to reduce the total number of resources from 30 to 12 agents while still having increased requests completed with shorter queue times compared to the AS-IS model results. The automation has streamlined the process and reduced the load on multiple teams by successfully triaging requests and providing solutions with no human intervention. Although the high-priority team utilization stands out and appears to be high compared to other resources, this can be justified as while the high-priority teams deal with fewer requests, their workflow involves more tasks that are more difficult than those of other teams (Table 1).

Table 1. AS-IS & TO-BE dynamic model's KPI comparison.

Key performance indicator	AS-IS model	TO-BE model
Average number of requests completed	8,222 requests	25,377 requests
Average queue time in the system	43,776 min	30,508 min
1st Level support agent utilization	94%	60%
2nd Level support agent utilization	86%	72%
3rd High Priority agent utilization	39%	78%

The application of discrete event simulation (DES) findings from service desk optimizations in a telecommunications company presents a compelling case for its extrapolation into broader, multiscale modeling and simulation endeavors. By enhancing the efficiency of service desk operations, we not only achieve immediate operational improvements but also lay a foundational framework for assessing and enhancing departmental and organizational-wide processes. This highlights the interconnected nature of organizational functions and the cascading effects of localized improvements.

4 Discussion

The decision to adopt a multiscale simulation approach stems from the recognition that improvements in service desk efficiency have the potential to impact various levels of an organization. The DES findings offer a micro-level view, which is crucial for understanding how changes can affect broader organizational dynamics, including interdepartmental workflows and overall customer satisfaction. Upon the successful optimization of service desk operations, characterized by significant reductions in queue times and improvements in resource utilization, the subsequent step involves the dissemination of these findings across various departments, for instance, IT and HR.

In the telecommunications company's internal service desk, the findings from the first phase of localized DES have direct implications for the processes within the IT and HR departments. For instance, the IT department can draw on these findings to refine the ticketing system, ensuring that it not only classifies queries by urgency but also directs them to the appropriate IT specialist based on the type of issue, which could range from software glitches to hardware malfunctions. This approach could be complemented by the introduction of a chatbot for initial troubleshooting, allowing IT professionals to focus on resolving more complex problems, especially during periods identified by the DES as having high incident reports.

Similarly, the HR department might implement a self-service portal that addresses the most common employee inquiries, freeing up HR personnel to deal with more sensitive or complex issues. Prioritization of issues could be particularly crucial during peak times, such as pay periods or benefits enrollment seasons, ensuring that HR resources are allocated effectively to manage the influx of employee queries.

Moreover, integrating DES findings within these departments can be enriched by employing other techniques like agent-based modeling and system dynamics, which offer a multi-scale perspective of the impact. Agent-based modeling can simulate the interactions of individual agents, such as IT and HR personnel, with each other and the system, capturing the mechanisms behind their choices. On the other hand, system dynamics provides a high-level view of the entire organization, highlighting how changes in one department can have ripple effects throughout the company. Together, these modeling techniques can offer a comprehensive view of the service desk's operations, supporting strategic decisions that enhance efficiency and employee satisfaction across the organization.

5 Conclusion

In conclusion, this study addresses the gap in the literature regarding optimizing organizational efficiency through multiscale modeling and simulation. While multiscale modeling has proven to be a transformative approach in other fields, its application within an organizational context remains relatively unexplored. This gap presents a unique opportunity for researchers to pioneer the adaptation and implementation of multiscale modeling and simulation techniques to uncover new insights into organizational structures and behaviors, thereby extending its benefits beyond the traditional domains.

This paper contributes to the existing literature by proposing a novel strategic framework for implementing multiscale simulation modeling within organizations through DES modeling. The second phase of the proposed strategic roadmap focuses on applying the insights gained from the localized DES models, such as reducing queuing times and optimizing resource utilization, to improve departmental efficiencies. This research has limitations, such as data gaps, as we were only provided with four months of operational data. Also, the research is based on some assumptions as not all the data could be provided due to the company's policies. In addition to this, the research assumed that resource availability was 100% daily, meaning, the agents would not take days off or fall sick.

Future research can extend the analysis to examine the cascading effects of service desk optimizations on the telecommunications company. Subsequent phases will investigate departmental interactions, overall organizational efficiency, the impact on customer satisfaction, and organizational resilience by employing a combination of ABMs, SD models, and potentially other multiscale simulation techniques. Furthermore, applying the suggested model to other organizational contexts may be necessary. Additional investigations into this model could refine the model's comprehensiveness and fulfill its purpose of enhancing organizational resilience.

References

1. Matthews, M.L., Marshall-Colón, A.: Multiscale plant modeling: from genome to phenome and beyond. Emerg. Topics Life Sci. **5**(2), 231–237 (2021)
2. Van Der Giessen, E., et al.: Roadmap on multiscale materials modeling. Model. Simul. Mater. Sci. Eng. **28**(4), 043001 (2020)
3. Antolínez, J.A.A., et al.: Predicting climate-driven coastlines with a simple and efficient multiscale model. J. Geophys. Res. Earth Surf. **124**(6), 1596–1624 (2019)
4. Duchek, S.: Organizational resilience: a capability-based conceptualization. Bus. Res. **13**, 215–246 (2020). https://doi.org/10.1007/s40685-019-0085-7
5. Knapp, D.: A Guide to Service Desk Concepts. 4. CENGAGE Learning. Yhdysvallat (2014)
6. Crawshaw, J.: AI in Telecom Operations: Opportunities and Obstacles. Heavy Reading, September 2018
7. Jäntti, M., Cater-Steel, A., Shrestha, A.: Towards an improved it service desk system and processes: a case study. Int. J. Adv. Syst. Measure. **5**(3&4), 203–215 (2012)
8. Taylor, S.J., Abbott, P., Young, T., Grocott-Mason, R.: Student modeling and simulation projects in healthcare: experiences with Hillingdon Hospital. In: Proceedings of the Winter Simulation Conference 2014, pp. 3650–3661. IEEE, 7 December 2014

Correction to: State Estimation of Partially Unknown Dynamical Systems with a Deep Kalman Filter

Erik Chinellato and Fabio Marcuzzi

Correction to:
Chapter 22 in: L. Franco et al. (Eds.): ***Computational Science – ICCS 2024*****, LNCS 14836,**
https://doi.org/10.1007/978-3-031-63775-9_22

In the originally published version of chapter 22, the references 3 and 12 had been rendered incorrectly. This has been corrected.

The updated version of this chapter can be found at
https://doi.org/10.1007/978-3-031-63775-9_22

L. Franco et al. (Eds.): ICCS 2024, LNCS 14836, p. C1, 2024.
https://doi.org/10.1007/978-3-031-63775-9_33

Author Index

L. Franco et al. (Eds.): ICCS 2024, LNCS 14836, pp. 437–438, 2024.
https://doi.org/10.1007/978-3-031-63775-9

Printed in the United States
by Baker & Taylor Publisher Services